Advances in Mechanics and Mathematics

Volume 33

More information about this series at http://www.springer.com/series/5613

Weimin Han • Stanisław Migórski • Mircea Sofonea
Editors

Advances in Variational and Hemivariational Inequalities

Theory, Numerical Analysis, and Applications

Springer

Editors
Weimin Han
University of Iowa
Iowa City, IA, USA

Xi'an Jiaotong University
Xi'an, China

Stanisław Migórski
Faculty of Mathematics
and Computer Science
Institute of Computer Science
Jagiellonian University
Kraków, Poland

Mircea Sofonea
Laboratoire de Mathématiques
et Physique (LAMPS)
Université de Perpignan Via Domitia
Perpignan, France

ISSN 1571-8689 ISSN 1876-9896 (electronic)
Advances in Mechanics and Mathematics
ISBN 978-3-319-14489-4 ISBN 978-3-319-14490-0 (eBook)
DOI 10.1007/978-3-319-14490-0

Library of Congress Control Number: 2015932094

Springer Cham Heidelberg New York Dordrecht London

Printed on acid-free paper

Springer International Publishing AG Switzerland is part of Springer Science+Business Media
(www.springer.com)

Preface

The theory of variational inequalities is a relatively young mathematical discipline. One of the bases for its development was the contribution of Fichera [5], who coined the term "Variational Inequality" in his paper on the solution of the frictionless contact problem between a linearly elastic body and a rigid foundation posed by Signorini [15]. The foundations of the mathematical theory of elliptic variational inequalities were laid by Stampacchia [16], Hartman and Stampacchia [7], Lions and Stampacchia [11], and others. Evolutionary variational inequalities have been preliminary treated by Brézis [2] who also connected the notion of variational inequality to convex subdifferential and maximal monotone operators. The theory of variational inequalities can be viewed as an important and significant extension of the variational principle of virtual work or power in inequality form, the origin of which can be traced back to Fermat, Euler, Bernoulli brothers, and Lagrange. The theory of variational inequalities and their applications represents the topics of several well-known classical monographs by Duvaut and Lions [4], Glowinski, Lions, and Trémoliéres [6], Kinderlehrer and Stampacchia [10], Baiocchi and Capelo [1], Kikuchi and Oden [9], and so on.

The notion of hemivariational inequality was first introduced by Panagiotopoulos [13] and is closely related to the development of the concept of the generalized gradient of a locally Lipschitz functional provided by Clarke [3]. Interest in hemivariational inequalities originated, similarly as in variational inequalities, in mechanical problems. From this point of view, the inequality problems in Mechanics can be divided into two main classes: that of variational inequalities, which is concerned with convex energy functions (potentials), and that of hemivariational inequalities, which is concerned with nonsmooth and nonconvex energy functions (superpotentials). Through the formulation of hemivariational inequalities, problems involving nonmonotone and multivalued constitutive laws and boundary conditions can be treated successfully mathematically and numerically. The theory of hemivariational inequalities and their applications was developed in several monographs by Panagiotopolous [13], Naniewicz and Panagiotopolous [12], and Haslinger, Mietten, and Panagiotopolous [8], among others.

During the last decades, variational and hemivariational inequalities were shown to be very useful across a wide variety of subjects, ranging from nonsmooth mechanics, physics, and engineering to economics. For this reason, there are a large number of problems which lead to mathematical models expressed in terms of variational and hemivariational inequalities. The mathematical literature dedicated to this field is growing rapidly, as illustrated by the list of references at the end of each chapter of this volume.

The purpose of this edited volume is to highlight recent advances in the field of variational and hemivariational inequalities with an emphasis on theory, numerical analysis, and applications. The theory includes existence and uniqueness results for various classes of nonlinear inclusions and variational and hemivariational inequalities. The numerical analysis addresses numerical methods and solution algorithms for solving variational and hemivariational inequalities and provides convergence results as well as error estimates. Finally, the applications illustrate the use of these results in the study of miscellaneous mathematical models which describe the contact between deformable bodies and a foundation. This includes the modeling, the variational and the numerical analysis of the corresponding contact processes.

This volume presents new and original results which have not been published before and have been obtained by recognized scholars in the area. It addresses to mathematicians, applied mathematicians, engineers, and scientists. Advanced graduate students can also benefit from the material presented in this book. Generally, the reader is expected to have background knowledge on nonlinear analysis, numerical analysis, partial differential equations, and mechanics of continua.

This volume is divided into three parts with 14 chapters. This division of the material is not strict and it is done only for the convenience of the reader. A brief description of each part is the following.

Part I, entitled *Theory*, is devoted to the study on abstract nonlinear evolutionary inclusions and hemivariational inequalities of the first and second order, an approximation method to solve nonsmooth problems and its application to variational–hemivariational inequalities, a bifurcation result for a nonlinear Dirichlet elliptic problem, and variational inequality problems on nonconvex sets.

Part II, entitled *Numerical Analysis*, deals with the numerical approximation of the hemivariational inequalities, extragradient algorithms for solving various classes of variational inequalities, the proximal methods for treating a nonlinear inverse problem in linearized elasticity relating to tumor identification, and discontinuous Galerkin methods for solving an elliptic variational inequality of the fourth order.

Part III, entitled *Applications*, is dedicated to the study of miscellaneous classes of problems issued from Contact Mechanics. Topics include the analysis of a dynamic contact model for Gao beam, an energy-consistent numerical model for the dynamic frictional contact between a hyperelastic body and a foundation, a nonclamped frictional contact problem in thermo-viscoelasticity, the large time asymptotics for contact problems for Navier–Stokes equation and antiplane elasticity, hemivariational inequalities, and history-dependent hemivariational inequalities in dynamic elastic-viscoplastic contact problems.

It is our pleasure to acknowledge the support of a Marie Curie International Research Staff Exchange Scheme Fellowship within the 7th European Community Framework Programme under Grant Agreement no. PIRSES-GA-2011-295118, of the National Science Centre of Poland within the Maestro Project under Grant Agreement no. DEC-2012/06/A/ST1/00262, of the Chang Jiang Scholars Program, and of the Simons Foundation under Grant Agreement No. 207052 and No. 228187. We thank the School of Mathematics and Statistics, Xi'an Jiaotong University, China, for the support of *International Workshop on Variational and Hemivariational Inequalities: Theory, Numerics, and Applications*, Xi'an, June 28–30, 2014. Many of the results in this volume were presented at this important and successful meeting. As editors, we are pleased to express our appreciation to all the authors for their valuable contributions and all the referees for their helpful suggestions. We extend our gratitude to Professor David Y. Gao for inviting us to make this contribution in the Springer book series on *Advances in Mechanics and Mathematics*.

Iowa City, IA, USA
Kraków, Poland
Perpignan, France
December 2014

Weimin Han
Stanisław Migórski
Mircea Sofonea

References

1. Baiocchi, C., Capelo, A.: Variational and Quasivariational Inequalities: Applications to Free-Boundary Problems. Wiley, Chichester (1984)
2. Brézis, H.: Problèmes unilatéraux. J. Math. Pures Appl. **51**, 1–168 (1972)
3. Clarke, F.H.: Generalized gradients and applications. Trans. Am. Math. Soc. **205**, 247–262 (1975)
4. Duvaut, G., Lions, J.-L.: Inequalities in Mechanics and Physics. Springer, Berlin (1976)
5. Fichera, G.: Problemi elastostatici con vincoli unilaterali, II. Problema di Signorini con ambique condizioni al contorno. Mem. Accad. Naz. Lincei VIII **7**, 91–140 (1964)
6. Glowinski, R., Lions, J.-L., Trémolières, R.: Numerical Analysis of Variational Inequalities. North-Holland, Amsterdam (1981)
7. Hartman, P., Stampacchia, G.: On some nonlinear elliptic differential functional equations. Acta Math. **15**, 271–310 (1966)
8. Haslinger, J., Miettinen, M., Panagiotopoulos, P.D.: Finite Element Method for Hemivariational Inequalities. Theory, Methods and Applications. Kluwer Academic, Boston (1999)
9. Kikuchi, N., Oden, J.T.: Contact Problems in Elasticity: A Study of Variational Inequalities and Finite Element Methods. SIAM, Philadelphia (1988)
10. Kinderlehrer, D., Stampacchia, G.: An Introduction to Variational Inequalities and their Applications. Academic, New York (1980); Classics in Applied Mathematics, vol. 31. SIAM, Philadelphia (2000)
11. Lions, J.-L., Stampacchia, G.: Variational inequalities. Commun. Pure Appl. Anal. **20**, 493–519 (1967)
12. Naniewicz, Z., Panagiotopoulos, P.D.: Mathematical Theory of Hemivariational Inequalities and Applications. Marcel Dekker, New York (1995)

13. Panagiotopoulos, P.D.: Nonconvex energy functions. Hemivariational inequalities and substationarity principles. Acta Mech. **42**, 160–183 (1983)
14. Panagiotopoulos, P.D.: Hemivariational Inequalities. Applications in Mechanics and Engineering. Springer, Berlin (1993)
15. Signorini, A.: Sopra alcune questioni di elastostatica. Atti della Società Italiana per il Progresso delle Scienze (1933)
16. Stampacchia, G.: Formes bilineaires coercitives sur les ensembles convexes. C. R. Acad. Sci. **258**, 4413–4416 (1964)

Contents

Part I
Theory

Chapter 1
Bifurcation Phenomena for Parametric Nonlinear Elliptic Hemivariational Inequalities

Leszek Gasiński and Nikolaos S. Papageorgiou

Abstract We consider a nonlinear Dirichlet parametric problem with discontinuous right hand side in which we have a competing effect of sub and superlinear nonlinearities. A bifurcation type result is studied when the parameter tends to zero.

Keywords Nonlinear elliptic hemivariational inequality • Concave-convex problem • Bifurcation • Mountain pass theorem

AMS Classification. 35J20, 35J60, 35J92

1.1 Introduction

Let $\Omega \subseteq \mathbb{R}^N$ be a bounded domain with a C^2-boundary $\partial\Omega$. In this paper we study the following nonlinear parametric Dirichlet problem

$$(A) \qquad \begin{cases} -\Delta_p u(z) = \lambda u(z)^{q-1} + f(z, u(z)) & \text{in } \Omega, \\ u|_{\partial\Omega} = 0, \ u > 0 \text{ in } \Omega, \ \lambda > 0, \end{cases}$$

where $1 < q < p$. Here Δ_p $(1 < p < +\infty)$ denotes the p-Laplace differential operator defined by

$$\Delta_p u = \operatorname{div}(\|\nabla u\|^{p-2}\nabla u) \quad \text{for all } u \in W_0^{1,p}(\Omega).$$

L. Gasiński (✉)
Faculty of Mathematics and Computer Science, Jagiellonian University, ul. Łojasiewicza 6, 30348 Krakow, Poland
e-mail: leszek.gasinski@uj.edu.pl

N.S. Papageorgiou
Department of Mathematics, National Technical University, Zografou Campus, Athens 15780, Greece
e-mail: npapg@math.ntua.gr

W. Han et al. (eds.), *Advances in Variational and Hemivariational Inequalities*, Advances in Mechanics and Mathematics 33, DOI 10.1007/978-3-319-14490-0_1

The interesting feature of our problem is that $f(z,\cdot)$ need not be continuous. It is only jointly measurable and the function $\zeta \mapsto f(z,\zeta)$ is L^∞-bounded on bounded sets. Then the potential function

$$F(z,\zeta) = \int_0^\zeta f(z,s)\,ds$$

is no longer smooth and it is only locally Lipschitz in the ζ-variable. The usual way to treat such discontinuous problems is to pass to an inclusion by "filling in the gaps" at the discontinuity points. This way we introduce a new form of variational expressions, known as hemivariational inequalities which arise in nonsmooth mechanics (see [17]). Hemivariational inequalities can be dealt effectively using the nonsmooth critical point theory (see [5]). The other important feature of our analysis is that the potential $F(z,\cdot)$ is p-superlinear near $+\infty$, but without satisfying the usual, in such cases, Ambrosetti–Rabinowitz condition. So, in the problem we have the competing effects of the concave (sublinear) term $\lambda\zeta^{q-1}$ and of the convex (superlinear) term $f(z,\zeta)$. We prove a bifurcation type result for small values of $\lambda > 0$. Such results were proved for semilinear equations (i.e., $p = 2$) by Ambrosetti et al. [1] and for nonlinear equations driven by the p-Laplacian by García Azorero et al. [4] and Guo and Zhang [12]. In these works the reaction has the special form

$$\lambda\zeta^{q-1} + \zeta^{r-1} \quad \text{for all } \zeta \geqslant 0,$$

with $1 < q < p < r < p^*$, where

$$p^* = \begin{cases} \dfrac{Np}{N-p} & \text{if } p < N, \\ +\infty & \text{if } N \leqslant p. \end{cases}$$

More general superlinear nonlinearities were employed by Hu and Papageorgiou [14] and Marano and Papageorgiou [15]. To the best of our knowledge, there are no such bifurcation results for problems with a nonsmooth potential.

A similar analysis is also conducted for the following parametric problem

$$(B) \qquad \begin{cases} -\Delta_p u(z) = \lambda f(z,u(z)) & \text{in } \Omega, \\ u|_{\partial\Omega} = 0, \ u > 0 \text{ in } \Omega, \ \lambda > 0. \end{cases}$$

Again $f(z,\cdot)$ may be discontinuous. So, we deal with this problem by passing to an inclusion obtained by filling in the gaps at the discontinuities (hemivariational inequality). We prove a bifurcation type theorem for large values of $\lambda > 0$ and with $f(z,\cdot)$ being $(p-1)$-sublinear near $+\infty$. Similar "smooth" equations were studied in [16, 20] (semilinear problems) and in [9, 11, 13, 19] (nonlinear equations driven by the p-Laplacian). Only [9] establishes the precise behaviour of the set of positive solutions as the parameter $\lambda > 0$ varies (bifurcation type result).

Our approach uses variational methods based on the nonsmooth critical point theory and suitable truncation and comparison techniques. In the next section for the convenience of the reader, we review the main mathematical tools which we will use in the sequel.

1.2 Mathematical Background

Let X be a Banach space and X^* its topological dual. By $\langle \cdot, \cdot \rangle$ we denote the duality brackets for the pair (X^*, X). Also by $\|\cdot\|_X$ we denote the norm of X and by $\|\cdot\|_{X^*}$ the norm of X^*. If $\varphi: X \to \mathbb{R}$ is a locally Lipschitz function, then the generalized directional derivative of φ at $x \in X$ in the direction $h \in X$ is defined by

$$\varphi^0(x;h) = \limsup_{\substack{y \to x \\ \lambda \searrow 0}} \frac{\varphi(y + \lambda h) - \varphi(y)}{\lambda}.$$

Then $\varphi^0(x;\cdot)$ is sublinear and continuous and so it is the support function of a nonempty, convex and w^*-compact set in X^* defined by

$$\partial\varphi(z) = \big\{x^* \in X^* : \langle x^*, h\rangle \leqslant \varphi^0(x;h) \text{ for all } h \in X\big\}.$$

The multifunction $\partial\varphi: X \to 2^{X^*} \setminus \{\emptyset\}$ is called the generalized subdifferential of φ. If φ is continuous, convex, then it is well known that φ is locally Lipschitz and the generalized subdifferential coincides with the subdifferential in the sense of convex analysis which is given by

$$\partial\varphi(x) = \big\{x^* \in X^* : \langle x^*, h\rangle \leqslant \varphi(x+h) - \varphi(x) \text{ for all } h \in X\big\}.$$

Moreover, if $\varphi \in C^1(X)$, then φ is locally Lipschitz and

$$\partial\varphi(x) = \{\varphi'(x)\}.$$

If $g: \Omega \times \mathbb{R} \to \mathbb{R}$ is a measurable function and for almost all $z \in \Omega$, the function $g(z,\cdot)$ is bounded on bounded sets, then the function $G(z,\cdot): \mathbb{R} \to \mathbb{R}$ defined by

$$G(z,\zeta) = \int_0^\zeta g(z,s)\,ds$$

is locally Lipschitz and

$$\partial G(z,\zeta) = \big[g_l(z,\zeta), g_u(z,\zeta)\big],$$

where

$$g_l(z,\zeta) = \lim_{\delta \searrow 0} \operatorname*{ess\,inf}_{|\zeta - s| \leqslant \delta} g(z,s) \quad \text{and} \quad g_u(z,\zeta) = \lim_{\delta \searrow 0} \operatorname*{ess\,sup}_{|\zeta - s| \leqslant \delta} g(z,s).$$

Moreover, if

$$|g(z,\zeta)| \leqslant a(z)(1 + |\zeta|^{r-1}) \quad \text{for almost all } z \in \Omega, \text{ all } \zeta > 0,$$

with $1 \leqslant r < +\infty$, $a \in L^\infty(\Omega)$, then the integral map $I_G : L^r(\Omega) \to \mathbb{R}$ defined by

$$I_G(u) = \int_\Omega G(z, u(z))\, dz$$

is locally Lipschitz and

$$\partial I_G(u) \subseteq \left\{ h \in L^{r'}(\Omega) : g_l(z, u(z)) \leqslant h(z) \leqslant g_u(z, u(z)) \text{ a.e. in } z \in \Omega \right\}$$

$(\frac{1}{r} + \frac{1}{r'} = 1)$. For details we refer to [3].

For a given locally Lipschitz function $\varphi : X \to \mathbb{R}$ we say that $x \in X$ is a critical point of φ, if $0 \in \partial\varphi(x)$. It is easy to see that when x is a local minimizer or a local maximizer of φ, then x is a critical point of φ.

For a given locally Lipschitz function $\varphi : X \to \mathbb{R}$, we set

$$m_\varphi(x) = \inf_{x^* \in \partial\varphi(x)} \|x^*\|_{X^*}.$$

We say that φ satisfies the Cerami condition at the level $c \in \mathbb{R}$, if for every sequence $\{x_n\}_{n \geqslant 1} \subseteq X$ such that $\varphi(x_n) \to c$ and $(1 + \|x_n\|_X)m_\varphi(x_n) \to 0$ has a strongly convergent subsequence. If this property holds at every level $c \in \mathbb{R}$, then we say that φ satisfies the Cerami condition. If $\varphi \in C^1(X)$, then we recover the classical definition of the Cerami condition (see e.g., [6]), since $m_\varphi(x_n) = \|\varphi'(x_n)\|_{X^*}$ for all $n \geqslant 1$.

Using this compactness type condition, one can prove a deformation theorem from which follows the minimax theory of the critical values of φ. One of the main results of this theory is the next theorem which is a nonsmooth version of the so called mountain pass theorem. For details we refer to [5].

Theorem 1.1. *If X is a Banach space, $\varphi : X \to \mathbb{R}$ is a locally Lipschitz functional which satisfies the Cerami condition, $x_0, x_1 \in X$ are such that $\|x_1 - x_0\| > r > 0$,*

$$\max\left\{\varphi(x_0), \varphi(x_1)\right\} < \inf\left\{\varphi(x) : \|x - x_0\|_X = r\right\} = m_r$$

and

$$c = \inf_{\gamma \in \Gamma} \max_{0 \leqslant t \leqslant 1} \varphi\big(\gamma(t)\big),$$

where

$$\Gamma = \{\gamma \in C([0,1];X) : \gamma(0) = x_0,\ \gamma(1) = x_1\},$$

then $c \geqslant m_r$ and c is a critical value of the functional φ.

In the analysis of the two parametric equations, in addition to the Sobolev space $W_0^{1,p}(\Omega)$, we will also use the Banach space

$$C_0^1(\overline{\Omega}) = \{u \in C^1(\overline{\Omega}) : u|_{\partial\Omega} = 0\}.$$

This is an ordered Banach space with positive cone

$$C_+ = \{u \in C_0^1(\overline{\Omega}) : u(z) \geqslant 0 \text{ for all } z \in \overline{\Omega}\}.$$

This cone has a nonempty interior given by

$$\operatorname{int} C_+ = \left\{u \in C_+ : u(z) > 0 \text{ for all } z \in \Omega,\ \frac{\partial u}{\partial n}(z) < 0 \text{ for all } z \in \partial\Omega\right\},$$

where $n(\cdot)$ is the outward unit normal on $\partial\Omega$.

The next theorem is a special case of a more general result of [8] and relates local C^1 and local $W^{1,p}$ minimizers for a large class of nonsmooth functionals.

So, let $j_0 \colon \Omega \times \mathbb{R} \to \mathbb{R}$ be a measurable function, which is locally Lipschitz in the $\zeta \in \mathbb{R}$ variable. Assume that

$$|u| \leqslant a(z)(1 + |\zeta|^{r-1}) \quad \text{for almost all } z \in \Omega,\ \text{all } \zeta \in \mathbb{R},\ \text{all } u \in \partial j_0(z,\zeta),$$

with $a \in L^\infty(\Omega)_+$ and $1 < r < p^*$. Let $\varphi_0 \colon W_0^{1,p}(\Omega) \to \mathbb{R}$ be the functional defined by

$$\varphi_0(u) = \frac{1}{p}\|\nabla u\|_p^p - \int_\Omega j_0(z,u(z))\,dz \quad \text{for all } u \in W_0^{1,p}(\Omega). \tag{1.1}$$

From [3, p. 83] we know that φ_0 is Lipschitz continuous on bounded sets, hence it is locally Lipschitz.

Theorem 1.2. *If φ_0 is defined by* (1.1) *and $u_0 \in W_0^{1,p}(\Omega)$ is a local $C_0^1(\overline{\Omega})$-minimizer of φ_0, i.e., there exists $\varrho_0 > 0$, such that*

$$\varphi_0(u_0) \leqslant \varphi_0(u_0 + h) \quad \textit{for all } h \in C_0^1(\overline{\Omega}),\ \|h\|_{C_0^1(\overline{\Omega})} \leqslant \varrho_0,$$

then $u_0 \in C_0^{1,\beta}(\overline{\Omega})$ for some $\beta \in (0,1)$ and u_0 is also a local $W_0^{1,p}(\Omega)$-minimizer of φ_0, i.e., there exists $\varrho_1 > 0$, such that

$$\varphi_0(u_0) \leqslant \varphi_0(u_0 + h) \quad \textit{for all } h \in W_0^{1,p}(\Omega),\ \|h\| \leqslant \varrho_1.$$

Hereafter by $\|\cdot\|$ we denote that norm of the Sobolev space $W_0^{1,p}(\Omega)$. By virtue of the Poincaré inequality, we have

$$\|u\| = \|\nabla u\|_p \quad \text{for all } u \in W_0^{1,p}(\Omega).$$

Let $A\colon W_0^{1,p}(\Omega) \to W^{-1,p'}(\Omega) = W_0^{1,p}(\Omega)^*$ (where $\frac{1}{p} + \frac{1}{p'} = 1$) be the nonlinear map defined by

$$\langle A(u), y\rangle = \int_\Omega \|\nabla u\|_{\mathbb{R}^N}^{p-2} (\nabla u, \nabla y)_{\mathbb{R}^N}\, dz \quad \text{for all } u, y \in W_0^{1,p}(\Omega). \tag{1.2}$$

The next proposition summarizes the main properties of the map A (see e.g., [7, Lemma 3.2] or [6, pp. 745–746]).

Proposition 1.3. *The map $A\colon W_0^{1,p}(\Omega) \to W^{-1,p'}(\Omega)$ defined by* (1.2) *is bounded (i.e., maps bounded sets to bounded sets), continuous, strictly monotone (hence maximal monotone too) and of type $(S)_+$, i.e., if $u_n \to u$ weakly in $W_0^{1,p}(\Omega)$ and* $\limsup_{n\to+\infty}\langle A(u_n), u_n - u\rangle \leqslant 0$, *then $u_n \to u$ in $W_0^{1,p}(\Omega)$.*

We will also need the following fact about ordered Banach spaces (see e.g., [15]).

Lemma 1.4. *If X is an ordered Banach space with order cone X_+,* $\operatorname{int} X_+ \neq \emptyset$, *$u_0 \in \operatorname{int} X_+$, then for every $u \in X$ there exists $t = t(u) > 0$ such that $t u_0 - u \in X_+$.*

In the following by $\hat{\lambda}_1$ we denote the first eigenvalue of $(-\Delta_p, W_0^{1,p}(\Omega))$. We know that $\hat{\lambda}_1 > 0$, it is isolated, simple and

$$\hat{\lambda}_1 = \inf\left\{\frac{\|\nabla u\|_p^p}{\|u\|_p^p} : u \in W_0^{1,p}(\Omega),\ u \neq 0\right\}. \tag{1.3}$$

In (1.3) the infimum is actually attained on the one dimensional eigenspace corresponding to $\hat{\lambda}_1 > 0$. From (1.3) it is clear that the elements of this eigenspace do not change sign. By $\hat{u}_1$ we denote the L^p-normalized (i.e., $\|\hat{u}_1\|_p = 1$) positive eigenfunction for $\hat{\lambda}_1 > 0$. From the nonlinear regularity theory and the nonlinear maximum principle (see e.g., [6, pp. 737–738]) we have that $\hat{u}_1 \in \operatorname{int} C_+$.

Finally let us fix our notation in this paper. For every $\zeta \in \mathbb{R}$, we set $\zeta^\pm = \max\{\pm\zeta, 0\}$. Then for $u \in W_0^{1,p}(\Omega)$ we set $u^\pm(\cdot) = u(\cdot)^\pm$. We know that

$$u^\pm \in W_0^{1,p}(\Omega), \quad u = u^+ - u^- \quad \text{and} \quad |u| = u^+ + u^-.$$

By $|\cdot|_N$ we denote the Lebesgue measure on $\mathbb{R}^N$.

1.3 Problem (A)

In this section we study problem (A).

Due to the discontinuous nature of the perturbation $f(z,\cdot)$, to study problem (A) we replace the equation by an inclusion, by filling in the gaps at the discontinuity points. So, we consider

$$(P_\lambda) \qquad \begin{cases} -\Delta_p u(z) - \lambda u(z)^{q-1} \in \partial F(z, u(z)) & \text{in } \Omega, \\ u|_{\partial\Omega} = 0, \ u > 0 \text{ in } \Omega, \ \lambda > 0. \end{cases}$$

By a positive solution of this problem we mean a function $u \in W_0^{1,p}(\Omega)$, $u \neq 0$, $u(z) \geqslant 0$ for almost all $z \in \Omega$ such that

$$\int_\Omega \|\nabla u\|_{\mathbb{R}^N}^{p-2} (\nabla u, \nabla h)_{\mathbb{R}^N}\, dz = \lambda \int_\Omega u^{q-1} h\, dz + \int_\Omega g^* h\, dz \quad \text{for all } h \in W_0^{1,p}(\Omega),$$

with $g^* \in L^{r'}(\Omega)$ (for some $r \in (p, p^*)$, $\frac{1}{r} + \frac{1}{r'} = 1$) and $g^*(z) \in \partial F(z, u(z)) = \left[f_l(z, u(z)), f_u(z, u(z))\right]$ for almost all $z \in \Omega$.

Our hypotheses on the perturbation f are the following.

H_1 $f : \Omega \times \mathbb{R} \to \mathbb{R}$ is a measurable function, such that for almost all $z \in \Omega$, $f(z, 0) = 0$, $f(z, \zeta) \geqslant 0$ for all $\zeta \geqslant 0$ and

(*i*) there exist $a \in L^\infty(\Omega)_+$ and $r \in (p, p^*)$ such that

$$f(z, \zeta) \leqslant a(z)(1 + |\zeta|^{r-1}) \quad \text{for almost all } z \in \Omega, \text{ all } \zeta \in \mathbb{R};$$

(*ii*) if $F(z, \zeta) = \int_0^\zeta f(z, s)\, ds$, then we have

$$\lim_{\zeta \to +\infty} \frac{F(z, \zeta)}{\zeta^p} = +\infty \quad \text{uniformly for almost all } z \in \Omega$$

and there exist $\mu \in (0, (r-p)\max\{\frac{N}{p}, 1\})$ and $\beta_0 > 0$ such that

$$\beta_0 \leqslant \liminf_{\zeta \to +\infty} \frac{f_l(z, \zeta)\zeta - pF(z, \zeta)}{\zeta^\mu} \quad \text{uniformly for almost all } z \in \Omega; \tag{1.4}$$

(*iii*) we have $\lim\limits_{\zeta \searrow 0} \dfrac{f_u(z, \zeta)}{\zeta^{p-1}} = 0$ uniformly for almost all $z \in \Omega$;

(*iv*) for every $\varrho > 0$, there exists $\xi_\varrho > 0$ such that for almost all $z \in \Omega$, the function $\zeta \mapsto F(z, \zeta) + \xi_\varrho \zeta^p$ is convex on $[0, \varrho]$.

Remark 1.5. Since we are looking for positive solutions and all the above hypotheses concern the positive semiaxis $(0, +\infty)$, without any loss of generality, we may assume that $f(z, \zeta) = 0$ for almost all $z \in \Omega$ and all $\zeta \leqslant 0$. Then $f_l(z, \zeta) = f_u(z, \zeta) = 0$ for almost all $z \in \Omega$ and all $\zeta < 0$, hence $\partial F(z, \zeta) = \{0\}$ for almost all $z \in \Omega$ and all $\zeta < 0$. Hypothesis $H_1(ii)$ implies that for almost all $z \in \Omega$, the potential $F(z, \cdot)$ is p-superlinear. Note that hypothesis $H_1(ii)$ implies that

$$\lim_{\zeta \to +\infty} \frac{f_l(z, \zeta)}{\zeta^{p-1}} = +\infty \text{ uniformly for almost all } z \in \Omega.$$

Nevertheless, we do not employ the usual, in such cases, Ambrosetti–Rabinowitz condition (the unilateral version since $f(z, \cdot)|_{(-\infty,0]} = 0$), which says that there exist $\tau > p$ and $M_1 > 0$ such that

$$0 < \tau F(z, \zeta) \leqslant f_l(z, \zeta)\zeta \quad \text{for almost all } z \in \Omega, \text{ all } \zeta \geqslant M_1 \tag{1.5}$$

and

$$\operatorname*{ess\,sup}_{\Omega} F(\cdot, M_1) > 0. \tag{1.6}$$

From (1.5)–(1.6) it follows that

$$c_1 \zeta^{\tau} \leqslant F(z, \zeta) \quad \text{for almost all } z \in \Omega, \text{ all } \zeta \geqslant M_1, \tag{1.7}$$

with $c_1 > 0$ (see [5, p. 298]). From (1.7) we infer the much weaker condition

$$\lim_{\zeta \to +\infty} \frac{F(z, \zeta)}{\zeta^p} = +\infty \quad \text{uniformly for almost all } z \in \Omega.$$

In our setting, this p-superlinear condition is coupled with (1.4), which is weaker than the Ambrosetti–Rabinowitz condition (1.5)–(1.6). Indeed, suppose that the Ambrosetti–Rabinowitz condition holds. We may assume that $\tau > (r - p) \max\{\frac{N}{p} - 1\}$. Then

$$\begin{aligned} \frac{f_l(z, \zeta)\zeta - pF(z, \zeta)}{\zeta^{\tau}} &= \frac{f_l(z, \zeta)\zeta - \tau F(z, \zeta)}{\zeta^{\tau}} + (\tau - p)\frac{F(z, \zeta)}{\zeta^{\tau}} \\ &\geqslant (\tau - p)c_1 \end{aligned}$$

(see (1.5)–(1.7)), so condition (1.4) is satisfied.

Note that (1.4) incorporates in our framework p-superlinear potentials with "slower" growth near $+\infty$ which fail to satisfy the Ambrosetti–Rabinowitz condition (see (1.5)–(1.6)). The next simple example illustrates this.

Example 1.6. The following locally Lipschitz potential F satisfies hypotheses H_1:

$$F(\zeta) = \begin{cases} 0 & \text{if } \zeta < 0, \\ \zeta^\tau - 2\zeta^\vartheta & \text{if } 0 \leqslant \zeta \leqslant 1, \\ \zeta^p(\ln \zeta + 1) & \text{if } 1 < \zeta, \end{cases}$$

with $1 < p < \tau < \vartheta < +\infty$. Note that F fails to satisfy Ambrosetti–Rabinowitz condition.

For every $\lambda > 0$, let $\varphi_\lambda \colon W_0^{1,p}(\Omega) \to \mathbb{R}$ be the energy functional for problem (P_λ) defined by

$$\varphi_\lambda(u) = \frac{1}{p}\|\nabla u\|_p^p - \frac{\lambda}{q}\|u^+\|_q^q - \int_\Omega F(z, u(z))\,dz \quad \text{for all } u \in W_0^{1,p}(\Omega).$$

From [3], we know that φ_λ is locally Lipschitz.

Proposition 1.7. *If hypotheses H_1 hold and $\lambda > 0$, then φ_λ satisfies the Cerami condition.*

Proof. Let $\{u_n\}_{n\geqslant 1} \subseteq W_0^{1,p}(\Omega)$ be a sequence such that

$$|\varphi_\lambda(u_n)| \leqslant M_2 \quad \text{for all } n \geqslant 1, \tag{1.8}$$

for some $M_2 > 0$ and

$$(1 + \|u_n\|)m_{\varphi_\lambda}(u_n) \to 0 \quad \text{as } n \to +\infty. \tag{1.9}$$

Since $\partial\varphi_\lambda(u_n) \subseteq W^{-1,p'}(\Omega)$ is weakly compact and the norm functional in a Banach space is weakly lower semicontinuous, by the Weierstrass theorem, we can find $u_n^* \in \partial\varphi_\lambda(u_n)$ such that

$$\|u_n^*\|_* = m_{\varphi_\lambda}(u_n) \quad \text{for all } n \geqslant 1.$$

From (1.9) we have

$$\left|\langle A(u_n), h\rangle - \lambda\int_\Omega (u_n^+)^{p-1}h\,dz - \int_\Omega g_n^* h\,dz\right| \leqslant \frac{\varepsilon_n\|h\|}{1 + \|u_n\|} \quad \text{for all } h \in W_0^{1,p}(\Omega), \tag{1.10}$$

with $\varepsilon \searrow 0$ and with

$$g_n^*(z) \in \big[f_l(z, u_n(z)), f_u(z, u_n(z))\big] \quad \text{for almost all } z \in \Omega.$$

Hence $g_n^* \in L^{r'}(\Omega)$ $(\frac{1}{r} + \frac{1}{r'} = 1)$ and $g_n^* \geqslant 0$ for all $n \geqslant 1$. In (1.10) we choose $h = -u_n^- \in W_0^{1,p}(\Omega)$. Then

$$\|\nabla u_n^-\|_p^p \leqslant \varepsilon_n \quad \text{for all } n \geqslant 1,$$

so

$$u_n^- \to 0 \quad \text{in } W_0^{1,p}(\Omega). \tag{1.11}$$

Next we show that the sequence $\{u_n^+\}_{n\geqslant 1} \subseteq W_0^{1,p}(\Omega)$ is bounded. To this end, in (1.10) we choose $h = u_n^+ \in W_0^{1,p}(\Omega)$. Then

$$-\|\nabla u_n^+\|_p^p + \lambda \|u_n^+\|_q^q + \int_\Omega g_n^* u_n^+ \, dz \leqslant \varepsilon_n \quad \text{for all } n \geqslant 1. \tag{1.12}$$

From (1.8) and (1.11), we have

$$\|\nabla u_n^+\|_p^p - \frac{\lambda p}{q}\|u_n^+\|_q^q - \int_\Omega pF(z, u_n^+)\, dz \leqslant pM_2 \quad \text{for all } n \geqslant 1. \tag{1.13}$$

We add (1.12) and (1.13) and obtain

$$\int_\Omega \big(g_n^* u_n^+ - pF(z, u_n^+)\big)\, dz \leqslant M_3 + \lambda\Big(\frac{p}{q} - 1\Big)\|u_n^+\|_q^q \quad \text{for all } n \geqslant 1, \tag{1.14}$$

for some $M_3 > 0$. By virtue of hypotheses $H_1(i)$ and (ii), we can find $\beta_1 \in (0, \beta_0)$ and $c_2 > 0$ such that

$$\beta_1 \zeta^\mu - c_2 \leqslant f_l(z, \zeta)\zeta - pF(z, \zeta) \quad \text{for almost all } z \in \Omega, \text{ all } \zeta \geqslant 0,$$

so

$$\beta_1 \zeta^\mu - c_2 \leqslant u\zeta - pF(z, \zeta) \quad \text{for almost all } z \in \Omega, \text{ all } \zeta \geqslant 0,\ u \in \partial j(z, \zeta). \tag{1.15}$$

Using (1.15) and (1.14), we obtain

$$\begin{aligned} \beta_1 \|u_n^+\|_\mu^\mu &\leqslant M_4 + \lambda\Big(\frac{p}{q} - 1\Big)\|u_n^+\|_q^q \\ &\leqslant c_3\Big(1 + \lambda\Big(\frac{p}{q} - 1\Big)\|u_n^+\|_\mu^q\Big) \quad \text{for all } n \geqslant 1, \end{aligned} \tag{1.16}$$

for some $M_4, c_3 > 0$ (recall that $\mu > q$; see hypothesis $H_1(ii)$). From (1.16) and since $\mu, p > q$, we infer that

$$\text{the sequence } \{u_n^+\}_{n\geqslant 1} \subseteq L^{\mu}(\Omega) \text{ is bounded.} \tag{1.17}$$

First assume that $N \neq p$. From hypothesis $H_1(ii)$, it is clear that without any loss of generality we may assume that $\mu \leqslant r < p^*$. So, we can find $t \in (0,1)$ such that

$$\frac{1}{r} = \frac{1-t}{\mu} + \frac{t}{p^*}.$$

Using the interpolation inequality (see e.g., [6, p. 905]) we have

$$\|u_n^+\|_r \leqslant \|u_n^+\|_{\mu}^{1-t} \|u_n^+\|_{p^*}^{t} \quad \text{for all } n \geqslant 1,$$

so

$$\|u_n^+\|_r^r \leqslant M_5 \|u_n^+\|^{tr} \quad \text{for all } n \geqslant 1,$$

for some $M_5 > 0$ (use (1.17) and the Sobolev embedding theorem). In (1.10) we choose $h = u_n^+ \in W_0^{1,p}(\Omega)$ and obtain

$$\|\nabla u_n^+\|_p^p - \lambda \|u_n^+\|_q^q - \int_{\Omega} g_n^* u_n^+ \, dz \leqslant \varepsilon_n \quad \text{for all } n \geqslant 1. \tag{1.18}$$

By virtue of hypothesis $H_1(i)$, we have

$$u\zeta \leqslant c_4(1+\zeta^r) \quad \text{for almost all } z \in \Omega, \text{ all } \zeta \geqslant 0, \ u \in \partial F(z,\zeta) \tag{1.19}$$

for some $c_4 > 0$. Using (1.19) in (1.18) and recalling that $g_n^*(z) = 0$ for almost all $z \in \{u_n < 0\}$ we have

$$\begin{aligned} \|\nabla u_n^+\|_p^p &\leqslant c_5 + \lambda \|u_n^+\|_q^q + c_6 \|u_n^+\|_r^r \\ &\leqslant c_5 + c_7(\lambda \|\nabla u_n^+\|_p^q + \|\nabla u_n^+\|_p^{tr}) \quad \text{for all } n \geqslant 1 \end{aligned} \tag{1.20}$$

for some $c_5, c_6, c_7 > 0$. The hypothesis on μ (see $H_1(ii)$) implies that $tr < p$. So, from (1.20) and since $q < p$, we infer that

$$\text{the sequence } \{u_n^+\}_{n\geqslant 1} \subseteq W_0^{1,p}(\Omega) \text{ is bounded.} \tag{1.21}$$

If $N = p$, then by definition $p^* = +\infty$, but from the Sobolev embedding theorem we know that $W_0^{1,p}(\Omega)$ is embedded (compactly) in $L^{\eta}(\Omega)$ for all $\eta \in [1, +\infty)$. So, in order for the above argument to work in the present setting, we have to replace p^* by $\eta > r$ big such that $tr = \frac{\eta(r-\mu)}{\eta-\mu} < p$ (recall that $r - \mu < p$). So, again we obtain (1.21).

From (1.11) and (1.21) it follows that the sequence $\{u_n\}_{n\geqslant 1} \subseteq W_0^{1,p}(\Omega)$ is bounded. So, passing to a subsequence if necessary, we may assume that

$$u_n \to u \quad \text{weakly in } W_0^{1,p}(\Omega), \tag{1.22}$$

$$u_n \to u \quad \text{in } L^r(\Omega). \tag{1.23}$$

In (1.10) we choose $h = u_n - u \in W_0^{1,p}(\Omega)$, pass to the limit as $n \to +\infty$ and use (1.22). We obtain

$$\lim_{n\to+\infty} \langle A(u_n), u_n - u\rangle = 0,$$

so

$$u_n \to u \quad \text{in } W_0^{1,p}(\Omega)$$

(see Proposition 1.3).

This proves that φ_λ satisfies the Cerami condition. □

In the next two propositions we show that for all $\lambda > 0$ small, the function φ_λ satisfies the mountain pass geometry (see Theorem 1.1).

Proposition 1.8. *If hypotheses H_1 hold, then there exists $\hat{\lambda} > 0$ such that for every $\lambda \in (0, \hat{\lambda})$ we can find $\varrho_\lambda > 0$ for which we have*

$$\inf\{\varphi_\lambda(u) : \|u\| = \varrho_\lambda\} = m_\lambda > 0.$$

Proof. By virtue of hypotheses $H_1(i)$ and (iii), for a given $\varepsilon > 0$ we can find $c_\varepsilon > 0$ such that

$$u \leqslant \varepsilon(\zeta^+)^{p-1} + c_\varepsilon(\zeta^+)^{r-1} \quad \text{for almost all } \zeta \in \mathbb{R} \text{ and all } u \in \partial F(z,\zeta)$$

(recall that $\partial F(z,\zeta) = \{0\}$ for almost all $z \in \Omega$, all $\zeta < 0$).

Using Lebourg's mean value theorem (see [3, p. 41]), we have

$$F(z,\zeta) \leqslant \frac{\varepsilon}{p}(\zeta^+)^p + \frac{c_\varepsilon}{r}(\zeta^+)^r \quad \text{for almost all } z \in \Omega, \text{ all } \zeta \in \mathbb{R}. \tag{1.24}$$

For every $u \in W_0^{1,p}(\Omega)$, we have

$$\begin{aligned}\varphi_\lambda(u) &= \frac{1}{p}\|\nabla u\|_p^p - \frac{\lambda}{q}\|u^+\|_q^q - \int_\Omega F(z,u)\,dz \\ &\geqslant \frac{1}{p}\Big(1 - \frac{\varepsilon}{\hat{\lambda}_1}\Big)\|\nabla u^+\|_p^p - \frac{\lambda}{q}c_8\|\nabla u^+\|_p^q - c_9\|\nabla u^+\|_p^r\end{aligned}$$

for some $c_8 > 0$, $c_9 = c_9(\varepsilon) > 0$ (see (1.3) and (1.24)). Choosing $\varepsilon \in (0, \hat{\lambda}_1)$, we have

$$\varphi_\lambda(u) \geqslant \big(c_{10} - (\frac{\lambda}{q} c_8 \|u^+\|^{q-p} + c_9 \|u^+\|^{r-p})\big) \|u^+\|^p \tag{1.25}$$

for some $c_{10} = c_{10}(\varepsilon) > 0$. Let

$$\xi(t) = \frac{\lambda}{q} c_8 t^{q-p} + c_9 t^{r-p} \quad \text{for all } t > 0.$$

Evidently $\xi \in C^1(0, +\infty)$ and since $q < p < r$, we have

$$\xi(t) \to +\infty \quad \text{as } t \searrow 0$$

and

$$\xi(t) \to +\infty \quad \text{as } t \to +\infty.$$

Therefore, we can find $t_0 > 0$ such that

$$\xi(t_0) = \min_{\mathbb{R}_+} \xi,$$

where $\mathbb{R}_+ = [0, +\infty)$. Hence $\xi'(t_0) = 0$, so

$$\frac{\lambda}{q}(q-p) c_8 t_0^{q-p-1} + (r-p) c_9 t_0^{r-p-1} = 0,$$

thus

$$(r-p) c_9 t_0^{r-q} = \frac{\lambda}{q}(p-q) c_8$$

and hence

$$t_0 = t_0(\lambda) = \big(\frac{\lambda}{q} \frac{p-q}{r-p} \frac{c_8}{c_9}\big)^{\frac{1}{r-q}}. \tag{1.26}$$

We see that $\xi(t_0(\lambda)) \to 0$ as $\lambda \searrow 0$ (see (1.26)). Therefore, we see that we can find $\hat{\lambda} > 0$ such that

$$\xi(t_0(\lambda)) < c_{10} \quad \text{for all } \lambda \in (0, \hat{\lambda}),$$

so

$$\varphi_\lambda(u) \geqslant m_\lambda > 0 \quad \text{for all } u \in W_0^{1,p}(\Omega),$$

with $\|u\| = t_0(\lambda) = \varphi_\lambda$ (see (1.25)). □

Hypothesis $H_1(iii)$ and the fact that $\hat{u}_1 \in \operatorname{int} C_+$ imply that the following result holds.

Proposition 1.9. *If hypotheses H_1 hold and $\lambda > 0$, then $\varphi_\lambda(t\hat{u}_1) \to -\infty$ as $t \to +\infty$.*

With Propositions 1.8 and 1.9 we have verified the mountain pass geometry for φ_λ when $\lambda \in (0, \hat{\lambda})$. This leads to the nonemptiness of the set

$$\mathcal{L} = \big\{\lambda > 0 : \text{ problem } (P_\lambda) \text{ has a positive solution}\big\}.$$

Proposition 1.10. *If hypotheses H_1 hold, then $\mathcal{L} \neq \emptyset$.*

Proof. Propositions 1.7–1.9 allow the use of the mountain pass theorem (see Theorem 1.1). So, for every $\lambda \in (0, \hat{\lambda}_1)$ we can find $u_0 \in W_0^{1,p}(\Omega)$ such that

$$\varphi_\lambda(0) = 0 < m_\lambda \leqslant \varphi_\lambda(u_0) \tag{1.27}$$

and

$$0 \in \partial\varphi_\lambda(u_0). \tag{1.28}$$

From (1.27) we see that $u_0 \neq 0$. From (1.28) we have

$$A(u_0) = \lambda(u_0^+)^{q-1} + g_0^*, \tag{1.29}$$

with $g_0^* \in L^{r'}(\Omega)$, $g_0^*(z) \in \partial F(z, u_0(z))$ for almost every $z \in \Omega$.

On (1.29) we act with $-u_0^- \in W_0^{1,p}(\Omega)$ and obtain

$$\|\nabla u_0^-\|_p^p \leqslant 0$$

(recall that $g_0^* \geqslant 0$; see hypothesis $H_1(i)$), so $u_0 \geqslant 0$, $u_0 \neq 0$. Then from (1.29) we have

$$-\Delta_p u_0(z) = \lambda u_0(z)^{q-1} + g_0^*(z) \quad \text{for almost all } z \in \Omega. \tag{1.30}$$

The nonlinear regularity theory (see e.g., [6, pp. 737–738]) implies that $u_0 \in C_+ \setminus \{0\}$. From (1.30) we have

$$\Delta_p u_0(z) \leqslant 0 \quad \text{for almost all } z \in \Omega,$$

so $u_0 \in \operatorname{int} C_+$ and $(0, \hat{\lambda}) \subseteq \mathcal{L}$ (see [6, p. 738]). □

Remark 1.11. Let $S_+(\lambda)$ denote the set of positive solutions for problem (P_λ). From the above proof, we see that for every $\lambda \in \mathcal{L}$, we have $S_+(\lambda) \subseteq \operatorname{int} C_+$.

Proposition 1.12. *If hypotheses H_1 hold, $\lambda_0 \in \mathcal{L}$ and $\lambda \in (0, \lambda_0)$, then $\lambda \in \mathcal{L}$.*

Proof. Since $\lambda_0 \in \mathcal{L}$, we can find $\overline{u}_0 \in \operatorname{int} C_+$ and $g_0^* \in L^{r'}(\Omega)$ such that $g_0^*(z) \in \partial F(z, \overline{u}_0(z))$ for almost all $z \in \Omega$, and

$$A(\overline{u}_0) = \lambda_0 \overline{u}_0^{q-1} + g_0^* \quad \text{in } W_0^{1,p}(\Omega). \tag{1.31}$$

We introduce the following measurable function

$$h_\lambda(z, \zeta) = \begin{cases} 0 & \text{if } \zeta < 0, \\ \lambda \zeta^{q-1} + f(z, \zeta) & \text{if } 0 \leqslant \zeta \leqslant \overline{u}_0(z), \\ \lambda \overline{u}_0(z)^{q-1} + g_0^*(z) & \text{if } \overline{u}_0(z) < \zeta. \end{cases} \tag{1.32}$$

Let

$$H_\lambda(z, \zeta) = \int_0^\zeta h_\lambda(z, s)\, ds$$

and consider the locally Lipschitz functional $\psi_\lambda : W_0^{1,p}(\Omega) \to \mathbb{R}$ defined by

$$\psi_\lambda(u) = \frac{1}{p} \|\nabla u\|_p^p - \int_\Omega H_\lambda(z, u(z))\, dz \quad \text{for all } u \in W_0^{1,p}(\Omega).$$

From (1.32) it is clear that ψ_λ is coercive. Also, it is sequentially weakly lower semicontinuous. So, we can find $u_0 \in W_0^{1,p}(\Omega)$ such that

$$\psi_\lambda(u_0) = \inf_{u \in W_0^{1,p}(\Omega)} \psi_\lambda(u). \tag{1.33}$$

Since $\overline{u}_0 \in \operatorname{int} C_+$, Lemma 1.4 implies that we can find $t \in (0, 1)$ small such that $t\hat{u}_1 \leqslant \overline{u}_0$. So we have

$$\psi_\lambda(t\hat{u}_1) \leqslant \frac{t^p}{p} \hat{\lambda}_1 - \frac{t^q}{q} \lambda \|\hat{u}_1\|_q^q$$

(see (1.32) and hypothesis $H_1(i)$). Recall that $1 < q < p$. So, choosing $t \in (0, 1)$ even smaller if necessary, we have $\psi_\lambda(t\hat{u}_1) < 0$, so

$$\psi_\lambda(u_0) < 0 = \psi_\lambda(0)$$

(see (1.33)), hence $u_0 \neq 0$. From (1.33) we have $\psi_\lambda'(u_0) = 0$, so

$$A(u_0) = \hat{v}^*, \tag{1.34}$$

with $\hat{v}^* \in L^{r'}(\Omega)$, $\hat{v}^*(z) \in \partial H_\lambda(z, u_0(z))$ for almost all $z \in \Omega$. On (1.34) we act with $-u_0^- \in W_0^{1,p}(\Omega)$. Using (1.32) we obtain

$$\|\nabla u_0^-\|_p^p = 0,$$

hence $u_0 \geqslant 0$, $u_0 \neq 0$. Also, on (1.34) we act with $(u_0 - \overline{u}_0)^+ \in W_0^{1,p}(\Omega)$. Then

$$\begin{aligned}\langle A(u_0), (u_0 - \overline{u}_0)^+\rangle &= \int_\Omega \hat{v}^*(u_0 - \overline{u}_0)^+ dz \\ &= \int_\Omega (\lambda \overline{u}_0^{q-1} + g_0^*)(u_0 - \overline{u}_0)^+ dz \\ &< \int_\Omega (\lambda_0 \overline{u}_0 + g_0^*)(u_0 - \overline{u}_0)^+ dz \\ &= \langle A(\overline{u}_0), (u_0 - \overline{u}_0)^+\rangle\end{aligned}$$

(see (1.32), (1.31) and recall that $\lambda < \lambda_0$, $\overline{u}_0 \in \operatorname{int} C_+$), so

$$\int_{\{u_0 > \overline{u}_0\}} \big(\|\nabla u_0\|_{\mathbb{R}^N}^{p-2} \nabla u_0 - \|\nabla \overline{u}_0\|_{\mathbb{R}^N}^{p-2} \nabla \overline{u}_0,\ \nabla u_0 - \nabla \overline{u}_0\big)_{\mathbb{R}^N} dz \leqslant 0,$$

thus $|\{u_0 > \overline{u}_0\}|_N = 0$, hence $u_0 \leqslant \overline{u}_0$. So, we have proved that $u_0 \in [0, \overline{u}_0] = \{u \in W_0^{1,p}(\Omega) : 0 \leqslant u(z) \leqslant \overline{u}_0(z)$ for almost all $z \in \Omega\}$.

Then from (1.32) and (1.34), we have

$$A(u_0) = \lambda u_0^{q-1} + \hat{g}^*,$$

with $\hat{g}^* \in L^{r'}(\Omega)$, $\hat{g}^*(z) \in \partial F(z, u_0(z))$ for almost all $z \in \Omega$, so

$$-\Delta_p u_0(z) = \lambda u_0(z)^{q-1} + \hat{g}^*(z) \quad \text{for almost all } z \in \Omega$$

and thus $u_0 \in S_+(\lambda) \subseteq \operatorname{int} C_+$, i.e., $\lambda \in \mathcal{L}$. ☐

Let $\lambda^* = \sup \mathcal{L}$.

Proposition 1.13. *If hypotheses H_1 hold, then $\lambda^* < +\infty$.*

Proof. By virtue of hypotheses $H_1(i)$ and (ii) and since $1 < q < p$, we can find $\overline{\lambda} > 0$ such that

$$\overline{\lambda}\zeta^{q-1} + f_l(z, \zeta) > \hat{\lambda}_1 \zeta^{p-1} \quad \text{for almost all } z \in \Omega,\ \text{all } \zeta > 0. \tag{1.35}$$

Suppose that $\overline{\lambda} \in \mathcal{L}$. Then we can find $\overline{u} \in S_+(\overline{\lambda}) \subseteq \operatorname{int} C_+$. Let $\vartheta > 0$ be the biggest positive real such that $\vartheta \hat{\lambda}_1 \leqslant \overline{u}$ (see Lemma 1.4). We have

$$-\Delta_p \overline{u}(z) = \overline{\lambda}\overline{u}(z)^{q-1} + \overline{g}^*(z) > \hat{\lambda}_1 \overline{u}(z)^{p-1}$$
$$\geqslant \hat{\lambda}_1 (\vartheta \hat{u}_1(z))^{p-1} = -\Delta_p(\vartheta \hat{u}_1(z)) \quad \text{for almost all } z \in \Omega,$$

(see (1.36) and using the fact that $\vartheta \hat{u}_1 \leqslant \overline{u}$), with $\overline{g}^* \in L^{r'}(\Omega)$, $\overline{g}^*(z) \in \partial F(z, \overline{u}(z))$ for almost all $z \in \Omega$. Invoking Proposition 2.2 of [10], we obtain

$$u - \vartheta \hat{u}_1 \in \operatorname{int} C_+$$

which contradicts the maximality of $\vartheta > 0$. This means that $\overline{\lambda} \notin \mathcal{L}$ and so $\lambda^* \leqslant \overline{\lambda} < +\infty$. □

Proposition 1.14. *If hypotheses H_1 hold and $\lambda \in (0, \lambda^*)$, then problem (P_λ) has at least two positive solutions.*

Proof. Let $\hat{\lambda} \in (\lambda, \lambda^*) \cap \mathcal{L}$ and let $\hat{u} \in S_+(\hat{\lambda}) \subseteq \operatorname{int} C_+$. We can find $\hat{g}^* \in L^{r'}(\Omega)$ with $\hat{g}^*(z) \in \partial F(z, \hat{u}(z))$ for almost all $z \in \Omega$ such that

$$-\Delta_p \hat{u}(z) = \hat{\lambda}\hat{u}(z) + \hat{g}^*(z) \quad \text{for almost all } z \in \Omega. \tag{1.36}$$

Reasoning as in the proof of Proposition 1.12, we can find $u_0 \in [0, \hat{u}] \cap S_+(\lambda)$. So, there exists $g_0^* \in L^{r'}(\Omega)$ with $g_0^*(z) \in \partial F(z, u_0(z))$ for almost all $z \in \Omega$ such that

$$-\Delta_p u_0(z) = \lambda(z) u_0(z)^{q-1} + g_0^*(z) \quad \text{for almost all } z \in \Omega. \tag{1.37}$$

Let $\varrho = \|\hat{u}\|_\infty$ and let $\xi_\varrho > 0$ be as postulated by hypothesis $H_1(iv)$. Then

$$\hat{g}^*(z) + \xi_\varrho \hat{u}(z)^{p-1} \geqslant g_0^*(z) + \xi_\varrho u_0(z)^{p-1} \quad \text{for almost all } z \in \Omega. \tag{1.38}$$

So, we have

$$\begin{aligned} -\Delta_p u_0(z) + \xi_\varrho u_0(z)^{p-1} &= \lambda u_0(z)^{q-1} + g_0^*(z) + \xi_\varrho u_0(z)^{p-1} \\ &\leqslant \lambda \hat{u}(z)^{q-1} + \hat{g}^*(z) + \xi_\varrho \hat{u}(z)^{p-1} \\ &< \hat{\lambda}\hat{u}(z)^{q-1} + \hat{g}^*(z) + \xi_\varrho \hat{u}(z)^{p-1} \\ &= -\Delta_p \hat{u}(z) + \xi_\varrho \hat{u}(z)^{p-1} \end{aligned}$$

(see (1.37), (1.38), (1.36) and recall that $u_0 \leqslant \hat{u}$, $\lambda < \hat{\lambda}$ and $\hat{u} \in \operatorname{int} C_+$).

From Proposition 2.6 of [2], we have

$$\hat{u}_0 - u_0 \in \operatorname{int} C_+. \tag{1.39}$$

Without loss of generality, we may assume that $S_+(\lambda) \cap [0, \hat{u}] = \{u_0\}$; otherwise, we already have a second positive solution and so we are done.

We introduce the following measurable function

$$\hat{g}_\lambda(z,\zeta) = \begin{cases} \lambda u_0(z)^{q-1} + g_0^*(z) & \text{if } \zeta \leqslant u_0(z), \\ \lambda \zeta^{q-1} + f_l(z,\zeta) & \text{if } u_0(z) < \zeta. \end{cases} \tag{1.40}$$

We set

$$\hat{G}_\lambda(z,\zeta) = \int_0^\zeta \hat{g}_\lambda(z,s)\,ds$$

and consider the locally Lipschitz function $\hat{\varphi}_\lambda \colon W_0^{1,p}(\Omega) \to \mathbb{R}$ defined by

$$\hat{\varphi}_\lambda(u) = \frac{1}{p}\|\nabla u\|_p^p - \int_\Omega \hat{G}_\lambda(z,u(z))\,dz \quad \text{for all } u \in W_0^{1,p}(\Omega).$$

Claim 1. $\hat{\varphi}_\lambda$ satisfies the Cerami condition.

Let $\{u_n\}_{n\geqslant 1} \subseteq W_0^{1,p}(\Omega)$ be a sequence such that

$$|\hat{\varphi}_\lambda(u_n)| \leqslant M_6 \quad \text{for all } n \geqslant 1 \tag{1.41}$$

for some $M_6 > 0$ and

$$(1 + \|u_n\|)m_{\hat{\varphi}_\lambda}(u_n) \to 0 \quad \text{as } n \to +\infty. \tag{1.42}$$

As before (see the proof of Proposition 1.7), we can find $u_n^* \in \partial\hat{\varphi}_\lambda(u_n)$ such that $m_{\hat{\varphi}_\lambda}(u_n) = \|u_n^*\|_*$ for all $n \geqslant 1$ and $u_n^* = A(u_n) - v_n^*$ with $v_n^* \in L^{r'}(\Omega)$, $v_n^*(z) \in \partial\hat{G}_\lambda(z, u_n(z))$ for almost all $z \in \Omega$, all $n \geqslant 1$.

From (1.41) we have

$$\|\nabla u_n\|_p^p - \int_\Omega p\hat{G}_\lambda(z,u_n)\,dz \leqslant pM_6 \quad \text{for all } n \geqslant 1. \tag{1.43}$$

Also, from (1.42) we have

$$|\langle u_n^*, h\rangle| \leqslant \frac{\varepsilon_n \|h\|}{1 + \|u_n\|} \quad \text{for all } h \in W_0^{1,p}(\Omega),$$

with $\varepsilon_n \searrow 0$, so

$$\left|\langle A(u_n), h\rangle - \int_\Omega v_n^* h\,dz\right| \leqslant \frac{\varepsilon_n \|h\|}{1 + \|u_n\|} \quad \text{for all } n \geqslant 1.$$

Choosing $h = u_n \in W_0^{1,p}(\Omega)$, we have

$$-\|\nabla u_n\|_p^p + \int_\Omega v_n^* u_n dz \leqslant \varepsilon_n \quad \text{for all } n \geqslant 1. \tag{1.44}$$

Adding (1.43) and (1.44), we obtain

$$\int_\Omega \big(v_n^* u_n - p\hat{G}_\lambda(z, u_n)\big)\, dz \leqslant M_7 \quad \text{for all } n \geqslant 1 \tag{1.45}$$

for some $M_7 > 0$. From [3, pp. 39 and 42], we have

$$\partial \hat{G}_\lambda(z,\zeta) \subseteq \begin{cases} \lambda u_0(z)^{q-1} + g_0^*(z) & \text{if } \zeta < u_0(z), \\ \{\tau(\lambda u_0(z)^{q-1} + \partial F(z, u_0(z)))\}_{\tau\in[0,1]} & \text{if } \zeta = u_0(z), \\ \lambda \zeta^{q-1} + \partial F(z,\zeta) & \text{if } u_0(z) < \zeta. \end{cases} \tag{1.46}$$

Recalling that $v_n^*(z) \in \partial \hat{G}_\lambda(z, u_n(z))$ for almost all $z \in \Omega$, from (1.45) and (1.46) it follows that we can find $g_n^* \in L^{r'}(\Omega)$ with $g_n^*(z) \in \partial F(z, u_n(z))$ for almost all $z \in \Omega$ for all $n \geqslant 1$ such that

$$\int_\Omega \big(g_n^* u_n - pF(z, u_n)\big)\, dz \leqslant M_8 + \lambda\Big(\frac{p}{q} - 1\Big)\|u_n\|_q^q \quad \text{for all } n \geqslant 1, \tag{1.47}$$

for some $M_8 > 0$. From (1.47) and reasoning as in the proof of Proposition 1.7 (see the part of the proof after (1.14)), we conclude that $\hat{\varphi}_\lambda$ satisfies the Cerami condition. This proves Claim 1.

Claim 2. $u_0 \in \operatorname{int} C_+$ is a local minimizer of $\hat{\varphi}_\lambda$.

We consider the following truncation of the nonlinearity $\hat{g}_\lambda(z,\zeta)$:

$$\tilde{g}_\lambda(z,\zeta) \subseteq \begin{cases} \tilde{g}_\lambda(z,\zeta) & \text{if } \zeta \leqslant \hat{u}_0(z), \\ \hat{g}_\lambda(z, \hat{u}_0(z)) & \text{if } \hat{u}_0(z) < \zeta. \end{cases} \tag{1.48}$$

This is a measurable function. We set

$$\tilde{G}_\lambda(z,\zeta) = \int_0^\zeta \tilde{g}_\lambda(z,s)\, ds$$

which is a potential function locally Lipschitz in the $\zeta \in \mathbb{R}$ variable. Let $\tilde{\varphi}_\lambda \colon W_0^{1,p}(\Omega) \to \mathbb{R}$ be the locally Lipschitz function defined by

$$\tilde{\varphi}_\lambda(u) = \frac{1}{p}\|\nabla u\|_p^p - \int_\Omega \tilde{G}_\lambda(z, u(z))\, dz \quad \text{for all } u \in W_0^{1,p}(\Omega).$$

Evidently $\tilde{\varphi}_\lambda$ is coercive (see (1.48)) and sequentially weakly lower semicontinuous. So, we can find $\tilde{u}_0 \in W_0^{1,p}(\Omega)$ such that

$$\tilde{\varphi}_\lambda(\tilde{u}_0) = \inf_{u \in W_0^{1,p}(\Omega)} \tilde{\varphi}_\lambda(u),$$

so $0 \in \partial\tilde{\varphi}_\lambda(\tilde{u}_0)$, thus

$$A(\tilde{u}_0) = \tilde{v}_0^*, \tag{1.49}$$

with $\tilde{v}_0^* \in L^{r'}(\Omega)$, $\tilde{v}_0^*(z) \in \partial\tilde{G}_\lambda(z, \tilde{u}_0(z))$ for almost all $z \in \Omega$.

On (1.49) we act with $(\tilde{u}_0 - \tilde{u}_0)^+ \in W_0^{1,p}(\Omega)$. Then

$$\begin{aligned}
\langle A(\tilde{u}_0), (\tilde{u}_0 - \hat{u}_0)^+\rangle &= \int_\Omega \tilde{v}_0^*(\tilde{u}_0 - \hat{u}_0)^+ dz \\
&= \int_\Omega \hat{g}_\lambda(z, \hat{u}_0)(\tilde{u}_0 - \hat{u}_0)^+ dz \\
&= \int_\Omega \big(\lambda\hat{u}_0^{q-1} + f_l(z, \hat{u}_0)\big)(\tilde{u}_0 - \hat{u}_0)^+ dz \\
&\leqslant \langle A(\hat{u}_0), (\tilde{u}_0 - \hat{u}_0)^+\rangle
\end{aligned}$$

(see (1.48), (1.40) and recall that $f_l(z, \hat{u}_0) = \inf \partial F(z, \hat{u}_0)$, $u_0 \leqslant \hat{u}_0$ and $\lambda < \hat{\lambda}$), so

$$\int_{\{\tilde{u}_0 > \hat{u}_0\}} \big(\|\nabla\tilde{u}_0\|_{\mathbb{R}^N}^{p-2}\nabla\tilde{u}_0 - \|\nabla\hat{u}_0\|_{\mathbb{R}^N}^{p-2}\nabla\hat{u}_0,\ \nabla\tilde{u}_0 - \nabla\hat{u}_0\big)_{\mathbb{R}^N} dz \leqslant 0,$$

thus

$$|\{\tilde{u}_0 > \hat{u}_0\}|_N = 0,$$

hence $\tilde{u}_0 \leqslant \hat{u}_0$. Also, on (1.49) we act with $(u_0 - \hat{u}_0)^+ \in W_0^{1,p}(\Omega)$. Then

$$\begin{aligned}
\langle A(\tilde{u}_0, (u_0 - \tilde{u}_0)^+\rangle &= \int_\Omega \hat{g}_\lambda(z, u_0)(u_0 - \tilde{u}_0)^+ dz \\
&= \int_\Omega \big(\lambda u_0^{q-1} - g_0^*\big)(u_0 - \tilde{u}_0)^+ dz \\
&= \langle A(u_0), (u_0 - \tilde{u}_0)^+\rangle
\end{aligned}$$

(see (1.48), (1.40), (1.37) and recall that $u_0 \leqslant \hat{u}_0$), so

$$\int_{\{u_0 > \tilde{u}_0\}} \big(\|\nabla\tilde{u}_0\|_{\mathbb{R}^N}^{p-2}\nabla\tilde{u}_0 - \|\nabla u_0\|_{\mathbb{R}^N}^{p-2}\nabla u_0,\ \nabla\tilde{u}_0 - \nabla u_0\big)_{\mathbb{R}^N} dz = 0,$$

thus $|\{u_0 > \tilde{u}_0\}|_N = 0$, hence $u_0 \leqslant \tilde{u}_0$. So, we have proved that $\tilde{u}_0 \in [u_0, \tilde{u}_0] = \{u \in W_0^{1,p}(\Omega) : u_0(z) \leqslant u(z) \leqslant \hat{u}_0(z)$ for almost all $z \in \Omega\}$. From (1.48) and (1.40) we see that (1.49) becomes

$$-\Delta_p \tilde{u}_0(z) = \lambda \tilde{u}_0(z)^{q-1} + f_l(z, \tilde{u}_0(z)) \quad \text{for almost all } z \in \Omega,$$

so $\tilde{u}_0 \in S_+(\lambda)$, hence $\tilde{u}_0 = u_0$ (recall that we have assumed $[0, \hat{u}_0] \cap S_+(\lambda) = \{u_0\}$). Note that

$$\tilde{\varphi}_\lambda|_{[0,\hat{u}_0]} = \hat{\varphi}_\lambda|_{[0,\hat{u}_0]} \tag{1.50}$$

(see (1.40) and (1.48)).

From (1.39) we know that $\hat{u}_0 - u_0 \in \operatorname{int} C_+$. Also recall that $u_0 \in S_+(\lambda) \subseteq \operatorname{int} C_+$. These facts and (1.50) imply that u_0 is a local $C_0^1(\overline{\Omega})$-minimizer of $\hat{\varphi}_\lambda$. Invoking Theorem 1.2 we infer that u_0 is a local $W_0^{1,p}(\Omega)$-minimizer of $\hat{\varphi}_\lambda$. This proves Claim 2.

Claim 2 implies that u_0 is a critical point of $\hat{\varphi}_\lambda$. Moreover, as in the proof of Claim 2, we can show that all the critical points u of $\hat{\varphi}_\lambda$ satisfy $u \geqslant u_0$. So, we may assume that u_0 is an isolated critical point of $\hat{\varphi}_\lambda$, or otherwise, we already have a whole sequence of distinct positive solutions of (P_λ) converging to u_0 and so we are done. Then, because u_0 is also a local minimizer of $\hat{\varphi}_\lambda$ (see Claim 2), as in [7, Proof of Theorem 2.12], we can find $\varrho \in (0, 1)$ small such that

$$\hat{\varphi}_\lambda(u_0) < \inf\left\{\hat{\varphi}_\lambda(u) : \ \|u - u_0\| = \varrho\right\} = \hat{m}_\varrho^\lambda. \tag{1.51}$$

Hypothesis $H_1(iii)$ implies that

$$\hat{\varphi}_\lambda(t\hat{u}) \to -\infty \quad \text{as } t \to +\infty. \tag{1.52}$$

From (1.51), (1.52) and Claim 1, we see that we can apply the mountain pass theorem (see Theorem 1.1) and find $\hat{u} \in W_0^{1,p}(\Omega)$, such that

$$\hat{\varphi}_\lambda(u_0) < \hat{m}_\varrho^\lambda \leqslant \hat{\varphi}_\lambda(\hat{u}) \tag{1.53}$$

and

$$0 \in \partial\hat{\varphi}_\lambda(\hat{u}). \tag{1.54}$$

From (1.53) we see that $\hat{u} \neq u_0$. From (1.54) we have $\hat{u} \geqslant u_0$ and so $\hat{u} \in S_+(\lambda)$. This is the second positive solution of problem (P_λ), $\lambda \in (0, \lambda^*)$. ☐

Finally, we examine what happens in the critical case $\lambda = \lambda^*$.

Proposition 1.15. *If hypotheses H_1 hold, then $\lambda^* \in \mathcal{L}$ and so $\mathcal{L} = (0, \lambda^*]$.*

Proof. Let $\{\lambda_n\}_{n\geqslant 1} \subseteq \mathcal{L}$ be such that $\lambda_n \nearrow \lambda$ as $n \to +\infty$. Then we can find $u_n \in S_+(\lambda_n) \subseteq \operatorname{int} C_+$ for all $n \geqslant 1$ such that

$$A(u_n) = \lambda_n u_n^{q-1} + g_n^* \quad \text{for all } n \geqslant 1, \tag{1.55}$$

and

$$\varphi_\lambda(u_n) < 0 \quad \text{for all } n \geqslant 1 \tag{1.56}$$

(see the proof of Proposition 1.12), where $g_n^* \in L^{r'}(\Omega)$, $g_n^*(z) \in \partial F(z, u_n(z))$ for almost all $z \in \Omega$. From (1.55) we have

$$\|\nabla u_n\|_p^p = \lambda_n \|u_n\|_q^q + \int_\Omega g_n^* u_n dz \quad \text{for all } n \geqslant 1. \tag{1.57}$$

From (1.56) we have

$$-\frac{\lambda_n p}{q}\|u_n\|_q^q - \int_\Omega pF(z, u_n)\, dz \leqslant -\|\nabla u_n\|_p^p \quad \text{for all } n \geqslant 1. \tag{1.58}$$

Adding (1.57) and (1.58) we obtain

$$\int_\Omega \big(g_n^* u_n - pF(z, u_n)\big)\, dz \leqslant \lambda_n\big(\frac{p}{q} - 1\big)\|u_n\|_q^q \quad \text{for all } n \geqslant 1. \tag{1.59}$$

From (1.59) and reasoning as in the proof of Proposition 1.7 (see the part of the proof after (1.22)), we infer that the sequence $\{u_n\}_{n\geqslant 1} \subseteq W_0^{1,p}(\Omega)$ is bounded. So, passing to a subsequence if necessary, we may assume that

$$u_n \to u^* \quad \text{weakly in } W_0^{1,p}(\Omega), \tag{1.60}$$

$$u_n \to u^* \quad \text{in } L^r(\Omega). \tag{1.61}$$

On (1.55) we act with $u_n - u^* \in W_0^{1,p}(\Omega)$, pass to the limit as $n \to$ and use (1.60). Then

$$\lim_{n\to+\infty} \langle A(u_n), u_n - u^*\rangle = 0,$$

so

$$u_n \to u^* \quad \text{in } W_0^{1,p}(\Omega) \tag{1.62}$$

(see Proposition 1.3). Also, $\{g_n^*\}_{n\geqslant 1} \subseteq L^{r'}(\Omega)$ and so we may assume that

$$g_n^* \to g^* \quad \text{weakly in } L^{r'}(\Omega), \tag{1.63}$$

so

$$g_n^*(z) \in \operatorname{conv} \limsup_{n \to +\infty} \{g_n^*(z)\} \quad \text{for almost all } z \in \Omega \tag{1.64}$$

(see [18, p. 521]).

Recall that $g_n^*(z) \in \partial F(z, u_n(z))$ for almost all $z \in \Omega$, all $n \geqslant 1$ and that $\zeta \mapsto \partial F(z, \zeta)$ is an upper semicontinuous multifunction (see [3]). So, from (1.64) it follows that

$$g^*(z) \in \partial F(z, u^*(z)) \quad \text{for almost all } z \in \Omega.$$

If in (1.55) we pass to the limit as $n \to +\infty$ and use (1.62) and (1.63), we obtain

$$A(u^*) = \lambda^*(u^*)^{q-1} + g^*,$$

so $u^* \in C_+$ and it solves problem (P_{λ^*}).

We need to show that $u^* \neq 0$ and then we will have $u^* \in S_+(\lambda^*)$, hence $\lambda^* \in \mathcal{L}$.

To this end let $\tilde{\lambda} < \lambda_1 < \lambda_2 < \ldots < \lambda_n \nearrow \lambda^*$. We consider the following auxiliary Dirichlet problem

$$\begin{cases} -\Delta_p u(z) = \tilde{\lambda} u^+(z)^{q-1} & \text{in } \Omega, \\ u|_{\partial\Omega} = 0. \end{cases} \tag{1.65}$$

The energy functional $\tilde{\xi}\colon W_0^{1,p}(\Omega) \to \mathbb{R}$ for problem (1.65) is given by

$$\tilde{\xi}(u) = \frac{1}{p}\|\nabla u\|_p^p - \frac{\tilde{\lambda}}{q}\|u^+\|_q^q \quad \text{for all } u \in W_0^{1,p}(\Omega).$$

Evidently $\tilde{\xi} \in C^1(W_0^{1,p}(\Omega))$ and it is coercive (since $1 < q < p$) and sequentially weakly lower semicontinuous. Therefore, by the Weierstrass theorem, we can find $\tilde{u} \in W_0^{1,p}(\Omega)$ such that

$$\tilde{\xi}(\tilde{u}) = \inf_{u \in W_0^{1,p}(\Omega)} \tilde{\xi}(u). \tag{1.66}$$

Since $q < p$, for $t \in (0,1)$ small, we have $\tilde{\xi}(t\hat{u}_1) < 0 = \tilde{\xi}(0)$, so

$$\tilde{\xi}(\tilde{u}) < 0 = \tilde{\xi}(0)$$

(see (1.66)), hence $\tilde{u} \neq 0$. From (1.66), we have $\tilde{\xi}'(\tilde{u}) = 0$, so

$$A(\tilde{u}) = \lambda^*(\tilde{u}^+)^{q-1}. \tag{1.67}$$

On (1.67) we act with $-\tilde{u}^- \in W_0^{1,p}(\Omega)$ and obtain that $\tilde{u} \geqslant 0$, $\tilde{u} \neq 0$. Moreover, the nonlinearity regularity theory and the nonlinear maximum principle (see e.g., [6, pp. 737–738]), imply that $\tilde{u} \in \operatorname{int} C_+$.

Since $u_0 \in \operatorname{int} C_+$, we can find $t_n > 0$ such that

$$t_n \tilde{u} \leqslant u_n \quad \text{for all } n \geqslant 1$$

(see Lemma 1.4). Let $t_n > 0$ be the biggest such positive real. Suppose that $t_n \in (0,1)$. Then

$$\begin{aligned} -\Delta_p u_n(z) &= \lambda_n u_n(z)^{q-1} + g_n^*(z) \geqslant \lambda_n u_n(z)^{q-1} \\ &\geqslant \lambda_n (t_n \tilde{u})^{q-1} > \tilde{\lambda} t_n^{p-1} \tilde{u}^{q-1} \\ &= -\Delta_p (t_n \tilde{u})(z) \quad \text{for almost all } z \in \Omega \end{aligned}$$

(since $g_n^* \geqslant 0$, $t_n \tilde{u} \leqslant u_n$ and $\tilde{\lambda} < \lambda_1 \leqslant \lambda_n$ for all $n \geqslant 1$, $t_n \in (0,1)$, $q < p$), so

$$u_n - t_n \tilde{u} \in \operatorname{int} C_+$$

(see [10]). But this contradicts the maximality of t_n. Therefore $t_n \geqslant 1$ for all $n \geqslant 1$ and so $\tilde{u} \leqslant u_n$ for all $n \geqslant 1$, thus $\tilde{u} \leqslant u^*$ (see (1.62)). Hence $u^* \neq 0$ and so $u^* \in S_+(\lambda^*) \subseteq \operatorname{int} C_+$, $\lambda^* \in \mathcal{L}$. □

Summarizing the situation for problem (P_λ), we can formulate the following bifurcation type result.

Theorem 1.16. *If hypotheses H_1 hold, then there exists $\lambda^* > 0$ such that*

(a) for all $\lambda \in (0, \lambda^*)$ problem (P_λ) has at least two positive solutions $u_0, \hat{u} \in \operatorname{int} C_+$;
(b) for $\lambda = \lambda^*$ problem (P_{λ^*}) has at least one positive solution $u^* \in \operatorname{int} C_+$;
(c) for all $\lambda > \lambda^*$ problem (P_λ) has no positive solutions.

1.4 Problem (B)

In this section we focus on problem (B) and again we provide a bifurcation type result, but now for large values of $\lambda > 0$.

Again, due to the discontinuous character of the reaction $f(z,\cdot)$, to study problem (B) we consider the elliptic inclusion

$$(Q_\lambda) \qquad \begin{cases} -\Delta_p u(z) \in \lambda \partial F(z, u(z)) & \text{in } \Omega, \\ u|_{\partial\Omega} = 0, \ u > 0 \text{ in } \Omega, \ \lambda > 0. \end{cases}$$

By a positive solution of this problem we mean a function $u \in W_0^{1,p}(\Omega)$, $u \neq 0$, $u(z) \geqslant 0$ for almost all $z \in \Omega$ such that

$$\int_\Omega \|\nabla u\|_{\mathbb{R}^N}^{p-2}(\nabla u, \nabla h)_{\mathbb{R}^N}\, dz = \lambda \int_\Omega g^* h\, dz \quad \text{for all } h \in W_0^{1,p}(\Omega),$$

with $g^* \in L^{p'}(\Omega)$, $g^*(z) \in \partial F(z, u(z)) = \big[f_l(z, u(z)), f_u(z, u(z))\big]$ for almost all $z \in \Omega$. So, again we fill in the gaps at the discontinuity points of $f(z, \cdot)$.

The hypotheses on the reaction f are the following.

H_2 $f : \Omega \times \mathbb{R} \to \mathbb{R}$ is a measurable function, such that for almost all $z \in \Omega$, $f(z, 0) = 0$, $f(z, \zeta) \geqslant 0$ for all $\zeta \geqslant 0$ and

(*i*) for every $\varrho > 0$, there exist $a_\varrho \in L^\infty(\Omega)_+$ such that

$$f(z, \zeta) \leqslant a_\varrho(z) \quad \text{for almost all } z \in \Omega, \text{ all } \zeta \in [0, \varrho];$$

(*ii*) $\lim\limits_{\zeta \to +\infty} \dfrac{f_u(z, \zeta)}{\zeta^{p-1}} = 0$ uniformly for almost all $z \in \Omega$;

(*iii*) $\lim\limits_{\zeta \to 0} \dfrac{f_u(z, \zeta)}{\zeta^{p-1}} = 0$ uniformly for almost all $z \in \Omega$;

(*iv*) for every $\varrho > 0$, there exists $\xi_\varrho > 0$ such that for almost all $z \in \Omega$, the function

$$\zeta \mapsto F(z, \zeta) + \xi_\varrho \zeta^p$$

is convex on $[0, \varrho]$ (recall that $F(z, \zeta) = \int_0^\zeta f(z, s)\, ds$);

(*v*) there exists $\tilde{v} \in L^1(\Omega)$ such that $\int_\Omega F(z, \tilde{v}(z))\, dz > 0$.

Remark 1.17. As before (see Sect. 1.3) without any loss of generality we assume that $f(z, \zeta) = 0$ for almost all $z \in \Omega$ and all $\zeta \leqslant 0$.

Example 1.18. The following locally Lipschitz potential F satisfies hypotheses H_2:

$$F(\zeta) = \begin{cases} 0 & \text{if } \zeta < 0, \\ \zeta^\tau - \zeta^\eta & \text{if } 0 \leqslant \zeta \leqslant 1, \\ \zeta^q \ln \zeta & \text{if } 1 < \zeta, \end{cases}$$

with $1 < q < p < \tau < \eta < +\infty$.

As before let

$$\hat{\mathcal{L}} = \{\lambda > 0 : \text{ problem } (Q_\lambda) \text{ admits a positive solution}\}.$$

Also for $\lambda > 0$ by $\hat{S}_+(\lambda)$ we denote the set of positive solutions of (Q_λ). Finally let

$$\lambda_* = \inf \hat{\mathcal{L}}$$

(if $\hat{\mathcal{L}} = \emptyset$, then $\lambda_* = +\infty$).

Proposition 1.19. *If hypotheses H_2 hold, then for every $\lambda \in \hat{\mathcal{L}}$, $\hat{S}_+(\lambda) \subseteq \operatorname{int} C_+$ and $\lambda_* > 0$.*

Proof. Let $\lambda \in \hat{\mathcal{L}}$ and $u \in \hat{S}_+(\lambda)$. Then

$$-\Delta_p u(z) = \lambda g^*(z) \quad \text{for almost all } z \in \Omega,$$

with $g^* \in L^{p'}(\Omega)$, $g^*(z) \in \partial F(z, u(z))$ for almost all $z \in \Omega$. Then nonlinear regularity theory implies $u \in C_+ \setminus \{0\}$ and we have

$$\Delta_p u(z) \leqslant 0 \quad \text{for almost all } z \in \Omega$$

(since $g^* \geqslant 0$), so $u \in \operatorname{int} C_+$ (by the nonlinear maximum principle; see e.g., [6, p. 738]), thus

$$\hat{S}_+(\lambda) \subseteq \operatorname{int} C_+.$$

By virtue of hypotheses $H_2(i), (ii)$ and (iii), we can find $c_{11} > 0$ such that

$$0 \leqslant f_u(z, \zeta) \leqslant c_{11} \zeta^p \quad \text{for almost all } z \in \Omega, \text{ all } \zeta \geqslant 0.$$

Let $\lambda \in (0, \frac{\hat{\lambda}_1}{c_{11}})$ and suppose that $\lambda \in \hat{\mathcal{L}}$. Then we can find $u_\lambda \in \hat{S}_+(\lambda) \subseteq \operatorname{int} C_+$ such that

$$-\Delta_p u_\lambda(z) = \lambda g^*_\lambda(z) \quad \text{for almost all } z \in \Omega,$$

with $g^*_\lambda \in L^{p'}(\Omega)$, $g^*_\lambda(z) \in \partial F(z, u_\lambda(z))$ for almost all $z \in \Omega$. Then

$$\|\nabla u_\lambda\|_p^p = \lambda \int_\Omega g^*_\lambda u_\lambda dz \leqslant \lambda c_{11} \|u_\lambda\|_p^p < \hat{\lambda}_1 \|u_\lambda\|_p^p,$$

which contradicts (1.3). Therefore $\lambda_* \geqslant \frac{\hat{\lambda}_1}{c_{11}} > 0$. ☐

Proposition 1.20. *If hypotheses H_2 hold, then $\hat{\mathcal{L}} \neq 0$ and if $\lambda \in \hat{\mathcal{L}}$ and $\mu > \lambda$, then $\mu \in \hat{\mathcal{L}}$.*

Proof. By virtue of hypotheses $H_2(i), (ii)$, for a given $\varepsilon > 0$ we can find $c_{12} = c_{12}(\varepsilon) > 0$ such that

$$F(z, \zeta) \leqslant \frac{\varepsilon}{p} (\zeta^+)^p + c_{12} \quad \text{for almost all } z \in \Omega, \text{ all } \zeta \in \mathbb{R}.$$

Then for all $\lambda > 0$ and all $u \in W^{1,p}(\Omega)$ we have

$$\varphi_\lambda(u) = \frac{1}{p}\|\nabla u\|_p^p - \lambda \int_\Omega F(z, u(z))\, dz \geqslant \frac{1}{p}\Big(1 - \frac{\varepsilon}{\hat{\lambda}_1}\Big)\|\nabla u\|_p^p - c_{12}|\Omega|_N$$

(see (1.47) and (1.22)).

So, choosing $\varepsilon \in (0, \hat{\lambda}_1)$ we see that φ_λ is coercive. Also, it is sequentially weakly lower semicontinuous. Thus, we can find $u_0 \in W_0^{1,p}(\Omega)$ such that

$$\varphi_\lambda(u_0) = \inf_{u \in W_0^{1,p}(\Omega)} \varphi_\lambda(u). \tag{1.68}$$

Consider the integral functional $K\colon L^p(\Omega) \to \mathbb{R}$ defined by

$$K(u) = \int_\Omega F(z, u(z)) \quad \text{for all } u \in L^p(\Omega).$$

Evidently K is continuous. Since $C_c^\infty(\Omega)$ is dense in $L^1(\Omega)$, using hypothesis $H_1(v)$, we see that we can find $\overline{v} \in C_c^\infty(\Omega)$ such that $K(\overline{v}) > 0$. Then we have

$$\varphi_\lambda(\overline{v}) = \frac{1}{p}\|\nabla \overline{v}\|_p^p - \lambda \int_\Omega F(z, \overline{v})\, dz.$$

Then for $\lambda > 0$ big we see that $\varphi_\lambda(\overline{v}) < 0$, so

$$\varphi_\lambda(u_0) < 0 = \varphi_\lambda(0)$$

(see (1.68)), hence $u_0 \neq 0$. From (1.68) we have $0 \in \partial\varphi_\lambda(u_0)$, so

$$A(u_0) = \lambda g_0^*, \tag{1.69}$$

with $g_0^* \in L^{r'}(\Omega)$, $g_0^*(z) \in \partial F(z, u_0(z))$ for almost all $z \in \Omega$. On (1.69) we act with $-u_0^- \in W_0^{1,p}(\Omega)$. Since $g_0^* \geqslant 0$, we obtain

$$\|\nabla u_0^-\|_p^p \leqslant 0,$$

hence $u_0 \geqslant 0$, $u_0 \neq 0$. We have

$$-\Delta_p u_0(z) = \lambda g_0^*(z) \quad \text{for almost all } z \in \Omega,$$

so

$$u_0 \in \hat{S}_+(\lambda) \quad \text{for all } \lambda > 0 \text{ big},$$

and thus

$$\hat{\mathcal{L}} \neq \emptyset.$$

Next let $\lambda \in \hat{\mathcal{L}}$ and $\mu > \lambda$. We can find $u_\lambda \in \hat{S}_+(\lambda)$. Then

$$-\Delta_p u_\lambda(z) = \lambda g^*(z) \quad \text{for almost all } z \in \Omega, \tag{1.70}$$

with $g^* \in L^{p'}(\Omega)$, $g^*(z) \in \partial F(z, u_\lambda(z))$ for almost all $z \in \Omega$.

We now introduce the following truncation of the reaction $f(z, \cdot)$:

$$h(z, \zeta) = \begin{cases} g^*(z) & \text{if } \zeta \leqslant u_\lambda(z), \\ f(z, \zeta) & \text{if } u_\lambda(z) < \zeta. \end{cases} \tag{1.71}$$

This is a measurable function and we set

$$H(z, \zeta) = \int_0^\zeta h(z, s)\, ds$$

which is locally Lipschitz in $\zeta \in \mathbb{R}$ variable. We introduce the locally Lipschitz function $\psi_\mu \colon W_0^{1,p}(\Omega) \to \mathbb{R}$ defined by

$$\psi_\mu(u) = \frac{1}{p}\|\nabla u\|_p^p - \mu \int_\Omega H(z, u(z)) \quad \text{for all } u \in W_0^{1,p}(\Omega).$$

Hypothesis $H_2(ii)$ implies that for a given $\varepsilon > 0$, we can find $M_9 = M_9(\varepsilon) > \|u_\lambda\|_\infty$ such that

$$f(z, \zeta) \leqslant \varepsilon \zeta^{p-1} \quad \text{for almost all } z \in \Omega, \text{ all } \zeta \geqslant M_9,$$

so

$$h(z, \zeta) \leqslant \varepsilon \zeta^{p-1} \quad \text{for almost all } z \in \Omega, \text{ all } \zeta \geqslant M_9 \tag{1.72}$$

(see (1.71)). From (1.72), (1.71) and hypothesis $H_2(i)$, it follows that

$$H(z, \zeta) \leqslant \frac{\varepsilon}{p}\zeta^p + c_{13} \quad \text{for almost all } z \in \Omega, \text{ all } \zeta \in \mathbb{R} \tag{1.73}$$

for some $c_{13} > 0$. Hence for all $u \in W_0^{1,p}(\Omega)$ we have

$$\psi_\mu(u) \geqslant \frac{1}{p}\Big(1 - \frac{\mu\varepsilon}{\hat{\lambda}_1}\Big)\|\nabla u\|_p^p - c_{14}$$

for some $c_{14} > 0$ (see (1.73) and 1.3). Thus for $\varepsilon \in (0, \frac{\hat{\lambda}_1}{\mu})$, ψ_μ is coercive. Also, it is sequentially weakly lower semicontinuous. So, we can find $u_\mu \in W_0^{1,p}(\Omega)$ such that

$$\psi_\mu(u_\mu) = \inf_{u \in W_0^{1,p}(\Omega)} \psi_\mu(u).$$

Then,

$$0 \in \partial \psi_\mu(u_\mu),$$

so

$$A(u_\mu) = \mu \hat{g}^*, \tag{1.74}$$

with $\hat{g}^* \in L^{p'}(\Omega)$, $\hat{g}^*(z) \in \partial H(z, u_\mu(z))$ for almost all $z \in \Omega$.

Recall that

$$\partial H(z,\zeta) \subseteq \begin{cases} g^*(z) & \text{if } \zeta < u_\lambda(z), \\ \{\tau \partial F(z, u_\lambda(z))\}_{\tau \in [0,1]} & \text{if } \zeta = u_\lambda(z), \\ \partial F(z,\zeta) & \text{if } u_\lambda(z) < \zeta \end{cases} \tag{1.75}$$

(see [3, p. 42]). So, acting with $(u_\lambda - u_\mu)^+ \in W_0^{1,p}(\Omega)$ in (1.74), we obtain

$$\begin{aligned} \langle A(u_\mu), (u_\lambda - u_\mu)^+ \rangle &= \int_\Omega \mu \hat{g}^* (u_\lambda - u_\mu)^+ dz \\ &\geqslant \int_\Omega \lambda g^* (u_\lambda - u_\mu)^+ dz \\ &= \langle A(u_\lambda), (u_\lambda - u_\mu)^+ \rangle \end{aligned}$$

(see (1.75) and (1.70)). So

$$\int_{\{u_\lambda > u_\mu\}} \left(\|\nabla u_\mu\|_{\mathbb{R}^N}^{p-2} \nabla u_\mu - \|\nabla u_\lambda\|_{\mathbb{R}^N}^{p-2} \nabla u_\lambda, \nabla u_\lambda - \nabla u_\mu \right)_{\mathbb{R}^N} \geqslant 0.$$

Therefore,

$$|\{u_\lambda > u_\mu\}|_N = 0;$$

hence $u_\lambda \leqslant u_\mu$. Then from (1.71) and (1.74), we have

$$-\Delta_p u_\mu(z) \leqslant \mu \partial F(z, u_\mu(z)) \quad \text{for almost all } z \in \Omega,$$

so

$$u_\mu \in \hat{S}_+(\mu)$$

and so $\mu \in \hat{\mathcal{L}}$. □

Proposition 1.21. *If hypotheses* H_2 *hold and* $\lambda > \lambda_*$, *then problem* (Q_λ) *admits at least two positive solutions*

$$u_0, \hat{u} \in \operatorname{int} C_+, \quad u_0 \neq \hat{u}.$$

Proof. Let $\vartheta \in (\lambda_*, \lambda) \cap \hat{\mathcal{L}}$. We can find $u_\vartheta \in \hat{S}_+(\vartheta) \subseteq \operatorname{int} C_+$. Reasoning as in the proof of Proposition 1.20, we can find $u_0 \in \hat{S}_+(\lambda) \in \operatorname{int} C_+$ such that

$$u_\vartheta \leqslant u_0 \quad \text{and} \quad \psi_\lambda(u_0) = \inf_{u \in W_0^{1,p}(\Omega)} \psi_\lambda(u). \tag{1.76}$$

Here $\psi_\lambda \colon W_0^{1,p}(\Omega) \to \mathbb{R}$ is the locally Lipschitz functional as in the proof of Proposition 1.20, resulting from the lower truncation of $f(z,\cdot)$ at $u_\vartheta(z)$ (see (1.71)).

Let

$$C(u_\vartheta) = \big\{u \in W_0^{1,p}(\Omega) : u_\vartheta(z) \leqslant u(z) \text{ for almost all } z \in \Omega\big\}.$$

From (1.71) we see that

$$\psi_\lambda|_{C(u_\vartheta)} = \varphi_\lambda|_{C(u_\vartheta)} - \hat{\xi}, \tag{1.77}$$

where

$$\hat{\xi} = \int_\Omega F(z, u_\vartheta(z))\,dz - \int_\Omega g^*(z) u_\vartheta(z)\,dz,$$

with $g^* \in L^{p'}(\Omega)$, $g^*(z) \in \partial F(z, u_\vartheta(z))$ for almost all $z \in \Omega$. Recall that $\varphi_\lambda \colon W_0^{1,p}(\Omega) \to \mathbb{R}$ is the locally Lipschitz energy functional for problem (Q_λ) defined by

$$\varphi_\lambda(u) = \frac{1}{p}\|\nabla u\|_p^p - \lambda \int_\Omega F(z, u(z))\,dz \quad \text{for all } u \in W_0^{1,p}(\Omega).$$

Let $\varrho = \|u_0\|_\infty$ and let $\xi_\varrho > 0$ be as postulated by hypothesis $H_2(iv)$. Then for some $g_0^*, g_\vartheta^* \in L^{p'}(\Omega)$ with $g_0^*(z) \in \partial F(z, u_0(z))$, $g_\vartheta^*(z) \in \partial F(z, u_\vartheta(z))$ for almost all $z \in \Omega$ and with $\hat{\xi}_\varrho > \xi_\varrho$, we have

$$\begin{aligned} -\Delta_p u_0(z) + \hat{\xi}_\varrho u_0(z)^{p-1} &= \lambda g_0^*(z) + \hat{\xi}_\varrho u_0(z)^{p-1} \\ &\geqslant \lambda g_\vartheta^*(z) + \hat{\xi}_\varrho u_\vartheta(z)^{p-1} \end{aligned}$$

$$\geqslant \vartheta g_{\vartheta}^{*}(z) + \hat{\xi}_{\varrho} u_{\vartheta}(z)^{p-1}$$
$$= -\Delta_p u_{\vartheta}(z) + \hat{\xi}_{\varrho} u_{\vartheta}(z)^{p-1} \quad \text{for almost all } z \in \Omega$$

(see hypothesis $H_2(iv)$), so

$$u_0 - u_{\vartheta} \in \operatorname{int} C_+ \tag{1.78}$$

(see [2]). From (1.76)–(1.78), it follows that u_0 is a local $C_0^1(\overline{\Omega})$-minimizer of φ_λ. Invoking Theorem 1.2, we conclude that u_0 is also a local $W_0^{1,p}(\Omega)$-minimizer of φ_λ.

By virtue of hypothesis $H_2(iii)$, for a given $\varepsilon > 0$, we can find $\delta = \delta(\varepsilon) > 0$ such that

$$f(z,\zeta) \leqslant \varepsilon \zeta^{p-1} \quad \text{for almost all } z \in \Omega, \text{ all } \zeta \in [0,\delta],$$

so

$$F(z,\zeta) \leqslant \frac{\varepsilon}{p}|\zeta|^p \quad \text{for almost all } z \in \Omega, \text{ all } |\zeta| \leqslant \delta \tag{1.79}$$

(recall that $f(z,\cdot)|_{(-\infty,0]} \equiv 0$).

Let $u \in C_0^1(\overline{\Omega})$ with $\|u\|_{C_0^1(\overline{\Omega})} \leqslant \delta$. We have

$$\varphi_\lambda(u) = \frac{1}{p}\|\nabla u\|_p^p - \lambda \int_\Omega F(z,u(z))\,dz \geqslant \frac{1}{p}\Big(1 - \frac{\lambda\varepsilon}{\hat{\lambda}_1}\Big)\|\nabla u\|_p^p \tag{1.80}$$

(see (1.79) and (1.22)).

Choosing $\varepsilon \in (0, \frac{\hat{\lambda}_1}{\lambda})$ from (1.80) we infer that $u = 0$ is a local $C_0^1(\overline{\Omega})$-minimizer of φ_λ, hence by Theorem 1.2, $u = 0$ is a local $W_0^{1,p}(\Omega)$-minimizer of φ_λ.

Without any loss of generality we may assume that $0 = \varphi_\lambda(0) \leqslant \varphi_\lambda(u_0)$ (the analysis is similar if the opposite inequality holds). Also, we may assume that u_0 is an isolated critical point of φ_λ. Indeed, if this not the case, we can find a sequence $\{u_n\}_{n\geqslant 1} \subseteq W_0^{1,p}(\Omega)$ such that

$$u_n \to u_0 \quad \text{in } W_0^{1,p}(\Omega) \quad \text{and} \quad 0 \in \partial\varphi_\lambda(u_n) \quad \text{for all } n \geqslant 1. \tag{1.81}$$

From (1.81) we have

$$A(u_n) = \lambda g_n^*,$$

with $g_n^* \in L^{p'}(\Omega)$, $g_n^*(z) \in \partial F(z, u_n(z))$ for almost all $z \in \Omega$. We have

$$\begin{cases} -\Delta_p u_n(z) = \lambda g_n^*(z) & \text{in } \Omega, \\ u_n|_{\partial\Omega} = 0, \end{cases}$$

for $n \geqslant 1$. From nonlinear regularity theory (see e.g., [6, pp. 737–738]), we know that we can find $\beta \in (0,1)$ and $M_{10} > 0$ such that

$$u_n \in C_0^{1,\beta}(\overline{\Omega}) \quad \text{and} \quad \|u_n\|_{C_0^{1,\beta}(\overline{\Omega})} \leqslant M_{10} \quad \text{for all } n \geqslant 1. \tag{1.82}$$

From (1.82) and since $C_0^{1,\beta}(\overline{\Omega})$ is embedded compactly in $C_0^1(\overline{\Omega})$, we infer that

$$u_n \to u_0 \quad \text{in } C_0^1(\overline{\Omega})$$

(see (1.81)). Since $u_0 \in \operatorname{int} C_+$, we see that we can find $n_0 \geqslant 1$ such that $\{u_n\}_{n \geqslant n_0} \subseteq \operatorname{int} C_+$ are all distinct positive solutions of (Q_λ) and we are done.

So, we assume that u_0 is an isolated critical point of φ_λ. Since u_0 is a local minimizer of φ_λ, as in [7] (see the proof of Theorem 2.12), we can find $\varrho > 0$ small such that

$$0 = \varphi_\lambda(0) \leqslant \varphi_\lambda(u_0) < \inf\left\{\varphi_\lambda(u) : \|u - u_0\| = \varrho\right\} = m_\varrho. \tag{1.83}$$

From the proof of Proposition 1.20 we know that φ_λ is coercive. So, it satisfies the Cerami condition. This fact and (1.83) permit the use of Theorem 1.1 and find $\hat{u} \in W_0^{1,p}(\Omega)$ such that

$$m_\varrho \leqslant \varphi_\lambda(\hat{u}) \quad \text{and} \quad 0 \in \partial\varphi_\lambda(\hat{u}). \tag{1.84}$$

From the inequality in (1.84) and (1.83) we have $\hat{u} \notin \{0, u_0\}$. From the inclusion in (1.84) it follows that $\hat{u} \in \hat{S}_+(\lambda) \subseteq \operatorname{int} C_+$. □

Finally we deal with the critical case $\lambda = \lambda_*$.

Proposition 1.22. *If hypotheses H_2 hold, then $\lambda_* \in \hat{\mathcal{L}}$, i.e., $\hat{\mathcal{L}} = [\lambda_*, +\infty)$.*

Proof. Let $\{\lambda_n\}_{n \geqslant 1} \subseteq \hat{\mathcal{L}}$ and $\lambda_n \searrow \lambda_*$. Let $u_n \in \hat{S}_+(\lambda_n)$ for $n \geqslant 1$. We can find $\{g_n^*\}_{n \geqslant 1} \subseteq L^{p'}(\Omega)$ such that $g_n^*(z) \in \partial F(z, u_n(z))$ for almost all $z \in \Omega$ and

$$A(u_n) = \lambda_n g_n^* \quad \text{for all } n \geqslant 1, \tag{1.85}$$

so

$$\|\nabla u_n\|_p^p \leqslant \lambda_1 \int_\Omega g_n^* u_n dz \quad \text{for all } n \geqslant 1. \tag{1.86}$$

Using hypotheses $H_2(i)$ and (ii), for a given $\varepsilon > 0$ we can find $c_{15} = c_{15}(\varepsilon) > 0$ such that

$$f_u(z,\zeta)\zeta \leqslant \varepsilon\zeta^p + c_{15} \quad \text{for almost all } z \in \Omega, \text{ all } \zeta \geqslant 0. \tag{1.87}$$

Returning to (1.86) and using (1.87) and (1.3) we obtain

$$\left(1-\frac{\lambda_1\varepsilon}{\hat{\lambda}_1}\right)\|\nabla u_n\|_p^p \leqslant \lambda_1 c_{15}|\Omega|_N. \tag{1.88}$$

Choosing $\varepsilon \in (0, \frac{\hat{\lambda}_1}{\lambda_1})$ from (1.88) we infer that the sequence $\{u_n\}_{n\geqslant 1} \subseteq W_0^{1,p}(\Omega)$ is bounded. So, we may assume that

$$u_n \to u_* \quad \text{weakly in } W_0^{1,p}(\Omega), \tag{1.89}$$

$$u_n \to u_* \quad \text{in } L^p(\Omega). \tag{1.90}$$

On (1.85) we act with $u_n - u_* \in W_0^{1,p}(\Omega)$, pass to the limit as $n \to +\infty$ and use (1.89). We obtain

$$\lim_{n\to+\infty} \langle A(u_n), u_n - u_*\rangle = 0,$$

so

$$u_n \to u_* \quad \text{in } W_0^{1,p}(\Omega) \tag{1.91}$$

(see Proposition 1.3). Also note that the sequence $\{g_n^*\}_{n\geqslant 1} \subseteq L^{p'}(\Omega)$ is bounded (see (1.87) and (1.91)). So, we may assume that

$$g_n^* \to g^* \quad \text{weakly in } L^{p'}(\Omega) \tag{1.92}$$

and as in the proof of Proposition 1.15, we have

$$g^*(z) \in \partial F(z, u_*(z)) \quad \text{for almost all } z \in \Omega. \tag{1.93}$$

Passing to the limit as $n \to +\infty$ in (1.85) and using (1.91) and (1.92), we obtain

$$A(u_*) = \lambda_* g^*$$

so u_* is a solution of (Q_λ) (see (1.93)).

We need to show that $u_* \neq 0$ in order to conclude that $\lambda^* \in \hat{\mathcal{L}}$.

Arguing indirectly, suppose that $u_* = 0$. We set $y_n = \frac{u_n}{\|u_n\|}$ for all $n \geqslant 1$. Then $\|u_n\| = 1$ for all $n \geqslant 1$ and so we may assume that

$$y_n \to y \quad \text{weakly in } W_0^{1,p}(\Omega),$$

$$y_n \to y \quad \text{in } L^p(\Omega).$$

Hypotheses $H_2(i)$–(iii) imply that we can find $c_{16} > 0$ such that

$$0 \leqslant f_u(z, \zeta) \leqslant c_{16}|\zeta|^{p-1} \quad \text{for almost all } z \in \Omega, \text{ all } \zeta \in \mathbb{R}. \tag{1.94}$$

Then, from (1.85) we have

$$A(y_n) = \lambda_n \frac{g_n^*}{\|u_n\|^{p-1}} \quad \text{for all } n \geqslant 1. \tag{1.95}$$

From (1.94) it is clear that the sequence $\left\{\frac{g_n^*}{\|u_n\|^{p-1}}\right\}_{n\geqslant 1} \subseteq L^{p'}(\Omega)$ is bounded. Using hypothesis $H_2(iii)$, we have (at least for a subsequence)

$$\frac{g_n^*}{\|u_n\|^{p-1}} \to 0 \quad \text{in } L^{p'}(\Omega). \tag{1.96}$$

Acting on (1.95) with $y_n - y \in W_0^{1,p}(\Omega)$, passing to the limit as $n \to +\infty$ and using (1.96), we obtain

$$\lim_{n\to+\infty} \langle A(y_n), y_n - y\rangle = 0,$$

so

$$y_n \to y \quad \text{in } W_0^{1,p}(\Omega) \tag{1.97}$$

(see Proposition 1.3), hence $\|y\| = 1$. Therefore, from (1.95) in the limit as $n \to +\infty$, we obtain

$$A(y) = 0$$

(see (1.96) and (1.97)), thus $y = 0$ (see Proposition 1.3) which contradicts (1.97). Therefore $u_* \neq 0$ and so $\lambda_* \in \hat{\mathcal{L}}$. □

Summarizing the situation for problem (Q_λ), we can formulate the following bifurcation type theorem for large values of $\lambda > 0$.

Theorem 1.23. *If hypotheses H_2 hold, then there exists λ_* such that*

(a) for all $\lambda > \lambda_*$ problem (Q_λ) has at least two positive solutions $u_0, \hat{u} \in \operatorname{int} C_+$, $u_0 \neq \hat{u}$;
(b) for $\lambda = \lambda_*$ problem (Q_{λ_*}) has at least one positive solution $u_* \in \operatorname{int} C_+$;
(c) for all $\lambda \in (0, \lambda_*)$ problem (Q_λ) has no positive solutions.

Acknowledgements This research was supported by the Marie Curie International Research Staff Exchange Scheme Fellowship within the 7th European Community Framework Programme under Grant Agreement No. 295118, the National Science Center of Poland under grant no. N N201

604640, the International Project co-financed by the Ministry of Science and Higher Education of Republic of Poland under grant no. W111/7.PR/2012, the National Science Center of Poland under Maestro Advanced Project no. DEC-2012/06/A/ST1/00262.

References

1. Ambrosetti, A., Brézis, H., Cerami, G.: Combined effects of concave and convex nonlinearities in some elliptic problems. J. Funct. Anal. **122**, 519–543 (1994)
2. Arcoya, D., Ruiz, D.: The Ambrosetti–Prodi problem for the p-Laplace operator. Commun. Partial Differ. Equ. **31**, 849–865 (2006)
3. Clarke, F.H.: Optimization and Nonsmooth Analysis. Wiley, New York (1983)
4. García Azorero, J., Manfredi, J., Peral Alonso, I.: Sobolev versus Hölder local minimizers and global multiplicity for some quasilinear elliptic equations. Commun. Contemp. Math. **2**, 385–404 (2000)
5. Gasiński, L., Papageorgiou, N.S.: Nonsmooth Critical Point Theory and Nonlinear Boundary Value Problems. Chapman and Hall/CRC, Boca Raton, FL (2005)
6. Gasiński, L., Papageorgiou, N.S.: Nonlinear Analysis. Chapman and Hall/CRC, Boca Raton, FL (2006)
7. Gasiński, L., Papageorgiou, N.S.: Nodal and multiple constant sign solutions for resonant p-Laplacian equations with a nonsmooth potential. Nonlinear Anal. **71**, 5747–5772 (2009)
8. Gasiński, L., Papageorgiou, N.S.: Multiple solutions for nonlinear coercive problems with a nonhomogeneous differential operator and a nonsmooth potential. Set-Valued Var. Anal. **20**, 417–443 (2012)
9. Gasiński, L., Papageorgiou, N.S.: Bifurcation-type results for nonlinear parametric elliptic equations. Proc. R. Soc. Edinb. Sect. A **142**, 595–623 (2012)
10. Guedda, M., Véron, L.: Quasilinear elliptic equations involving critical Sobolev exponents. Nonlinear Anal. **13**, 879–902 (1989)
11. Guo, Z.: Some existence and multiplicity results for a class of quasilinear elliptic eigenvalue problems. Nonlinear Anal. **18**, 957–971 (1992)
12. Guo, Z., Zhang, Z.: $W^{1,p}$ versus C^1 local minimizers and multiplicity results for quasilinear elliptic equations. J. Math. Anal. Appl. **286**, 32–50 (2003)
13. Hu, S., Papageorgiou, N.S.: Multiple positive solutions for nonlinear eigenvalue problems with the p-Laplacian. Nonlinear Anal. **69**, 4286–4300 (2008)
14. Hu, S., Papageorgiou, N.S.: Multiplicity of solutions for parametric p-Laplacian equations with nonlinearity concave near the origin. Tôhoku Math. J. **62**, 137–162 (2010)
15. Marano, S.A., Papageorgiou, N.S.: Positive solutions to a Dirichlet problem with p-Laplacian and concave-convex nonlinearity depending on a parameter. Commun. Pure Appl. Anal. **12**, 815–829 (2013)
16. Maya, C., Shivaji, R.: Multiple positive solutions for a class of semilinear elliptic boundary value problems. Nonlinear Anal. **38**, 497–504 (1999)
17. Panagiotopoulos, P.D.: Hemivariational Inequalities: Applications to Mechanics and Engineering. Springer, Berlin (1993)
18. Papageorgiou, N.S., Kyritsi, S.: Handbook of Applied Analysis. Springer, New York (2009)
19. Perera, K.: Multiple positive solutions of a class of quasilinear elliptic boundary value problems. Electron. J. Differ. Equ. **7**, 1–5 (2003)
20. Rabinowitz, P.H.: Pairs of positive solutions of nonlinear elliptic partial differential equations. Indiana Univ. Math. J. **23**, 173–186 (1973)

Chapter 2
Evolutionary Inclusions and Hemivariational Inequalities

Stanisław Migórski, Anna Ochal, and Mircea Sofonea

Abstract We consider a class of abstract nonlinear evolutionary inclusions of first order with a multivalued Clarke subgradient term. We use a surjectivity result for pseudomonotone multivalued operators in order to prove existence and uniqueness of solutions. Next, we use the Banach fixed point theorem and establish the unique solvability to evolutionary inclusion with history-dependent operators. We apply this result to second order evolutionary inclusions governed by two history-dependent operators, which depend on the solution and its time derivative, respectively. Finally, we specify existence and uniqueness results for nonlinear first and second order hemivariational inequalities with or without history-dependent operators.

Keywords Evolutionary inclusion • Hemivariational inequality • Clarke subdifferential • Pseudomonotone operator • History-dependent operator • Weak solution

AMS Classification. 34G25, 35L86, 47J20, 47J35, 35L90, 47J22

2.1 Introduction

In this chapter we study abstract evolutionary inclusions of first and second order involving a multivalued term in the form of the Clarke subdifferential of a locally Lipschitz functional. We also investigate the first and second order evolutionary hemivariational inequalities with integral functionals defined on the boundary of a

---

S. Migórski (✉) • A. Ochal
Faculty of Mathematics and Computer Science, Jagiellonian University in Krakow, ul. Łojasiewicza 6, 30348 Krakow, Poland
e-mail: stanislaw.migorski@uj.edu.pl; ochal@ii.uj.edu.pl

M. Sofonea
Laboratoire de Mathématiques et Physique (LAMPS), Université de Perpignan Via Domitia, 52 Avenue Paul Alduy, 66860 Perpignan, France
e-mail: sofonea@univ-perp.fr

W. Han et al. (eds.), *Advances in Variational and Hemivariational Inequalities*, Advances in Mechanics and Mathematics 33, DOI 10.1007/978-3-319-14490-0_2

given domain. Our main results concern the existence and uniqueness of solutions to several classes of nonlinear inclusions and hemivariational inequalities.

We recall that hemivariational inequalities, introduced and studied in the early 1980s by Panagiotopoulos in [19, 20], are closely related to nonlinear inclusions of subdifferential type. During the last three decades the number of contributions to this area was enormous, both in the theory and applications of hemivariational inequalities, see, e.g., [4,7,8,10–18,21,22] to mention only a few, and the references therein.

The first order evolutionary inclusions we study in this chapter are governed by a time dependent pseudomonotone operator and are considered in the framework of evolution triple of spaces. First, we provide results on the existence and uniqueness of solutions to the Cauchy problem for this class of inclusions. Existence is established by employing a surjectivity result for multivalued L-pseudomonotone operator (cf. Proposition 2.2). The uniqueness of a solution is obtained in a case when the operator is strongly monotone, the subdifferential of the superpotential satisfies a relaxed monotonicity condition, and a smallness hypothesis holds. In comparison to our earlier results on the first order evolutionary subdifferential inclusions, the main result we present here, Theorem 2.6, does not require the introduction of an additional intermediate space to the problem. Next, using a fixed point argument, we provide a result on the unique solvability of the class of first order evolutionary subdifferential inclusions with history-dependent operators. Subsequently, based on our results for first order problems, we derive several results on the unique solvability of the Cauchy problems for second order evolutionary inclusions involving history-dependent operators. In this way we obtain some generalizations of the results obtained in [10, 12–14, 16, 22]. Finally, we give results on the existence and uniqueness of solutions to various evolutionary hemivariational inequalities of first and second order with or without history-dependent operators.

The results of this chapter find many applications in carrying out the variational analysis of various contact models of mechanics. The systematic studies of problems in Contact Mechanics by exploiting results on inclusions and hemivariational inequalities can be found in recent monograph [16]. Applications of results of this chapter to the study of nonlinear problems which describe the contact between a deformable body and a foundation are illustrated in Chap. 14 of this volume.

The chapter is organized as follows. In Sect. 2.2 we recall some notation and present some auxiliary material. In Sects. 2.3 and 2.4 we treat, respectively, abstract evolutionary inclusions of first and second order both without and with history-dependent operators. Results on existence and uniqueness of solutions to hemivariational inequalities of first and second order are delivered in Sect. 2.5.

2.2 Notation and Preliminaries

In this section we present the notation and recall some definitions from nonlinear analysis needed in the sequel. For further details, we refer e.g. to [6, 24].

Given a reflexive Banach space X we denote by $\|\cdot\|_X$ its norm, by X^* its topological dual and by $\langle\cdot,\cdot\rangle_{X^*\times X}$ the duality pairing of X and X^*. For a set $U \subset X$ we define $\|U\|_X = \sup\{\|u\|_X \mid u \in U\}$, and we denote by $\mathcal{L}(X, Y)$ a space of linear and bounded operators between the Banach space X with values in the Banach space Y with the usual norm $\|\cdot\|_{\mathcal{L}(X,Y)}$.

Let $A\colon X \to 2^{X^*}$ be a multivalued operator. We say it is *pseudomonotone*, if the following conditions are satisfied:

(a) for all $u \in X$ the set Au is a nonempty, bounded, closed, and convex, subset of X^*.
(b) A is upper semicontinuous from each finite dimensional subspace of X to X^* endowed with the weak topology.
(c) if $\{u_n\} \subset X$, $u_n \to u$ weakly in X and $u_n^* \in Au_n$ is such that the following inequality holds $\limsup \langle u_n^*, u_n - u\rangle_{X^*\times X} \le 0$, then for every $y \in X$, there exists $u^*(y) \in Au$ such that

$$\langle u^*(y), u - y\rangle_{X^*\times X} \le \liminf \langle u_n^*, u_n - y\rangle_{X^*\times X}.$$

An operator $A\colon X \to 2^{X^*}$ is called *bounded*, if it maps bounded sets into bounded ones. It is called *coercive* if either the domain $D(A)$ of A is bounded or $D(A)$ is unbounded and

$$\lim_{\|u\|_X\to\infty,\ u\in D(A)} \frac{\inf\{\langle u^*, u\rangle_{X^*\times X} \mid u^* \in Au\}}{\|u\|_X} = +\infty.$$

The following version of the notion of pseudomonotonicity of multivalued operators will be useful in what follows. Let $L\colon D(L) \subset X \to X^*$ be a linear, maximal monotone operator. We say that $A\colon X \to 2^{X^*}$ is *pseudomonotone with respect to* $D(L)$ or L-*pseudomonotone*, if the conditions (a) and (b) hold and, in addition,

(d) if $\{u_n\} \subset D(L)$, $u_n \to u$ weakly in X, $Lu_n \to Lu$ weakly in X^*, $u_n^* \in Au_n$ is such that $u_n^* \to u^*$ weakly in X^* and, in addition, $\limsup \langle u_n^*, u_n - u\rangle_{X^*\times X} \le 0$, then $u^* \in Au$ and $\langle u_n^*, u_n\rangle_{X^*\times X} \to \langle u^*, u\rangle_{X^*\times X}$.

The class of multivalued bounded L-pseudomonotone operators is closed under addition of mappings. We have the following result which corresponds to Proposition 2 of [4].

Proposition 2.1. *Let X be a reflexive Banach space, $L\colon D(L) \subset X \to X^*$ a linear, maximal monotone operator and A_1, $A_2\colon X \to 2^{X^*}$ multivalued L-pseudomonotone operators. If A_1 or A_2 is bounded, then $A_1 + A_2$ is L-pseudomonotone.*

We also need the following surjectivity result (Theorem 1.3.73 of [6]) for operators which are pseudomonotone with respect to $D(L)$.

Proposition 2.2. *Let X be a reflexive and strictly convex Banach space, let $L: D(L) \subset X \to X^*$ be a linear and maximal monotone operator. If $A: X \to 2^{X^*}$ is bounded, coercive, and pseudomonotone with respect to $D(L)$, then $L + A$ is surjective, i.e., $(L + A)(D(L)) = X^*$.*

Next, we recall some definitions for single-valued operators. A single-valued operator $A: X \to X^*$ is said to be *pseudomonotone* if it is bounded (i.e., it maps bounded subsets of X into bounded subsets of X^*) and satisfies the inequality

$$\langle Au, u - v\rangle_{X^* \times X} \le \liminf \langle Au_n, u_n - v\rangle_{X^* \times X} \text{ for all } v \in X,$$

whenever $\{u_n\}$ converges weakly in X towards u with

$$\limsup \langle Au_n, u_n - u\rangle_{X^* \times X} \le 0.$$

Let $L: D(L) \subset X \to X^*$ be a linear, maximal monotone operator. An operator $A: X \to X^*$ is said to be *L-pseudomonotone*, if for any sequence $\{u_n\}$ in $D(L)$ with $u_n \to u$ weakly in X, $Lu_n \to Lu$ weakly in X^* and $\limsup \langle Au_n, u_n - u\rangle_{X^* \times X} \le 0$, it follows that $Au_n \to Au$ weakly in X^* and $\langle Au_n, u_n\rangle_{X^* \times X} \to \langle Au, u\rangle_{X^* \times X}$. For characterizations of pseudomonotonicity we refer to [6, 24].

Given an operator $A: (0, T) \times X \to X^*$, its *Nemitsky* (superposition) operator is the operator $\mathcal{A}: L^2(0, T; X) \to L^2(0, T; X^*)$ defined by $(\mathcal{A}v)(t) = A(t, v(t))$ for $v \in L^2(0, T; X)$ and $t \in (0, T)$.

Next, we provide the definitions of the generalized directional derivative and the generalized gradient of Clarke for a locally Lipschitz function $h: E \to \mathbb{R}$, where E is a Banach space (see [3, 5, 16]). The generalized directional derivative of h at $x \in E$ in the direction $v \in E$, denoted by $h^0(x; v)$, is defined by

$$h^0(x; v) = \limsup_{y \to x,\, t \downarrow 0} \frac{h(y + tv) - h(y)}{t}.$$

The generalized gradient of h at $x \in E$, denoted by $\partial h(x)$, is a subset of a dual space E^* given by

$$\partial h(x) = \{\zeta \in E^* \mid h^0(x; v) \ge \langle \zeta, v\rangle_{E^* \times E} \text{ for all } v \in E\}.$$

The locally Lipschitz function h is called regular (in the sense of Clarke) at $x \in E$ if for all $v \in E$ the one-sided directional derivative $h'(x; v)$ exists and satisfies $h^0(x; v) = h'(x; v)$ for all $v \in E$.

Finally, we recall the following fixed point type result (see Lemma 7 in [9] or Proposition 3.1 in [23]).

Lemma 2.3. *Let E be a Banach space and $0 < T < \infty$. Let $\Lambda\colon L^2(0,T;E) \to L^2(0,T;E)$ be an operator such that*

$$\|(\Lambda\eta_1)(t) - (\Lambda\eta_2)(t)\|_E^2 \le c \int_0^t \|\eta_1(s) - \eta_2(s)\|_E^2 \, ds$$

for every η_1, $\eta_2 \in L^2(0,T;E)$, a.e. $t \in (0,T)$ with a constant $c > 0$. Then Λ has a unique fixed point in $L^2(0,T;E)$, i.e., there exists a unique $\eta^* \in L^2(0,T;E)$ such that $\Lambda\eta^* = \eta^*$.

Lemma 2.3 is a consequence of the Banach Contraction Principle applied to the nth iteration of operator Λ for sufficiently large $n \in \mathbb{N}$. This lemma will be used in various places in the rest of this chapter.

2.3 First Order Inclusions

In this section we present an existence result for an abstract inclusion of first order. We treat the inclusion within the setting of an evolution triple of spaces.

Let $V \subset H \subset V^*$ be an evolution triple of spaces, i.e., V is a reflexive separable Banach space, H is a separable Hilbert space, the embedding $V \subset H$ is continuous, and V is dense in H. Let $0 < T < +\infty$. We set $\mathcal{V} = L^2(0,T;V)$ and introduce the space $\mathcal{W}$ defined by $\mathcal{W} = \{w \in \mathcal{V} \mid w' \in \mathcal{V}^*\}$, where the time derivative $w' = \partial w/\partial t$ is understood in the sense of vector-valued distributions and $\mathcal{V}^* = L^2(0,T;V^*)$ is the dual space to $\mathcal{V}$. Recall that for any Banach space Y the space $L^2(0,T;Y)$ of vector-valued functions consists of all measurable functions $u\colon (0,T) \to Y$ for which $\int_0^T \|u(t)\|_Y^2 \, dt$ is finite. It is well known that the space $\mathcal{W}$ endowed with the graph norm $\|w\|_{\mathcal{W}} = \|w\|_{\mathcal{V}} + \|w'\|_{\mathcal{V}^*}$ is a Banach space which is separable and reflexive, due to the separability and reflexivity of $\mathcal{V}$ and $\mathcal{V}^*$. Furthermore, let $\mathcal{H} = L^2(0,T;H)$. Identifying $\mathcal{H}$ with its dual, we have the following continuous embeddings $\mathcal{W} \subset \mathcal{V} \subset \mathcal{H} \subset \mathcal{V}^*$. It is also well known that the embedding $\mathcal{W} \subset C(0,T;H)$ is continuous, where $C(0,T;H)$ denotes the space of continuous functions on $[0,T]$ with values in H. If, in addition, we suppose that the embedding $V \subset H$ is compact, then by the Lions–Aubin lemma (cf. Theorem 3.4.13 of [6]), we know that the embedding $\mathcal{W} \subset \mathcal{H}$ is also compact. The duality pairing between $\mathcal{V}^*$ and $\mathcal{V}$ is given by

$$\langle w, v \rangle_{\mathcal{V}^* \times \mathcal{V}} = \int_0^T \langle w(t), v(t) \rangle_{V^* \times V} \, dt \quad \text{for } w \in \mathcal{V}^*,\ v \in \mathcal{V},$$

where $\langle \cdot, \cdot \rangle_{V^* \times V}$ stands for the duality brackets for the pair (V^*, V). Moreover, let X be a separable and reflexive Banach space.

Let $A\colon (0,T) \times V \to V^*$ be a nonlinear operator, $M\colon V \to X$ be a linear and continuous operator, and $J\colon (0,T) \times X \to \mathbb{R}$ be a functional. We denote by M^* the adjoint operator to M and by ∂J the Clarke generalized subdifferential of the functional J with respect to its second variable. Denote by $v_0 \in V$ an initial value. With these data we consider the following evolutionary inclusion.

Problem 2.4. Find $w \in \mathcal{W}$ such that

$$\left.\begin{aligned} & w'(t) + A(t, w(t)) + M^*\partial J(t, Mw(t)) \ni f(t) \ \text{ a.e. } t \in (0,T), \\ & w(0) = v_0. \end{aligned}\right\}$$

In the study of Problem 2.4 we introduce the following definition.

Definition 2.5. A function $w \in \mathcal{W}$ is called a solution of Problem 2.4 if there exists $\zeta \in \mathcal{V}^*$ such that

$$\left.\begin{aligned} & w'(t) + A(t, w(t)) + \zeta(t) = f(t) \ \text{ a.e. } \ t \in (0,T), \\ & \zeta(t) \in M^*\partial J(t, Mw(t)) \ \text{ a.e. } \ t \in (0,T), \\ & w(0) = v_0. \end{aligned}\right\}$$

We consider the following the hypotheses on the data.

$$\left.\begin{aligned} & A\colon (0,T) \times V \to V^* \text{ is such that} \\ & \text{(a) } A(\cdot, v) \text{ is measurable on } (0,T) \text{ for all } v \in V. \\ & \text{(b) } A(t,\cdot) \text{ is pseudomonotone on } V \text{ for a.e. } t \in (0,T). \\ & \text{(c) } \|A(t,v)\|_{V^*} \le a_0(t) + a_1\|v\|_V \text{ for all } v \in V, \text{ a.e. } t \in (0,T) \\ & \qquad \text{with } a_0 \in L^2(0,T), a_0 \ge 0 \text{ and } a_1 > 0. \\ & \text{(d) } \langle A(t,v), v\rangle_{V^*\times V} \ge \alpha\|v\|_V^2 \text{ for all } v \in V, \text{ a.e. } t \in (0,T) \\ & \qquad \text{with } \alpha > 0. \\ & \text{(e) } A(t,\cdot) \text{ is strongly monotone for a.e. } t \in (0,T), \text{ i.e., there} \\ & \qquad \text{is } m_1 > 0 \text{ such that for all } v_1, v_2 \in V, \text{ a.e. } t \in (0,T) \\ & \qquad \langle A(t,v_1) - A(t,v_2), v_1 - v_2\rangle_{V^*\times V} \ge m_1\|v_1 - v_2\|_V^2. \end{aligned}\right\} \tag{2.1}$$

$$\left.\begin{aligned} & M \in \mathcal{L}(V,X) \text{ is such that its Nemytski operator} \\ & \qquad \mathcal{M}\colon \mathcal{W} \subset L^2(0,T;V) \to L^2(0,T;X) \text{ is compact.} \end{aligned}\right\} \tag{2.2}$$

$$\left.\begin{array}{l} J\colon(0,T)\times X\to\mathbb{R} \text{ is such that} \\ \text{(a) } J(\cdot,v) \text{ is measurable on } (0,T) \text{ for all } v\in X. \\ \text{(b) } J(t,\cdot) \text{ is locally Lipschitz on } X \text{ for a.e. } t\in(0,T). \\ \text{(c) } \|\partial J(t,v)\|_{X^*}\le c_0(t)+c_1\|v\|_X \text{ for all } v\in X, \\ \quad \text{a.e. } t\in(0,T) \text{ with } c_0\in L^2(0,T),\, c_0,c_1\ge 0. \\ \text{(d) } \langle z_1-z_2, v_1-v_2\rangle_{X^*\times X}\ge -m_2\|v_1-v_2\|_X^2 \text{ for all} \\ \quad z_i\in\partial J(t,v_i),\ z_i\in X^*,\ v_i\in X,\ i=1,2, \text{ a.e. } t\in(0,T) \\ \quad \text{with } m_2\ge 0. \end{array}\right\} \tag{2.3}$$

$$\left.\begin{array}{l} \text{One of the following conditions is satisfied} \\ \text{(a) } \alpha>2\sqrt{2}\,c_1\,\|M\|^2, \text{ where } \|M\|=\|M\|_{\mathcal{L}(V,X)}. \\ \text{(b) } J^0(t,v;-v)\le d_0\,(1+\|v\|_X) \text{ for all } v\in X, \text{a.e. } t\in(0,T) \\ \quad \text{with } d_0\ge 0. \end{array}\right\} \tag{2.4}$$

$$m_1\ge m_2\,\|M\|^2. \tag{2.5}$$

$$f\in\mathcal{V}^*,\ v_0\in V. \tag{2.6}$$

We have the following existence and uniqueness result.

Theorem 2.6. *Assume that hypotheses* (2.1)(a)–(d), (2.2), (2.3)(a)–(c), (2.4) *and* (2.6) *hold. Then Problem* 2.4 *has at least one solution. If, in addition, conditions* (2.1)(e), (2.3)(d) *and* (2.5) *hold, then the solution to Problem* 2.4 *is unique.*

Proof. We begin with the proof of the existence part. To this end, we start by providing an equivalent form to Problem 2.4 which is given by an operator inclusion. Let $\mathcal{A}\colon\mathcal{V}\to\mathcal{V}^*$ and $\mathcal{N}\colon\mathcal{V}\to 2^{\mathcal{V}^*}$ be the Nemitsky (superposition) operators corresponding to the translations of A and $M^*\circ\partial J(t,M\,\cdot)$, i.e.,

$$(\mathcal{A}w)(t)=A(t,w(t)+v_0),$$

$$(\mathcal{N}w)(t)=\{\,\zeta\in\mathcal{V}^*\mid\zeta(t)\in M^*\left(\partial J(t,M(w(t)+v_0))\right)\}$$

for $w\in\mathcal{V}$ and a.e. $t\in(0,T)$. Using these operators, we formulate the inclusion

$$\left.\begin{array}{l} w'+\mathcal{A}w+\mathcal{N}w\ni f, \\ w(0)=0. \end{array}\right\} \tag{2.7}$$

We note that $w \in \mathcal{W}$ is a solution to Problem 2.4 if and only if $w - v_0 \in \mathcal{W}$ solves inclusion (2.7).

Next, we introduce the operator $L: D(L) \subset \mathcal{V} \to \mathcal{V}^*$ given by

$$Lv = v' \quad \text{for all} \quad v \in D(L)$$

with its domain defined by $D(L) = \{w \in \mathcal{W} \mid w(0) = 0\}$. Recall that the operator L is linear and maximal monotone (cf. Proposition 32.10 in [24]). Let the operator $\mathcal{T}: \mathcal{V} \to 2^{\mathcal{V}^*}$ be given by $\mathcal{T}w = \mathcal{A}w + \mathcal{N}w$ for $w \in \mathcal{V}$. Then, problem (2.7) takes the form

$$\text{find } w \in D(L) \text{ such that } Lw + \mathcal{T}w \ni f. \tag{2.8}$$

In order to show the existence of a solution to problem (2.8), we consider the following claims that are state below and proved later in this section.

Claim 1. $\mathcal{T}$ is a bounded operator.

Claim 2. $\mathcal{T}$ is coercive.

Claim 3. $\mathcal{T}$ is L-pseudomonotone.

From these claims, by using Proposition 2.2, it follows that problem (2.8) has a solution $w \in D(L)$, so $w - v_0$ solves (2.7). This concludes the proof of the existence part in Theorem 2.6.

To prove the uniqueness part we assume that w_1, w_2 are solutions to Problem 2.4. Then, by Definition 2.5, there exist $\zeta_1, \zeta_2 \in \mathcal{V}^*$ such that

$$\left.\begin{array}{l} w_i'(s) + A(s, w_i(s)) + \zeta_i(s) = f(s) \ \text{ a.e. } s \in (0,T), \\ \zeta_i(s) \in M^* \partial J(s, M w_i(s)) \ \text{ a.e. } s \in (0,T), \\ w_i(0) = v_0 \end{array}\right\} \tag{2.9}$$

for $i = 1, 2$. Subtracting the two equations in (2.9), taking the result in duality with $w_1(s) - w_2(s)$ and integrating by parts, we obtain

$$\frac{1}{2}\|w_1(t) - w_2(t)\|_H^2 + \int_0^t \langle A(s, w_1(s)) - A(s, w_2(s)), w_1(s) - w_2(s)\rangle_{V^* \times V} ds$$
$$+ \int_0^t \langle \zeta_1(s) - \zeta_2(s), w_1(s) - w_2(s)\rangle_{V^* \times V} ds = 0$$

for all $t \in [0, T]$. We also have $\zeta_i(s) = M^* z_i(s)$ with $z_i(s) \in \partial J(s, M w_i(s))$ for a.e. $s \in (0, T)$ and $i = 1, 2$. Therefore, using condition (2.3)(d) yields

$$\int_0^t \langle \zeta_1(s) - \zeta_2(s), w_1(s) - w_2(s)\rangle_{V^* \times V} ds$$

$$
\begin{aligned}
&= \int_0^t \langle z_1(s) - z_2(s), Mw_1(s) - Mw_2(s)\rangle_{X^*\times X} ds \\
&\geq -m_2 \int_0^t \|Mw_1(s) - Mw_2(s)\|_X^2 ds \\
&\geq -m_2 \|M\|^2 \int_0^t \|w_1(s) - w_2(s)\|_V^2 ds
\end{aligned}
\tag{2.10}
$$

for all $t \in [0, T]$. We combine now (2.10) with hypothesis (2.1)(e) to obtain

$$
\begin{aligned}
&\frac{1}{2} \|w_1(t) - w_2(t)\|_H^2 + m_1 \int_0^t \|w_1(s) - w_2(s)\|_V^2 \, ds \\
&\qquad - m_2 \|M\|^2 \int_0^t \|w_1(s) - w_2(s)\|_V^2 \, ds \leq 0
\end{aligned}
$$

for all $t \in [0, T]$. From this inequality and hypothesis (2.5), we deduce that $w_1 = w_2$ on $[0, T]$. This completes the proof of the uniqueness part in Theorem 2.6. □

We turn now to prove of the three claims used in the proof of Theorem 2.6.

Proof of Claim 1. We prove that $\mathcal{T}$ is a bounded operator. Let $w \in \mathcal{V}$ and $w^* \in \mathcal{T}w$. By the definition, we have $w^* \in \mathcal{A}w + \mathcal{N}w$. From hypothesis (2.1)(a)–(d) and Lemma 11 in [10], it follows that $\|\mathcal{A}w\|_{\mathcal{V}^*} \leq \overline{a}_0 + \overline{a}_1 \|w\|_{\mathcal{V}}$ with $\overline{a}_0 \geq 0$ and $\overline{a}_1 > 0$. By an argument of Lemma 13 in [10], we have $\|\zeta\|_{\mathcal{V}^*} \leq \overline{c}_0 + \overline{c}_1 \|w\|_{\mathcal{V}}$ for all $\zeta \in \mathcal{N}w$ with $\overline{c}_0, \overline{c}_1 \geq 0$. Hence, we deduce that $\|w^*\|_{\mathcal{V}^*} \leq \overline{b}_0 + \overline{b}_1 \|w\|_{\mathcal{V}}$ with $\overline{b}_0, \overline{b}_1 \geq 0$. This inequality entails the boundedness of the operator $\mathcal{T}$. □

Proof of Claim 2. We prove that $\mathcal{T}$ is coercive. First, we assume that condition (2.4)(a) holds. Let $w \in \mathcal{V}$ and $w^* \in \mathcal{T}w$, i.e., $w^* = \mathcal{A}w + \zeta$ with $\zeta \in \mathcal{N}w$. From (2.1)(c), (d) and the inequality $|a + b|^2 \geq \frac{1}{2}a^2 - b^2$, valid for $a, b \in \mathbb{R}$, we have

$$
\begin{aligned}
&\langle \mathcal{A}w, w\rangle_{\mathcal{V}^*\times\mathcal{V}} \\
&= \int_0^T \Big(\langle A(t, w(t) + v_0), w(t) + v_0\rangle_{V^*\times V} - \langle A(t, w(t) + v_0), v_0\rangle_{V^*\times V} \Big) dt \\
&\geq \alpha \int_0^T \left(\frac{1}{2} \|w(t)\|_V^2 - \|v_0\|_V^2 \right) dt - T \|v_0\|_V \|a_0\|_{L^2(0,T)} \\
&\qquad - a_1 \|v_0\|_V \int_0^T \|w(t) + v_0\|_V dt \\
&\geq \frac{\alpha}{2} \|w\|_{\mathcal{V}}^2 - \alpha_1 \|w\|_{\mathcal{V}} - \alpha_2
\end{aligned}
\tag{2.11}
$$

with α_1, $\alpha_2 > 0$. Since $\zeta \in \mathcal{N}w$, we have $\zeta(t) = M^* z(t)$ with $z(t) \in \partial J(t, M(w(t) + v_0))$ for a.e. $t \in (0, T)$. From (2.3)(c), using the inequality $(a+b)^2 \le 2(a^2 + b^2)$ for $a, b \in \mathbb{R}$, we obtain

$$\|z(t)\|_{X^*}^2 \le 2\,c_1^2\,\|M\|^2\,\|w(t)\|_V^2 + 2\,(c_0(t) + c_1\,\|M\|\,\|v_0\|_V)^2$$

and

$$\|z\|_{L^2(0,T;X^*)}^2 = \int_0^T \|z(t)\|_{X^*}^2\,dt \le 2\,c_1^2\,\|M\|^2\,\|w\|_{\mathcal{V}}^2 + d^2$$

with $d \ge 0$. Using this inequality, we deduce that

$$\begin{aligned} |\langle \zeta, w\rangle_{\mathcal{V}^*\times\mathcal{V}}| &\le \|\zeta\|_{\mathcal{V}^*}\,\|w\|_{\mathcal{V}} \le \|M^*\|\,\|z\|_{L^2(0,T;X^*)}\,\|w\|_{\mathcal{V}} \\ &\le c_1\,\sqrt{2}\,\|M\|^2\,\|w\|_{\mathcal{V}}^2 + d\,\|w\|_{\mathcal{V}} \end{aligned}$$

which implies that

$$\langle \zeta, w\rangle_{\mathcal{V}^*\times\mathcal{V}} \ge -c_1\,\sqrt{2}\,\|M\|^2\,\|w\|_{\mathcal{V}}^2 - d\,\|w\|_{\mathcal{V}}. \tag{2.12}$$

The coercivity of $\mathcal{T}$ is now a consequence of (2.11), (2.12) and hypothesis (2.4)(a), that is

$$\begin{aligned} \langle w^*, w\rangle_{\mathcal{V}^*\times\mathcal{V}} &= \langle \mathcal{A}w, w\rangle_{\mathcal{V}^*\times\mathcal{V}} + \langle \zeta, w\rangle_{\mathcal{V}^*\times\mathcal{V}} \\ &\ge \left(\frac{\alpha}{2} - c_1\sqrt{2}\,\|M\|^2\right)\|w\|_{\mathcal{V}}^2 - (\alpha_1 + d)\,\|w\|_{\mathcal{V}} - \alpha_2. \end{aligned}$$

Secondly, we suppose condition (2.4)(b). As before, let $w \in \mathcal{V}$ and $w^* \in \mathcal{T}w$, which means that $w^* = \mathcal{A}w + \zeta$ with $\zeta \in \mathcal{N}w$. Hence $\zeta(t) = M^* z(t)$ with $z(t) \in \partial J(t, M(w(t) + v_0))$ for a.e. $t \in (0, T)$. By (2.4)(b), we obtain

$$\begin{aligned} -\langle z(t), M(w(t) + v_0)\rangle_{X^*\times X} &\le J^0(t, M(w(t) + v_0); -M(w(t) + v_0)) \\ &\le d_0\,(1 + \|M\|\,\|v_0\|_V + \|M\|\,\|w(t)\|_V) \end{aligned}$$

for a.e. $t \in (0, T)$. On the other hand, from (2.3)(c), we get

$$\begin{aligned} \langle z(t), M v_0\rangle_{X^*\times X} &\le \|z(t)\|_{X^*}\,\|M v_0\|_X \\ &\le \|M\|\|v_0\|_V\,(c_0(t) + c_1\|M(w(t) + v_0)\|_X) \\ &\le \|M\|\|v_0\|_V\,(c_0(t) + c_1\|M\|\|v_0\|_V + c_1\|M\|\|w(t)\|_V)\,. \end{aligned}$$

Therefore,

$$\langle \zeta, w\rangle_{\mathcal{V}^*\times\mathcal{V}} = \int_0^T \langle M^*z(t), w(t)\rangle_{V^*\times V} dt$$

$$= \int_0^T \Big(\langle z(t), M(w(t) + v_0)\rangle_{X^*\times X} - \langle z(t), M v_0\rangle_{X^*\times X}\Big) dt$$

$$\geq \int_0^T \Big(- d_0 (1 + \|M\| \|v_0\|_V) - d_0\|M\| \|w(t)\|_V$$

$$- \|M\| \|v_0\|_V (c_0(t) + c_1 \|M\| \|v_0\|_V) - c_1 \|v_0\|_V \|M\|^2 \|w(t)\|_V\Big) dt$$

$$\geq -d_1 \|w\|_{\mathcal{V}} - d_2$$

with $d_1, d_2 \geq 0$. Hence and from inequality (2.11), we deduce

$$\langle w^*, w\rangle_{\mathcal{V}^*\times\mathcal{V}} = \langle \mathcal{A}w, w\rangle_{\mathcal{V}^*\times\mathcal{V}} + \langle \zeta, w\rangle_{\mathcal{V}^*\times\mathcal{V}}$$
$$\geq \frac{\alpha}{2} \|w\|_{\mathcal{V}}^2 - (\alpha_1 + d_1) \|w\|_{\mathcal{V}} - (\alpha_2 + d_2)$$

which implies the coercivity of $\mathcal{T}$ and concludes the proof of Claim 2. □

Proof of Claim 3. We prove that $\mathcal{T}$ is L-pseudomonotone. We start with the properties of the operator $\mathcal{N}$ and show that it is L-pseudomonotone.

It is well known (see Proposition 2.1.2 of Clarke [3]) that the values of $\partial J(t, \cdot)$ are nonempty, weakly compact and convex subsets of X^* for a.e. $t \in (0, T)$. Hence for every $w \in \mathcal{V}$ the set $\mathcal{N}w$ is nonempty and convex in $\mathcal{V}^*$. To show that $\mathcal{N}w$ is weakly compact in $\mathcal{V}^*$, we prove that it is closed in $\mathcal{V}^*$. Let $\{\zeta_n\} \subset \mathcal{N}w$, $\zeta_n \to \zeta$ in $\mathcal{V}^*$. Then, passing to a subsequence if necessary, we have $\zeta_n(t) \to \zeta(t)$ in V^* for a.e. $t \in (0, T)$. Since for every $n \in \mathbb{N}$, $\zeta_n(t) \in M^*\partial J(t, M(w(t) + v_0))$ for a.e. $t \in (0, T)$ and the latter is a closed subset of V^*, we deduce that $\zeta(t) \in M^*\partial J(t, M(w(t) + v_0))$ for a.e. $t \in (0, T)$. Hence $\zeta \in \mathcal{N}w$. Consequently, the set $\mathcal{N}w$ is closed in $\mathcal{V}^*$ and convex, so it is also weakly closed in $\mathcal{V}^*$. Since $\mathcal{N}w$ is a bounded set in a reflexive Banach space $\mathcal{V}^*$, we obtain that $\mathcal{N}w$ is weakly compact in $\mathcal{V}^*$.

Now we prove that $\mathcal{N}$ is upper semicontinuous from $\mathcal{V}$ into $2^{\mathcal{V}^*}$ where $\mathcal{V}^*$ is endowed with the weak topology. For this purpose (cf. Proposition 4.1.4 of [5]), we show that if a set D is weakly closed in $\mathcal{V}^*$, then the set $\mathcal{N}^-(D) = \{w \in \mathcal{V} \mid \mathcal{N}w \cap D \neq \emptyset\}$ is closed in $\mathcal{V}$. Let $\{w_n\} \subset \mathcal{N}^-(D)$ be such that $w_n \to w$ in $\mathcal{V}$. Then, we may assume that $w_n(t) \to w(t)$ in V for a.e. $t \in (0, T)$. So, we can find $\zeta_n \in \mathcal{N}w_n \cap D$ for $n \in \mathbb{N}$. Since $\{w_n\}$ is bounded in $\mathcal{V}$ and $\mathcal{N}$ is a bounded map (cf. Claim 1), we have that the sequence $\{\zeta_n\}$ is bounded in $\mathcal{V}^*$. Therefore, we may suppose that

$$\zeta_n \to \zeta \text{ weakly in } \mathcal{V}^* \tag{2.13}$$

and, since D is weakly closed in $\mathcal{V}^*$, we have $\zeta \in D$. From relation $\zeta_n \in \mathcal{N} w_n$, we have

$$\zeta_n(t) = M^* z_n(t) \text{ a.e. } t \in (0,T) \tag{2.14}$$

and

$$z_n(t) \in \partial J(t, M(w_n(t) + v_0)) \text{ a.e. } t \in (0,T). \tag{2.15}$$

Next, similarly as in Claim 2, by (2.3)(c), we have

$$\|z_n\|_{L^2(0,T;X^*)} \le c_1 \sqrt{2}\, \|M\|\, \|w_n\|_{\mathcal{V}} + d,$$

where $d \ge 0$. Passing to a subsequence, if necessary, we may assume that

$$z_n \to z \text{ weakly in } L^2(0,T;X^*). \tag{2.16}$$

Note that $M(w_n(t) + v_0) \to M(w(t) + v_0)$ in X for a.e. $t \in (0,T)$ and the generalized gradient $\partial J(t,\cdot)$ is upper semicontinuous from X to X^* endowed with the weak topology (cf. Proposition 5.6.10 of [5]) with convex values for a.e. $t \in (0,T)$. Therefore, from the Convergence Theorem (cf. Theorem 5.4 in [1]), due to (2.15) and (2.16), we have

$$z(t) \in \partial J(t, M(w(t) + v_0)) \text{ a.e. } t \in (0,T).$$

Exploiting (2.13) and (2.16), we pass to the limit in (2.14) to get $\zeta(t) = M^* z(t)$ for a.e. $t \in (0,T)$. Subsequently, we obtain $\zeta \in \mathcal{N} w \cap D$, i.e., $w \in \mathcal{N}^-(D)$. This shows that $\mathcal{N}^-(D)$ is closed in $\mathcal{V}$ and proves the upper semicontinuity of $\mathcal{N}$ from $\mathcal{V}$ into the subsets of $\mathcal{V}^*$ equipped with the weak topology.

To show that $\mathcal{N}$ is L-pseudomonotone, it remains to check condition (d) on page 41. Let $\{w_n\} \subset D(L)$, $w_n \to w$ weakly in $\mathcal{W}$, $\zeta_n \in \mathcal{N} w_n$, $\zeta_n \to \zeta$ weakly in $\mathcal{V}^*$ and assume that $\limsup \langle \zeta_n, w_n - w\rangle_{\mathcal{V}^* \times \mathcal{V}} \le 0$. Since $\mathcal{N}\colon \mathcal{V} \to 2^{\mathcal{V}^*}$ is a bounded map (cf. Claim 1), we infer that $\{\zeta_n\}$ belongs to a bounded subset of $\mathcal{V}^*$, where

$$\zeta_n(t) = M^* z_n(t) \text{ a.e. } t \in (0,T) \tag{2.17}$$

and

$$z_n(t) \in \partial J(t, M(w_n(t) + v_0)) \text{ a.e. } t \in (0,T). \tag{2.18}$$

From hypothesis (2.2), we have $\mathcal{M}(w_n + v_0) \to \mathcal{M}(w + v_0)$ in $L^2(0, T; X)$. This entails that, passing to a subsequence again denoted by $\{w_n\}$, we have

$$M(w_n(t) + v_0) \to M(w(t) + v_0) \text{ in } X \text{ for a.e. } t \in (0, T). \tag{2.19}$$

Using now condition (2.3)(c) and (2.18), we deduce that $\{z_n\}$ is bounded in $L^2(0, T; X^*)$ and so we may suppose that

$$z_n \to z \text{ weakly in } L^2(0, T; X^*). \tag{2.20}$$

As before, using (2.20) and the convergence $\zeta_n \to \zeta$ weakly in $\mathcal{V}^*$, from (2.17) we obtain that $\zeta(t) = M^* z(t)$ for a.e. $t \in (0, T)$. Moreover, taking into account (2.19) and (2.20), we apply again the Convergence Theorem of [1] to inclusion (2.18). We get

$$z(t) \in \partial J(t, M(w(t) + v_0)) \text{ a.e. } t \in (0, T).$$

Therefore, $\zeta \in \mathcal{N} w$. Combining convergence (2.20) and $\mathcal{M} w_n \to \mathcal{M} w$ in $L^2(0, T; X)$, we have

$$\begin{aligned}
\langle \zeta_n, w_n \rangle_{\mathcal{V}^* \times \mathcal{V}} &= \int_0^T \langle M^* z_n(t), w_n(t) \rangle_{V^* \times V} dt \\
&= \langle z_n, \mathcal{M} w_n \rangle_{L^2(0,T;X^*) \times L^2(0,T;X)} \\
&\to \langle z, \mathcal{M} w \rangle_{L^2(0,T;X^*) \times L^2(0,T;X)} \\
&= \int_0^T \langle M^* z(t), w(t) \rangle_{V^* \times V} \, dt = \langle \zeta, w \rangle_{\mathcal{V}^* \times \mathcal{V}}.
\end{aligned}$$

This completes the proof that $\mathcal{N}$ is L-pseudomonotone.

Next, from hypothesis (2.1) it follows by Theorem 2 of Berkovits and Mustonen [2] that the operator $\mathcal{A}\colon \mathcal{V} \to \mathcal{V}^*$ is L-pseudomonotone. Exploiting the boundedness of $\mathcal{A}$ and L-pseudomonotonicity of both $\mathcal{A}$ and $\mathcal{N}$, we deduce from Proposition 2.1 that the operator $\mathcal{T} = \mathcal{A} + \mathcal{N}\colon \mathcal{V} \to 2^{\mathcal{V}^*}$ is L-pseudomonotone. This completes the proof of Claim 3. □

We are now in a position to study the following perturbation of the inclusion in Problem 2.4.

Problem 2.7. Find $w \in \mathcal{W}$ such that

$$\left.\begin{aligned}
&w'(t) + A(t, w(t)) + (\mathcal{S} w)(t) + M^* \partial J(t, M w(t)) \ni f(t) \text{ a.e. } t \in (0, T), \\
&w(0) = v_0.
\end{aligned}\right\}$$

We assume that the perturbation operator satisfies the following hypothesis.

$$\left.\begin{array}{l} \mathcal{S}\colon \mathcal{V} \to \mathcal{V}^* \text{ is such that} \\ \qquad \|(\mathcal{S}v_1)(t) - (\mathcal{S}v_2)(t)\|_{V^*} \le L_{\mathcal{S}} \displaystyle\int_0^t \|v_1(s) - v_2(s)\|_V \, ds \\ \text{for all } v_1,\, v_2 \in \mathcal{V},\ \text{a.e. } t \in (0,T) \text{ with } L_{\mathcal{S}} > 0. \end{array}\right\} \tag{2.21}$$

Note that condition (2.21) is satisfied for the operator $\mathcal{S}\colon \mathcal{V} \to \mathcal{V}^*$ given by

$$(\mathcal{S}v)(t) = R\left(\int_0^t v(s)\,ds + v_0\right) \quad \text{for all } v \in \mathcal{V},\ \text{a.e. } t \in (0,T), \tag{2.22}$$

where $R\colon V \to V^*$ is a Lipschitz continuous operator and $v_0 \in V$. It is also satisfied for the *Volterra operator* $\mathcal{S}\colon \mathcal{V} \to \mathcal{V}^*$ given by

$$(\mathcal{S}v)(t) = \int_0^t R(t-s)\,v(s)\,ds \quad \text{for all } v \in \mathcal{V},\ \text{a.e. } t \in (0,T), \tag{2.23}$$

where now $R \in L^2(0,T;\mathcal{L}(V,V^*))$. Clearly, in the case of the operators (2.22) and (2.23) the current value $(\mathcal{S}v)(t)$ at the moment t depends on the history of the values of v at the moments $0 \le s \le t$ and, therefore, we refer the operators of form (2.22) or (2.23) as *history-dependent operators*. We extend this definition to all the operators $\mathcal{S}\colon \mathcal{V} \to \mathcal{V}^*$ which satisfy condition (2.21) and, for this reason, we say that the subdifferential inclusion in Problem 2.7 represents a *history-dependent subdifferential inclusion*. Its main feature consists in the fact that they contain operators which, at any moment $t \in (0,T)$ depend on the *history* of the solution up to the moment t, see the term $(\mathcal{S}u)(t)$. This feature makes the difference with respect to the time-dependent subdifferential inclusions studied in literature in which, usually, the operators involved in are assumed to depend on the *current* value of the solution, $u(t)$.

We have the following existence and uniqueness result.

Theorem 2.8. *Assume that hypotheses* (2.1)–(2.6) *and* (2.21) *hold. Then Problem* 2.7 *has a unique solution.*

Proof. The proof is carried out in three steps and it is based on Theorem 2.6 combined with a fixed-point argument.

Step 1. We fix $\eta \in \mathcal{V}^*$ and consider the following intermediate problem.

$$\left.\begin{array}{l} \text{Find } w_\eta \in \mathcal{W} \text{ such that} \\ w_\eta'(t) + A(t, w_\eta(t)) + M^*\,\partial J(t, M w_\eta(t)) \ni f(t) - \eta(t) \\ \hfill \text{a.e. } t \in (0,T), \\ w_\eta(0) = v_0. \end{array}\right\} \tag{2.24}$$

By Theorem 2.6, problem (2.24) has a unique solution $w_\eta \in \mathcal{W}$.
Step 2. Let $\Lambda\colon \mathcal{V}^* \to \mathcal{V}^*$ be the operator defined by

$$\Lambda\eta = \mathcal{S}w_\eta \quad \text{for all } \eta \in \mathcal{V}^*,$$

where $w_\eta \in \mathcal{W}$ is the unique solution to (2.24). We prove that the operator Λ has a unique fixed point. To this end, let $\eta_1, \eta_2 \in \mathcal{V}^*$ and let $w_i = w_{\eta_i}$, for $i = 1, 2$ be the corresponding solutions to (2.24). We have

$$w_1'(s) + A(s, w_1(s)) + \zeta_1(s) = f(s) - \eta_1(s) \quad \text{a.e. } s \in (0, T), \tag{2.25}$$

$$w_2'(s) + A(s, w_2(s)) + \zeta_2(s) = f(s) - \eta_2(s) \quad \text{a.e. } s \in (0, T), \tag{2.26}$$

$$\zeta_i(s) \in M^*\partial J(s, Mw_i(s)) \quad \text{a.e. } s \in (0, T),\ i = 1, 2, \tag{2.27}$$

$$w_1(0) = w_2(0) = v_0. \tag{2.28}$$

Subtracting (2.26) from (2.25), multiplying the result in duality by $w_1(s) - w_2(s)$ and integrating by parts with initial conditions (2.28), we obtain

$$\begin{aligned}
&\frac{1}{2}\|w_1(t) - w_2(t)\|_H^2 \\
&\quad + \int_0^t \langle A(s, w_1(s)) - A(s, w_2(s)), w_1(s) - w_2(s)\rangle_{V^*\times V}\, ds \\
&\quad + \int_0^t \langle \zeta_1(s) - \zeta_2(s), w_1(s) - w_2(s)\rangle_{V^*\times V}\, ds \\
&= \int_0^t \langle \eta_2(s) - \eta_1(s), w_1(s) - w_2(s)\rangle_{V^*\times V}\, ds
\end{aligned} \tag{2.29}$$

for all $t \in [0, T]$. Note also that from (2.27) we have $\zeta_i(s) = M^*z_i(s)$ with $z_i(s) \in \partial J(s, Mw_i(s))$ for a.e. $s \in (0, t)$ and $i = 1, 2$. Therefore, by using hypothesis (2.3)(d), similarly as in (2.10), we have

$$\begin{aligned}
&\int_0^t \langle \zeta_1(s) - \zeta_2(s), w_1(s) - w_2(s)\rangle_{V^*\times V}\, ds \\
&\qquad\qquad \geq -m_2\|M\|^2 \int_0^t \|w_1(s) - w_2(s)\|_V^2\, ds
\end{aligned} \tag{2.30}$$

for all $t \in [0, T]$. We combine now (2.29), (2.30), use hypotheses (2.1)(e) and (2.5) to obtain

$$\frac{1}{2}\|w_1(t)-w_2(t)\|_H^2+\tilde{c}\int_0^t\|w_1(s)-w_2(s)\|_V^2 ds$$

$$\leq \int_0^t \|\eta_1(s)-\eta_2(s)\|_{V^*}\|w_1(s)-w_2(s)\|_V ds$$

for all $t\in[0,T]$ with $\tilde{c}=m_1-m_2\|M\|^2>0$. Hence, by the Hölder inequality, we have

$$\tilde{c}\,\|w_1-w_2\|^2_{L^2(0,t;V)}\leq\|\eta_1-\eta_2\|_{L^2(0,t;V^*)}\,\|w_1-w_2\|_{L^2(0,t;V)}$$

for all $t\in[0,T]$, which implies that

$$\|w_1-w_2\|_{L^2(0,t;V)}\leq\frac{1}{\tilde{c}}\,\|\eta_1-\eta_2\|_{L^2(0,t;V^*)} \tag{2.31}$$

for all $t\in[0,T]$.

Next, by the definition of the operator $\mathcal{S}$, hypothesis (2.21), the Hölder inequality and (2.31), we have

$$\begin{aligned}\|(\Lambda\eta_1)(t)-(\Lambda\eta_2)(t)\|^2_{V^*}&=\|(\mathcal{S}w_1)(t)-(\mathcal{S}w_2)(t)\|^2_{V^*}\\&\leq L_{\mathcal{S}}^2\,T\int_0^t\|w_1(s)-w_2(s)\|_V^2 ds\\&\leq\frac{L_{\mathcal{S}}^2\,T}{\tilde{c}^2}\int_0^t\|\eta_1(s)-\eta_2(s)\|^2_{V^*}ds\end{aligned}$$

for a.e. $t\in(0,T)$. Applying Lemma 2.3 we deduce that the operator Λ has a unique fixed point $\eta^*\in\mathcal{V}^*$ such that $\eta^*=\Lambda\eta^*$.

Step 3. Let $\eta^*\in\mathcal{V}^*$ be the unique fixed point of Λ. Then w_{η^*} is a solution to Problem 2.7, which concludes the proof of the existence part of the theorem.

To prove the uniqueness part let $w\in\mathcal{W}$ be a solution to Problem 2.7 and define the element $\eta\in\mathcal{V}^*$ by

$$\eta=\mathcal{S}w.$$

It follows that w is the solution to problem (2.24) and, by the uniqueness of solution to (2.24), we obtain $w=w_\eta$. This implies $\Lambda\eta=\eta$ and, by the uniqueness of the fixed point of Λ, we have $\eta=\eta^*$. Therefore, $w=w_{\eta^*}$, which concludes the proof of the uniqueness part of the theorem. □

2.4 Second Order Inclusions

In this section we study second order evolutionary inclusions for which we provide results on the unique solvability. Consider the following problem.

Problem 2.9. Find $u \in \mathcal{V}$ such that $u' \in \mathcal{W}$ and

$$\left.\begin{array}{l} u''(t) + A(t, u'(t)) + (\mathcal{S}u')(t) + M^* \partial J(t, Mu'(t)) \ni f(t) \text{ a.e. } t \in (0, T), \\ u(0) = u_0, \ \ u'(0) = v_0. \end{array}\right\}$$

We need the following hypothesis on the initial condition.

$$u_0 \in V. \tag{2.32}$$

We have the following existence and uniqueness result.

Theorem 2.10. *Assume that hypotheses* (2.1)–(2.6), (2.21) *and* (2.32) *hold. Then Problem* 2.9 *has a unique solution.*

Proof. We note that if $u \in \mathcal{V}$ with $u' \in \mathcal{W}$ is a solution to Problem 2.9, then $w = u'$ solves Problem 2.7. Vice versa, if $w \in \mathcal{W}$ solves Problem 2.7 and u_0 satisfies condition (2.32), then the function u defined by

$$u(t) = u_0 + \int_0^t w(s)\, ds \qquad \text{for all } \ t \in (0, T)$$

is a solution to Problem 2.9. Hence, Theorem 2.10 is a direct consequence of Theorem 2.8. □

In the following problem we consider the second order evolutionary inclusion with two history-dependent operators.

Problem 2.11. Find $u \in \mathcal{V}$ such that $u' \in \mathcal{W}$ and

$$\left.\begin{array}{l} u''(t) + A(t, u'(t)) + (\mathcal{P}u')(t) + (\mathcal{R}u)(t) + M^* \partial J(t, Mu'(t)) \ni f(t) \\ \qquad\qquad\qquad\qquad\qquad\qquad\qquad\qquad \text{a.e. } t \in (0, T), \\ u(0) = u_0, \ \ u'(0) = v_0. \end{array}\right\}$$

In the study of Problem 2.11 we consider the following hypothesis.

$$\left.\begin{array}{l} \mathcal{P}, \mathcal{R} \colon \mathcal{V} \to \mathcal{V}^* \text{ are operators which satisfy (2.21)} \\ \text{with positive constants } L_{\mathcal{P}} \text{ and } L_{\mathcal{R}}, \text{ respectively.} \end{array}\right\} \tag{2.33}$$

We have the following existence and uniqueness result.

Theorem 2.12. *Assume that hypotheses* (2.1)–(2.6), (2.32) *and* (2.33) *hold. Then Problem* 2.11 *has a unique solution.*

Proof. We define the operator $\mathcal{S}\colon \mathcal{V} \to \mathcal{V}^*$ by

$$(\mathcal{S}w)(t) = (\mathcal{P}w)(t) + \mathcal{R}\left(\int_0^t w(s)\,ds + u_0\right) \quad \text{for all } w \in \mathcal{V}, \text{ a.e. } t \in (0,T).$$

Note that hypothesis (2.33) implies that $\mathcal{S}$ is a history-dependent operator, i.e., it satisfies (2.21). Indeed, for all $w_1, w_2 \in \mathcal{V}$ and a.e. $t \in (0,T)$, we have

$$\begin{aligned}
&\|(\mathcal{S}w_1)(t) - (\mathcal{S}w_2)(t)\|_{V^*} = \|(\mathcal{P}w_1)(t) - (\mathcal{P}w_2)(t)\|_{V^*} \\
&\quad + \|\mathcal{R}\left(\int_0^t w_1(s)\,ds + u_0\right) - \mathcal{R}\left(\int_0^t w_2(s)\,ds + u_0\right)\|_{V^*} \\
&\quad \le L_{\mathcal{P}} \int_0^t \|w_1(s) - w_2(s)\|_V ds \\
&\quad + L_{\mathcal{R}} \int_0^t \|\int_0^s w_1(\tau)\,d\tau + u_0 - \int_0^s w_2(\tau)\,d\tau - u_0\|_V ds \\
&\quad \le L_{\mathcal{P}} \int_0^t \|w_1(s) - w_2(s)\|_V ds + L_{\mathcal{R}} \int_0^t \int_0^s \|w_1(\tau) - w_2(\tau)\|_V d\tau ds \\
&\quad \le L_{\mathcal{P}} \int_0^t \|w_1(s) - w_2(s)\|_V\, ds + L_{\mathcal{R}} t \int_0^t \|w_1(\tau) - w_2(\tau)\|_V d\tau \\
&\quad \le (L_{\mathcal{P}} + T L_{\mathcal{R}}) \int_0^t \|w_1(s) - w_2(s)\|_V ds.
\end{aligned}$$

Hence, it follows that for $u \in \mathcal{V}$ such that $u' \in \mathcal{W}$, we have $\mathcal{S}u' = \mathcal{P}u' + \mathcal{R}u$. The conclusion of the theorem follows from Theorem 2.10. □

We conclude this section with two particular cases of Problem 2.11. First, we consider the following second order inclusion which involves a Volterra-type operator.

Problem 2.13. Find $u \in \mathcal{V}$ such that $u' \in \mathcal{W}$ and

$$\left.\begin{aligned}
& u''(t) + A(t, u'(t)) + B(t, u(t)) + \int_0^t C(t-s)u(s)\,ds \\
& \qquad\qquad + M^* \partial J(t, M u'(t)) \ni f(t) \ \text{ a.e. } t \in (0,T), \\
& u(0) = u_0, \ \ u'(0) = v_0.
\end{aligned}\right\}$$

We consider the following hypotheses.

$$\left.\begin{array}{l} B\colon (0,T)\times V\to V^* \text{ is such that}\\ \text{(a) } B(\cdot,v) \text{ is measurable on } (0,T) \text{ for all } v\in V.\\ \text{(b) } B(t,\cdot) \text{ is Lipschitz continuous with constant } L_B>0\\ \quad\text{for a.e. } t\in(0,T). \end{array}\right\} \tag{2.34}$$

$$C\in L^2(0,T;\mathcal{L}(V,V^*)). \tag{2.35}$$

We have the following existence and uniqueness result.

Theorem 2.14. *Assume that hypotheses* (2.1)–(2.6), (2.32), (2.34) *and* (2.35) *hold. Then Problem* 2.13 *has a unique solution.*

Proof. Let us consider two operators $\mathcal{P},\mathcal{R}\colon\mathcal{V}\to\mathcal{V}^*$ given by

$$(\mathcal{P}w)(t)=B\left(t,\int_0^t w(s)\,ds+u_0\right)+\int_0^t C(t-s)\left(\int_0^s w(\tau)\,d\tau+u_0\right)ds$$

for all $w\in\mathcal{V}$, a.e. $t\in(0,T)$ and $\mathcal{R}\equiv 0$. It is clear from (2.34) and (2.35) that

$$\|(\mathcal{P}w_1)(t)-(\mathcal{P}w_2)(t)\|_{V^*}\le\overline{c}\int_0^t\|w_1(s)-w_2(s)\|_V ds$$

for all $w_1,w_2\in\mathcal{V}$, a.e. $t\in(0,T)$ with $\overline{c}=L_B+\sqrt{T}\,\|C\|_{L^2(0,T;\mathcal{L}(V,V^*))}$, i.e., hypothesis (2.33) holds. Note that for $u\in\mathcal{V}$ such that $u'\in\mathcal{W}$, we have

$$(\mathcal{P}u')(t)=B(t,u(t))+\int_0^t C(t-s)u(s)\,ds\quad\text{a.e. } t\in(0,T).$$

We now use Theorem 2.12 to complete the proof. □

Note that Theorem 2.14 represents an extension of Theorem 5.17 in [16]. There, the operator B was assumed to be time-independent, linear, continuous, monotone and symmetric.

Finally, we consider the following problem.

Problem 2.15. Find $u\in\mathcal{V}$ such that $u'\in\mathcal{W}$ and

$$\left.\begin{array}{l} u''(t)+A(t,u'(t))+B(t,u(t))+M^*\partial J(t,Mu'(t))\ni f(t)\ \text{ a.e. } t\in(0,T),\\ u(0)=u_0,\ \ u'(0)=v_0. \end{array}\right\}$$

The following existence and uniqueness result represents a direct consequence of Theorem 2.14.

Corollary 2.16. *Assume that hypotheses* (2.1)–(2.6), (2.32) *and* (2.34) *hold. Then Problem* 2.15 *has a unique solution.*

2.5 Hemivariational Inequalities

In this section we formulate results on existence and uniqueness of solutions to hemivariational inequalities of first and second order. These results represent a consequence of the existence and uniqueness results proved in Sects. 2.3 and 2.4, in the study of evolutionary inclusions of first and second order, respectively.

Let $\Omega \subset \mathbb{R}^d$ be a bounded domain with a Lipschitz boundary Γ and let Γ_C be a measurable part of Γ, $\Gamma_C \subseteq \Gamma$. Let V be a closed subspace of $H^1(\Omega;\mathbb{R}^d)$ and $H = L^2(\Omega;\mathbb{R}^d)$. It is well known that $V \subset H \subset V^*$ form an evolution triple of spaces, cf. e.g., Section 3.4 of [6]. We introduce the trace operator $\gamma\colon V \to L^2(\Gamma;\mathbb{R}^d)$ and its adjoint $\gamma^*\colon L^2(\Gamma;\mathbb{R}^d) \to V^*$.

We consider the following hemivariational inequality of first order.

$$\left.\begin{aligned}
&\text{Find } w \in \mathcal{W} \text{ such that}\\
&\langle w'(t) + A(t,w(t)), v\rangle_{V^*\times V} + \int_{\Gamma_C} j^0(x,t,\gamma w(t);\gamma v)\,d\Gamma\\
&\qquad\qquad \geq \langle f(t), v\rangle_{V^*\times V} \quad \text{for all } v \in V, \text{ a.e. } t \in (0,T),\\
&w(0) = v_0.
\end{aligned}\right\} \tag{2.36}$$

In the study of this hemivariational inequality we consider the following hypotheses on the data.

$$\left.\begin{aligned}
&j\colon \Gamma_C \times (0,T) \times \mathbb{R}^d \to \mathbb{R} \text{ is such that}\\
&\text{(a) } j(\cdot,\cdot,\xi) \text{ is measurable for all } \xi \in \mathbb{R}^d \text{ and there exists}\\
&\qquad e \in L^2(\Gamma_C;\mathbb{R}^d) \text{ such that } j(\cdot,\cdot,e(\cdot)) \in L^1(\Gamma_C \times (0,T)).\\
&\text{(b) } j(x,t,\cdot) \text{ is locally Lipschitz for a.e. } (x,t) \in \Gamma_C \times (0,T).\\
&\text{(c) } \|\partial j(x,t,\xi)\|_{\mathbb{R}^d} \leq b_0(x,t) + b_1\|\xi\|_{\mathbb{R}^d} \text{ for all } \xi \in \mathbb{R}^d,\\
&\qquad \text{a.e. } (x,t) \in \Gamma_C \times (0,T) \text{ with } b_0 \in L^2(\Gamma_C \times (0,T)),\\
&\qquad b_0, b_1 \geq 0.\\
&\text{(d) } (\zeta_1 - \zeta_2, \xi_1 - \xi_2)_{\mathbb{R}^d} \geq -\overline{m}_2\,\|\xi_1 - \xi_2\|^2_{\mathbb{R}^d}\\
&\qquad \text{for all } \zeta_i \in \partial j(x,t,\xi_i),\ \xi_i \in \mathbb{R}^d,\ i = 1,2 \text{ with } \overline{m}_2 \geq 0.
\end{aligned}\right\} \tag{2.37}$$

$$\left.\begin{array}{l}\text{One of the following conditions is satisfied}\\ \quad \text{(a) } \alpha > 2\sqrt{2}\, b_1 \, \|\gamma\|^2, \text{ where } \|\gamma\| = \|\gamma\|_{\mathcal{L}(V,L^2(\Gamma;\mathbb{R}^d))}.\\ \quad \text{(b) } j^0(x,t,\xi;-\xi) \le \overline{d}_0 \, (1 + \|\xi\|_{\mathbb{R}^d}) \text{ for all } \xi \in \mathbb{R}^d, \text{ a.e.}\\ \quad\quad (x,t) \in \Gamma_C \times (0,T) \text{ with } \overline{d}_0 \ge 0.\end{array}\right\} \tag{2.38}$$

$$m_1 \ge \overline{m}_2 \, \|\gamma\|^2. \tag{2.39}$$

$$\left.\begin{array}{l} j\colon \Gamma_C \times (0,T) \times \mathbb{R}^d \to \mathbb{R} \text{ is such that}\\ \quad \text{either } j(x,t,\cdot) \text{ or } -j(x,t,\cdot) \text{ is regular on } \mathbb{R}^d\\ \quad \text{for a.e. } (x,t) \in \Gamma_C \times (0,T).\end{array}\right\} \tag{2.40}$$

Note that in hypotheses (2.37) and (2.38) the symbols ∂j and j^0 denote the Clarke generalized gradient of $j(x,t,\cdot)$ and its directional derivative, respectively.

We consider the integral functional $J\colon (0,T) \times L^2(\Gamma_C;\mathbb{R}^d) \to \mathbb{R}$ defined by

$$J(t,v) = \int_{\Gamma_C} j(x,t,v(x))\, d\Gamma \quad \text{for } v \in L^2(\Gamma_C;\mathbb{R}^d), \text{ a.e. } t \in (0,T). \tag{2.41}$$

We recall the following result whose proof can be found in Theorem 3.47 of [16].

Lemma 2.17. *Assume that* (2.37) *holds. Then the functional* J *given by* (2.41) *satisfies* (2.3) *and for all* u, $v \in L^2(\Gamma_C;\mathbb{R}^d)$, *a.e.* $t \in (0,T)$, *we have*

$$J^0(t,u;v) \le \int_{\Gamma_C} j^0(x,t,u(x);v(x))\, d\Gamma, \tag{2.42}$$

where $J^0(t,u;v)$ *denotes the directional derivative of* $J(t,\cdot)$ *at a point* $u \in L^2(\Gamma_C;\mathbb{R}^d)$ *in the direction* $v \in L^2(\Gamma_C;\mathbb{R}^d)$. *If, in addition,* $j\colon \Gamma_C \times (0,T) \times \mathbb{R}^d \to \mathbb{R}$ *satisfies condition* (2.38)(b), *then condition* (2.4)(b) *holds.*

We now use Theorem 2.6 and Lemma 2.17 to obtain the following existence and uniqueness result.

Theorem 2.18. *Assume that hypotheses* (2.1)(a)–(d), (2.6), (2.37)(a)–(c) *and* (2.38) *hold. Then the hemivariational inequality* (2.36) *has at least one solution. If, in addition, conditions* (2.1)(e), (2.37)(d), (2.39) *and* (2.40) *hold, then the solution to* (2.36) *is unique.*

Proof. We begin with the existence part. Let $X = L^2(\Gamma_C;\mathbb{R}^d)$ and $M = \gamma\colon V \to X$. We introduce the Nemitsky operator corresponding to the trace operator γ and denote it by the same symbol $\gamma\colon \mathcal{V} = L^2(0,T;V) \to L^2(0,T;X)$. This operator satisfies hypothesis (2.2). Indeed, let $v_n \to v$ weakly in $\mathcal{W}$. Since $\mathcal{W}$ is compactly embedded in $L^2(0,T;H^{1-\delta}(\Omega))$ with $\delta \in (0,1/2)$ (which is a consequence of the compact embedding $H^1(\Omega) \subset H^{1-\delta}(\Omega)$, cf. [6, 24]), we have $v_n \to v$ in $L^2(0,T;H^{1-\delta}(\Omega))$. Next, from the fact that

$$\gamma: L^2(0,T;H^{1-\delta}(\Omega)) \to L^2(0,T;H^{1-\delta-1/2}(\Gamma)) \subset L^2(0,T;L^2(\Gamma;\mathbb{R}^d))$$

is linear and continuous, we deduce that $\gamma v_n \to \gamma v$ in $L^2(0,T;X)$.

We denote by $w \in \mathcal{W}$ the solution of Problem 2.4 with the functional J given by (2.41). The existence and uniqueness of this solution is guaranteed by Theorem 2.6 combined with Lemma 2.17. According to Definition 2.5, we have

$$w'(t) + A(t,w(t)) + \zeta(t) = f(t) \ \text{ for a.e. } t \in (0,T), \tag{2.43}$$

where $\zeta(t) = \gamma^* z(t) \in V^*$ and $z(t) \in \partial J(t,\gamma w(t))$ for a.e. $t \in (0,T)$. The last inclusion is equivalent to

$$\langle z(t), \overline{w}\rangle_{L^2(\Gamma_C;\mathbb{R}^d)} \le J^0(t,\gamma w(t);\overline{w}) \tag{2.44}$$

for all $\overline{w} \in L^2(\Gamma_C;\mathbb{R}^d)$ and a.e. $t \in (0,T)$. We now combine (2.42)–(2.44) to obtain

$$\begin{aligned}\langle f(t) - w'(t) - A(t,w(t)), v\rangle_{V^*\times V} &= \langle \zeta(t), v\rangle_{V^*\times V}\\ &= \langle z(t), \gamma v\rangle_{L^2(\Gamma_C;\mathbb{R}^d)}\\ &\le J^0(t,\gamma w(t);\gamma v)\\ &\le \int_{\Gamma_C} j^0(x,t,\gamma w(t);\gamma v)\,d\Gamma\end{aligned}$$

for all $v \in V$, a.e. $t \in (0,T)$. It follows from the last inequality that w is a solution to (2.36), which concludes the proof of existence part of the theorem.

We now proceed with the proof of the uniqueness part and, to this end, we denote by $w \in \mathcal{W}$ a solution to (2.36) obtained in the first part of the theorem. It is well known (cf. Theorem 2.7.2 of [3]) that under the regularity hypothesis (2.40), that either $J(t,\cdot)$ or $-J(t,\cdot)$ is regular for a.e. $t \in (0,T)$, respectively, and (2.42) holds with equality. Therefore, using the equality in (2.42), we have

$$\langle w'(t) + A(t,w(t)) - f(t), v\rangle_{V^*\times V} + J^0(t,\gamma w(t);\gamma v) \ge 0$$

for all $v \in V$ and a.e. $t \in (0,T)$. Also, by Proposition 3.37 in [16], we obtain

$$\langle f(t) - w'(t) - A(t,w(t)), v\rangle_{V^*\times V} \le (J\circ\gamma)^0(t,w(t);v)$$

for all $v \in V$ and a.e. $t \in (0, T)$. Using now the definition of the subdifferential and Proposition 3.37 in [16], it is straightforward to see that the previous inequality implies that

$$f(t) - w'(t) - A(t, w(t)) \in \partial(J \circ \gamma)(t, w(t)) = \gamma^* \partial J(t, \gamma w(t))$$

for a.e. $t \in (0, T)$. Therefore, we find that w is a solution to Problem 2.4. The uniqueness of solution to (2.36) follows now from the uniqueness part in Theorem 2.6, which concludes the proof. □

We consider now the following hemivariational inequality of first order with the history-dependent operator.

$$\left.\begin{aligned} &\text{Find } w \in \mathcal{W} \text{ such that} \\ &\langle w'(t) + A(t, w(t)) + (\mathcal{S}w)(t), v\rangle_{V^*\times V} + \int_{\Gamma_C} j^0(x, t, \gamma w(t); \gamma v)\, d\Gamma \\ &\qquad \geq \langle f(t), v\rangle_{V^*\times V} \quad \text{for all } v \in V, \text{ a.e. } t \in (0, T), \\ &w(0) = v_0. \end{aligned}\right\} \tag{2.45}$$

Exploiting the argument used in the proof of Theorem 2.18, from Theorem 2.8 and Lemma 2.17, we deduce the following result.

Corollary 2.19. *Assume that* (2.1), (2.6), (2.21), *and* (2.37)–(2.39) *hold. Then the hemivariational inequality* (2.45) *has at least one solution. If, in addition* (2.40) *holds, then the solution to* (2.45) *is unique.*

Next, we pass to the hemivariational inequalities of second order. We consider the following problem.

$$\left.\begin{aligned} &\text{Find } u \in \mathcal{V} \text{ such that } u' \in \mathcal{W} \text{ and} \\ &\langle u''(t) + A(t, u'(t)) + (\mathcal{S}u')(t), v\rangle_{V^*\times V} + \int_{\Gamma_C} j^0(x, t, \gamma u'(t); \gamma v)\, d\Gamma \\ &\qquad \geq \langle f(t), v\rangle_{V^*\times V} \quad \text{for all } v \in V, \text{ a.e. } t \in (0, T), \\ &u(0) = u_0, \ \ u'(0) = v_0. \end{aligned}\right\} \tag{2.46}$$

Using Theorem 2.10, Lemma 2.17 and Theorem 2.18, we deduce the following existence and uniqueness result.

Corollary 2.20. *Assume that* (2.1), (2.6), (2.21), (2.32), *and* (2.37)–(2.39) *hold. Then the hemivariational inequality* (2.46) *has at least one solution. If, in addition* (2.40) *holds, then the solution to* (2.46) *is unique.*

Subsequently, we consider hemivariational inequalities with two history-dependent operators. The first problem in which our interest is can be formulated as follows.

$$\left.\begin{array}{l}\text{Find } u \in \mathcal{V} \text{ such that } u' \in \mathcal{W} \text{ and} \\ \langle u''(t) + A(t, u'(t)) + (\mathcal{P}u')(t) + (\mathcal{R}u)(t), v\rangle_{V^* \times V} \\ \qquad + \displaystyle\int_{\Gamma_C} j^0(x, t, \gamma u'(t); \gamma v)\, d\Gamma \geq \langle f(t), v\rangle_{V^* \times V} \\ \qquad\qquad\qquad \text{for all } v \in V, \text{ a.e. } t \in (0, T), \\ u(0) = u_0, \ u'(0) = v_0. \end{array}\right\} \quad (2.47)$$

Using Theorem 2.12, Lemma 2.17 and arguments similar to those used in the proof of Theorem 2.18, we obtain the following result.

Corollary 2.21. *Assume that* (2.1), (2.6), (2.32), (2.33), *and* (2.37)–(2.39) *hold. Then the hemivariational inequality* (2.47) *has at least one solution. If, in addition* (2.40) *holds, then the solution to* (2.47) *is unique.*

Next, we consider hemivariational inequalities involving a Volterra-type operator. The problem can be formulated as follows.

$$\left.\begin{array}{l}\text{Find } u \in \mathcal{V} \text{ such that } u' \in \mathcal{W} \text{ and} \\ \langle u''(t) + A(t, u'(t)) + B(t, u(t)) + \displaystyle\int_0^t C(t-s)u(s)\, ds, v\rangle_{V^* \times V} \\ \qquad + \displaystyle\int_{\Gamma_C} j^0(x, t, \gamma u'(t); \gamma v)\, d\Gamma \geq \langle f(t), v\rangle_{V^* \times V} \\ \qquad\qquad\qquad \text{for all } v \in V, \text{ a.e. } t \in (0, T), \\ u(0) = u_0, \ u'(0) = v_0. \end{array}\right\} \quad (2.48)$$

From Theorem 2.14, Lemma 2.17 and Theorem 2.18, we deduce the following existence and uniqueness result.

Corollary 2.22. *Assume that* (2.1), (2.6), (2.32), (2.34), (2.35), *and* (2.37)–(2.39) *hold. Then the hemivariational inequality* (2.48) *has at least one solution. If, in addition* (2.40) *holds, then the solution to* (2.48) *is unique.*

We end this chapter with the study of a particular case of the hemivariational inequality (2.48). The problem under consideration is the following

$$\left.\begin{array}{l}\text{Find } u \in \mathcal{V} \text{ such that } u' \in \mathcal{W} \text{ and} \\ \langle u''(t) + A(t, u'(t)) + B(t, u(t)), v\rangle_{V^* \times V} \\ \qquad + \displaystyle\int_{\Gamma_C} j^0(x, t, \gamma u'(t); \gamma v)\, d\Gamma \geq \langle f(t), v\rangle_{V^* \times V} \\ \qquad\qquad \text{for all } v \in V, \text{ a.e. } t \in (0, T), \\ u(0) = u_0, \ u'(0) = v_0. \end{array}\right\} \quad (2.49)$$

The following result represents a direct consequence of Corollaries 2.16 and 2.22.

Corollary 2.23. *Assume that* (2.1), (2.6), (2.32), (2.34), *and* (2.37)–(2.39) *hold. Then the hemivariational inequality* (2.49) *has at least one solution. If, in addition* (2.40) *holds, then the solution to* (2.49) *is unique.*

The existence and uniqueness results in Theorem 2.18 and Corollaries 2.19–2.23 are useful in the study of various dynamic or quasistatic contact problems with viscoelastic and viscoplastic materials, as illustrated in Chap. 14 of this book. Indeed, a large number of such problems leads to evolutionary hemivariational inequalities of first and second order, in which the unknown is the displacement or the velocity field.

Acknowledgements This research was supported by the Marie Curie International Research Staff Exchange Scheme Fellowship within the 7th European Community Framework Programme under Grant Agreement No. 295118, the National Science Center of Poland under grant no. N N201 604640, the International Project co-financed by the Ministry of Science and Higher Education of Republic of Poland under grant no. W111/7.PR/2012, the National Science Center of Poland under Maestro Advanced Project no. DEC-2012/06/A/ST1/00262, and the project Polonium "Mathematical and Numerical Analysis for Contact Problems with Friction" 2014/15 between the Jagiellonian University in Krakow and Université de Perpignan Via Domitia.

References

1. Aubin, J.-P., Cellina, A.: Differential Inclusions. Springer, Berlin/Heidelberg/New York/Tokyo (1984)
2. Berkovits, J., Mustonen, V.: Monotonicity methods for nonlinear evolution equations. Nonlinear Anal. **27**, 1397–1405 (1996)
3. Clarke, F.H.: Optimization and Nonsmooth Analysis. Wiley Interscience, New York (1983)
4. Denkowski, Z., Migórski, S.: A system of evolution hemivariational inequalities modeling thermoviscoelastic frictional contact. Nonlinear Anal. **60**, 1415–1441 (2005)
5. Denkowski, Z., Migórski, S., Papageorgiou, N.S.: An Introduction to Nonlinear Analysis: Theory. Kluwer Academic/Plenum Publishers, Boston/Dordrecht/London/New York (2003)
6. Denkowski, Z., Migórski, S., Papageorgiou, N.S.: An Introduction to Nonlinear Analysis: Applications. Kluwer Academic/Plenum Publishers, Boston/Dordrecht/London/New York (2003)
7. Goeleven, D., Miettinen, M., Panagiotopoulos, P.D.: Dynamic hemivariational inequalities and their applications. J. Optim. Theory Appl. **103**, 567–601 (1999)
8. Haslinger, J., Miettinen, M., Panagiotopoulos, P.D.: Finite Element Method for Hemivariational Inequalities. Theory, Methods and Applications. Kluwer Academic Publishers, Boston/Dordrecht/London (1999)
9. Kulig, A., Migórski, S.: Solvability and continuous dependence results for second order nonlinear evolution inclusions with a Volterra-type operator. Nonlinear Anal. Theory Methods Appl. **75**, 4729–4746 (2012)
10. Migórski, S.: Dynamic hemivariational inequality modeling viscoelastic contact problem with normal damped response and friction. Appl. Anal. **84**, 669–699 (2005)
11. Migórski, S.: Evolution hemivariational inequalities with applications. In: Gao, D.Y., Motreanu, D. (eds.) Handbook of Nonconvex Analysis and Applications, pp. 409–473. International Press, Boston (2010)
12. Migórski, S., Ochal, A.: Boundary hemivariational inequality of parabolic type. Nonlinear Anal. **57**, 579–596 (2004)

13. Migórski, S., Ochal, A.: A unified approach to dynamic contact problems in viscoelasticity. J. Elast. **83**, 247–275 (2006)
14. Migórski, S., Ochal, A., Sofonea, M.: Integrodifferential hemivariational inequalities with applications to viscoelastic frictional contact. Math. Models Methods Appl. Sci. **18**, 271–290 (2008)
15. Migórski, S., Ochal, A., Sofonea, M.: History-dependent subdifferential inclusions and hemivariational inequalities in contact mechanics. Nonlinear Anal. Real World Appl. **12**, 3384–3396 (2011)
16. Migórski, S., Ochal, A., Sofonea, M.: Nonlinear Inclusions and Hemivariational Inequalities. Models and Analysis of Contact Problems. Advances in Mechanics and Mathematics, vol. 26. Springer, New York (2013)
17. Naniewicz, Z., Panagiotopoulos, P.D.: Mathematical Theory of Hemivariational Inequalities and Applications. Marcel Dekker, New York, Basel/Hong Kong (1995)
18. Ochal, A.: Existence results for evolution hemivariational inequalities of second order. Nonlinear Anal. **60**, 1369–1391 (2005)
19. Panagiotopoulos, P.D.: Inequality Problems in Mechanics and Applications. Convex and Nonconvex Energy Functions. Birkhäuser, Boston (1985)
20. Panagiotopoulos, P.D.: Hemivariational Inequalities, Applications in Mechanics and Engineering. Springer, Berlin (1993)
21. Panagiotopoulos, P.D., Pop, G.: On a type of hyperbolic variational-hemivariational inequalities. J. Appl. Anal. **5**, 95–112 (1999)
22. Park, J.Y., Ha, T.G.: Existence of antiperiodic solutions for hemivariational inequalities. Nonlinear Anal. **68**, 747–767 (2008)
23. Sofonea, M., Matei, A.: Mathematical Models in Contact Mechanics. London Mathematical Society Lecture Note Series, vol. 398. Cambridge University Press, Cambridge (2012)
24. Zeidler, E.: Nonlinear Functional Analysis and its Applications II. Springer, New York (1990)

Chapter 3
Location Results for Variational–Hemivariational Inequalities

Dumitru Motreanu and Viorica Venera Motreanu

Abstract The chapter presents a general method, based on approximation of spaces and operators, to solve certain nonsmooth problems. The method allows us to obtain location properties of the solutions, for instance the inclusion of the solutions in prescribed sets. This is achieved through an approximation approach by means of sequences of associated problems formulated in a simpler setting, possibly on finite dimensional spaces. The abstract results are applied to various classes of hemivariational and variational–hemivariational inequalities. An essential tool is represented by pseudomonotone operators.

Keywords Variational–hemivariational inequality • Hemivariational inequality • Generalized gradient • Pseudomonotone operator • Approximation method • Location result

AMS Classification. 47H05, 47J20, 49J52, 41A65

3.1 Introduction

The aim of this chapter is to present a general method for studying certain nonsmooth problems. Consider a general problem of the form

$$\begin{cases} \text{Find } u \in C \text{ such that} \\ \langle T(u), v - u \rangle + f(u, v - u) \geq 0 \quad \forall\, v \in C, \end{cases} \tag{3.1}$$

D. Motreanu (✉)
Laboratoire de Mathématiques et Physique, Université de Perpignan Via Domitia, 52 Avenue Paul Alduy, 66860 Perpignan, France
e-mail: motreanu@univ-perp.fr

V.V. Motreanu
Department of Mathematics, Ben Gurion University of the Negev, P.O.B. 653, Be'er Sheva 84105, Israel
e-mail: motreanu@post.bgu.ac.il

W. Han et al. (eds.), *Advances in Variational and Hemivariational Inequalities*, Advances in Mechanics and Mathematics 33, DOI 10.1007/978-3-319-14490-0_3

involving the following data:

- $(X, \|\cdot\|)$ is a real reflexive Banach space, its dual space is denoted by $(X^*, \|\cdot\|_*)$, and the duality pairing (X^*, X) is denoted by $\langle\cdot,\cdot\rangle$;
- $C \subset X$ is a nonempty subset;
- $f : C \times X \to \mathbb{R}$ is a given function and $T : C \to X^*$ a given operator.

A solution of this problem is necessarily located in the set C. In fact, our method enables us to derive some more location properties of the solutions. For instance, our setting incorporates variational–hemivariational inequalities of the form

$$\begin{cases} \text{Find } u \in \partial C \text{ such that} \\ \langle T(u), v-u\rangle + \int_\Omega j^0(x,u;v-u)\,dx \ge 0 \quad \forall\, v \in C, \end{cases} \tag{3.2}$$

and

$$\begin{cases} \text{Find } u \in C \text{ such that} \\ \langle T(u), v\rangle + \int_\Omega j^0(x,u;v)\,dx \ge 0 \quad \forall\, v \in X. \end{cases} \tag{3.3}$$

Problems (3.2) and (3.3) involve the following data:

- $\Omega \subset \mathbb{R}^N$ is a bounded domain;
- the real reflexive Banach space $(X, \|\cdot\|)$ is compactly embedded in $L^q(\Omega)$, for some $q \in (1,+\infty)$; for instance $X = W^{1,p}(\Omega)$ (or $X = W_0^{1,p}(\Omega)$) with $p \in (1,+\infty)$ such that $q < p^*$, where $p^* \in (1,+\infty]$ stands for the Sobolev critical exponent for the Sobolev space $W^{1,p}(\Omega)$;
- $C \subset X$ is nonempty, closed, convex; for instance $C = \{u \in X : \|u\| \le \rho\}$ ($\rho \in (0,+\infty)$) or $C = \{u \in X : u_- \le u \le u_+ \text{ a.e. in } \Omega\}$ with $u_-, u_+ \in L^q(\Omega)$ (provided C is nonempty); by ∂C we denote the boundary of C;
- $T : C \to X^*$ is a bounded pseudomonotone operator; for instance, in the case where $X = W^{1,p}(\Omega)$ (or $X = W_0^{1,p}(\Omega)$), the operator T can be the negative p-Laplacian $-\Delta_p u = -\mathrm{div}(|\nabla u|^{p-2}\nabla u)$;
- $j : \Omega \times \mathbb{R} \to \mathbb{R}$ is a function satisfying the following conditions:

  (i) $j(\cdot,s) \in L^1(\Omega)$ for all $s \in \mathbb{R}$ and $j(x,\cdot)$ is locally Lipschitz for a.a. $x \in \Omega$;
  (ii) (growth condition) $|z| \le k(x) + c|s|^{p-1}$ for a.e. $x \in \Omega$, all $s \in \mathbb{R}$, all $z \in \partial j(x,s)$.

Hereafter, by $j^0(x,s;t)$ we denote the generalized directional derivative of $j(x,\cdot)$ at s in the direction t, and by $\partial j(x,s)$ we denote the generalized gradient of $j(x,\cdot)$ at s, in the sense of Clarke [2] (see (3.18)–(3.19) below). In the case where $j(x,\cdot)$ is continuously differentiable, $j^0(x,s;t)$ coincides with the usual directional derivative while $\partial j(x,s)$ is the singleton $\{j'(x,\cdot)(s)\}$.

Problem (3.2) can be seen as a location problem with respect to the variational–hemivariational inequality

$$\begin{cases} \text{Find } u \in C \text{ such that} \\ \langle T(u), v - u\rangle + \int_\Omega j^0(x, u; v - u)\, dx \geq 0 \quad \forall\, v \in C, \end{cases} \tag{3.4}$$

whereas problem (3.3) can be viewed as a location problem with respect to the hemivariational inequality

$$\begin{cases} \text{Find } u \in X \text{ such that} \\ \langle T(u), v\rangle + \int_\Omega j^0(x, u; v)\, dx \geq 0 \quad \forall\, v \in X. \end{cases}$$

Our result shows the following alternative.

Theorem 3.1. *Let Ω, X, C, T, j be as above. Then one of the problems* (3.2) *and* (3.3) *admits a solution.*

For instance, assume that $X = W_0^{1,p}(\Omega)$ with $p \in (1, +\infty)$, $T = -\Delta_p$, $j(x, s) = -\frac{\lambda}{p}|s|^p$ where $\lambda \in \mathbb{R}$ is not an eigenvalue of $-\Delta_p$, and $0 \notin C$. Then, problem (3.3) reads as the eigenvalue problem

$$\begin{cases} \text{Find } u \in C \text{ such that} \\ -\Delta_p u = \lambda |u|^{p-2} u \ \text{ in } (W_0^{1,p}(\Omega))^*, \end{cases}$$

which has no solution (due to the assumption on λ). Therefore, in this situation, we can conclude from Theorem 3.1 that problem (3.2) has a solution.

Theorem 3.1 is proved in Sect. 3.5. In fact, our results go beyond the setting of Theorem 3.1. Specifically, our aim is to develop an abstract setting in which (3.1) can be solved by approximating it with a sequence of related problems. The abstract setting is presented in Sects. 3.2–3.3. The abstract result that we obtain (established in Sect. 3.4) will be shown to lead to various existence and approximation results for classes of hemivariational and variational–hemivariational inequalities (see Sects. 3.5–3.8).

3.2 Abstract Setting

Let us describe our abstract setting. For every $n \in \mathbb{N}$, we consider a subset $C_n \subset X$, a function $f_n : C_n \times X \to \mathbb{R}$, and an operator $T_n : C_n \to X^*$, and we formulate the corresponding problem

$$\begin{cases} \text{Find } u_n \in C_n \text{ such that} \\ \langle T_n(u_n), v - u_n\rangle + f_n(u_n, v - u_n) \geq 0 \quad \forall\, v \in C_n. \end{cases} \tag{3.5}$$

Then, assuming that problem (3.5) admits a solution u_n for every $n \in \mathbb{N}$, our goal is to study the existence of a solution of problem (3.1) obtained as a weak limit of the sequence (u_n) (or at least of a subsequence). Specifically, our main abstract result (formulated in Theorem 3.14) provides an abstract setting where this approximation principle is fulfilled. This abstract setting consists of the following set of hypotheses:

(H$_1$) (i) for every subsequence (C_{n_k}) of (C_n) and every sequence (u_k) with $u_k \in C_{n_k}$, $\forall\, k \in \mathbb{N}$, and $u_k \rightharpoonup u$ in X, we have $u \in C$;

(ii) there exist a subset $\mathcal{C}$ dense in C and a sequence of mappings $r_n : \mathcal{C} \to C_n$, $n \in \mathbb{N}$, with

$$\lim_{n\to\infty} \|r_n(v) - v\| = 0 \quad \forall\, v \in \mathcal{C};$$

(H$_2$) (i) for every $u \in C$, the function $f(u,\cdot) : X \to \mathbb{R}$ is subadditive and there exists a constant $a = a(u) > 0$ such that

$$f(u,v) \le a\|v\| \quad \forall\, v \in X;$$

(ii) for every subsequence (f_{n_k}) of (f_n) and every sequence (u_k) with $u_k \in C_{n_k}$, $\forall k \in \mathbb{N}$, and $u_k \rightharpoonup u$ in X, we have

$$\limsup_{k\to\infty} f_{n_k}(u_k, r_{n_k}(v) - u_k) \le f(u, v-u) \quad \forall\, v \in \mathcal{C};$$

(H$_3$) (i) for every subsequence (T_{n_k}) of (T_n) and every bounded sequence (u_k) with $u_k \in C_{n_k}$, $\forall k \in \mathbb{N}$, the sequence $(T_{n_k}(u_k))$ is bounded;

(ii) for every subsequence (T_{n_k}) of (T_n) and every sequence (u_k) with $u_k \in C_{n_k}$, $\forall k \in \mathbb{N}$, and $u_k \rightharpoonup u$ in X and

$$\limsup_{k\to\infty} \langle T_{n_k}(u_k), u_k - u\rangle \le 0,$$

we have

$$\langle T(u), u - v\rangle \le \limsup_{k\to\infty} \langle T_{n_k}(u_k), u_k - v\rangle \quad \forall\, v \in \mathcal{C}.$$

Hereafter, the notation $\rightharpoonup$ stands for the weak convergence in X. We also consider a further hypothesis under which we will be able to guarantee the *strong* convergence of a sequence of solution of (3.5) to a solution of (3.1):

(H$_4$) for every subsequence (T_{n_k}) of (T_n) and every sequence (u_k) with $u_k \in C_{n_k}$, $\forall k \in \mathbb{N}$, and $u_k \rightharpoonup u$, for some u, and

$$\limsup_{k\to\infty} \langle T_{n_k}(u_k), u_k - u\rangle \le 0$$

we have $u_k \to u$ in X.

The setting corresponding to (H_1)–(H_4) is quite natural, in particular, it is well-suited for studying hemivariational and variational–hemivariational inequalities. This is explained in Sect. 3.3.

3.3 Preliminary Observations

This section contains some preliminaries concerning the hypotheses stated previously for a nonempty subset C of X, a sequence of subsets C_n of X, $n \in \mathbb{N}$, functions $f : C \times X \to \mathbb{R}$ and $f_n : C_n \times X \to \mathbb{R}$, and operators $T : C \to X^*$ and $T_n : C_n \to X^*$.

Hypothesis (H_1) is fulfilled under usual conditions in approximation of variational inequalities (see, e.g., [3, Section I.4.4]). The next result gives an example where (H_1) holds true.

Lemma 3.2. *Assume that C is a (nonempty) separable, closed and convex subset of X. Let a countable dense subset $\{w_n\}_{n\in\mathbb{N}}$ of C and denote*

$$C_n = C \cap \operatorname{span}\{w_1, w_2, \dots, w_n\} \quad \forall\, n \in \mathbb{N}.$$

Then assumption (H_1) *is satisfied with $\mathcal{C} = C$ and the mappings $r_n : C \to C_n$ characterized by*

$$\|r_n(u) - u\| = \min_{w\in C_n} \|w - u\| \quad \forall\, u \in C. \tag{3.6}$$

Proof. As the set C is convex and closed, it is weakly closed in X, hence condition (H_1) (i) is verified. To check (H_1) (ii) with $\mathcal{C} = C$, let $u \in C$ and $\varepsilon > 0$. Since $C = \overline{\cup_{m=1}^{\infty} C_m}$, we find some $v \in \cup_{m=1}^{\infty} C_m$ with $\|v - u\| < \varepsilon$. Let $m_0 \in \mathbb{N}$ be such that $v \in C_{m_0}$. Then, for every $n \geq m_0$ we have $v \in C_n$, thus

$$\|r_n(u) - u\| = \min_{w\in C_n} \|w - u\| \leq \|v - u\| < \varepsilon.$$

Consequently, (H_1) (ii) is fulfilled, which completes the proof. □

Next, we discuss conditions (H_2) and (H_3). The following lemma is immediate.

Lemma 3.3. *If $f : C \times X \to \mathbb{R}$ satisfies* (H_2) (i), *then for every $u \in C$, the function $f(u,\cdot) : X \to \mathbb{R}$ is Lipschitz continuous with the Lipschitz constant $a = a(u)$.*

Proof. From (H_2) (i), for every $v, w \in X$ we have

$$f(u,w) - f(u,v) \leq f(u,w - v) \leq a\|w - v\|.$$

Reversing the roles of v and w we infer that $|f(u,w) - f(u,v)| \leq a\|w - v\|$. □

Lemma 3.4. *Let $f : C \times X \to \mathbb{R}$ be a function satisfying*

(H$_5$) *for each $u \in C$, $f(u,\cdot)$ is subadditive, positively homogeneous, continuous and $f(u,0) = 0$.*

Then (H$_2$) (i) *is satisfied.*

Proof. Since $f(u,\cdot)$ is convex and continuous, it is locally Lipschitz on X, so there exist constants $r = r(u) > 0$ and $a = a(u) > 0$ such that

$$f(u,v) \leq |f(u,v) - f(u,0)| \leq a\|v\| \quad \forall\, v \in X,\ \|v\| \leq r.$$

Taking into account that $f(u,\cdot)$ is positively homogeneous, we obtain that $f(u,v) \leq a\|v\|$, $\forall\, v \in X$, whence (H$_2$) (i). □

Lemma 3.5. *Assume that $T : C \to X^*$ and $T_n : C_n \to X^*$ ($n \in \mathbb{N}$) satisfy the following property:*

(H$_6$) *for every subsequence (T_{n_k}) of (T_n) and every sequence (u_k) with $u_k \in C_{n_k}$, $\forall\, k \in \mathbb{N}$, and $u_k \to u$, we have $T_{n_k}(u_k) \rightharpoonup T(u)$.*

If (T_n) satisfies (H$_4$), *then* (H$_3$) (ii) *holds.*

Proof. Let (T_{n_k}) be an arbitrary subsequence of (T_n) and let (u_k) be a sequence such that $u_k \in C_{n_k}$, $\forall k \in \mathbb{N}$, and satisfying $u_k \rightharpoonup u$ in X and $\limsup_{k\to\infty}\langle T_{n_k}(u_k), u_k - u\rangle \leq 0$. By (H$_4$) and (H$_6$), we see that $u_k \to u$ and $T_{n_k}(u_k) \rightharpoonup T(u)$ in X. Then $\lim_{k\to\infty} \langle T_{n_k}(u_k), u_k - v\rangle = \langle T(u), u - v\rangle$, $\forall v \in C$, whence (H$_3$) (ii). □

We recall that $T : C \to X^*$ is bounded if it maps bounded subsets of C into bounded sets of X^*. The operator $T : C \to X^*$ is said to be pseudomonotone if for every sequence $(u_n) \subset C$ such that $u_n \rightharpoonup u$, for some $u \in C$, and $\limsup_{n\to\infty}\langle T(u_n), u_n - u\rangle \leq 0$ we have

$$\langle T(u), u - v\rangle \leq \liminf_{n\to\infty}\langle T(u_n), u_n - v\rangle \quad \forall\, v \in X.$$

The operator $T : C \to X^*$ satisfies condition $(S)_+$ if every sequence $(u_n) \subset C$ such that $u_n \rightharpoonup u$ and $\limsup_{n\to\infty}\langle T(u_n), u_n - u\rangle \leq 0$, for some $u \in C$, is strongly convergent to u in X.

Recall that a function $f : C \times X \to \mathbb{R}$ is said to be sequentially weakly upper semicontinuous if for every sequences $(u_n) \subset C$ and $(v_n) \subset X$ such that $u_n \rightharpoonup u$, $v_n \rightharpoonup v$, for some $u \in C$, $v \in X$, we have $\limsup_{n\to\infty} f(u_n, v_n) \leq f(u,v)$.

In the case of constant sequences of functions, we have:

Lemma 3.6. (a) *Assume that the sets C and (C_n) satisfy* (H$_1$) (ii) *and assume that $f = f_n : (C \cup (\cup_{m=1}^{\infty} C_m)) \times X \to \mathbb{R}$ (for all $n \in \mathbb{N}$). If either*

$$f \text{ is sequentially weakly upper semicontinuous}$$

or

$$f \text{ is upper semicontinuous, } C_n \subset C, \ \forall\, n \in \mathbb{N}, \text{ and } C \text{ is compact,}$$

then (H$_2$) (ii) *is satisfied.*

(b) *If* $T = T_n : C \cup (\cup_{m=1}^{\infty} C_m) \to X^*$ *(for all* $n \in \mathbb{N}$*) is bounded and pseudomonotone, then* T *satisfies the properties* (i) *and* (ii) *stated in* (H$_3$)*, respectively. In addition,* T *is demicontinuous.*

(c) *If* $T_n = T : C \cup (\cup_{m=1}^{\infty} C_m) \to X^*$ *(for all* $n \in \mathbb{N}$*), then property* (H$_4$) *reduces to condition* $(S)_+$ *for* T*, while property* (H$_6$) *reduces to the demicontinuity of* T*.*

Now, we study some situations related to which problem (3.5) has a solution. First we recall from [4] the following result extending [8, Corollary 1] (for the sake of completeness we give the proof). This situation is done in the context of finite dimensional Banach spaces.

Lemma 3.7. *Assume that* C *is a nonempty, compact, convex subset of a real finite dimensional Banach space* X *and let* $f : C \times X \to \mathbb{R}$ *be an upper semicontinuous function such that* $f(u, \cdot)$ *is convex and* $f(u, 0) = 0$ *for all* $u \in C$*. If* $T : C \to X^*$ *is continuous, then the problem*

$$\begin{cases} \text{Find } u \in C \text{ such that} \\ \langle T(u), v - u\rangle + f(u, v - u) \geq 0 \quad \forall\, v \in C \end{cases}$$

has at least one solution.

Proof. Arguing by contradiction, assume that for every $u \in C$ we can find $v(u) \in C$ such that

$$\langle T(u), v(u) - u\rangle + f(u, v(u) - u) < 0. \tag{3.7}$$

Given $v \in C$, set

$$N(v) = \{u \in C : \langle T(u), v - u\rangle + f(u, v - u) < 0\}.$$

By (3.7) we know that

$$C \subset \bigcup_{v \in C} N(v).$$

As f is upper semicontinuous and T is continuous, each set $N(v)$ is open in C. Since C is compact, we find $v_1, v_2, \ldots, v_k \in C$ such that

$$C \subset \bigcup_{j=1}^{k} N(v_j). \tag{3.8}$$

For all $j = 1, \ldots, k$ and $u \in C$, let $\rho_j(u) = \mathrm{dist}(u; C \setminus N(v_j))$ and set

$$\psi_j(u) = \frac{\rho_j(u)}{\sum_{i=1}^k \rho_i(u)}.$$

Relation (3.8) ensures that the map $\psi_j : C \to \mathbb{R}$ is well defined and continuous. Let $p : C \to C$ be defined by

$$p(u) = \sum_{j=1}^k \psi_j(u) v_j \quad \forall\, u \in C.$$

The mapping p takes values in C (since C is convex) and is continuous. Applying Brouwer's fixed point theorem, there exists an $u_0 \in C$ such that $p(u_0) = u_0$.

Next, we define a function $q : C \to \mathbb{R}$ by

$$q(u) = \langle T(u), p(u) - u \rangle + f(u, p(u) - u) \quad \forall\, u \in C.$$

Since $f(u, \cdot)$ is convex, we derive

$$q(u) \leq \sum_{j=1}^k \psi_j(u) \big(\langle T(u), v_j - u \rangle + f(u, v_j - u) \big) \quad \forall\, u \in C.$$

By (3.8) and the definition of $N(v_j)$, we infer that $q(u) < 0$, $\forall u \in C$. On the other hand, we have $q(u_0) = 0$ because $p(u_0) = u_0$ (since $f(u_0, 0) = 0$). This contradiction completes the proof. □

Another way to guarantee the solvability of problem (3.5) is through the minimization method. Here we consider a real Banach space $(X, \|\cdot\|)$ which is compactly embedded in another real Banach space Z. Let C_0 be a nonempty, closed, convex subset of X. Let $J : X \to \mathbb{R}$ be a Gâteaux differentiable functional and $F : Z \to \mathbb{R}$ be a locally Lipschitz function. Denote by J' the Gâteaux differential of J and by F^0 the generalized directional derivative of F. Recall that $F^0(u; v)$ is the generalized directional derivative of F at u in the direction v, i.e.,

$$F^0(u; v) = \limsup_{\substack{w \to u \\ t \to 0^+}} \frac{F(w + tv) - F(w)}{t}$$

(see Clarke [2]). We consider the following variational–hemivariational inequality:

$$\begin{cases} \text{Find } u \in C_0 \text{ such that} \\ \langle J'(u), v - u \rangle + F^0(u; v - u) \geq 0 \quad \forall\, v \in C_0. \end{cases} \tag{3.9}$$

Let $\Phi : X \to \mathbb{R}$ be defined by

$$\Phi = F|_X + J. \tag{3.10}$$

We associate to (3.9) the minimization problem

$$\begin{cases} \text{Find } u \in C_0 \text{ such that} \\ \Phi(u) \le \Phi(v) \quad \forall\, v \in C_0. \end{cases} \tag{3.11}$$

Lemma 3.8. *If u is a solution of* (3.11)*, then it is a solution of* (3.9).

Proof. If u is a solution of (3.11), then

$$F(u + t(v-u)) - F(u) + J(u + t(v-u)) - J(u) \ge 0 \quad \forall\, v \in C_0,\ t \in (0,1],$$

and thus $F^0(u; v-u) + \langle J'(u), v-u\rangle \ge 0$, $\forall\, v \in C_0$. □

We assume that J satisfies the following alternative:

(H_7) Either

$$J \text{ is convex}$$

or

$$J \in C^1(X,\mathbb{R}) \text{ with } J' : X \to X^* \text{ pseudomonotone, bounded.}$$

Remark 3.9. If the Gâteaux differentiable functional $J : X \to X^*$ satisfies (H_7), then:

(a) J is sequentially weakly lower semicontinuous on X;
(b) J is continuous on X.

Indeed, if J is convex and Gâteaux differentiable, part (a) is straightforward. In the case where J is continuously differentiable with J' pseudomonotone and bounded, part (a) follows from [10, Proposition 25.21]. Part (b) is a consequence of part (a) (because either J is continuously differentiable or it is convex, hence locally Lipschitz).

Remark 3.10. If the Gâteaux differentiable functional $J : X \to X^*$ satisfies (H_7), then:

(a) Φ is sequentially weakly lower semicontinuous on X;
(b) Φ is continuous on X.

Indeed, since J is sequentially weakly lower semicontinuous on X [see Remark 3.9 (a)] and since X is compactly embedded in Z, we infer that F is sequentially weakly lower semicontinuous on X, so Φ is. Moreover, since J is continuous on X (see Remark 3.9 (b)) and F is locally Lipschitz on Z, we infer that Φ is continuous on X.

Remark 3.11. Assume that the Gâteaux differentiable functional $J : X \to X^*$ satisfies the following condition:

(H$_8$) Either

$$J \text{ is convex with } J' \text{ hemicontinuous} \tag{3.12}$$

or

$$J \in C^1(X, \mathbb{R}) \text{ with } J' : X \to X^* \text{ pseudomonotone.} \tag{3.13}$$

Then:

(a) J' satisfies (H$_3$) (ii) (with $T = J'|_C$ and $T_n = J'|_{C_n}$).
(b) If, in addition, J' is bounded, then J' satisfies (H$_3$) (with $T = J'|_C$ and $T_n = J'|_{C_n}$) and (H$_7$).

Indeed, we first observe that hypothesis (H$_8$) implies that $J' : X \to X^*$ is a pseudomonotone operator. To see this, we note that this is clear if (3.13) holds. In the case where (3.12) holds, the convexity of J ensures that $J' : X \to X^*$ is a monotone operator, which in conjunction with the hemicontinuity of J' guarantees that $J' : X \to X^*$ is pseudomonotone (see, e.g., [10, Proposition 27.6 (a)]). Hence, $J'|_{C\cup(\cup_{m=1}^{\infty} C_m)}$ is pseudomonotone, so hypothesis (H$_3$) (ii) is fulfilled (see Lemma 3.6 (b)), which shows assertion (a). Part (b) is a consequence of part (a) and Lemma 3.6 (b).

We have the following existence result of a solution of (3.11) (and, a fortiori, of (3.9)).

Lemma 3.12. *Assume that* (H$_1$) *holds and that* Φ *in* (3.10) *is sequentially weakly lower semicontinuous and coercive in the sense that*

$$\Phi(v) \to +\infty \quad as\ \|v\| \to \infty.$$

Then problem (3.11) *has at least one solution* $u \in C_0$. *In particular, u is a solution of* (3.9).

Proof. The sequential weak lower semicontinuity and the coercivity of Φ ensure the first part of the conclusion. The second part of the conclusion follows from Lemma 3.8. □

Remark 3.13. Lemma 3.12 also holds true if we assume that the set C_0 is bounded in place of the coercivity of Φ.

3.4 Abstract Existence Result

Let X be a real reflexive Banach space, let a nonempty subset C of X, a sequence of subsets C_n of X, $n \in \mathbb{N}$, functions $f : C \times X \to \mathbb{R}$ and $f_n : C_n \times X \to \mathbb{R}$, and operators $T : C \to X^*$ and $T_n : C_n \to X^*$. We have the following convergence result.

Theorem 3.14. *Assume* (H$_1$)–(H$_3$) *and that problem* (3.5) *admits a solution u_n for each $n \in \mathbb{N}$.*

(a) *If the sequence (u_n) is bounded, then it admits a subsequence (u_{n_k}) that is weakly convergent to a solution $u \in C$ of problem* (3.1).
(b) *If, in addition, (T_n) satisfies* (H$_4$), *then the subsequence (u_{n_k}) strongly converges to u in X.*
(c) *Under the assumptions in part* (b), *if the solution of problem* (3.1) *is unique, then the whole sequence (u_n) is strongly convergent to u in X.*

Proof. The assumption that the sequence (u_n) is bounded implies that we can find a subsequence (u_{n_k}) of (u_n) such that $u_{n_k} \rightharpoonup u$, for some $u \in X$ (using that X is reflexive). Hypothesis (H$_1$) (i) implies that $u \in C$.

We claim that

$$\limsup_{k\to\infty} \langle T_{n_k}(u_{n_k}), u_{n_k} - u \rangle \le 0. \tag{3.14}$$

To see this, by (H$_2$) (i) we have that

$$f(u, w) \le a\|w\| \quad \forall\, w \in X.$$

Using (H$_3$) (i), we find a constant $a_0 > 0$ such that

$$\|T_{n_k}(u_{n_k})\|_* \le a_0 \quad \forall\, k \in \mathbb{N}. \tag{3.15}$$

Let $\varepsilon > 0$. Invoking the density of the subset $\mathcal{C} \subset C$ [cf. (H$_1$) (ii)], there exists some $v \in \mathcal{C}$ such that

$$\|v - u\| < \min\left\{\frac{\varepsilon}{2a_0}, \frac{\varepsilon}{2a}\right\}.$$

Using (3.15) and the fact that u_{n_k} is a solution of problem (3.5), we obtain

$$\begin{aligned}
\langle T_{n_k}(u_{n_k}), u_{n_k} - u \rangle &= \langle T_{n_k}(u_{n_k}), u_{n_k} - r_{n_k}(v) \rangle + \langle T_{n_k}(u_{n_k}), r_{n_k}(v) - v \rangle \\
&\quad + \langle T_{n_k}(u_{n_k}), v - u \rangle \\
&\le f_{n_k}(u_{n_k}, r_{n_k}(v) - u_{n_k}) + a_0\|r_{n_k}(v) - v\| \\
&\quad + a_0\|v - u\| \quad \forall\, k \in \mathbb{N}.
\end{aligned}$$

Passing to the upper limit in the above inequality, by (H_1) (ii) and (H_2) (ii), it follows

$$\limsup_{k\to\infty}\langle T_{n_k}(u_{n_k}), u_{n_k} - u\rangle \le \limsup_{k\to\infty} f_{n_k}(u_{n_k}, r_{n_k}(v) - u_{n_k}) + a_0\|v - u\|$$
$$\le f(u, v - u) + \frac{\varepsilon}{2} \le a\|v - u\| + \frac{\varepsilon}{2} < \varepsilon.$$

As $\varepsilon > 0$ is arbitrary, (3.14) holds true, as claimed.

Let us show that u is a solution of problem (3.1). To this end, take $v \in \mathcal{C}$ arbitrary. Applying (H_3) (ii) [due to (3.14)] and using the fact that u_{n_k} is a solution of (3.5), (H_2) (ii), (3.15) and (H_1) (ii), it follows that

$$\begin{aligned}
\langle T(u), u - v\rangle &\le \limsup_{k\to\infty}\langle T_{n_k}(u_{n_k}), u_{n_k} - v\rangle \\
&\le \limsup_{k\to\infty}\big(\langle T_{n_k}(u_{n_k}), u_{n_k} - r_{n_k}(v)\rangle + \langle T_{n_k}(u_{n_k}), r_{n_k}(v) - v\rangle\big) \\
&\le \limsup_{k\to\infty} f_{n_k}(u_{n_k}, r_{n_k}(v) - u_{n_k}) \\
&\quad + \limsup_{k\to\infty}\langle T_{n_k}(u_{n_k}), r_{n_k}(v) - v\rangle \\
&\le f(u, v - u) + a_0 \lim_{k\to\infty}\|r_{n_k}(v) - v\| \\
&= f(u, v - u).
\end{aligned} \tag{3.16}$$

As v is arbitrary in the dense subset $\mathcal{C} \subset C$, taking Lemma 3.3 into account, we conclude from (3.16) that u solves problem (3.1). This proves part (a).

Assume now that (T_n) satisfies (H_4). Then from (3.14) and the fact that $u_{n_k} \rightharpoonup u$ we obtain that $u_{n_k} \to u$ in X, which proves part (b).

If, in addition, the solution of problem (3.1) is unique, then the sequence (u_n) is strongly convergent to u since the above reasoning can be done for any subsequence of (u_n), hence part (c) holds true. □

Remark 3.15. The setting in [4] (and in particular [4, Theorem 2.1]) is recovered for $f_n = f : (C \cup (\cup_{m=1}^{\infty} C_m)) \times X \to \mathbb{R}$ and $T_n = T : C \cup (\cup_{m=1}^{\infty} C_m) \to X^*$ (for all $n \in \mathbb{N}$) and when the elements u and u_k ($k \in \mathbb{N}$) in the hypotheses are taken in the whole domain $C \cup (\cup_{m=1}^{\infty} C_m)$.

Next, we give a consequence of Theorem 3.14 that will be applied in Sect. 3.5 to hemivariational inequalities (see [7]).

Let X be a real, reflexive Banach space and let C be a nonempty, bounded, closed, convex, separable subset of X. Let $\{w_n\}_{n\in\mathbb{N}}$ be a countable dense subset of C and let

$$X_n = \text{span}\{w_1, w_2, \dots, w_n\} \quad \text{and} \quad C_n = C \cap X_n, \ \ n \in \mathbb{N}.$$

For functions $f : C \times X \to \mathbb{R}$ and $f_n : C_n \times X \to \mathbb{R}$ $(n \in \mathbb{N})$, we formulate the following condition:

(H$_9$) for every subsequence (f_{n_k}) of (f_n) and every sequences (u_k), (v_k) with $u_k, v_k \in C_{n_k}$, $\forall\, k \in \mathbb{N}$, and $u_k \rightharpoonup u$ in X, $v_k \rightharpoonup v$ in X, we have

$$\limsup_{k\to\infty} f_{n_k}(u_k, v_k - u_k) \le f(u, v-u).$$

Corollary 3.16. *Let X, C and X_n be as above, let functions $f : C \times X \to \mathbb{R}$ and $f_n : C_n \times X \to \mathbb{R}$ $(n \in \mathbb{N})$, let an operator $T : C \to X^*$ and let continuous operators $T_n : C_n \to X^*$ $(n \in \mathbb{N})$. Assume that* (H$_3$), (H$_9$) *hold, that f satisfies* (H$_5$) *(in Lemma 3.4), and that $f_n : C_n \times X \to \mathbb{R}$ $(n \in \mathbb{N})$ are upper semicontinuous functions such that $f_n(u,\cdot)$ is convex and $f_n(u,0) = 0$ for all $u \in C_n$. Then:*

(a) *Problem*

$$\begin{cases} \text{Find } u \in C \text{ such that} \\ \langle T(u), v-u\rangle + f(u, v-u) \ge 0 \quad \forall\, v \in C \end{cases} \tag{3.17}$$

has a solution. Moreover, for each $n \in \mathbb{N}$, there is a solution u_n of (3.5), *and the sequence (u_n) admits a subsequence (u_{n_k}) that is weakly convergent to a solution u of problem* (3.17).

(b) *If, in addition, (T_n) satisfies* (H$_4$), *then the subsequence (u_{n_k}) strongly converges to u in X.*

(c) *Under the assumptions in part* (b), *if the solution of problem* (3.17) *is unique, then the whole sequence (u_n) is strongly convergent to u in X.*

Proof. We aim to apply Theorem 3.14. By Lemma 3.2, propriety (H$_1$) is satisfied with $\mathcal{C} = C$ and r_n defined by means of (3.6). Lemma 3.4 guarantees that (H$_2$) (i) is satisfied, whereas (H$_2$) (ii) follows from (H$_9$). Condition (H$_3$) is satisfied by hypothesis. For each $n \in \mathbb{N}$, we obtain that (3.5) has a solution $u_n \in C_n$ by applying Lemma 3.7 with (X_n, C_n) in place of (X, C), $f|_{C_n \times X_n}$ and $i_n^* T|_{C_n}$ in place of f and T, where i_n^* stands for the dual mapping of the inclusion map $i_n : X_n \to X$. The sequence (u_n) is bounded (since C is bounded), hence Theorem 3.14 can be applied and the proof is complete. □

Remark 3.17. Assume that $f = f_n : C \times X \to \mathbb{R}$ (for all $n \in \mathbb{N}$) is a function such that $f(u,\cdot)$ is subadditive, positively homogeneous, continuous and $f(u,0) = 0$ whenever $u \in C$. In addition, suppose that either

$$f \text{ is a sequentially weakly upper semicontinuous on } C \times X,$$

or

$$f \text{ is upper semicontinuous on } C \times X \text{ and the set } C \text{ is compact.}$$

By Lemma 3.6, we see that Corollary 3.16 applies [providing a solution of problem (3.17)] when each of the above cases is combined with the case where $T = T_n : C \to X^*$ (for all $n \in \mathbb{N}$) is a bounded and pseudomonotone (thus, demicontinuous) operator (note that then T is continuous on the compact set C_n).

3.5 Application to Hemivariational Inequalities

Our goal is to apply Corollary 3.16 to study the existence of solutions for hemivariational inequalities.

Let Ω be a bounded domain in $\mathbb{R}^N$ ($N \geq 1$). Given $M \geq 1$, we consider functions $j, j_n : \Omega \times \mathbb{R}^M \to \mathbb{R}$ ($n \in \mathbb{N}$) with the following properties:

(H$_{10}$) (i) $j(\cdot, y), j_n(\cdot, y) : \Omega \to \mathbb{R}$ are measurable for all $y \in \mathbb{R}^M$ and satisfy that $j(\cdot, 0), j_n(\cdot, 0) \in L^1(\Omega)$;

(ii) $j(x, \cdot), j_n(x, \cdot) : \mathbb{R}^M \to \mathbb{R}$ is locally Lipschitz for a.e. $x \in \Omega$;

(iii) there exist $q \in (1, +\infty)$, $c \geq 0$, and $k \in L^{\frac{q}{q-1}}(\Omega)$ such that

$$|z| \leq k(x) + c|y|^{q-1} \quad \text{for a.e. } x \in \Omega, \forall\, y \in \mathbb{R}^M, \forall\, z \in \partial j(x, y) \cup \partial j_n(x, y)$$

for all $n \in \mathbb{N}$.

Here, for $x \in \Omega$, we denote by $\partial j(x, y)$ the generalized gradient of the locally Lipschitz function $j(x, \cdot)$ at $y \in \mathbb{R}^M$, that is,

$$\partial j(x, y) = \{z \in \mathbb{R}^M : \langle z, \xi \rangle \leq j^0(x, y; \xi) \quad \forall\, \xi \in \mathbb{R}^M\}, \tag{3.18}$$

where $j^0(x, y; \xi)$ is the generalized directional derivative of $j(x, \cdot)$ at y in the direction ξ, i.e.,

$$j^0(x, y; \xi) = \limsup_{\substack{\eta \to y \\ t \to 0+}} \frac{j(x, \eta + t\xi) - j(x, \eta)}{t}. \tag{3.19}$$

As before, let X be a real, reflexive Banach space and let $C \subset X$ be a nonempty, bounded, closed, convex, separable subset. Let $\{w_n\}_{n \in \mathbb{N}}$ be a countable dense subset of C and let

$$X_n = \text{span}\{w_1, w_2, \ldots, w_n\} \quad \text{and} \quad C_n = C \cap X_n, \quad n \in \mathbb{N}.$$

Let a linear continuous operator $L : X \to L^q(\Omega; \mathbb{R}^M)$ (for instance, X can be the Sobolev space $W^{1,p}(\Omega)$ with $p \in (1, +\infty)$, $M = 1$, and L the inclusion $W^{1,p}(\Omega) \hookrightarrow L^q(\Omega)$ (which is continuous whenever $q < p^*$, where p^* is the Sobolev critical exponent); or, $X = W^{1,q}(\Omega)$, $M = N$, and $Lu = \nabla u$). We assume moreover that

(H_{11}) for every subsequence (j_{n_k}) of (j_n) and every sequences (u_k), (v_k) with $u_k, v_k \in C_{n_k}$, $\forall k \in \mathbb{N}$, and $u_k \rightharpoonup u$ in X, $v_k \rightharpoonup v$ in X, we have

$$\limsup_{k\to\infty} j_{n_k}^0(x, Lu_k; L(v_k - u_k)) \le j^0(x, Lu; L(v - u)).$$

Given (as before) an operator $T : C \to X^*$, we formulate the following problem in terms of a hemivariational inequality

$$\begin{cases} \text{Find } u \in C \text{ such that} \\ \langle T(u), v - u\rangle + \displaystyle\int_\Omega j^0(x, Lu; Lv - Lu)\, dx \ge 0 \quad \forall\, v \in C. \end{cases} \tag{3.20}$$

We have the following existence result for problem (3.20).

Theorem 3.18. *Let X, C, X_n, C_n, Ω, L, j, j_n be as above and satisfy* (H_{10}), (H_{11}). *Let $T : C \to X^*$ and $T_n : C_n \to X^*$ $(n \in \mathbb{N})$ be continuous operators satisfying* (H_3). *Then:*

(a) *Problem* (3.20) *has a solution. Moreover, for all $n \in \mathbb{N}$, the problem*

$$\begin{cases} \textit{Find } u_n \in C_n \textit{ such that} \\ \langle T_n(u_n), v - u_n\rangle + \displaystyle\int_\Omega j_n^0(x, Lu_n; Lv - Lu_n)\, dx \ge 0 \quad \forall\, v \in C_n, \end{cases} \tag{3.21}$$

has a solution u_n, and the sequence (u_n) admits a subsequence (u_{n_k}) which is weakly convergent to a solution u of problem (3.20).

(b) *If, in addition, (T_n) satisfies* (H_4), *then the subsequence (u_{n_k}) strongly converges to u in X.*

(c) *Under the assumptions in part* (b), *if the solution of problem* (3.20) *is unique, then the whole sequence (u_n) is strongly convergent to u in X.*

Proof. We aim to apply Corollary 3.16 for the functions

$$f(u, v) = \int_\Omega j^0(x, Lu; Lv)\, dx \quad \forall\, u \in C,\ \forall\, v \in X,$$

and

$$f_n(u, v) = \int_\Omega j_n^0(x, Lu; Lv)\, dx \quad \forall\, u \in C_n,\ \forall\, v \in X$$

and the considered operators T, T_n. The hypotheses (H_{10}) and (H_{11}) imply that the hypotheses of Corollary 3.16 are satisfied. The conclusions thus follow from Corollary 3.16. □

Remark 3.19. (a) Assume that $j = j_n : \Omega \times \mathbb{R}^M \to \mathbb{R}$ (for all $n \in \mathbb{N}$) is a function satisfying (H_{10}), that $T = T_n : C \to X^*$ (for all $n \in \mathbb{N}$) is a bounded and pseudomonotone operator. If in addition we assume that either C is a compact set or L is a compact operator, then (using that $j^0(x,\cdot;\cdot)$ is upper semicontinuous) (H_{11}) is satisfied, so that Theorem 3.18 applies providing a solution of problem (3.20).

(b) Theorem 3.18 is an existence result of Hartman–Stampacchia type for hemivariational inequalities. It extends the corresponding result in [8].

Proof of Theorem 3.1. Applying Theorem 3.18 in the setting of Theorem 3.1, we obtain that problem (3.4) has a solution $u \in C$. This solution either belongs to the boundary of C (in which case problem (3.2) admits a solution), or to the interior of C. In the latter case, we can find an open set $U \subset X$ such that $u \in U \subset C$. Choosing the test function $v = u + tw$ in (3.4), for $w \in X$ and $t > 0$ small enough so that $v \in U$, we obtain

$$\langle T(u), tw\rangle + \int_\Omega j^0(x, u; tw)\, dx \geq 0.$$

Since $\langle T(u), \cdot\rangle$ is linear and $j^0(x, u; \cdot)$ is positively homogeneous, we conclude that

$$\langle T(u), w\rangle + \int_\Omega j^0(x, u; w)\, dx \geq 0 \quad \forall\, w \in X.$$

This shows that u is a solution of (3.3). □

3.6 Application to Variational–Hemivariational Inequalities

Let $(X, \|\cdot\|)$ be a real reflexive Banach space, which is continuously embedded in another real Banach space Z. Let C, C_n $(n \in \mathbb{N})$ be nonempty, closed, convex subsets of X satisfying hypothesis (H_1). Let $J, J_n : X \to \mathbb{R}$ $(n \in \mathbb{N})$ be Gâteaux differentiable functionals and $F, F_n : Z \to \mathbb{R}$ $(n \in \mathbb{N})$ be locally Lipschitz functions.

Consider the following inequality problem

$$\begin{cases} \text{Find } u \in C \text{ such that} \\ \langle J'(u), v - u\rangle + F^0(u; v - u) \geq 0 \quad \forall\, v \in C. \end{cases} \tag{3.22}$$

The inequality is a variational–hemivariational inequality in the sense of Motreanu–Panagiotopoulos [5]. Let $\Phi, \Phi_n : X \to \mathbb{R}$ be defined by

$$\Phi = F|_X + J, \quad \Phi_n = F_n|_X + J_n \quad (n \in \mathbb{N}). \tag{3.23}$$

Consider the following conditions on the functions Φ, Φ_n in (3.23):

(H_{12}) (i) for each $n \in \mathbb{N}$, Φ_n is coercive in the sense that

$$\Phi_n(v) \to +\infty \quad \text{as } \|v\| \to \infty;$$

(ii) for every subsequence (C_{n_k}) of (C_n) and every sequence (u_k) with $u_k \in C_{n_k}$, $\forall k \in \mathbb{N}$, and $\|u_k\| \to +\infty$, we have $\Phi_{n_k}(u_k) \to +\infty$;

(iii) for every subsequence (C_{n_k}) of (C_n) and every sequence (u_k) with $u_k \in C_{n_k}$, $\forall k \in \mathbb{N}$, and $u_k \rightharpoonup u$, for some u, we have $\Phi(u) \leq \limsup_{k\to\infty} \Phi_{n_k}(u_k)$;

(iv) for every sequence (u_n) with $u_n \in C_n$, $\forall\, n \in \mathbb{N}$, and $u_n \to u$, for some u, we have $\Phi_n(u_n) \to \Phi(u)$.

We consider the following condition on the generalized directional derivatives F^0, F_n^0 of the functions F, F_n, respectively:

(H_{13}) for every subsequence (F_{n_k}) of (F_n) and every sequences (u_k), (v_k) with $u_k, v_k \in C_{n_k}$, $\forall\, k \in \mathbb{N}$, and $u_k \rightharpoonup u$ in X, $v_k \rightharpoonup v$ in X, we have

$$\limsup_{k\to\infty} F_{n_k}^0(u_k; v_k - u_k) \leq F^0(u; v - u).$$

We also consider the following condition on the Gâteaux differentials J', J_n' of the functions J, J_n, respectively:

(H_{14}) (i) for every subsequence (J_{n_k}) of (J_n) and every bounded sequence (u_k) with $u_k \in C_{n_k}$, $\forall\, k \in \mathbb{N}$, the sequence $(J_{n_k}'(u_k))$ is bounded;

(ii) for every subsequence (J_{n_k}) of (J_n) and every sequence (u_k) with $u_k \in C_{n_k}$, $\forall\, k \in \mathbb{N}$, $u_k \rightharpoonup u$ in X, and

$$\limsup_{k\to\infty} \langle J_{n_k}'(u_k), u_k - u\rangle \leq 0,$$

we have

$$\langle J'(u), u - v\rangle \leq \limsup_{k\to\infty} \langle J_{n_k}'(u_k), u_k - v\rangle \quad \forall\, v \in \mathcal{C},$$

where $\mathcal{C} \subset C$ is as in (H_1).

We formulate the minimization problem

$$\begin{cases} \text{Find } u \in C \text{ such that} \\ \Phi(u) \leq \Phi(v) \quad \forall\, v \in C. \end{cases} \tag{3.24}$$

We have the following existence and approximation result of a solution u of (3.24) [and, a fortiori, of (3.22)]. An essential aspect of this result is that the solution u and its approximations u_n are required to belong to the prescribed closed and convex sets C and C_n, respectively.

Theorem 3.20. *Let $X, Z, C, C_n, F, F_n, J, J_n$ be as above. Assume that* (H_1) *holds. Assume moreover that J, J_n ($n \in \mathbb{N}$) satisfy* (H_7)*, that is, for every $K \in \{J, J_n\ (n \in \mathbb{N})\}$, we have either*

$$K \text{ is convex} \tag{3.25}$$

or

$$K \in C^1(X, \mathbb{R}) \text{ with } K' : X \to X^* \text{ pseudomonotone and bounded.} \tag{3.26}$$

Assume that (H_{12}) (i) *holds. Then:*

(a) *For each $n \in \mathbb{N}$, problem*

$$\begin{cases} \text{Find } u_n \in C_n \text{ such that} \\ \Phi_n(u_n) \leq \Phi_n(v) \quad \forall\, v \in C_n \end{cases} \tag{3.27}$$

has at least one solution u_n. Moreover, if (H_{12}) (ii), (iv) *hold, then the sequence (u_n) is bounded in X.*

(b) *If* (H_{12}) *holds, then every subsequence of (u_n) possesses a subsequence which is weakly convergent to a solution u of* (3.24)*. In particular, u is a solution of problem* (3.22) *(see Lemma 3.8).*

(c) *If* (H_{12})–(H_{14}) *and* (H_4) *hold and if the solution of* (3.24) *is unique, then the whole sequence (u_n) is strongly convergent to u in X.*

Proof. By Lemma 3.12 [using Remark 3.10 and the facts that X_n is of finite dimension and that J_n satisfies (3.25) or (3.26)], for each $n \in \mathbb{N}$ problem (3.27) has a solution u_n. To show that the sequence (u_n) is bounded, suppose by contradiction that there exists a subsequence (u_{n_k}) of (u_n), with $\|u_{n_k}\| \to +\infty$ as $k \to \infty$. Hypothesis (H_{12}) (ii) implies that $\Phi_{n_k}(u_{n_k}) \to +\infty$ as $k \to \infty$. Fix $v_0 \in \mathcal{C}$, where $\mathcal{C}$ is the set entering (H_1) (ii). Since u_{n_k} solves problem (3.27) we have $\Phi_{n_k}(u_{n_k}) \leq \Phi_{n_k}(r_{n_k}(v_0))$, $\forall k \in \mathbb{N}$, which gives

$$\Phi_{n_k}(r_{n_k}(v_0)) \to +\infty \ \text{ as } k \to \infty. \tag{3.28}$$

On the other hand, by (H_1) (ii), it is known that $r_n(v_0) \to v_0$. Then hypothesis (H_{12}) (iv) ensures that $\Phi_n(r_n(v_0)) \to \Phi(v_0)$ as $n \to \infty$, contradicting (3.28). Hence part (a) is verified.

To justify part (b), let (u_{n_k}) be an arbitrary subsequence of the sequence (u_n) consisting of solutions u_n of (3.27). By the reflexivity of X, we find a subsequence of (u_{n_k}), denoted again by (u_{n_k}), converging weakly to some $u \in X$. By hypothesis (H_1) (i) we have that $u \in C$. By (H_{12}) (iii), the fact that u_{n_k} solves problem (3.27), (H_1) (ii), and (H_{12}) (iv), we obtain

$$\Phi(u) \leq \limsup_{k\to\infty} \Phi_{n_k}(u_{n_k}) \leq \limsup_{k\to\infty} \Phi_{n_k}(r_{n_k}(v)) = \Phi(v) \quad \forall\, v \in \mathcal{C}. \tag{3.29}$$

Taking into account that Φ is continuous on X (since J is continuous by virtue of Remark 3.9) and $\mathcal{C}$ is dense in C, from (3.29) we derive that $\Phi(u) \leq \Phi(v), \forall\, v \in C$, hence u solves problem (3.24). Therefore part (b) holds true.

Now let us show part (c). Since u_n is a solution of problem (3.27) we have

$$\langle J_n'(u_n), v - u_n\rangle + F_n^0(u_n; v - u_n) \geq 0 \quad \forall\, v \in C_n, \tag{3.30}$$

for each $n \in \mathbb{N}$ (see Lemma 3.8). In view of part (b), every arbitrary subsequence (u_{n_k}) of the sequence (u_n) admits a subsequence that weakly converges to the solution u of (3.24) (which is assumed to be unique by hypothesis). It follows that the whole sequence (u_n) weakly converges to u.

We prove that the hypotheses of Theorem 3.14 with $f = F^0|_{C\times X}$, $f_n = F^0|_{C_n\times X}$, $T = J'|_C$, $T_n = J'|_{C_n}$ are satisfied. Assumption (H_1) is satisfied by hypothesis. Note that (H_2) (i) is verified due to Lemma 3.4 and the properties of the generalized directional derivative F^0. Conditions (H_2) (ii) and (H_3) follow from (H_{13}) and (H_{14}), respectively. Since u_n solves (3.30), we can apply Theorem 3.14 which guarantees that (u_n) is strongly convergent to u. The proof is complete. □

Remark 3.21. Theorem 3.20 holds true if we assume that $\cup_{n=1}^{\infty} C_n$ is bounded in place of hypotheses (H_{12}) (i), (ii) (see Remark 3.13).

Remark 3.22. (a) If $F_n = F : Z \to \mathbb{R}$ is locally Lipschitz and $J_n = J : X \to X^*$ (for all $n \in \mathbb{N}$) is Gâteaux differentiable and satisfies (H_7), then (H_{12}) (iv) holds true (since $\Phi = \Phi_n$ (for all $n \in \mathbb{N}$) is continuous as noted in Remark 3.10).

(b) If $\Phi_n = \Phi$ (for all $n \in \mathbb{N}$) is coercive on X, then (H_{12}) (i), (ii) hold true.

(c) If the embedding $X \subset Z$ is compact and $F_n = F : Z \to \mathbb{R}$ is locally Lipschitz, then (H_{13}) is verified (by the properties of the generalized directional derivative F^0).

(d) If $J_n = J : X \to \mathbb{R}$ (for all $n \in \mathbb{N}$) is Gâteaux differentiable and satisfies (H_8) and J' is bounded and satisfies condition $(S)_+$, then (H_4) and (H_{14}) hold true (see Lemma 3.6 and Remark 3.11).

3.7 Extension of Main Theorem on Pseudomonotone Operators

Let X be a real, reflexive, separable Banach space. Let $\{w_n\}_{n\in\mathbb{N}}$ be a countable dense subset of X and, for each $n \in \mathbb{N}$, denote

$$X_n = \text{span}\{w_1, w_2, \ldots, w_n\}.$$

For a sequence of operators $T_n : X_n \to X^*$ ($n \in \mathbb{N}$), we consider the following condition:

(H_{15}) for every sequence (u_n) with $u_n \in X_n$, $\forall n \in \mathbb{N}$, and $\|u_n\| \to +\infty$ we have

$$\lim_{n\to\infty} \frac{\langle T_n(u_n), u_n\rangle}{\|u_n\|} = +\infty.$$

Theorem 3.23. *Let X, X_n be as above and let an operator $T : X \to X^*$. Let $T_n : X_n \to X^*$ ($n \in \mathbb{N}$) be demicontinuous operators which are coercive in the sense that*

$$\lim_{\|u\|\to+\infty} \frac{\langle T_n(u), u\rangle}{\|u\|} = +\infty.$$

Assume that (H$_3$) *(with $C_n = X_n$) and* (H$_{15}$) *hold. Then:*

(a) *For each $b \in X^*$ the problem*

$$\begin{cases} \text{Find } u \in X \text{ such that} \\ T(u) = b \end{cases} \tag{3.31}$$

has a solution.

(b) *For fixed $b \in X^*$ and each $n \in \mathbb{N}$, the problem*

$$\begin{cases} \text{Find } u_n \in X_n \text{ such that} \\ \langle T_n(u_n), v\rangle = \langle b, v\rangle \quad \forall\, v \in X_n \end{cases} \tag{3.32}$$

has a solution u_n and the sequence (u_n) is bounded in X. Moreover, there exists a subsequence (u_{n_k}) of (u_n) that is weakly convergent to a solution u of problem (3.31).

(c) *If (T_n) satisfies property* (H$_4$) *(with $C_n = X_n$), then (u_{n_k}) strongly converges to u in X. In this case, if the solution of problem* (3.31) *is unique, then the whole sequence (u_n) is strongly convergent to u in X.*

Proof. We aim to apply Theorem 3.14 with $C = X$ and $C_n = X_n$, $n \in \mathbb{N}$. Lemma 3.2 ensures that condition (H$_1$) is satisfied with $\mathcal{C} = X$ and r_n, $n \in \mathbb{N}$, as in (3.6). Let the function $f : X \times X \to \mathbb{R}$ be given by

$$f(u, v) = -\langle b, v\rangle \quad \forall\, u, v \in X.$$

Clearly, (H$_2$) is satisfied with the above f and $f_n = f|_{X_n}$, $n \in \mathbb{N}$, whereas (H$_3$) is satisfied by hypothesis. Moreover, problem (3.5) becomes problem (3.32), which is equivalent to finding $u_n \in X_n$ such that

$$i_n^* T_n(u_n) = i_n^* b\,,$$

where i_n^* stands for the dual mapping of the inclusion map $i_n : X_n \to X$. Since X_n is finite dimensional, the demicontinuity of $T_n : X_n \to X^*$ implies the continuity of $i_n^* T_n : X_n \to X_n^*$. Since $i_n^* T_n$ is also coercive, by a standard consequence of the Brouwer's fixed point theorem, problem (3.32) has at least one solution.

We also observe that any sequence of solutions u_n of (3.32), $n \in \mathbb{N}$, is bounded in X due to (H_{15}) and the inequality

$$\frac{\langle T_n(u_n), u_n\rangle}{\|u_n\|} = \frac{\langle b, u_n\rangle}{\|u_n\|} \leq \|b\|_* \text{ whenever } u_n \neq 0.$$

It suffices to apply Theorem 3.14 to complete the proof. □

Remark 3.24. (a) If $T_n = T : X \to X^*$ (for all $n \in \mathbb{N}$) is coercive, that is,

$$\lim_{\|v\|\to\infty} \frac{\langle T(v), v\rangle}{\|v\|} = +\infty,$$

then (H_{15}) is satisfied.

(b) In the case where $T_n = T : X \to X^*$ (for all $n \in \mathbb{N}$) in Theorem 3.23 is bounded and pseudomonotone, we have that T satisfies hypothesis (H_3) (see Lemma 3.6) and is demicontinuous (see [10, Proposition 27.7 (b)]). Therefore Theorem 3.23 extends the main theorem on pseudomonotone operators due to Brezis [1] (see also Zeidler [10, Theorem 27.A]), which deals with the case where $T = T_n : X \to X^*$ (for all $n \in \mathbb{N}$).

3.8 Extension of Skrypnik's Result for Equations with Odd Operator

We start with a slightly modified version of [9, Lemma 5.1] that we recall from [4] (for the sake of completeness we include the proof). The notation $B(0, R)$ and $S(0, R)$ mean the open ball centered at the origin of radius R and its boundary in the underlying space, respectively.

Lemma 3.25. *Let Z be a finite dimensional real Banach space, $A : \overline{B(0, R)} \subset Z \to Z^*$ be a continuous mapping and $y \in Z^*$ which satisfy*

(i) $A(-z) = -A(z) \quad \forall\, z \in S(0, R)$,
(ii) $A(z) \neq \alpha y \quad \forall\, \alpha \in [0, 1],\ \forall\, z \in S(0, R)$.

Then the equation $A(x) = y$ has a solution $x \in B(0, R)$.

Proof. As Z is finite dimensional, we may consider a linear isomorphism $\Phi : Z^* \to Z$ and introduce $\tilde{A} : \overline{B(0, R)} \to Z$ by $\tilde{A} = \Phi A$. The mapping $h : [0, 1] \times \overline{B(0, R)} \to Z$ given by

$$h(\alpha, z) = \tilde{A}(z) - \alpha\Phi(y) \quad \forall\, (\alpha, z) \in [0, 1] \times \overline{B(0, R)}$$

is a homotopy between $\tilde{A}$ and $\tilde{A} - \Phi(y)$. By (ii), Brouwer's topological degree $\deg(h(\alpha, \cdot), B(0, R), 0)$ is well defined for all $\alpha \in [0, 1]$ (see, e.g., [6, Section 4.1]).

The homotopy invariance property of Brouwer's degree implies

$$\deg(\tilde{A} - \Phi(y), B(0, R), 0) = \deg(\tilde{A}, B(0, R), 0).$$

Hypothesis (i) enables us to apply Borsuk's theorem (see, e.g., [6, Corollary 4.13]), which implies that $\deg(\tilde{A} - \Phi(y), B(0, R), 0)$ is an odd number. Thus the equation $\tilde{A}(x) = \Phi(y)$ is solvable in $B(0, R)$ (by the properties of Brouwer's degree, see, e.g., [6, Theorem 4.5 (e)]). The proof is complete. □

Now, let X be a real, reflexive, separable Banach space. Let $\{w_n\}_{n\in\mathbb{N}}$ be a countable dense subset of $\overline{B(0, R)}$ and, for each $n \in \mathbb{N}$, denote

$$X_n = \operatorname{span}\{w_1, w_2, \ldots, w_n\}.$$

Theorem 3.26. *Let X, X_n be as above and an operator $T : \overline{B(0, R)} \to X^*$. Let $T_n : \overline{B(0, R)} \cap X_n \to X^*$ ($n \in \mathbb{N}$) be demicontinuous operators such that*

$$T_n(-v) = -T_n(v) \quad \forall\, v \in S(0, R) \cap X_n,\ \forall\, n \in \mathbb{N} \tag{3.33}$$

and satisfying (H$_3$) (i), (H$_6$) *(in Lemma 3.5) and* (H$_4$) *(with $C_{n_k} = \overline{B(0, R)} \cap X_{n_k}$). Fix $b \in X^*$ with*

$$\|b\|_* < \|T(v)\|_* \quad \forall\, v \in S(0, R). \tag{3.34}$$

Then:

(a) *The problem*

$$\begin{cases} \text{Find } u \in \overline{B(0, R)} \text{ such that} \\ T(u) = b \end{cases} \tag{3.35}$$

has a solution in $B(0, R)$.

(b) *For each $n \in \mathbb{N}$, the problem*

$$\begin{cases} \text{Find } u_n \in \overline{B(0, R)} \cap X_n \text{ such that} \\ \langle T_n(u_n), v\rangle = \langle b, v\rangle \quad \forall\, v \in X_n \end{cases} \tag{3.36}$$

has a solution $u_n \in B(0, R) \cap X_n$. Furthermore, there exists a subsequence (u_{n_k}) of (u_n) which is weakly convergent to a solution u of problem (3.35).

(c) *If (T_n) satisfies property* (H$_4$) *(with $C_n = \overline{B(0, R)} \cap X_n$), then the subsequence (u_{n_k}) strongly converges to u in X. In this case, if the solution of* (3.35) *is unique, then the whole sequence (u_n) strongly converges to u in X.*

Proof. We aim to apply Theorem 3.14 with $C = \overline{B(0, R)}$ and $C_n = \overline{B(0, R)} \cap X_n$, $n \in \mathbb{N}$. Let $b \in X^*$ satisfy (3.34). By Lemma 3.2, we see that hypothesis (H$_1$) is satisfied with $\mathcal{C} = \overline{B(0,R)}$ and r_n defined as in (3.6). Let $f : \overline{B(0,R)} \times X \to X$ be given by

$$f(u,v) = -\langle b, v\rangle \quad \forall\, u \in \overline{B(0,R)},\ \forall\, v \in X.$$

It is clear that f satisfies (H$_2$) with $f_n = f|_{\overline{B(0,R)}\cap X_n}$, $n \in \mathbb{N}$. Condition (H$_3$) (i) is satisfied by hypothesis, whereas condition (H$_3$) (ii) follows from (H$_6$) and (H$_4$) (see Lemma 3.5).

Let us show that Lemma 3.25 can be applied with $Z = X_n$, $A = i_n^* T_n|_{\overline{B(0,R)}\cap X_n}$ and $y = i_n^* b$, where i_n^* stands for the dual mapping of the inclusion map $i_n : X_n \to X$. Since X_n is finite dimensional, the demicontinuity of $T_n : \overline{B(0,R)} \cap X_n \to X^*$ implies the continuity of $i_n^* T_n : \overline{B(0,R)} \cap X_n \to X_n^*$. Condition (i) in Lemma 3.25 follows from (3.33). We prove that condition (ii) in Lemma 3.25 holds for all n sufficiently large, that is, there exists $n_0 \in \mathbb{N}$ such that (ii) in Lemma 3.25 is true for $n \geq n_0$. Arguing by contradiction, assume that we can find sequences $\alpha_k \in [0,1]$, $u_k \in S(0,R) \cap X_{n_k}$ such that

$$\langle T_{n_k}(u_k), v\rangle = \alpha_k \langle b, v\rangle \quad \forall\, v \in X_{n_k},\ k \in \mathbb{N}. \tag{3.37}$$

Taking into account that $\|u_k\| = R$, passing to a relabelled subsequence, we have $u_k \rightharpoonup u_0$ in X and $\alpha_k \to \alpha_0$ as $k \to \infty$, for some $u_0 \in \overline{B(0,R)}$ and $\alpha_0 \in [0,1]$. Then from (3.37), (H$_1$) (ii) and (H$_3$) (i) we obtain

$$\begin{aligned}\langle T_{n_k}(u_k), u_k - u_0\rangle &= \langle T_{n_k}(u_k), r_{n_k}(u_0) - u_0\rangle + \langle T_{n_k}(u_k), u_k - r_{n_k}(u_0)\rangle \\ &= \langle T_{n_k}(u_k), r_{n_k}(u_0) - u_0\rangle + \alpha_k \langle b, u_k - r_{n_k}(u_0)\rangle \to 0\end{aligned}$$

as $k \to \infty$. Combining with hypothesis (H$_4$), this yields $u_k \to u_0$ in X, hence $u_0 \in S(0,R)$. Let $v \in \cup_{k\in\mathbb{N}} X_{n_k}$. Then $v \in X_{n_k}$ for all k sufficiently large, so from (3.37) we have $\langle T_{n_k}(u_k), v\rangle = \alpha_k \langle b, v\rangle$. Letting $k \to \infty$ and using that $u_k \to u_0$ in X and hypothesis (H$_6$), we obtain $\langle T(u_0), v\rangle = \alpha_0 \langle b, v\rangle$. The density of $\cup_{k\in\mathbb{N}} X_{n_k}$ in X implies

$$\langle T(u_0), v\rangle = \alpha_0 \langle b, v\rangle \quad \forall\, v \in X.$$

It follows that $T(u_0) = \alpha_0 b$, which contradicts the choice of b in (3.34), so condition (ii) of Lemma 3.25 is satisfied.

By Lemma 3.25 it follows that for every $n \geq n_0$ problem (3.36) has a solution $u_n \in B(0,R) \cap X_n$. The conclusions in the theorem follow by applying Theorem 3.14, while the fact that $u \in B(0,R)$ is a consequence of (3.34). The proof is complete. □

Remark 3.27. (a) If $T_n = T : X \to X^*$ (for all $n \in \mathbb{N}$) is bounded, demicontinuous and satisfies condition $(S)_+$, then (H$_3$) (i), (H$_6$) and (H$_4$) are satisfied.

(b) Theorem 3.26 extends the existence result for equations with odd operator in [9, Theorem 5.1], which deals with the case $T = T_n : \overline{B(0, R)} \to X^*$ (for all $n \in \mathbb{N}$).

Acknowledgements The second author is supported by a Marie Curie Intra-European Fellowship for Career Development within the European Community's 7th Framework Program (Grant Agreement No. PIEF-GA-2010-274519).

References

1. Brezis, H.: Équations et inéquations non linéaires dans les espaces vectoriels en dualité. Annales de l'Institut Fourier (Grenoble) **18**, 115–175 (1968)
2. Clarke, F.H.: Optimization and Nonsmooth Analysis. Wiley, New York (1983)
3. Glowinski, R., Lions, J.-L., Trémolières, R.: Analyse Numérique des Inéquations Variationnelles, I, Théorie Générale, Premières Applications. Dunod, Paris (1976)
4. Jebelean, P., Motreanu, D., Motreanu, V.V.: A unified approach for a class of problems involving a pseudomonotone operator. Math. Nachr. **281**, 1283–1293 (2008)
5. Motreanu, D., Panagiotopoulos, P.D.: Minimax Theorems and Qualitative Properties of the Solutions of Hemivariational Inequalities. Kluwer Academic Publishers, Dordrecht (1999)
6. Motreanu, D., Motreanu, V.V., Papageorgiou, N.S.: Topological and Variational Methods with Applications to Nonlinear Boundary Value Problems. Springer, New York (2014)
7. Naniewicz, Z., Panagiotopoulos, P.D.: Mathematical Theory of Hemivariational Inequalities and Applications. Marcel Dekker, New York (1995)
8. Panagiotopoulos, P.D., Fundo, M., Rădulescu, V.: Existence theorems of Hartman-Stampacchia type for hemivariational inequalities and applications. J. Glob. Optim. **15**, 41–54 (1999)
9. Skrypnik, I.V.: Methods for Analysis of Nonlinear Boundary Value Problems, translated by D. Pascali. Translations of Mathematical Monographs, vol. 139. American Mathematical Society, Providence (1994)
10. Zeidler, E.: Nonlinear Functional Analysis and its Applications, *II/B*: Nonlinear Monotone Operators. Springer, New York (1990)

Chapter 4
Nonconvex Variational Inequalities

Si-sheng Yao and Nan-jing Huang

Abstract In this chapter, we consider a nonconvex variational inequality defined on proximally smooth set and star-shaped set. Based on arguments of the fixed point theorem and the proximal normal method, some existence and uniqueness results concerning the nonconvex variational inequality are proved under suitable conditions. Some iterative algorithms for approximating the solutions of the nonconvex variational inequality are constructed and the convergence results for the iterative sequences generated by the algorithms are also given.

Keywords Nonconvex variational inequality • Proximally smooth set • Star-shaped set

AMS Classification. 47J20, 49J40.

4.1 Introduction

Variational inequalities, introduced by Hartman and Stampacchia [13] in the early 1960s, are a very powerful tool of the current mathematical technology. These have been extended and generalized to the study of a wide class of problems arising in mechanics, physics, optimization and control, nonlinear programming, economics, transportation equilibrium and engineering sciences, etc. The development trace of variational inequality theory can be counted as the process to reveal some important problems and the tool for developing highly efficient algorithms for solving the relevant applied problems. Those numerical methods, such as the projection method

S. Yao
Department of Mathematics, Kunming University, Kunming, Yunnan 650221, People's Republic of China
e-mail: yaosisheng@outlook.com

N. Huang (✉)
Department of Mathematics, Sichuan University, Chengdu, Sichuan 610064, People's Republic of China
e-mail: nanjinghuang@hotmail.com

W. Han et al. (eds.), *Advances in Variational and Hemivariational Inequalities*, Advances in Mechanics and Mathematics 33, DOI 10.1007/978-3-319-14490-0_4

and its variants, the finite element method, Wiener-Hopf skill and the auxiliary principle, solve abundant interesting and important problems, see [11, 14].

The projection type technique, initially introduced by Lions and Stampacchia [17] and developed later by many authors, is an important tool for finding the approximating solutions of various kinds of variational inequalities (see, for example, [20,29]). The main idea in this technique is to establish the equivalence between a variational inequality and a fixed point problem [2, 22]. In addition, the projection method plays a significant role in the numerical solution of the problems mentioned above. For more work concerned with the theory, algorithms and applications of the projection type technique, we refer to [10, 29] and the references therein.

It is well known that the variational inequality problem and many other mathematical problems, such as optimization, control theory, dynamical systems and differential inclusions etc., both the numerical algorithms and their theoretical analysis rely heavily on the assumption of convexity. For example, many classical algorithms in optimization use the fact that a local optimum is a global optimum; consequently, almost all the techniques are based on the properties of the operators over convex sets.

Clarke et al. [7,8] and Poliquin et al. [26] have introduced and studied two classes of nonconvex sets, which are called prox-regular sets and proximally smooth sets. It is known that proximally smooth sets may or may not be convex. This class of proximally smooth sets has many useful properties and has played an important role in many nonconvex applications. In 2003, Bounkhel et al. [4] used these proximally smooth sets to consider the variational inequalities. They have shown that nonconvex variational inequalities are equivalent to nonconvex variational problems. Noor [19] introduced nonconvex variational inequalities based on the proximally smooth sets. Moreover, he discussed the existence and algorithms of the solution for nonconvex variational inequalities and nonconvex mixed variational inequalities, and his research shows that the projection technique can be extended to nonconvex sets. In 2010, Alimohammady et al. [2] constructed some new perturbed finite step projection iterative algorithms with mixed errors for approximating the solutions of a new class of general nonconvex set-valued variational inequalities. Noor et al. [23] used the extragradient projection technique to solve the nonconvex variational inequalities. In 2011, Wen [31] modified the projection methods to a generalized system of nonconvex variational inequalities with different nonlinear operators.

Since Brunn [5] introduced star-shaped sets in 1913, star-shaped sets have been used naturally in many application fields, including integral geometry, computational geometry, mixed integer programming problem (see, for example, [3, 12]). Because of the importance of those sets in structural and mechanical systems, a considerable effort has been made in their theory and numerical analysis (see, for example, [6] and the references therein). It is worth mentioning that some authors discussed the developed mathematical modeling which plays an important role for some practical problems. Naniewicz [18] used the hemivariational inequality approach to establish the existence of solutions for a large class of star-shaped with respect to a ball constrained problems in a reflexive Banach space. And some applications to nonconvex constrained variational problems were considered. In 2003, Lin et al. [16] suggested and analyzed a homotopy continuation method for solving fixed points of self-mappings in a class of star-shaped subsets.

Motivated and inspired by the work mentioned above, in this chapter we introduce and study the nonconvex variational inequality. By using the arguments of Banach's fixed point theorem, star-shaped set theory and proximally smooth set methods, we prove the existence and uniqueness of a solution for the nonconvex variational inequality under some suitable conditions. Furthermore, we consider algorithms of the corresponding mathematical problem and give a convergence result. The results presented in this chapter generalize and improve some known results presented in [2, 20, 27].

This chapter is structured as follows. In Sect. 4.2, we list the preliminaries on nonconvex sets, including proximally smooth sets and star-shaped sets, and some useful results on the nonconvex variational inequality. We prove the existence and uniqueness of a solution to the nonconvex variational inequality in Sect. 4.3. Finally, in Sect. 4.4, we construct some iterative algorithms for finding the approximate solutions of the nonconvex variational inequality and prove the convergence of iterative sequence generated by the algorithms.

4.2 Preliminaries

Let X be a real Hilbert space which is equipped with an inner product $\langle \cdot, \cdot \rangle$ and corresponding norm $\|\cdot\|$, respectively. Let K be a nonempty subset of X and $CB(X)$ denote the family of all closed and bounded subsets of X.

Assume that C is a nonempty closed convex subset of X. For a given nonlinear operator $T : X \to X$, consider the problem of finding $u \in C$ such that

$$\langle Tu, v - u \rangle \geq 0 \quad \forall\, v \in C, \tag{4.1}$$

which is called a variational inequality. This problem was introduced and studied by Stampacchia in 1964 [30].

Now we recall some well known concepts and auxiliary results in nonsmooth analysis [8, 26].

Definition 4.1 ([8]). Let K be a nonempty subset of X. The proximal normal cone of K at $u \in X$ is given by

$$N_K^P(u) = \{\xi \in X \mid \exists\, t > 0 \text{ such that } d_K(u + t\xi) = t\|\xi\|\},$$

where t is a constant and $d_K : X \to \mathbb{R}$ is a distance function defined by

$$d_K(u) = \inf_{v \in K} \|v - u\|.$$

Definition 4.2 ([8]). Let $u \in X$ be a point not lying in K, a nonempty subset of X. A point $v \in K$ is called a closest point or a projection of u onto K if $d_K(u) = \|v - u\|$. The set of all such closest points is denoted by $\mathcal{P}_K(u)$, that is,

$$\mathcal{P}_K(u) = \{v \in K \mid d_K(u) = \|u - v\|\}.$$

Lemma 4.3 ([8]). *Let K be a nonempty subset in X. Then $\xi \in N_K^P(u)$ if and only if there exists a constant $\delta > 0$ such that*

$$\langle \xi, v - u \rangle \le \delta \|v - u\|^2 \quad \forall\, v \in K.$$

Definition 4.4. A closed set K is prox-regular at x if and only if there exists $\varepsilon > 0$ and $\rho > 0$ such that whenever $y \in K$ and $\xi \in N_K^P(y)$ with $\|y - x\| < \varepsilon$ and $\|\xi\| < \varepsilon$, one has

$$\langle \xi, x' - y \rangle - \frac{\rho}{2}\|x' - y\|^2 \le 0 \quad \forall\, x' \in K \text{ with } \|x' - x\| < \varepsilon.$$

Definition 4.5 ([7]). For a given $r \in (0, +\infty]$, a subset K is said to be normalized uniformly r-prox-regular if and only if every nonzero proximal normal to K can be realized by an r-ball, that is, for all $u \in K$ and $0 \neq \xi \in N_K^P(u)$ with $\|\xi\| = 1$, one has

$$\left\langle \frac{\xi}{\|\xi\|}, v - u \right\rangle \le \frac{1}{2r}\|v - u\|^2 \quad \forall\, v \in K.$$

Remark 4.6. There are several interesting equivalence statements to characterize the uniformly r-prox-regular sets. From [26], one can know that a closed set K is proximally smooth with associated tube $U_K(r)$ if and only if the set K is uniformly prox-regular with constant $\frac{1}{r'}$ for every $0 < r' < r$. Thus, we do not distinguish the two definitions "uniformly r-prox-regular" and "proximally smooth" in this chapter.

(i) Following the definition of Clarke et al. [7], a closed set K is said to be proximally smooth if d_K is (norm-to-norm-) continuously differentiable on an open "tube" of the type

$$U_K(r) = \{u \in X \mid 0 < d_K(u) < r\}$$

for some $r > 0$;

(ii) Clarke et al. [7] showed that K is proximally smooth if and only if there exists $r > 0$ such that, for all $u \in U_K(r)$, the projection $\mathcal{P}_K(u)$ is nonempty and each of its elements x belongs also to $\mathcal{P}_K(x + v)$ for $v = r\frac{u-x}{\|u-x\|}$;

(iii) Clarke et al. [7] showed that a weakly closed set K is proximally smooth if and only if $\mathcal{P}_K$ is single-valued on a tube $U_K(r)$.

Lemma 4.7 ([26]). *For a weakly closed set $K \subset H$ and any point $x \in K$, the following statements are equivalent.*

(i) *K is prox-regular at x;*
(ii) *$\mathcal{P}_K$ is single valued around x.*

It is clear that the class of proximally smooth sets is sufficiently large to include the class of convex sets, p-convex sets, $C^{1,1}$ submanifolds (possibly with boundary) of X, the images under a $C^{1,1}$ diffeomorphism of convex sets and many other nonconvex sets (see [26]). Obviously, if $r = +\infty$, then proximal smoothness of K is equivalent to the convexity of C. It is known that the proximal normal cone N_K^P is a set-valued mapping with closed values if K is a proximally smooth set. This class of proximally smooth sets have played an important part in many nonconvex applications such as optimization, dynamic systems and differential inclusions [21,32].

Lemma 4.8 ([7,26]). *Let K be a nonempty closed and proximally smooth subset of X and $r \in (0, +\infty]$. Then the following statements hold:*

(i) *For all $u \in K$, $\mathcal{P}_K(u) \neq \emptyset$;*
(ii) *For all $r' \in (0, r)$, $\mathcal{P}_K$ is Lipschitz continuous with constant $L = \frac{r}{r-r'}$ on $U_K(r')$;*
(iii) *The proximal normal cone is a set-valued mapping with closed values;*
(iv) *If K is weakly closed, then one can get the following to the list of equivalent properties: K is proximally smooth with associated tube $U_K(r) \Leftrightarrow \mathcal{P}_K$ is single-valued on $U_K(r)$.*

Let x be a point of a vector space Z with $x \neq 0$ and $R_x = \{\lambda x \in Z \mid \lambda \geq 0\}$. We recall the definition of the star-shaped set as follows.

Definition 4.9 ([32]). Let A be a nonempty subset of Z.

(i) The set *kern* A of all $a \in A$ such that

$$\{a \in A, 0 \leq \lambda \leq 1\} \Rightarrow a + \lambda(x - a) \in A$$

is called the kernel of A;
(ii) The subset A is called a star-shaped set if *kern* $A \neq \emptyset$.

The totality of all star-shaped sets with respect to zero will be denoted by $\mathcal{U}$. Letting $U \in \mathcal{U}$, the function defined by

$$\mu_U(x) = \inf\{\lambda > 0 \mid x \in \lambda U\} \quad \forall\, x \in Z$$

is called the Minkowski gauge of the set U.

Lemma 4.10 ([32]). *Let $p : Z \to \overline{\mathbb{R}}$ and $U = \{x \in Z \mid p(x) \leq 1\}$. Then the following statements are equivalent.*

(i) *p is positively homogeneous, nonnegative, l.s.c. and $p(0) = 0$;*
(ii) *U is a nonempty star-shaped with respect to zero set and $p = \mu_U$.*

It is well known that the union of two disjoint intervals $[a, b]$ and $[c, d]$ is a prox-regular set with $r = \frac{c-b}{2}$ and it is not a star-shaped set [1, 6]. Moreover, let

$$K = \{(t, z) \in \mathbb{R}^2 \mid t^2 + (z-2)^2 \geq 4, -2 \leq t \leq 2, z \geq -2\}$$

be a subset of the Euclidean plane, then K is a proximally smooth set and K is not a star-shaped set [21].

Next we provide more examples of the nonconvex sets.

Example 4.11. Let $K = \{(\rho, \theta) \in \mathbb{R}^2 \mid \rho \leq 1 - \cos(\theta), r \in [0, 2], \theta \in [0, 2\pi]\}$ be a subset of $\mathbb{R}^2$. By the definition, it is easy to see that K is a star-shaped set but it is without proximally smoothness at $(0, 0)$.

Example 4.12. Let

$$K = \left\{(p, q) \in \mathbb{R}^2 \,\middle|\, \frac{(p-0.5)^2}{0.25} + \frac{q^2}{(0.5p + 0.001)^2} \leq 1, 0 \leq p \leq 2, 0 \leq q \leq 1\right\}.$$

Then we know that K is a star-shaped set. Moreover, it follows from Remark 4.6 (ii) that K is also a proximally smooth set.

If K is a nonconvex set, the classical variation inequality (4.1) can be extended to the following form (4.2), which is called the nonconvex variational inequality [19, 22].

For a nonempty closed subset K in X and a given nonlinear operator $T : K \to K$, consider the problem of finding $u \in K$ such that

$$\langle Tu, v - u \rangle + \delta \|v - u\|^2 \geq 0 \quad \forall\, v \in K. \tag{4.2}$$

To provide a concrete working example for later developments, we consider an elastoplastic deformation model for geomaterials. In the past decades, some researchers studied problems about geomaterials, and got many useful results. Piccolroaz and Bigoni [25] proposed a new yield criteria within the class of isotropic functions of the stress tensor, and studied the nonconvex field surface which is the consequences of the new yield criteria. In 2010, Laydi [15] studied the yield conditions of phase transformation problem in which he gave a counter-example to show the convexity condition is not fulfilled for some field criteria case.

Convexity assumption, as a useful mathematical property, is not always supported by experiments in practical problem and some newly theoretical analysis process [15, 25]. All these works show that the classical method for solving elastoplastic deformation problems, especially in geomaterials constitutive models, will lead to a result that likely violates thermomechanics laws under some stress paths. In this chapter, we will not pay more attention to the fundamental details. The fundamental assumptions and postulates of the theory of thermomechanics are well known and can be found in [9, 33].

The symbol δ presents the state increment, δQ is the dissipation increment, which defined per unit volume. And Q is a function of the inner state variables (the state variables of the elastoplastic materials can usually be described as the following four parameters: stress tensor σ, strain tensor ε, entropy density η, plastic increment tensor ε^p). These four variables are related by the generalized Hooke's law and the laws of thermomechanics; only two variables can be the independent state variables [33, 34].

Firstly, we give an example to show that the yield surface may be a nonconvex set.

Example 4.13. Assume that the dissipative function is a simplified form of the case (18) in Collins and Hilder [9], that is,

$$\delta Q = \left[(ap + bp_0)^2(d\varepsilon_v^p)^2 + (cp + dp_0)^2(d\varepsilon_\gamma^p)^2\right]^{\frac{1}{2}},$$

where a, b, c, d are some experiment parameters, p_0 is a known function of the plastic volume strain and p, q are stress invariants. ε_v^p is the plastic volume strain increment and ε_γ^p is the plastic shear strain increment. Then the yield surface can be derived from the yield function by the method mentioned above as follows:

$$\frac{(p - bp_0)^2}{(ap + bp_0)^2} + \frac{q^2}{(cp + dp_0)^2} = 1.$$

The yield surface in the dissipative stress space is convex, while it is nonconvex in the true stress space, see Fig. 4.1.

Let $\mathbb{R}^d$ be a d-dimensional Euclidean space and $\mathbb{S}^d$ be the space of second order symmetric tensors on $\mathbb{R}^d$. The canonical inner products and corresponding norms on $\mathbb{R}^d$ and $\mathbb{S}^d$ are defined as follows:

$$u \cdot v = u_i v_i, \quad \|v\| = (v \cdot v)^{1/2} \quad \forall\, u, v \in \mathbb{R}^d,$$
$$\sigma : \tau = \sigma_{ij}\tau_{ij}, \quad |\tau| = (\tau : \tau)^{1/2} \quad \forall\, \sigma, \tau \in \mathbb{S}^d.$$

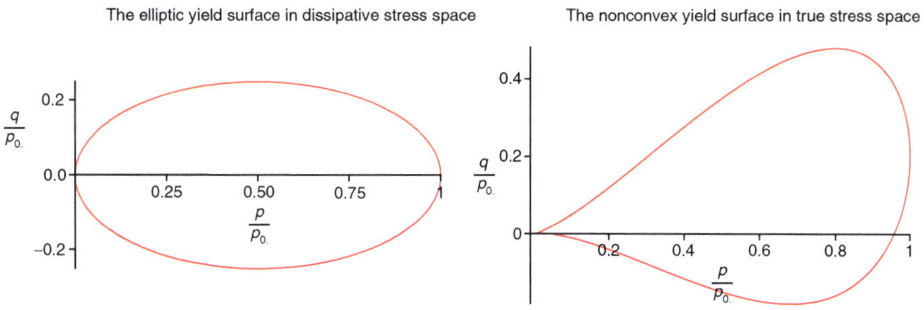

Fig. 4.1 Convex and nonconvex yield surfaces in stress space

Everywhere in the sequel the index i and j run between 1 and d and the summation convention over repeated indices is implied.

Now, we consider an elastoplastic geomaterial deformation problem. Suppose that the geomaterials occupies an open, connected and bounded domain $\Omega \subset \mathbb{R}^d$ with a Lipschitz boundary $\Gamma = \partial\Omega$. Let the time interval of interest be $[0, T]$ with $T > 0$. Since the boundary is Lipschitz continuous, the outward normal vector exists on Γ a.e. and we denote it by ν. Assume that a volume force $f(x,t)$ with $f(x,0) = 0$ acts in $\Omega \times [0, T]$. We denote by u the displacement field, ε the strain tensor, ε^e the elastic strain tensor, ε^p the plastic strain tensor and σ the stress tensor.

Since the yield surface may be nonconvex, the deformation body obeys a non-normality plastic flow rule (see [33]). By Examples 4.12 and 4.13, we know that the nonconvex set K is proximally smooth and star-shaped, which is called the star-shaped set with the proximal smoothness.

Let

$$d\varepsilon^q \in N_K^P(\sigma), \quad K = \{\sigma \in \mathbb{S}^d \mid \overline{f}(\sigma, \varepsilon^p) \le 0\}, \tag{4.3}$$

where K is the nonconvex closed set and $d\varepsilon^q$ is the normal increment of the plastic strain on the yield surface K, which is a nonlinear function of the variables $\varepsilon(u)$, $\varepsilon^p(u)$ and $d\varepsilon^p(u)$ under some suitable hypotheses. The boundary of K, denoted by $\mathcal{K}$, is the yield surface, the elastic region is

$$\mathcal{E} = \left\{\sigma \in \mathbb{S}^d \mid \overline{f}(\sigma, \varepsilon^p) < 0\right\}.$$

From Lemma 4.3, we can rewrite (4.3) as follows: there exists a positive number $\delta > 0$ such that

$$d\varepsilon^q : (\sigma' - \sigma) - \delta|\sigma' - \sigma|^2 \le 0 \quad \forall \sigma' \in K, \tag{4.4}$$

which give the variational formulation of the elastoplastic geomaterials deformation problem.

The nonconvexness of yield surface K implies that the considered elastoplastic geomaterials deformation problem can be solved by employing the nonconvex variational inequality method.

To solve the nonconvex variational inequality, we also need the following lemmas and definitions.

Lemma 4.14 ([1]). *Let X be a reflexive Banach space and $\{x_n\}$ be a bounded sequence of X. Then $\omega_w(x_n) \neq \emptyset$, where*

$$\omega_w(x_n) = \{x \in H \mid x_{n_j} \rightharpoonup x, \ \{n_j\} \subset \{n\}\}.$$

Lemma 4.15 ([8]). *Let K be a nonempty subset of X. For every $x \in X$, a point $u \in \mathcal{P}_K(x)$ if and only if u is a solution of variational inequality*

$$\langle x - u, v - u\rangle \le \frac{1}{2}\|v - u\|^2 \quad \forall v \in K. \tag{4.5}$$

Definition 4.16. Let K be a nonempty closed subset of X. A mapping $T : K \to K$ is said to be

(i) ζ-strongly monotone if there exists a constant $\zeta > 0$ such that

$$\langle Tu - Tv, u - v\rangle \geq \zeta\|u - v\|^2 \quad \forall\, u, v \in K;$$

(ii) χ-Lipschitz continuous if there exists a constant $\chi > 0$ such that

$$\|Tu - Tv\| \leq \chi\|u - v\| \quad \forall\, u, v \in K;$$

(iii) (γ, ζ)-relaxed co-coercive if there exist constants $\gamma, \zeta > 0$ such that

$$\langle Tu - Tv, u - v\rangle \geq -\gamma\|Tu - Tv\|^2 + \zeta\|u - v\|^2 \quad \forall\, u, v \in K.$$

Remark 4.17. It is easy to see that a ζ-strongly monotone mapping is (γ, ζ)-relaxed co-coercive, but the converse is not true in general.

4.3 Main Results

In this section, we use the fixed point theorem, projection technique, star-shaped set and proximally smooth set method to suggest and analyze some properties of the projection $\mathcal{P}_K$ when K is nonconvex. The existence and uniqueness results for nonconvex variational inequality (4.2) are also given.

Theorem 4.18. *Let K be a nonempty subset of X. For a given nonlinear operator $T : K \to K$, $u \in K$ is a solution of the nonconvex variational inequality* (4.2) *if and only if $u \in K$ satisfies the relation $u \in \mathcal{P}_K(u - \rho Tu)$, where $\mathcal{P}_K$ is the projection of X onto K.*

Proof. "$\Rightarrow$" Suppose that (4.2) has a solution $u \in K$. Then there exists $\delta > 0$ such that

$$\langle Tu, v - u\rangle + \delta\|v - u\|^2 \geq 0 \quad \forall\, v \in K$$

and so

$$\langle u - (u - \rho Tu), v - u\rangle + \frac{1}{2}\|v - u\|^2 \geq 0 \quad \forall\, v \in K,$$

where $\rho\delta = \frac{1}{2}$. By Lemma 4.15, we have

$$u \in \mathcal{P}_K(u - \rho Tu).$$

"$\Leftarrow$" Let $u \in \mathcal{P}_K(u - \rho Tu)$. Then it follows from Lemma 4.15 that

$$\langle (u - \rho Tu) - u, v - u\rangle - \frac{1}{2}\|v - u\|^2 \leq 0 \quad \forall\, v \in K,$$

which implies that

$$\rho\langle Tu, v-u\rangle + \frac{1}{2}\|v-u\|^2 \geq 0 \quad \forall\, v \in K$$

and so

$$\langle Tu, v-u\rangle + \delta\|v-u\|^2 \geq 0 \quad \forall\, v \in K,\ \delta = \frac{1}{2\rho}.$$

Thus, $u \in K$ is a solution of (4.2). This completes the proof. □

For $u \in \mathcal{P}_K(u - \rho Tu)$, by Lemma 4.8 and Theorem 4.18, we know that $0 < d_K(u - \rho Tu) < r$ when $\rho < \frac{r}{1+\|Tu\|}$. Thus, we have the following result.

Corollary 4.19. *Let K be a nonempty proximal smoothness on $U_K(r)$. For a given nonlinear operator $T : K \to K$, $u \in K$ is a solution of the nonconvex variational inequality* (4.2) *if and only if $u \in K$ satisfies the relation $u = \mathcal{P}_K(u - \rho Tu)$, where $\rho < \frac{r}{1+\|Tu\|}$.*

For $u \in \mathcal{P}_K(u - \rho Tu)$, by Lemma 4.7 and Theorem 4.18, if K is prox-regular at u, $\mathcal{P}_K$ is single valued around u, that is, there exists $\gamma > 0$ such that $\mathcal{P}_K$ is single on $B(u, \gamma)$ (the neighbourhood of u). It is easy to see that $0 < d_K(u - \rho Tu) < \gamma$ when $\rho < \frac{\gamma}{1+\|Tu\|}$. Thus, we can get the following result.

Corollary 4.20. *Let K be a prox-regular set at u. For a given nonlinear operator $T : K \to K$, $u \in K$ is a solution of the nonconvex variational inequality* (4.2) *if and only if there exists $\gamma = \gamma(u)$ such that $u \in K$ satisfies the relation $u = \mathcal{P}_K(u - \rho Tu)$, where $\rho < \frac{\gamma}{1+\|Tu\|}$.*

Remark 4.21. If K is a nonempty closed convex subset of X, for a given nonlinear operator T, it is well known that $u \in K$ is a solution of the variational inequality (4.1) if and only if $u \in K$ satisfies the relation $u = \mathcal{P}_K(u - \rho Tu)$ for all $\rho > 0$. Thus, Theorem 4.18, Corollaries 4.19 and 4.20 can be considered as the extension of the classical result from the nonempty closed convex subset of X to the nonempty subset of X, the nonempty proximally smooth subset of X and the nonempty prox-regular set of X, respectively.

On the other hand, the Opial property is an abstract property of Banach spaces that plays an important role in the study of weak convergence of iterates of mappings in Banach spaces. The property is named after the Polish mathematician Opial. Let $(X, \|\cdot\|)$ be a Banach space. We say that X has the Opial property if, whenever $\{x_n\}$ is a sequence in X converging weakly to some $x_0 \in X$ and $x \neq x_0$, it follows that

$$\liminf_{n\to\infty} \|x_n - x_0\| < \liminf_{n\to\infty} \|x_n - x\|.$$

Lemma 4.22. *Let X be a reflexive Banach space which satisfies Opial property, K be a nonempty bounded weak closed subset of X and $T : K \to K$ be nonexpansive.*

Then the mapping $I - T$ is demiclosed on K, that is, if $u_n \rightharpoonup u$ with $(I - T)u_n \to w$, then $(I - T)u = w$.

Proof. Suppose that $\{u_n\}$ in K satisfies $u_n \rightharpoonup u$ and $(I - T)u_n \to w$ as $n \to \infty$. Since K is a weak closed set, we know that $u \in K$. If we replace the mapping T by the mapping T_w which defined by $T_w x = Tx + w$, then $\|u_n - T_w u_n\| \to 0$ and T_w is also nonexpansive.

Next we show that $(I - T)u = w$. If $(I - T)u \neq w$, by Opial property, we have

$$\begin{aligned}\liminf_{n\to\infty} \|u_n - u\| &< \liminf_{n\to\infty} \|u_n - T_w u\| \\ &= \liminf_{n\to\infty} \|u_n - [u_n - T_w u_n] - T_w u\| \\ &= \liminf_{n\to\infty} \|T_w u_n - T_w u\| \\ &\leq \liminf_{n\to\infty} \|u_n - u\|,\end{aligned}$$

which is a contradiction and so $(I - T)u = w$. This completes the proof of Lemma 4.22. □

It is well known that every Hilbert space has the Opial property [24].

Theorem 4.23. *Let X be a Hilbert space, K be a nonempty weak closed bounded star-shaped set with the proximal smoothness on $U_K(r)$ and T be (γ, ζ)-relaxed co-coercive and χ-Lipschitz continuous. For some $r' \in (0, r)$, if the following conditions hold:*

$$\beta > 0, \quad r' \leq r(1 - \beta^2), \quad \rho < \frac{r'}{1 + \|Tu\|} \qquad \forall\, u \in K, \tag{4.6}$$

with

$$\beta = 1 - 2\rho\zeta + \rho^2\chi^2 + 2\rho\gamma\chi^2,$$

then the nonconvex variational inequality (4.2) *has a solution $u^* \in K$.*

Proof. For any $u \in K$, by using $\rho < \frac{r'}{1+\|Tu\|}$ for some $r' \in (0, r)$, we have

$$d_K(u - \rho Tu) \leq d_K(u) + \rho\|Tu\| < \frac{r'\|Tu\|}{1 + \|Tu\|} < r'.$$

It follows that, for $u, v \in K$, $u - \rho Tu \in U_K(r')$ and $v - \rho Tv \in U_K(r')$. Now Lemma 4.8 shows that

$$\begin{aligned}\|\mathcal{P}_K(I - \rho T)(u) - \mathcal{P}_K(I - \rho T)(v)\| &\leq \frac{r}{r - r'}\|u - \rho Tu - (v - \rho Tv)\| \\ &= \frac{r}{r - r'}\|(u - v) - \rho(Tu - Tv)\|.\end{aligned}$$

Since T is (γ, ζ)-relaxed co-coercive and χ-Lipschitz continuous, we have

$$\begin{aligned}\|(u-v)-\rho(Tu-Tv)\|^2 &= \|u-v\|^2 - 2\rho\langle Tu-Tv, u-v\rangle + \rho^2\|Tu-Tv\|^2\\ &\le (1-2\rho\zeta)\|u-v\|^2 + (\rho^2+2\rho\gamma)\|Tu-Tv\|^2\\ &\le (1-2\rho\zeta+\rho^2\chi^2+2\rho\gamma\chi^2)\|u-v\|^2\end{aligned}$$

and so

$$\|\mathcal{P}_K(I-\rho T)(u) - \mathcal{P}_K(I-\rho T)(v)\| \le \alpha\|u-v\|, \tag{4.7}$$

where

$$\alpha = \frac{r}{r-r'}\left(1-2\rho\zeta+\rho^2\chi^2+2\rho\gamma\chi^2\right)^{\frac{1}{2}}.$$

(I) If $0<\alpha<1$, we know that $\beta>0$ and $r'<r-r\beta^2$ with

$$\beta = 1-2\rho\zeta+\rho^2\chi^2+2\rho\gamma\chi^2$$

and so $\mathcal{P}_K(I-\rho T)$ is a contraction mapping. It follows that there exists a unique point $u^*\in K$ such that $\mathcal{P}_K(u^*+\rho Tu^*)=u^*$. Now, Corollary 4.19 guarantees that $u^*\in K$ is a solution of the nonconvex variational inequality (4.2).

(II) If $\alpha = 1$, we know that $r' = r - r\beta^2$ and so (4.7) implies that G is nonexpansive mapping. In order to get the solution of the nonconvex variational inequality (4.2), we need the proximal smoothness and the star-shapeness of the nonempty bounded weakly closed set K. The star-shapeness of K guarantees that there exists a $u\in K$ such that $u\in \mathit{kern}\, K$. For every $n\ge 1$, let $G=\mathcal{P}_K(I-\rho T)$ and

$$G_n(x) = G\left[\frac{1}{n}u + (1-\frac{1}{n})x\right] \quad \forall x\in K.$$

From the (γ,ζ)-relaxed co-coerciveness and χ-Lipschitz continuity of T, for every $x, y\in K$, Lemma 4.8 shows that

$$\begin{aligned}&\|G_n(x)-G_n(y)\|\\ &= \left\|\mathcal{P}_K(I-\rho T)\left(\frac{1}{n}u+(1-\frac{1}{n})x\right) - \mathcal{P}_K(I-\rho T)\left(\frac{1}{n}u+(1-\frac{1}{n})y\right)\right\|\\ &\le \frac{r}{r-r'}\left\|(I-\rho T)\left(\frac{1}{n}u+(1-\frac{1}{n})x\right) - (I-\rho T)\left(\frac{1}{n}u+(1-\frac{1}{n})y\right)\right\|\end{aligned}$$

$$\leq \frac{r}{r-r'} \cdot (1-\frac{1}{n})\|(x-y)-\rho(Tx-Ty)\|$$

$$\leq \frac{r}{r-r'} \cdot (1-\frac{1}{n})(1-2\rho\zeta+\rho^2\chi^2+2\rho\gamma\chi^2)^{\frac{1}{2}}\|x-y\|,$$

where $\alpha = 1$. This shows that $(1-\frac{1}{n})\alpha < 1$ and so G_n is a contraction mapping for each $n \in \mathbb{N}$. Thus, for every $n \in \mathbb{N}$, there exists a unique $x_n \in K$ such that $G_n(x_n) = x_n$.

On the other hand,

$$\begin{aligned} \|(I-G)x_n\| &= \|G_n x_n - G x_n\| \\ &= \left\| G\left(\frac{1}{n}u + (1-\frac{1}{n})x_n\right) - Gx_n \right\| \\ &\leq \frac{r}{n(r-r')}\|(u-x_n)-\rho(Tu-Tx_n)\| \\ &\leq \frac{\alpha}{n}\|u-x_n\| \to 0 \qquad (\text{as } n\to\infty). \end{aligned}$$

By $\{x_n\} \subset K$ and Lemma 4.14, we have $\omega_w(x_n) \neq \emptyset$. Taking $\overline{x} \in \omega_w(x_n)$, there exists a subsequence $\{x_{n_j}\} \subset \{x_n\}$ such that $x_{n_j} \rightharpoonup \overline{x}(j\to\infty)$. Since K is weak closed, we know that $\overline{x} \in K$. Now Lemma 4.22 shows that $G\overline{x} = \overline{x}$. Thus, that exists $\overline{x} \in K$ such that

$$\mathcal{P}_K(\overline{x}+\rho T\overline{x}) = \overline{x}.$$

By Corollary 4.19, $\overline{x} \in K$ is a solution of the nonconvex variational inequality (4.2).

Combining the cases of (I) and (II), we know that the nonconvex variational inequality (4.2) has a solution when (4.6) holds, which completes the proof of Theorem 4.23. □

Taking the particular cases of the set K and the mapping T in Theorem 4.23, it is easy to show the following result.

Corollary 4.24. *Let K be a nonempty bounded weak closed star-shaped set of a Hilbert space X and T be a nonexpansive mapping. Then $F(T)$, the fixed point set of T, is nonempty.*

Corollary 4.25. *Let X be a Hilbert space, K be a nonempty weak closed bounded set with the proximal smoothness on $U_K(r)$ and T be (γ,ζ)-relaxed co-coercive and χ-Lipschitz continuous. For some $r' \in (0,r)$, if the following conditions hold:*

$$\beta > 0, \quad r' < r(1-\beta^2), \quad \rho < \frac{r'}{1+\|Tu\|} \quad \forall u \in K, \tag{4.8}$$

with

$$\beta = 1 - 2\rho\zeta + \rho^2\chi^2 + 2\rho\gamma\chi^2,$$

then the nonconvex variational inequality (4.2) has a unique solution $u^* \in K$.

Proof. For any $u \in K$, by using $\rho < \frac{r'}{1+\|Tu\|}$ for some $r' \in (0, r)$, we have

$$d_K(u - \rho Tu) \le d_K(u) + \rho\|Tu\| < \frac{r'\|Tu\|}{1 + \|Tu\|} < r'.$$

It follows that, for $u, v \in K$, $u - \rho Tu \in U_K(r')$ and $v - \rho Tv \in U_K(r')$. Now Lemma 4.8 shows that

$$\begin{aligned}\|\mathcal{P}_K(I - \rho T)(u) - \mathcal{P}_K(I - \rho T)(v)\| &\le \frac{r}{r - r'}\|u - \rho Tu - (v - \rho Tv)\| \\ &= \frac{r}{r - r'}\|(u - v) - \rho(Tu - Tv)\|.\end{aligned}$$

Since T is (γ, ζ)-relaxed co-coercive and χ-Lipschitz continuous, one has

$$\begin{aligned}\|(u - v) - \rho(Tu - Tv)\|^2 &= \|u - v\|^2 - 2\rho\langle Tu - Tv, u - v\rangle + \rho^2\|Tu - Tv\|^2 \\ &\le (1 - 2\rho\zeta)\|u - v\|^2 + (\rho^2 + 2\rho\gamma)\|Tu - Tv\|^2 \\ &\le (1 - 2\rho\zeta + \rho^2\chi^2 + 2\rho\gamma\chi^2)\|u - v\|^2\end{aligned}$$

and so

$$\|\mathcal{P}_K(I - \rho T)(u) - \mathcal{P}_K(I - \rho T)(v)\| \le \alpha\|u - v\|, \tag{4.9}$$

where

$$\alpha = \frac{r}{r - r'}(1 - 2\rho\zeta + \rho^2\chi^2 + 2\rho\gamma\chi^2)^{\frac{1}{2}}.$$

By the condition (4.8), we get $0 < \alpha < 1$ and so $\mathcal{P}_K(I - \rho T)$ is a contraction mapping. It follows that there exists a unique point $u^* \in K$ such that $\mathcal{P}_K(u^* + \rho Tu^*) = u^*$. Now, Corollary 4.19 guarantees that $u^* \in K$ is a solution of the nonconvex variational inequality (4.2), which completes the proof of Corollary 4.25. □

It is well known every nonexpansive mapping has a fixed point in the nonempty bounded closed convex subset of Hilbert space X [28]. However, this conclusion is not true for the nonconvex subset of X. Furthermore, the existence theorem of solutions for the nonconvex variational inequality (4.2) is meaningful for computing the iterative sequences. In 2007, Singh [27] proved some fixed point theorems for a class of generalized set-valued contraction mapping on nonconvex sets. Recently, Alimohammady et al. [2] provided some existence theorems of solutions for an extended nonconvex variational inequality problem by using the Hausdorff pseudo-metric technique.

Remark 4.26. Corollary 4.24 generalizes the result Corollary 2.7 of Singh [27] and Theorem 4.23 generalizes and improves Lemma 3.1 of Noor [20]. Moreover, we would like to point out that the proof method of Theorem 4.23 and Corollary 4.25 can be also used in the one of Theorem 4.2 of Alimohammady et al. [2].

Theorem 4.27. *Let X be a Hilbert space, K be a nonempty weak closed bounded star-shaped set and T be (γ, ζ)-relaxed co-coercive and χ-Lipschitz continuous. If $F(T) \neq \emptyset$, then $F(T)$ is closed.*

Proof. Let $\{u_n\}_{n=1}^{\infty} \subset F(T)$ be a sequence which converge to a point $u \in K$. Since T is (γ, ζ)-relaxed co-coercive, one has

$$\langle Tu - Tu_n, u_n - u\rangle \leq \gamma \|Tu - Tu_n\|^2 - \zeta \|u - u_n\|^2.$$

It is easy to check that

$$2\langle Tu - Tu_n, u_n - u\rangle = \|Tu - u\|^2 - \|Tu - Tu_n\|^2 - \|u - u_n\|^2$$

and so

$$\begin{aligned} \|Tu - u\|^2 &= 2\langle Tu - Tu_n, u_n - u\rangle + \|Tu - Tu_n\|^2 + \|u - u_n\|^2 \\ &\leq (1 + 2\gamma)\|Tu - Tu_n\|^2 + (1 - 2\zeta)\|u - u_n\|^2 \\ &\leq \alpha_1 \|u - u_n\|^2, \end{aligned}$$

where

$$\alpha_1 = 2\gamma\chi^2 + \chi^2 + 1 - 2\zeta.$$

From the fact that $\|u_n - u\| \to 0$ as $n \to \infty$, we know that

$$\lim_{n\to\infty} \|Tu - u\| = 0.$$

Thus, $u \in F(T)$, which completes the proof of Theorem 4.27. □

4.4 Algorithms and Convergence

In this section, let X be a Hilbert space, K be a nonempty bounded weak closed star-shaped subset with the proximal smoothness on $U_K(r)$. We shall construct some projection iterative algorithms for solving the nonconvex variational inequality (4.2). We also establish the convergence of the iterative sequences generated by the iterative algorithms.

Algorithm 4.1. *For a given $u_0 \in K$, compute the approximate solution u_{n+1} by the iterative scheme*

$$\langle \rho T u_n + u_{n+1} - u_n, v - u_{n+1} \rangle + \rho\delta \|v - u_{n+1}\|^2 \geq 0 \quad \forall\, v \in K \tag{4.10}$$

for $n = 0, 1, \cdots$.

Algorithm 4.1 is called the proximal point algorithm for solving the nonconvex variational inequality (4.2). Note that, if $r = +\infty$, then the proximally smooth set K becomes to a convex set and Algorithm 4.1 reduces to an algorithm for solving the convex variational inequality.

We now give the convergence criteria of Algorithm 4.1 as follows.

Theorem 4.28. *Let K be a nonempty bounded weak closed star-shaped subset with the proximal smoothness on $U_K(r)$, the mapping $T : K \to K$ be (γ, ζ)-relaxed co-coercive and χ-Lipschitz continuous and $\{u_{n+1}\}$ be the approximate solution sequences generated by Algorithm 4.1. If $u \in K$ is a solution of (4.2) and there exists $r' \in (0, r)$ such that*

$$\rho < \frac{r'}{1 + \|Tu\|} \quad \forall\, u \in K, \quad \beta > 0, \quad r' \leq r - r\beta^2, \tag{4.11}$$

with

$$\beta = 1 - 2\rho\zeta + \rho^2\chi^2 + 2\rho\gamma\chi^2,$$

then $\lim_{n\to\infty} u_n = u$.

Proof. Let $u \in K$ be a solution of (4.2). Then

$$\langle Tu, v - u \rangle + \delta\|v - u\|^2 \geq 0 \quad \forall\, v \in K$$

and Corollary 4.19 implies that

$$u = \mathcal{P}_K(u - \rho Tu).$$

From (4.10), Lemmas 4.8 and 4.15, we get

$$u_{n+1} = \mathcal{P}_K(u_n - \rho T u_n)$$

and so

$$\begin{aligned} \|u_{n+1} - u\| &= \|\mathcal{P}_K(u_n - \rho T u_n) - \mathcal{P}_K(u - \rho T u)\| \\ &\le \frac{r}{r - r'} \|u_n - u - (\rho T u_n - \rho T u)\|, \end{aligned}$$

where $\rho < \frac{r'}{1+\|Tu\|}$ for all $u \in K$. It follows from the (γ, ζ)-relaxed co-coerciveness and χ-Lipschitz continuity of T that

$$\begin{aligned} &\|u_{n+1} - u\| \\ &\le \frac{r}{r - r'} \cdot \left[(\|u_n - u\|^2 - 2\rho\langle Tu_n - Tu, u_n - u\rangle + \rho^2 \|Tu_n - Tu\|^2) \right]^{\frac{1}{2}} \\ &= \alpha \|u_n - u\|, \end{aligned}$$

where

$$\alpha = \frac{r}{r - r'} \left[(1 - 2\rho\zeta + \rho^2\chi^2 + 2\rho\gamma\chi^2) \right]^{\frac{1}{2}}.$$

It is easy to see that

$$\|u_{n+1} - u\| \le \alpha^n \|u_0 - u\|.$$

By (4.9), we know that $\lim_{n\to\infty} u_n = u$. This completes the proof. □

It is well known that to apply the proximal point method, one has to compute the approximate solution by a iterative scheme implicitly, we suggest the implicit algorithm as follows.

Algorithm 4.2. For a given $u_0 \in K$, calculate the approximate solution u_{n+1} by the iterative scheme

$$\langle \rho T u_{n+1} + u_{n+1} - u_n, v - u_{n+1}\rangle + \rho\delta\|v - u_{n+1}\|^2 \ge 0 \quad \forall\, v \in K.$$

We now give the convergence criteria of Algorithm 4.2 as follows.

Theorem 4.29. *Let K be a nonempty totally bounded weak closed proximally smooth subset of a Hilbert space X, the mapping $T : K \to K$ be (γ, ζ)-relaxed co-coercive and χ-Lipschitz continuous and $\{u_{n+1}\}$ be the approximate solution sequences generated by Algorithm 4.2. If $u \in K$ is a solution of* (4.2) *and there exist $r' \in (0, r)$ such that*

$$\rho < \frac{r'}{1 + \|Tu\|} \;\forall\, u \in K, \quad \rho(\zeta - \gamma\chi^2) = 1, \quad \beta > 0, \quad r' < r - r\beta^2, \qquad (4.12)$$

with

$$\beta = 1 - 2\rho\zeta + \rho^2\chi^2 + 2\rho\gamma\chi^2,$$

then $\lim_{n\to\infty} u_n = u$.

Proof. Let $u \in K$ be a solution of (4.2). Then

$$\langle Tu, v-u\rangle + \delta\|v-u\|^2 \geq 0 \quad \forall\, v \in K.$$

Since T is (γ, ζ)-relaxed co-coercive, by condition (4.12), we have

$$-\langle Tv, u-v\rangle - \delta\|u-v\|^2 \geq \langle Tu, v-u\rangle + \delta\|u-v\|^2$$

and so

$$-\langle Tv, u-v\rangle - \delta\|u-v\|^2 \geq 0 \quad \forall\, v \in K. \tag{4.13}$$

Taking $v = u_{n+1}$ in (4.13), one has

$$-\langle Tu_{n+1}, u-u_{n+1}\rangle - \delta\|u-u_{n+1}\|^2 \geq 0. \tag{4.14}$$

Setting $v = u$ in Algorithm 4.2, it follows from (4.14) that

$$\begin{aligned}\langle u_{n+1}-u_n, u-u_{n+1}\rangle &\geq -\rho\langle Tu_{n+1}, u-u_{n+1}\rangle - \rho\delta\|u-u_{n+1}\|^2\\ &\geq 0.\end{aligned} \tag{4.15}$$

It is easy to see that

$$2\langle u, v\rangle = \|u+v\|^2 - \|u\|^2 - \|v\|^2 \qquad \forall u, v \in H. \tag{4.16}$$

Taking $v = u - u_{n+1}$ and $u = u_{n+1} - u_n$ in (4.16), we get

$$\begin{aligned}2\langle u_{n+1}-u_n, u-u_{n+1}\rangle = \|u-u_n\|^2 - \|u_n-u_{n+1}\|^2\\ -\|u-u_{n+1}\|^2.\end{aligned} \tag{4.17}$$

It follows from (4.15) and (4.17) that

$$\|u-u_{n+1}\|^2 \leq \|u-u_n\|^2 - \|u_n-u_{n+1}\|^2.$$

Thus, we know that the sequence $\{u_n\}$ is bounded and

$$\sum_{n=0}^{\infty} \|u_n-u_{n+1}\|^2 \leq \|u_0-u\|^2.$$

This implies that

$$\lim_{n\to\infty} \|u_n - u_{n+1}\| = 0. \tag{4.18}$$

Since K is totally bounded and weak closed, it is easy to know that K is complete and totally bounded and so K is compact. Let $\{u_{n_i}\}$ be a subsequence of $\{u_n\}$ such that $u_{n_i} \to \tilde{u} \in K$. Taking limit $n_i \to \infty$ in Algorithm 4.2 for the subsequence $\{u_{n_i}\}$ and using (4.18), we have

$$\langle T\tilde{u}, v - \tilde{u}\rangle + \delta\|v - \tilde{u}\|^2 \geq 0 \quad \forall\, v \in K,$$

which shows that $\tilde{u}$ is a solution of the nonconvex variational inequality (4.2) and so

$$\|\tilde{u} - u_{n+1}\|^2 \leq \|\tilde{u} - u_n\|^2.$$

Since $u_{n_i} \to \tilde{u}$, we know that $\lim_{n\to\infty} u_n = \tilde{u}$. Now Corollary 4.25 guarantees that $u = \tilde{u}$, which completes the proof of Theorem 4.29. □

Acknowledgements This research was partially supported by the Chinese National Science Foundation (Grant No. 11171237).

References

1. Agarwal, R.P., Regan, O.D., Sahu, D.R.: Iterative construction of fixed points of nearly asymptotically nonexpansive mappings. J. Nonlinear Convex Anal. **8**, 61–79 (2007)
2. Alimohammadya, M., Balooee, J., Cho, Y.J., Roohi, M.: Iterative algorithms for a new class of extended general nonconvex set-valued variational inequalities. Nonlinear Anal. **73**, 3907–3923 (2010)
3. Boltyanski, V., Martini, H., Soltan, P.S.: Star-shaped sets in normed spaces. Discrete Comput. Geom. **15**, 63–71 (1996)
4. Bounkhel, M., Tadj, L., Hamdi, A.: Iterative schemes to solve nonconvex variational problems. J. Inequal. Pure Appl. Math. **4**, 1–14 (2003)
5. Brunn, H.: Uber Kerneigebiete. Math. Ann. **73**, 436–440 (1913)
6. Chepoi, V.: On starshapedness in products of interval spaces. Arch. Math. **64**, 264–268 (1995)
7. Clarke, F.H., Stern, R.J., Wolenski, P.R.: Proximal smoothness and the lower-C^2 property. J. Convex Anal. **2**, 117–144 (1995)
8. Clarke, F.H., Ledyaev, Y.S., Wolenski, P.R.: Nonsmooth Analysis and Control Theory. Springer, Berlin (1998)
9. Collins, I.F., Hilder, T.: A theoretical framework for constructing elastic/plastic constitutive models of triaxial tests. Int. J. Numer. Anal. Methods Geom. **26**, 1313–1347 (2002)
10. Facchinei, F., Pang, J.S.: Finite-Dimensional Variational Inequalities and Complementarity Problems. Springer, New York (2003)
11. Glowinski, R., Lions, J.L., Tremolieres, R.: Numerical Analysis of Variational Inequalities. Elsevier, Amsterdam (1981)
12. Goeleven, D.: Noncoercive Variational Problems and Related Results. Addison Wesley-Longman, Harlow (1996)

13. Hartman, P., Stampacchia, G.: On some nonlinear elliptic differential functional equations. Acta Math. **15**, 271–310 (1966)
14. Kinderlehrer, D., Stampacchia, G.: An Introduction to Variational Inequalities and Their Applications. SIAM, Philadelphia (2000)
15. Laydi, M.R., Lexcellent, C.: Yield criteria for shape memory materials: convexity conditions and surface transport. Math. Mech. Solids **15**, 165–208 (2010)
16. Lin, Z.H., Yu, B., Zhu, D.L.: A continuation method for solving fixed points of self-mappings in general nonconvex sets. Nonlinear Anal. **52**, 905–915 (2003)
17. Lions, J.L., Stampacchia, G.: Variational inequalities. Commun. Pure Appl. Anal. **20**, 493–519 (1967)
18. Naniewicz, Z.: Hemivariational inequality approach to constrained problems for star-shaped admissible sets. J. Optim. Theory Appl. **83**, 97–112 (1994)
19. Noor, M.A.: Iterative schemes for nonconvex variational inequalities. J. Optim. Theory Appl. **121**, 385–395 (2004)
20. Noor, M.A.: Projection methods for nonconvex variational Inequalities. Optim Lett. **3**, 411–418 (2009)
21. Noor, M.A.: On an implicit method for nonconvex variational inequalities. J. Optim. Theory Appl. **147**, 411–417 (2010)
22. Noor, M.A.: On nonconvex bifunction variational inequalities. Optim. Lett. **6**, 1477–1484 (2012)
23. Noor, M.A., Al-Said, E., Noor, K.I., Yao, Y.H.: Extragradient methods for solving nonconvex variational inequalities. J. Concr. Appl. Math. **235**, 3104–3108 (2011)
24. Opial, Z.: Weak convergence of the sequence of successive approximations for nonexpansive mappings. Bull. Am. Math. Soc. **73**, 591–597 (1967)
25. Piccolroaz, A., Bigoni, D.: Yield criteria for quasibrittle and frictional materials: a generalization to surfaces with corners. Int. J. Solids Struct. **46**, 3587–3596 (2009)
26. Poliquin, R.A., Rockafellar, R.T., Thibault, L.: Local differentiability of distance functions. Trans. Am. Math. Soc. **352**, 5231–5249 (2000)
27. Singh, N.: Fixed point theorems for multifunctions on nonconvex sets. Int. J. Math. Anal. **1**, 1383–1387 (2007)
28. Smaïl, D., Karima, H.: Fixed point theorems for nonexpansive maps in Banach spaces. Nonlinear Anal. **73**, 3440–3449 (2010)
29. Solodov, M.V., Svaiter, B.F.: A new projection method for variational inequality problems. SIAM J. Control Optim. **37**, 765–776 (1999)
30. Stampacchia, G.: Formes bilineaires coercitives sur les ensembles convexes. C. R. Acad. Sci. **258**, 4413–4416 (1964)
31. Wen, D.J.: Projection methods for a generalized system of nonconvex variational inequalities with different nonlinear operators. Nonlinear Anal. **7**, 2292–2297 (2010)
32. Yang, X.Q., Alistair, I.M., Fisher, M., Jennings, L.: Progress in Optimization: Contributions from Australasia. Kluwer Academic, Dordrecht (2000)
33. Ziegler, H.: An Introduction to Thermomechanics, 2nd edn. North-Holland, Amsterdam (1983)
34. Ziegler, H., Wehrli, C.: The derivation of constitutive relations from the free energy and the dissipation function. Adv. Appl. Mech. **25**, 183–238 (1987)

Part II
Numerical Analysis

Chapter 5
Numerical Methods for Evolution Hemivariational Inequalities

Krzysztof Bartosz

Abstract We consider numerical methods of solving evolution subdifferential inclusions of nonmonotone type. In the main part of the chapter we apply Rothe method for a class of second order problems. The method consists in constructing a sequence of piecewise constant and piecewise linear functions being a solution of approximate problem. Our main result provides a weak convergence of a subsequence to a solution of exact problem. Under some more restrictive assumptions we obtain also uniqueness of exact solution and a strong convergence result. Next, for the reference class of problems we apply a semi discrete Faedo-Galerkin method as well as a fully discrete one. For both methods we present a result on optimal error estimate.

Keywords Evolution hemivariational inequalities • Rothe method • Faedo-Galerkin method • Weak convergence • Strong convergence • Error estimate

AMS Classification. 34G25, 35L70, 35L85, 74M10, 74M15, 65P99, 65G99

5.1 Introduction

In this chapter we present recent results in numerical methods addressed to evolution subdifferential inclusions of evolution type referred to as evolution hemivariational inequalities (HVIs). The methods under consideration can be classified according to several criteria.

- *Type of the reference problem.* We distinguish between two general classes of evolution problems related to the order of time derivative of unknown function. They are first and second order problems, known as parabolic and hyperbolic

K. Bartosz (✉)
Faculty of Mathematics and Computer Science, Jagiellonian University, ul. Łojasiewicza 6, 30348 Krakow, Poland
e-mail: bartosz@ii.uj.edu.pl

W. Han et al. (eds.), *Advances in Variational and Hemivariational Inequalities*, Advances in Mechanics and Mathematics 33, DOI 10.1007/978-3-319-14490-0_5

HVIs, respectively. The order of considered problem depends on the dynamics of modelled physical or mechanical problems and has a crucial influence on the choice of suitable numerical methods.

- *Discretization method.* There are two levels of discretization for time dependent problems based on semidiscrete and a fully discrete schemes, respectively. The semi-discrete schemes rely on replacing one of two variables (time or spatial) by its discrete approximation, while the second variable is kept continuous. The semi-discrete strategy consisting in time discretization is referred to as Rothe method and the one based on space discretization is known as Faedo-Galerkin method. Finally, a fully discrete scheme consists in replacing both time and spatial variables by their discrete approximations.
- *Convergence rate.* Constructing a sequence of approximate solution for the reference problem one can expect several possibilities concerning a quality of convergence. A typical result provides a weak convergence of a subsequence to a solution of the exact problem. Moreover, under some suitable assumptions (for example S_+ property of involved operator) we can get also a strong convergence of the sequence. Finally, the best which one can expect is to estimate an error understood as a norm of difference between an exact solution and an approximate one. In that case not only strong convergence is provided, but its rate can be expressed in terms of spatial and time discretization mesh's size.

Although the theory of HVIs has been developed broadly in last years, there are still very few publications devoted to numerical methods for evolution HVIs. The earliest one is probably [12], where Finite Element Method has been adopted for parabolic HVIs. Those results have been extended into hyperbolic case in [8], which has become the basic handbook for numerical methods for HVIs. Recently the Rothe method has been applied for parabolic HVIs in [10]. The result obtained there has been generalized in [11] by applying the θ scheme. Moreover in [9] an error estimate result has been obtained. Independently in [14] and [15], the time discretization method has been applied for parabolic doubly nonlinear problems. Moreover we refer to [4] for a result concerning Rothe method for evolution variational-hemivariational inequalities of parabolic type.

In this chapter we present new results concerning semidiscrete and fully discrete approach to second order evolution HVIs, as an example of typical strategy used in that types of problems. In Sect. 5.4 we study the Rothe method, which relies on constructing a sequence of piecewise constant and piecewise linear functions, which converges weakly to a solution of second order HVIs. In this case the existence of the solution is a consequence of applied method, so in fact we provide an alternative existence result for whose obtained in [13]. In Sect. 5.5 we consider a semidiscrete scheme based on Faedo-Galerkin approximation as well as a fully discrete scheme. For both schemes we provide an optimal error rate bounds. The results obtained in Sect. 5.5 come from [2] and they are generalization of [3]. In Sect. 5.6 we describe a concrete mechanical problem, for which the results obtained in Sects. 5.4 and 5.5 are applicable.

5.2 Preliminaries

In this section we recall some definitions and propositions which we will refer to in the sequel. We start with the definitions of Clarke directional derivative and Clarke subdifferential. Let X be a Banach space, X^* its dual and let $J: X \to \mathbb{R}$ be a locally Lipschitz functional.

Definition 5.1. Generalized directional derivative in the sense of Clarke at the point $x \in X$ in the direction $v \in X$, is defined by

$$J^0(x, v) = \limsup_{y \to x, \lambda \searrow 0} \frac{J(y + \lambda v) - J(y)}{\lambda}. \tag{5.1}$$

Definition 5.2. Clarke subdifferential of J at $x \in X$ is defined by

$$\partial J(x) = \{\xi \in X^* \mid J^0(x, v) \geq \langle \xi, v \rangle_{X^* \times X} \text{ for all } v \in X\}.$$

Now, we pass to the definition of a pseudomonotone operator.

Definition 5.3 (See [17], Chapter 27). Let X be a Banach space. A single valued operator $A: X \to X^*$ is called pseudomonotone, if for any sequence $\{v_n\}_{n=1}^{\infty} \subset X$ such that $v_n \to v$ weakly in X and $\limsup_{n\to\infty} \langle Av_n, v_n - v \rangle \leq 0$ we have $\langle Av, v - y \rangle \leq \liminf_{n\to\infty} \langle Av_n, v_n - y \rangle$ for every $y \in X$.

Definition 5.4. Let X be a real Banach space. The multivalued operator $A: X \to 2^{X^*}$ is called pseudomonotone if the following conditions hold:

1) A has values which are nonempty, bounded, closed and convex,
2) A is usc from every finite dimensional subspace of X into X^* furnished with weak topology,
3) if $v_n \to v$ weakly in X and $v_n^* \in Av_n$ is such that $\limsup_{n\to\infty} \langle v_n^*, v_n - v \rangle \leq 0$ then for every $y \in X$ there exists $u(y) \in A(v)$ such that $\langle u(y), v - y \rangle \leq \liminf_{n\to\infty} \langle v_n^*, v_n - y \rangle$.

The following result can be found, for example, in [13] (see Proposition 3.58).

Proposition 5.5. *Let X be a real reflexive Banach space, and assume that $A : X \to 2^{X^*}$ satisfies the following conditions*

1) *for each $v \in X$ we have that Av is a nonempty, closed and convex subset of X^*,*
2) *A is bounded, i.e., it maps bounded sets into bounded ones,*
3) *If $v_n \to v$ weakly in X and $v_n^* \to v^*$ weakly in X^* with $v_n^* \in Av_n$ and if $\limsup_{n\to\infty} \langle v_n^*, v_n - v \rangle \leq 0$, then $v^* \in Av$ and $\langle v_n^*, v_n \rangle \to \langle v^*, v \rangle$.*

Then the operator A is pseudomonotone.

The next proposition provides pseudomonotonicity of a multivalued operator corresponding to a superposition of Clarke subdifferential with a compact operator.

Proposition 5.6. *Let X and U be two reflexive Banach spaces, $\iota\colon X \to U$ be a linear, continuous and compact operator and $\iota^* : U^* \to X^*$ be its adjoint operator. Let $J\colon U \to \mathbb{R}$ be a locally Lipschitz functional and its Clarke subdifferential satisfies*

$$\|\xi\|_{U^*} \leq c(1 + \|v\|_U) \text{ for all } \xi \in \partial J(v) \tag{5.2}$$

with $c > 0$. Then the multivalued operator $M : X \to 2^{X^*}$ defined by

$$M(v) = \iota^* \partial J(\iota v) \quad \forall v \in X \tag{5.3}$$

is pseudomonotone.

Proof. We use Proposition 5.5. To this end we recall (see [6]) that the Clarke subdifferential of a locally Lipschitz functional is a nonempty, weakly closed and convex set. So from the linearity of ι we conclude that the values of M are nonempty, closed and convex subsets of X^*. It follows from (5.2) that M is bounded. Now we prove that M satisfies condition 3) of Proposition 5.5. Let $v_n \to v$ weakly in U and $v_n^* \to v^*$ weakly in U^* with $v_n^* \in \iota^* \partial J(\iota v_n)$. Thus we have $v_n^* = \iota^* \eta_n$ where $\eta_n \in \partial J(\iota v_n)$ and for all $x \in X$ we obtain

$$\langle \eta_n, \iota x\rangle_{U^* \times U} = \langle \iota^* \eta_n, x\rangle_{X^* \times X} = \langle v_n^*, x\rangle_{X^* \times X} \to \langle v^*, x\rangle_{X^* \times X}. \tag{5.4}$$

Since the sequence $\{v_n\}$ converges weakly, it is bounded and so is $\{\eta_n\}$. Thus, from the reflexivity of U^*, for a subsequence $\{\eta_{n_k}\}$, we have $\eta_{n_k} \to \eta$ weakly in U^* with $k \to \infty$ so in particular for all $x \in X$, we get $\langle \eta_{n_k}, \iota x\rangle_{U^* \times U} \to \langle \eta, \iota x\rangle_{U^* \times U}$. Next, from (5.4) we have $\langle \eta_{n_k}, \iota x\rangle_{U^* \times U} \to \langle v^*, x\rangle_{X^* \times X}$ and from the uniqueness of limit we have $\langle v^*, x\rangle_{X^* \times X} = \langle \eta, \iota x\rangle_{U^* \times U} = \langle \iota^* \eta, x\rangle_{X^* \times X}$ for all $x \in X$, so $v^* = \iota^* \eta$. Numerating the sequence n_k by n again, we get $v_n \to v$ weakly in X and $\eta_n \to \eta$ weakly in U^* with $\eta_n \in \partial J(\iota v_n)$. Since $\{v_n\}$ is bounded and ι is compact we have for a subsequence $\iota v_{n_k} \to u$ strongly in U. Since ι is linear and continuous, it is also weakly continuous so $\iota v_{n_k} \to \iota v$ weakly in U. From the uniqueness of weak limit we have that $\iota v_{n_k} \to \iota v$. Thus from the closedness of graph of the Clarke subdifferential in $U_{strong} \times U^*_{weak}$ topology we deduce that $\eta \in \partial J(\iota v)$. Since $v^* = \iota^* \eta$, it follows that $v^* \in M(v)$. It remains to show that $\langle v_n^*, v_n\rangle_{X^* \times X} \to \langle v^*, v\rangle_{X^* \times X}$. Let us take any subsequence of $\{\langle v_n^*, v_n\rangle_{X^* \times X}\}$ still numerated by n. From the previous part of the proof, we can find a subsequence $\langle v_{n_k}^*, v_{n_k}\rangle_{X^* \times X}$ such that $v_{n_k}^* = \iota^* \eta_{n_k}$ and $\eta_{n_k} \to \eta$ weakly in U^* with $v^* = \iota\eta$ and $\iota v_{n_k} \to \iota v$ strongly in X. Thus, we have

$$\begin{aligned}\langle v_{n_k}^*, v_{n_k}\rangle_{X^* \times X} &= \langle \iota^* \eta_{n_k}, v_{n_k}\rangle_{X^* \times X} = \langle \eta_{n_k}, \iota v_{n_k}\rangle_{U^* \times U} \\ &\to \langle \eta, \iota v\rangle_{U^* \times U} = \langle \iota\eta, v\rangle_{X^* \times X} = \langle v^*, v\rangle_{X^* \times X}.\end{aligned}$$

We conclude from here that whole sequence $\{\langle v_n^*, v_n\rangle_{X^* \times X}\}$ converges to $\langle v^*, v\rangle_{X^* \times X}$ as $n \to \infty$, which completes the proof. □

The next proposition deals with a superposition of pseudomonotone operator with an affine one.

Proposition 5.7. *Let X be a reflexive Banach space and $A\colon X \to 2^{X^*}$ be a pseudomonotone operator. Then for a given $v_0 \in X$ and $\lambda > 0$ the operator $M : X \to 2^{X^*}$ defined by $Mv = A(v_0+\lambda v)$ for all $v \in X$ is also pseudomonotone.*

Proof. Let $v_n \to v$ weakly in X and let $v_n^* \in Mv_n$ with $\limsup_{n\to\infty}\langle v_n^*, v_n - v\rangle_{X^*\times X} \le 0$. Taking $w_n := v_0 + \lambda v_n$ and $w := v_0 + \lambda v$ we have $w_n \to w$ weakly in X, $v_n^* \in Aw_n$ and

$$\limsup_{n\to\infty}\langle v_n^*, w_n - w\rangle_{X^*\times X} = \lambda \limsup_{n\to\infty}\langle v_n^*, v_n - v\rangle_{X^*\times X} \le 0.$$

Since A is pseudomonotone, it implies that for all $z \in X$ there exists $\overline{u}(z) \in Aw$ such that $\langle \overline{u}(z), w - z\rangle_{X^*\times X} \le \liminf_{n\to\infty}\langle v_n^*, w_n - z\rangle_{X^*\times X}$. Lets take any $y \in X$ and put $z = v_0 + \lambda v - v + y$. Taking $u(y) = \overline{u}(z)$ we have $u(y) \in A(w) = M(v)$. Moreover,

$$\begin{aligned}
\langle u(y), v - y\rangle_{X^*\times X} &= \langle \overline{u}(z), w - z\rangle_{X^*\times X} \\
&\le \liminf_{n\to\infty}\langle v_n^*, w_n - z\rangle_{X^*\times X} \\
&= \liminf_{n\to\infty}\langle v_n^*, \lambda(v_n - v) + v - y\rangle_{X^*\times X} \\
&\le \lambda \limsup_{n\to\infty}\langle v_n^*, v_n - v\rangle_{X^*\times X} + \liminf_{n\to\infty}\langle v_n^*, v - y\rangle_{X^*\times X} \\
&\le \liminf_{n\to\infty}\langle v_n^*, v - y\rangle_{X^*\times X}.
\end{aligned}$$

This shows that M is a pseudomonotone operator. □

We also recall a well known property for the sum of two pseudomonotone operators.

Proposition 5.8. *Let X be a reflexive Banach space. If $A_1, A_2 : X \to 2^{X^*}$ are pseudomonotone then so is $A_1 + A_2$.*

In what follows we introduce the notion of coercivity.

Definition 5.9. Let X be a Banach space and $A\colon X \to 2^{X^*}$ be an operator. We say that A is coercive if either $D(A)$ is bounded or $D(A)$ is unbounded and

$$\lim_{\|v\|_X\to\infty\ v\in D(A)} \frac{\inf\{\langle v^*, v\rangle_{X^*\times X} \mid v^* \in Av\}}{\|v\|_X} = +\infty.$$

The following is the main surjectivity result for pseudomonotone and coercive operator.

Proposition 5.10. *Let X be a reflexive Banach space and $A: X \to 2^{X^*}$ be pseudomonotone and coercive. Then A is surjective, i.e., for all $b \in X^*$ there exists $v \in X$ such that $Av = b$.*

Let X be a Banach space and $(0, T)$ be a time interval. We introduce the space $BV(0, T; X)$ of functions of bounded total variation on $(0, T)$. Let π denotes any finite partition of $(0, T)$ by a family of disjoint subintervals $\{\sigma_i = (a_i, b_i)\}$ such that $[0, T] = \cup_{i=1}^n \overline{\sigma}_i$. Let $\mathcal{F}$ denotes the family of all such partitions. Then for a function $x: (0, T) \to X$ we define its total variation by

$$\|x\|_{BV(0,T;X)} = \sup_{\pi \in \mathcal{F}} \left\{ \sum_{\sigma_i \in \pi} \|x(b_i) - x(a_i)\|_X \right\}.$$

As a generalization of the above definition, for $1 \leq q < \infty$, we define a seminorm

$$\|x\|^q_{BV^q(0,T;X)} = \sup_{\pi \in \mathcal{F}} \left\{ \sum_{\sigma_i \in \pi} \|x(b_i) - x(a_i)\|^q_X \right\}$$

and the space

$$BV^q(0, T; X) = \{x: (0, T) \to X \mid \|x\|_{BV^q(0,T;X)} < \infty\}.$$

For $1 \leq p \leq \infty$, $1 \leq q < \infty$ and Banach spaces X, Z such that $X \subset Z$, we introduce the vector space

$$M^{p,q}(0, T; X, Z) = L^p(0, T; X) \cap BV^q(0, T; Z).$$

Then, it is easy to see that $M^{p,q}(0, T; X, Z)$ is also a Banach space, endowed with the norm $\|\cdot\|_{L^p(0,T;X)} + \|\cdot\|_{BV^q(0,T;Z)}$.

The following proposition will play the crucial role for the convergence of Rothe functions which will be constructed later. For its proof, we refer to [10].

Proposition 5.11. *Let $1 \leq p, q < \infty$. Let $X_1 \subset X_2 \subset X_3$ be real Banach spaces such that X_1 is reflexive, the embedding $X_1 \subset X_2$ is compact and the embedding $X_2 \subset X_3$ is continuous. Then the embedding $M^{p,q}(0, T; X_1; X_3) \subset L^p(0, T; X_2)$ is compact.*

The following version of Aubin-Celina convergence theorem (see [1]) will be used in what follows.

Proposition 5.12. *Let X and Y be Banach spaces, and let $F: X \to 2^Y$ be a multifunction such that*

(a) the values of F are nonempty closed and convex subsets of Y
(b) F is upper semicontinuous from X into $w - Y$.

Let $x_n: (0,T) \to X$, $y_n: (0,T) \to Y$, $n \in \mathbb{N}$, be measurable functions such that x_n converges almost everywhere on $(0,T)$ to a function $x: (0,T) \to X$ and y_n converges weakly in $L^1(0,T;Y)$ to $y : (0,T) \to Y$. If $y_n(t) \in F(x_n(t))$ for all $n \in \mathbb{N}$ and almost all $t \in (0,T)$ then $y(t) \in F(x(t))$ for a.e. $t \in (0,T)$.

For the convergence result obtained in next section we will apply Lemma 1 of [10]. We present it without proof.

Lemma 5.13. *Let $A: V \to V^*$ satisfies $H(A)$ (see below) and let $\mathcal{A}$ be its Nemytskii operator defined by $(\mathcal{A}v)(t) = Av(t)$. Let a sequence $\{v_n\} \subset \mathcal{V}$ be bounded in $M^{2,2}(0,T;V,V^*)$. If $v_n \to v$ weakly in $\mathcal{V}$ and* $\limsup_{n\to\infty} \langle \mathcal{A}v_n, v_n - v\rangle_{\mathcal{V}^* \times \mathcal{V}} \le 0$, *then $\mathcal{A}v_n \to \mathcal{A}v$ weakly in $\mathcal{V}^*$.*

5.3 Problem Formulation

Let V be a reflexive and separable Banach space, V^* its dual and H a separable Hilbert space. Identifying H with its dual we consider an evolution triple $V \subset H \subset V^*$ with dense, continuous and compact embeddings. We denote by $\langle \cdot, \cdot \rangle$ the duality of V and V^*, by $(\cdot,\cdot)$ the scalar product in H. Let $i : V \to H$ be an embedding operator (for $v \in V$ we will denote $iv \in H$ again by v). For all $u \in H$ and $v \in V$ we have $\langle u, v\rangle = (u, v)$. We denote by $\|\cdot\|$ and $|\cdot|$ the norms in V and H, respectively. We also introduce a reflexive Banach space U and a linear, continuous operator $\iota: V \to U$. By $\|i\|$ and $\|\iota\|$ we always mean $\|i\|_{\mathcal{L}(V,H)}$ and $\|\iota\|_{\mathcal{L}(V,U)}$, respectively. For $T > 0$ we define the spaces $\mathcal{V} = L^2(0,T;V)$, $\mathcal{V}^* = L^2(0,T;V^*)$, $\mathcal{H} = L^2(0,T;H)$, $\mathcal{U} = L^2(0,T;U)$ and $\mathcal{W} = \{v \in \mathcal{V} \mid v' \in \mathcal{V}^*\}$, where the derivative is understood in the sense of distributions. We consider two problems denoted by (P^1) and (P^2) which read as follows:

$$(P^m) \quad \begin{cases} \text{Find } u \in \mathcal{V} \text{ with } u' \in \mathcal{W} \text{ such that} \\ u''(t) + Au'(t) + Bu(t) + \iota^* \partial J(\iota z(t)) \ni f(t) \quad \text{for a.e. } t \in (0,T), \\ u(0) = u_0, \quad u'(0) = u_1, \end{cases}$$

where $m = 1, 2$, $A: V \to V^*$, $B: V \to V^*$, $f : (0,T) \to V^*$, $J: U \to \mathbb{R}$, $\partial J(\cdot)$ denotes its Clarke subdifferential and we put $z = u$ in the problem (P^1) while $z = u'$ in the Problem (P^2), respectively. We remark that each solution of the problem (P^m) is a solution of a corresponding second order hemivariational inequality of the form

$$\langle u''(t) + Au'(t) + Bu(t), v\rangle + J^\circ(\iota z(t); \iota v) \ge \langle f(t), v\rangle \ \forall\, v \in V, \text{ a.e. } t \in (0,T).$$

In fact if u is a solution of (P^m) then for a.e. $t \in (0,T)$ there exists $\eta \in \partial J(\iota z(t))$ such that

$$\langle f(t) - u''(t) - Au'(t) - Bu(t), v\rangle = \langle \iota^*\eta, v\rangle_{U^*\times U} = \langle \eta, \iota v\rangle \le J^\circ(\iota z(t); \iota v)$$

for all $v \in V$ for a.e. $t \in (0,T)$.

In what follows we will refer to the following equivalent formulation of Problem (P^m).

$$(P^m)_* \quad \begin{cases} \text{Find } (u,w,\eta) \in \mathcal{V}\times\mathcal{W}\times\mathcal{U} \text{ such that} & \\ u'(t) = w(t) & \text{for a.e. } t \in (0,T), \\ w'(t) + Aw(t) + Bu(t) + \iota^*\eta(t) = f(t) & \text{for a.e. } t \in (0,T), \\ \eta(t) \in \partial J(\iota z(t)) & \text{for a.e. } t \in (0,T), \\ u(0) = u_0, \quad w(0) = u_1, & \end{cases}$$

where $z = u$ in the problem $(P^1)_*$ while $z = w$ in the Problem $(P^2)_*$, respectively.

Remark 5.14. Note that time derivatives in $(P^m)_*$ are understood in the sense of distributions. In particular (see Proposition 23.20 of [17]), the second equation of $(P^m)_*$ is equivalent to

$$\int_0^T (w(t), h)\varphi'(t)dt = \int_0^T \langle Aw(t) + Bu(t) + \iota^*\eta(t) - f(t), h\rangle\varphi(t)dt$$

for all $h \in V$, $\varphi \in C_0^\infty(0,T)$.

We consider the following hypotheses on the data of the problem (P^m).

$H(A)$: The operator $A\colon V \to V^*$ satisfies

(i) A is pseudomonotone;
(ii) $\|Av\|_{V^*} \le a + b\|v\|$ for all $v \in V$ with $a \ge 0$, $b > 0$;
(iii) $\langle Av, v\rangle \ge \alpha\|v\|^2 - \beta|v|^2 - \gamma$ for all $v \in V$ with $\alpha > 0$, $\beta, \gamma \ge 0$.

$H(B)$: The operator $B\colon V \to V^*$ is bounded, linear, monotone and symmetric, i.e. $B \in \mathcal{L}(V,V^*)$, $\langle Bv, v\rangle \ge 0$ for all $v \in V$, $\langle Bv, w\rangle = \langle Bw, v\rangle$ for all $v, w \in V$.

$H(J)$: The functional $J\colon U \to \mathbb{R}$ satisfies

(i) J is locally Lipschitz;
(ii) $\|\eta\|_{U^*} \le c(1 + \|w\|_U)$ for all $\eta \in \partial J(w)$, $w \in U$ with $c > 0$;

H_0: $f \in \mathcal{V}^*$, $u_0 \in V$, $u_1 \in H$.

$H(\iota)$: The operator $\iota \in \mathcal{L}(V,U)$ is compact and its associated Nemytskii operator $\bar{\iota} : M^{2,2}(0,T;V,V^*) \to \mathcal{U}$ defined by $(\bar{\iota}v)(t) = \iota(v(t))$ for all $v \in V$, a.e. $t \in (0,T)$ is also compact.

H_{aux}: The triple of spaces V, H and U and the operator ι satisfy: for all $\varepsilon > 0$ there exists $C(\varepsilon) > 0$ such that we have

$$\|\iota u\|_U \leq \varepsilon \|u\| + C(\varepsilon)|u| \text{ for all } u \in V.$$

These hypotheses will be used in the next section, in the analysis of the semidiscrete schemes associated to the problem (P^m).

5.4 Rothe Method

In this section we consider a semidiscrete approach to Problem $(P^m)_*$, $m = 1, 2$ based on a time discretization method known as the Rothe method. To this end, we divide the time interval $(0, T)$ by means of a sequence $\{t_k\}_{k=0}^{N_n} \subset (0, T)$ defined as follows:

$$t_k = k\tau_n \text{ where } \tau_n = T/N_n \text{ for } k = 0, \ldots, N_n.$$

In the above notation N_n denotes the number of time steps in n-th division of $[0, T]$, so we have $N_n \to \infty$ and $\tau_n \to 0$ as $n \to \infty$. For the convenience we will omit the subscript n and write N, τ instead of N_n, τ_n. We define the piecewise constant interpolant of f by

$$f_\tau(t) = f_\tau^k := \frac{1}{\tau} \int_{(k-1)\tau}^{k\tau} f(s)\, ds \text{ for } t \in ((k-1)\tau, k\tau], \quad k = 1, \ldots, N.$$

By Lemma 3.3 from [5], we know that $f_\tau \to f$ strongly in $\mathcal{V}^*$ as $\tau \to 0$. Finally, we approximate the initial conditions u_0 and u_1 by elements of V. Namely, let $\{u_\tau^0\}, \{u_\tau^1\} \subset V$ be sequences such that $u_\tau^0 \to u_0$ weakly in V and $u_\tau^1 \to u_1$ strongly in H, and $\|u_\tau^1\| \leq C/\sqrt{\tau}$ for some constant $C > 0$.

For a given $\tau > 0$ and $k = 1, \ldots, N$ we formulate the following two Rothe problems denoted by (P_τ^1) and (P_τ^2):

$$(P_\tau^m) \begin{cases} \text{Find } \{u_\tau^k\}_{k=0}^N \subset V, \ \{w_\tau^k\}_{k=0}^N \subset V, \text{ and } \{\eta_\tau^k\}_{k=0}^N \subset U^* \text{ such that} \\ u_\tau^k = u_\tau^0 + \tau \sum_{i=1}^k w_\tau^i & \text{for } k = 1, \ldots, N, \\ \frac{1}{\tau}(w_\tau^k - w_\tau^{k-1}, v) + \langle A w_\tau^k, v \rangle \\ +\langle B u_\tau^k, v \rangle + \langle \eta_\tau^k, \iota v \rangle_{U^* \times U} = \langle f_\tau^k, v \rangle & \text{for all } v \in V, \ k = 1, \ldots, N, \\ \eta_\tau^k \in \partial J(\iota z_\tau^k) & \text{for } k = 1, \ldots, N, \\ w_\tau^0 = u_\tau^1. \end{cases}$$

In Problem (P_τ^1) we take $z_\tau^k = u_\tau^k$ while in Problem (P_τ^2) $z_\tau^k = w_\tau^k$ for $k = 1, \ldots, N$. Now, we study the existence of solutions to Problems (P_τ^1) and (P_τ^2). First we define two mappings $L_1, L_2 : V \to 2^{V^*}$ by

$$L_1 w := \frac{1}{\tau} i^* i w + Aw + \tau Bw + \iota^* \partial J(\iota(u_\tau^0 + \tau \sum_{i=1}^{k-1} w_\tau^i + \tau w)) \quad \text{for } w \in V, \tag{5.5}$$

$$L_2 w := \frac{1}{\tau} i^* i w + Aw + \tau Bw + \iota^* \partial J(\iota w) \quad \text{for } w \in V. \tag{5.6}$$

We remark that for existence of solution to the problems (P_τ^1) and (P_τ^2), it is enough to examine the surjectivity of L_1 and L_2 respectively. To this end we use the following lemmas.

Lemma 5.15. *If the assumptions $H(A)$, $H(B)$, $H(J)$ hold then there exists $\tau_0 > 0$ such that for all $0 < \tau < \tau_0$ the mapping L_1 is coercive. Moreover if H_{aux} holds then L_2 is also coercive.*

Proof. We deal with the operator L_1 first. Let $w \in V$ and $w^* \in L_1 w$. It means that $w^* = \frac{1}{\tau} i^* i w + Aw + Bw + \iota^* \zeta$, where

$$\eta \in \partial J(\iota(u_\tau^0 + \tau \sum_{i=1}^{k-1} w_\tau^i + \tau w)). \tag{5.7}$$

Using $H(A)(iv)$ and $H(B)$ we obtain

$$\begin{aligned} \langle w^*, w \rangle &= \frac{1}{\tau} |w|^2 + \langle Aw, w \rangle + \tau \langle Bw, w \rangle + \langle \eta, \iota w \rangle_{U^* \times U} \\ &\geq \frac{1}{\tau} |w|^2 + \alpha \|w\|^2 - \beta |w|^2 - \gamma + \langle \eta, \iota w \rangle_{U^* \times U}. \end{aligned} \tag{5.8}$$

Using $H(J)(ii)$, we estimate the last term of (5.8) from below,

$$\begin{aligned} \langle \eta, \iota w \rangle_{U^* \times U} &\geq -\|\iota\| \|\eta\|_{U^*} \|w\| \\ &\geq -c \left(1 + \|\iota\| (\|u_0\| + \tau \sum_{i=1}^{k-1} \|w_\tau^i\| + \tau \|w\|) \right) \|\iota\| \|w\| \\ &\geq -c_k \|w\| - c\tau \|\iota\|^2 \|w\|^2, \end{aligned}$$

where $c_k > 0$. Combining the latter with (5.8), we obtain

$$\langle w^*, w \rangle \geq (\alpha - c\tau \|\iota\|) \|w\|^2 - c_k \|w\| + \left(\frac{1}{\tau} - \beta \right) |w|^2 - \gamma.$$

Let us take $\tau_1 := \alpha/c\|\iota\|$ and $\tau_2 := 1/\beta$. It is clear that L_1 is coercive for $\tau < \tau_0 = \min\{\tau_1, \tau_2\}$. For the operator L_2 we take $w \in V$, $w^* \in L_2 w$ and obtain (5.8) again. Instead of (5.7), we now assume that $\eta \in \partial J(\iota w)$. From $H(J)(ii)$ and H_{aux}, we infer that for any $\varepsilon > 0$ we can find $C(\varepsilon) > 0$ such that

$$\langle \eta, \iota w\rangle_{U^* \times U} \geq -\varepsilon\|w\|^2 - C(\varepsilon)|w|^2 - C(\varepsilon). \tag{5.9}$$

From (5.8) and (5.9), we get

$$\langle w^*, w\rangle \geq (\alpha - \varepsilon)\|w\|^2 + \left(\frac{1}{\tau} - \beta - C(\varepsilon)\right)|w|^2 - \gamma - C(\varepsilon).$$

Taking $\varepsilon = \frac{1}{2}\alpha$, we see that L_2 is coercive for τ small enough. This completes the proof. □

Now we formulate an existence results for the Rothe problems $(P_\tau^m)_{m=1,2}$.

Theorem 5.16. *If the assumptions $H(A)$, $H(B)$, $H(J)$ and $H(\iota)$ hold then, for $\tau > 0$ small enough, there exists a solution of problem (P_τ^1). Moreover, if H_{aux} holds then, for $\tau > 0$ small enough, there exists a solution of problem (P_τ^2).*

Proof. From $H(A)$ and $H(B)$ we deduce that the mapping $V \ni w \to \frac{1}{\tau} i^* i w + A_\tau^k w + \tau B w \in V^*$ is pseudomonotone. By $H(\iota)$ and Proposition 5.6, we know that the mapping $V \ni w \to \iota^* \partial J(\iota w) \in 2^{V^*}$ is pseudomonotone, and by $H(\iota)$, Propositions 5.6 and 5.7 so is the mapping

$$V \ni w \to \iota^* \partial J(\iota(u_\tau^0 + \tau \sum_{i=1}^{k-1} w_\tau^i + \tau w)) \in 2^{V^*}.$$

Thus from Proposition 5.8 both operators L_1 and L_2 are pseudomonotone. From Lemma 5.15, they are coercive for $\tau > 0$ small enough. Finally, from Proposition 5.10 it is clear that L_1 and L_2 are surjective. It means that, for any given $w_\tau^0, w_\tau^1, \ldots w_\tau^{k-1}$, we can find w_τ^k such that $(P_\tau^m)_{m=1,2}$ hold. So starting from a given w_τ^0, we can construct the whole Rothe sequence. □

5.4.1 A Priori Estimate

In this subsection we study an a priori estimate for the Rothe sequences being the solutions of Problems (P_τ^1) and (P_τ^2). We formulate the following lemma.

Lemma 5.17. *Let the assumptions $H(A)$, $H(B)$, $H(J)$, $H(\iota)$ and H_{aux} hold. Let the triple $\left(\{u_\tau^k\}_{k=0}^N, \{w_\tau^k\}_{k=0}^N, \{\eta_\tau^k\}_{k=0}^N\right)$ be a solution of Problem (P_τ^1) or (P_τ^2). Then we have*

$$\max_{k=1,\ldots,N} |w_\tau^k| \le const, \tag{5.10}$$

$$\sum_{k=1}^{N} |w_\tau^k - w_\tau^{k-1}|^2 \le const, \tag{5.11}$$

$$\tau \sum_{k=1}^{N} \|w_\tau^k\|^2 \le const, \tag{5.12}$$

$$\max_{k=1,\ldots,N} \|u_\tau^k\| \le const, \tag{5.13}$$

where const denotes a positive constant independent on τ.

Proof. Consider Problems $(P_\tau^m)_{m=1,2}$ and chose $v = w_\tau^k$. By a property of scalar product in Hilbert space, we have

$$(w_\tau^k - w_\tau^{k-1}, w_\tau^k) = \frac{1}{2}|w_\tau^k|^2 - \frac{1}{2}|w_\tau^{k-1}|^2 + \frac{1}{2}|w_\tau^k - w_\tau^{k-1}|^2. \tag{5.14}$$

From the assumption $H(A)(iii)$, we obtain

$$\langle A w_\tau^k, w_\tau^k \rangle \ge \alpha \|w_\tau^k\|^2 - \beta |w_\tau^k|^2 - \gamma. \tag{5.15}$$

By the assumptions $H(B)$, we obtain

$$\begin{aligned} \tau \langle B u_\tau^k, w_\tau^k \rangle &= \langle B u_\tau^k, u_\tau^k - u_\tau^{k-1} \rangle \\ &= \frac{1}{2}\langle B u_\tau^k, u_\tau^k \rangle - \frac{1}{2}\langle B u_\tau^{k-1}, u_\tau^{k-1} \rangle + \frac{1}{2}\langle B(u_\tau^k - u_\tau^{k-1}), u_\tau^k - u_\tau^{k-1} \rangle \\ &\ge \frac{1}{2}\langle B u_\tau^k, u_\tau^k \rangle - \frac{1}{2}\langle B u_\tau^{k-1}, u_\tau^{k-1} \rangle. \end{aligned} \tag{5.16}$$

Moreover, for any $\delta > 0$, we have

$$\langle f_\tau^k, w_\tau^k \rangle \le \|f_\tau^k\|_{V*} \|w_\tau^k\| \le \delta \|w_\tau^k\|^2 + \frac{1}{4\delta} \|f_\tau^k\|_{V*}^2 \tag{5.17}$$

Let $\eta_\tau^k \in \partial J(\iota z_\tau^k)$. Summing up the equations $(P_\tau^m)_{m=1,2}$ with $v = w_\tau^k$ for $k = 1, \ldots n \le N$ and using (5.14)–(5.17), we obtain

$$\frac{1}{2}|w_\tau^n|^2 + \frac{1}{2}\sum_{k=1}^{n} |w_\tau^k - w_\tau^{k-1}|^2 + \sum_{k=1}^{n} \tau(\alpha - \delta)\|w_\tau^k\|^2$$

$$+ \frac{1}{2}\tau \langle B u_\tau^n, u_\tau^n \rangle + \tau \sum_{k=1}^{n} \langle \eta_\tau^k, \iota w_\tau^k \rangle_{U^* \times U}$$

$$
\leq \frac{1}{2}|w_\tau^0|^2 + \frac{1}{2}\langle Bu_\tau^0, u_\tau^0\rangle + \tau\sum_{k=1}^n \beta|w_\tau^k|^2 + \tau\sum_{k=1}^n \frac{1}{4\delta}\|f_\tau^k\|_{V^*}^2 + \tau n\gamma
$$

$$
\leq \frac{1}{2}|w_\tau^0|^2 + \frac{1}{2}\langle Bu_\tau^0, u_\tau^0\rangle + \tau\sum_{k=1}^n \beta|w_\tau^k|^2 + \frac{1}{4\delta}\|f_\tau\|_{\mathcal{V}^*}^2 + T\gamma. \tag{5.18}
$$

Now we estimate the term $\tau\sum_{k=1}^n\langle\eta_\tau^k, \iota w_\tau^k\rangle_{U^*\times U}$ from below. To this end, we consider two cases.

- **Case 1:** $z_\tau^k = w_\tau^k$. We use $H(J)(ii)$ and (5.9) with $\varepsilon = \delta = \frac{1}{4}\alpha$. Then from (5.18), we obtain

$$
(1-2\tau(\beta+C(\varepsilon)))\,|w_\tau^n|^2 + \sum_{k=1}^n|w_\tau^k - w_\tau^{k-1}|^2 + \sum_{k=1}^n \tau\alpha\|w_\tau^k\|^2
$$

$$
\leq |w_\tau^0|^2 + \langle Bu_\tau^0, u_\tau^0\rangle + 2\tau\sum_{k=1}^{n-1}(\beta + C(\varepsilon))|w_\tau^k|^2
$$

$$
+ \frac{2}{\alpha}\|f_\tau\|_{\mathcal{V}^*}^2 + 2T\gamma + C(\varepsilon). \tag{5.19}
$$

- **Case 2:** $z_\tau^k = u_\tau^k$. In this case, using $H(J)(ii)$, we have

$$
\tau\sum_{k=1}^n|\langle\eta_\tau^k, \iota w_\tau^k\rangle_{U^*\times U}| \leq \tau\sum_{k=1}^n c(1+\|\iota u_\tau^k\|_U)\|\iota w_\tau^k\|_U
$$

$$
\leq \tau\sum_{k=1}^n c\left(1 + \|\iota u_0\|_U + \tau\sum_{i=1}^k\|\iota w_\tau^i\|\right)\|\iota w_\tau^k\|_U
$$

$$
\leq \tau\sum_{k=1}^n c\|\iota w_\tau^k\|_U + \tau\sum_{k=1}^n c\|\iota u_0\|_U\,\|\iota w_\tau^k\|_U
$$

$$
+ \tau^2\left(\sum_{k=1}^n\|\iota w_\tau^k\|_U\right)^2. \tag{5.20}
$$

In what follows, we use the inequality $ab \leq \frac{1}{2}(a^2+b^2)$ for $a, b \in \mathbb{R}$ and the fact that $\tau N = T$.

$$
\tau\sum_{k=1}^n c\|\iota w_\tau^k\|_U \leq \tau\sum_{k=1}^n\left(\frac{1}{2}c^2 + \frac{1}{2}\|\iota w_\tau^k\|_U^2\right) \leq \frac{1}{2}c^2T + \frac{1}{2}\tau\sum_{k=1}^n\|\iota w_\tau^k\|_U^2.
$$

Analogously we deduce that

$$\tau \sum_{i=1}^{n} c\|\iota u_0\| \|\iota w_\tau^k\|_U \le \frac{1}{2} c^2 T \|\iota u_0\|_U^2 + \frac{1}{2}\tau \sum_{k=1}^{n} \|\iota w_\tau^k\|_U^2 .$$

Moreover, for the last term of (5.20) we use inequality $(a_1 + \ldots + a_n)^2 \le n(a_1^2 + \ldots + a_n^2)$ for $a_i \in \mathbb{R}$, $i = 1 \ldots N$, $n \in \mathbb{N}$ and obtain

$$\tau^2 \left(\sum_{k=1}^{n} \|\iota w_\tau^k\|_U \right)^2 \le \tau^2 n \sum_{k=1}^{n} \|\iota w_\tau^k\|_U^2 \le \tau T \sum_{k=1}^{n} \|\iota w_\tau^k\|_U^2 .$$

Thus, from (5.20) and H_{aux}, for any $\varepsilon > 0$ we get

$$\begin{aligned} \tau \sum_{k=1}^{n} \langle \eta_\tau^k, \iota w_\tau^k \rangle_{U^* \times U} &\ge -\frac{1}{2} c^2 T (1 + \|\iota u_0\|_U^2) - \tau(1+T) \sum_{k=1}^{n} \|\iota w_\tau^k\|_U^2 \\ &\ge -\frac{1}{2} c^2 T (1 + \|\iota u_0\|_U^2) - \tau(1+T) \sum_{k=1}^{n} \varepsilon \|w_\tau^k\|^2 \\ &\quad - \tau(1+T) \sum_{k=1}^{n} C(\varepsilon) |w_\tau^k|^2 . \end{aligned}$$

Then from (5.18), we obtain

$$\begin{aligned} &(1 - 2\tau(\beta + (1+T)C(\varepsilon)))\, |w_\tau^n|^2 + \sum_{k=1}^{n} |w_\tau^k - w_\tau^{k-1}|^2 \\ &+ 2\tau \sum_{k=1}^{n} (\alpha - \delta - (1+T)\varepsilon) \|w_\tau^k\|^2 \\ &\quad \le |w_\tau^0|^2 + \langle B u_\tau^0, u_\tau^0 \rangle + 2\tau \sum_{k=1}^{n-1} (\beta + (1+T)C(\varepsilon)) |w_\tau^k|^2 \\ &\quad + \frac{1}{4\delta} \|f_\tau\|_{\mathcal{V}^*}^2 + 2T\gamma + \frac{1}{2} c^2 T (1 + \|\iota u_0\|_U^2). \end{aligned} \tag{5.21}$$

In Case 1 we take $\tau < \frac{1}{2}(\beta + C(\varepsilon))^{-1}$. In Case 2 we take $\delta < \alpha$, $\varepsilon < \frac{\alpha - \delta}{1+T}$ and $\tau < \frac{1}{2}(\beta + (1+T)C(\varepsilon))^{-1}$. Since $f_\tau \to f$ strongly in $\mathcal{V}^*$, it is bounded in $\mathcal{V}^*$. Thus, we can apply a discrete Gronwall lemma for the sequence $\{|w_\tau^k|^2\}_{k=1,\ldots,n}$ and, since $n = 1, \dots, N$ is arbitrary, we deduce (5.10). From (5.10), (5.19) and (5.21) we immediately get (5.11) and (5.12). From the definition of u_τ^k in (P_τ^m), we obtain

$$\|u_\tau^k\|_V^2 \le 2\|u_0\|^2 + 2\tau^2 N \sum_{i=1}^{k} \|w_\tau^i\|^2 \le 2\|u_0\|^2 + 2\tau T \sum_{i=1}^{N} \|w_\tau^i\|^2,$$

which together with (5.12) gives (5.13). □

5.4.2 Convergence of Rothe Method

In this subsection we construct sequences of time dependent piecewise constant and piecewise linear functions built on the solution of Rothe problem. Next we study their convergence to the solutions of Problems $(P^1)_*$ and $(P^2)_*$. Let the triple $(\{u_\tau^k\}_{k=0}^N, \{w_\tau^k\}_{k=0}^N, \{\eta_\tau^k\}_{k=0}^N)$ be a solution of Problem (P_τ^1) or (P_τ^2).

We define the functions $w_\tau, \overline{w}_\tau, u_\tau, \overline{u}_\tau \colon [0, T] \to V$ by the formulas

$$\overline{w}_\tau(t) = \begin{cases} w_\tau^k, \ t \in ((k-1)\tau, k\tau], \\ w_\tau^0 = u_\tau^1, \ t = 0, \end{cases}$$

$$w_\tau(t) = w_\tau^k + \left(\frac{t}{\tau} - k\right)(w_\tau^k - w_\tau^{k-1}) \ \text{for} \ t \in ((k-1)\tau, k\tau], \quad k = 1, \dots, N,$$

$$\overline{u}_\tau(t) = \begin{cases} u_\tau^k, \ t \in ((k-1)\tau, k\tau], \\ u_\tau^0, \ t = 0, \end{cases}$$

$$u_\tau(t) = u_\tau^k + \left(\frac{t}{\tau} - k\right)(u_\tau^k - u_\tau^{k-1}) \ \text{for} \ t \in ((k-1)\tau, k\tau], \quad k = 1, \dots, N.$$

The piecewise constant function $\overline{\eta}_\tau \colon (0, T] \to U^*$ is given by

$$\overline{\eta}_\tau(t) = \eta_\tau^k \ \text{for} \ t \in ((k-1)\tau, k\tau], \quad k = 1, \dots, N.$$

We observe that $\overline{w}_\tau$ is the distributional derivative of u_τ, namely $u_\tau'(t) = \overline{w}_\tau(t)$ for a.e. $t \in (0, T)$. Moreover, the distributional derivative of w_τ is given by $w_\tau'(t) = \frac{w_\tau^k - w_\tau^{k-1}}{\tau}$ for a.e. $t \in (0, T)$, $k = 1, \dots, N$. Thus, (P_τ^m) is equivalent to

$$(w_\tau'(t), v) + \langle A\overline{w}_\tau(t), v\rangle + \langle B\overline{u}_\tau(t), v\rangle + \langle \overline{\eta}_\tau(t), \iota v\rangle_{U^*\times U} = \langle f_\tau(t), v\rangle$$
$$\text{for all } v \in V, \ \text{for all } t \in (0, T), \tag{5.22}$$

$$\overline{\eta}_\tau(t) \in \partial J(\iota z_\tau(t)) \quad \text{for all } t \in (0, T). \tag{5.23}$$

where $z_\tau = \overline{u}_\tau$ in case $m = 1$ and $z_\tau = \overline{w}_\tau$ in case $m = 2$. We define the Nemytskii operators $\mathcal{A}, \mathcal{B} : \mathcal{V} \to \mathcal{V}^*$ by $(\mathcal{A}v)(t) = A(v(t))$, $(\mathcal{B}v)(t) = B(v(t))$ for $v \in \mathcal{V}$ and $\bar{\iota} : \mathcal{V} \to \mathcal{U}$ by $(\bar{\iota}v)(t) = \iota v(t)$ for $v \in \mathcal{V}$ and observe that problem (5.22), (5.23) is now equivalent to

$$(w'_\tau, v)_\mathcal{H} + \langle \mathcal{A}\overline{w}_\tau, v\rangle_{\mathcal{V}^*\times\mathcal{V}} + \langle \mathcal{B}\overline{u}_\tau, v\rangle_{\mathcal{V}^*\times\mathcal{V}} + \langle \overline{\eta}_\tau, \bar{\iota}v\rangle_{\mathcal{U}^*\times\mathcal{U}} = \langle f_\tau, v\rangle_{\mathcal{V}^*\times\mathcal{V}} \quad \text{for all } v \in \mathcal{V}, \tag{5.24}$$

$$\overline{\eta}_\tau(t) \in \partial J((\bar{\iota} z_\tau)(t)) \quad \text{for all } t \in (0, T). \tag{5.25}$$

In what follows we deal with a priori estimates for the piecewise constant and piecewise linear functions built on the solution of the Rothe problem.

Lemma 5.18. *Under assumptions $H(A)$, $H(B)$, $H(J)$, $H(\iota)$, H_0 and H_{aux}, there exists $\tau_0 > 0$ such that for all $\tau \in (0, \tau_0)$, we have*

$$\|\overline{u}_\tau\|_{L^\infty(0,T;V)} \le const, \tag{5.26}$$

$$\|u_\tau\|_\mathcal{V} \le const, \tag{5.27}$$

$$\|\overline{w}_\tau\|_{L^\infty(0,T;H)} \le const, \tag{5.28}$$

$$\|w_\tau\|_{C(0,T;H)} \le const, \tag{5.29}$$

$$\|\overline{w}_\tau\|_\mathcal{V} \le const, \tag{5.30}$$

$$\|w_\tau\|_\mathcal{V} \le const, \tag{5.31}$$

$$\|\mathcal{A}\overline{w}_\tau\|_{\mathcal{V}^*} \le const, \tag{5.32}$$

$$\|\overline{\eta}_\tau\|_{\mathcal{U}^*} \le const, \tag{5.33}$$

$$\|w'_\tau\|_{\mathcal{V}^*} \le const, \tag{5.34}$$

$$\|\overline{u}_\tau\|_{M^{2,2}(0,T;V,V^*)} \le const, \tag{5.35}$$

$$\|\overline{w}_\tau\|_{M^{2,2}(0,T;V,V^*)} \le const, \tag{5.36}$$

with a constant independent of τ.

Proof. The estimate (5.26) follows directly from (5.13), while from (5.10), we easily get (5.28) and (5.29). Moreover, $\|\overline{w}_\tau\|^2_\mathcal{V} = \tau\sum_{k=1}^N \|w^k_\tau\|^2$ so, from (5.12), we obtain (5.30). The simple calculation shows that $\|w_\tau\|^2_\mathcal{V} \le \tau\sum_{k=0}^N \|w^k_\tau\|^2$. Thus using (5.12) and the fact that $\|w^0_\tau\| \le C/\sqrt{\tau}$, we get (5.31). Analogously, we have

$$\|u_\tau\|^2_\mathcal{V} \le \tau\sum_{k=0}^N \|u^k_\tau\|^2 \le \tau N \max_{k=0,\ldots,N} \|u^k_\tau\|^2 = T \max_{k=0,\ldots,N} \|u^k_\tau\|^2,$$

which, together with (5.13), gives (5.27). Using $H(A)(ii)$, we calculate

$$\|\mathcal{A}\overline{w}_\tau\|_{\mathcal{V}^*} = \left(\int_0^T \|A\overline{w}_\tau(t)\|_{V^*}^2\,dt\right)^{\frac{1}{2}} \le \left(\int_0^T 2a^2 + 2b^2\|\overline{w}_\tau(t)\|^2\,dt\right)^{\frac{1}{2}}$$
$$\le a\sqrt{2T} + \sqrt{2}b\|\overline{w}_\tau\|_{\mathcal{V}},$$

and from (5.30) we obtain (5.32). Next, using $H(J)(ii)$ we get

$$\|\overline{\eta}_\tau\|_{\mathcal{U}^*} = \left(\int_0^T \|\overline{\eta}_\tau(t)\|_{U^*}^2 dt\right)^{\frac{1}{2}} \le \left(\int_0^T 2c^2 + 2\|\iota\|^2\|z_\tau(t)\|_{U^*}^2\,dt\right)^{\frac{1}{2}}$$
$$\le c\sqrt{2T} + \sqrt{2}\|\iota\|\,\|z_\tau\|_{\mathcal{V}},$$

and we obtain (5.33) from (5.26) in case $z_\tau = \overline{u}_\tau$ and from (5.30) in case $z_\tau = \overline{w}_\tau$. We also have

$$\|\mathcal{B}\overline{u}_\tau\|_{\mathcal{V}^*} = \left(\int_0^T \|B\overline{u}_\tau(t)\|_{V^*}^2 dt\right)^{\frac{1}{2}} \le \left(\int_0^T \|B\|_{\mathcal{L}(V,V^*)}\|\overline{u}_\tau(t)\|_{V^*}^2\,dt\right)^{\frac{1}{2}}$$
$$= \|B\|_{\mathcal{L}(V,V^*)}\|\overline{u}\|_{\mathcal{V}} \le \|B\|_{\mathcal{L}(V,V^*)}\sqrt{T}\|\overline{u}\|_{L^\infty(0,T;V)}. \tag{5.37}$$

From (5.24), we get

$$\|w_\tau'\|_{\mathcal{V}^*} \le \|\mathcal{A}\overline{w}_\tau\|_{\mathcal{V}^*} + \|\mathcal{B}\overline{u}_\tau\|_{\mathcal{V}^*} + \|\overline{\eta}_\tau\|_{\mathcal{U}^*} + \|f_\tau\|_{\mathcal{V}^*}$$

and, using (5.32), (5.33), (5.37) and boundedness of f_τ in $\mathcal{V}^*$, we obtain (5.34). Finally in order to prove (5.36), let us assume that the seminorm in $BV^2(0,T;V^*)$ of piecewise constant function $\overline{w}_\tau$ is realized by some division $0 = a_0 < a_1 < \ldots < a_n = T$, and each a_i is in different interval $((m_i-1)\tau, m_i\tau]$, such that $\overline{w}_\tau(a_i) = w_\tau^{m_i}$ with $m_0 = 0$, $m_n = N$ and $m_{i+1} > m_i$ for $i = 1,\ldots,N-1$. Thus

$$\|\overline{w}_\tau\|_{BV^2(0,T;V^*)}^2 = \sum_{i=1}^n \|w_\tau^{m_i} - w_\tau^{m_{i-1}}\|_{V^*}^2$$
$$\le \sum_{i=1}^n \left((m_i - m_{i-1}) \sum_{k=m_{i-1}+1}^{m_i} \|w_\tau^k - w_\tau^{k-1}\|_{V^*}^2\right)$$
$$\le \left(\sum_{i=1}^n (m_i - m_{i-1})\right)\left(\sum_{i=1}^n \sum_{k=m_{i-1}+1}^{m_i} \|w_\tau^k - w_\tau^{k-1}\|_{V^*}^2\right)$$
$$= N\sum_{k=1}^N \|w_\tau^k - w_\tau^{k-1}\|_{V^*}^2 = T\tau\sum_{k=1}^N \left\|\frac{w_\tau^k - w_\tau^{k-1}}{\tau}\right\|_{V^*}^2$$
$$= T\int_0^T \|w_\tau'(t)\|_{V^*}^2\,dt.$$

The last term is bounded by (5.34) which, together with (5.30), completes the proof of (5.36). Analogously we prove (5.35). □

Theorem 5.19. *Let assumptions $H(A)$, $H(B)$, $H(J)$, $H(\iota)$, H_0 and H_{aux} hold and let $u_\tau, \overline{u}_\tau, w_\tau, \overline{w}_\tau$ and $\overline{\eta}_\tau$ be piecewise linear and piecewise constant functions built on a solution of Rothe problem (P_τ^m) with $m = 1, 2$. Then, there exists a triple (u, w, η) which is a solution of the corresponding Problem $(P^m)_*$. Moreover, for a subsequence, we have $\overline{u}_\tau \to u$ weakly* in $L^\infty(0,T;V)$, $u_\tau \to u$ weakly in $\mathcal{V}$, $\overline{w}_\tau \to w$ weakly in $\mathcal{V}$ and weakly* in $L^\infty(0,T;H)$, $w_\tau \to w$ weakly in $\mathcal{W}$ and weakly* in $L^\infty(0,T;H)$ and $\overline{\eta}_\tau \to \eta$ weakly in $\mathcal{U}^*$.*

Proof. From (5.26)–(5.34) and reflexivity of spaces $\mathcal{V}, \mathcal{V}^*, \mathcal{U}$ and $\mathcal{U}^*$, we can assume, passing to the subsequence if necessary, that the following convergences hold:

$$\overline{u}_\tau \to \overline{u} \quad \text{weakly* in } L^\infty(0,T;V), \tag{5.38}$$

$$u_\tau \to u \quad \text{weakly in } \mathcal{V}, \tag{5.39}$$

$$\overline{w}_\tau \to \overline{w} \quad \text{weakly in } \mathcal{V} \text{ and weakly* in } L^\infty(0,T;H), \tag{5.40}$$

$$w_\tau \to w \quad \text{weakly in } \mathcal{V} \text{ and weakly* in } L^\infty(0,T;H), \tag{5.41}$$

$$w'_\tau \to w_1 \quad \text{weakly in } \mathcal{V}^*, \tag{5.42}$$

$$\mathcal{A}\overline{w}_\tau \to \zeta \quad \text{weakly in } \mathcal{V}^*, \tag{5.43}$$

$$\overline{\eta}_\tau \to \eta \quad \text{weakly in } \mathcal{U}^*. \tag{5.44}$$

First we show that $\overline{u} = u$. To this end we calculate

$$\|\overline{u}_\tau - u_\tau\|_{\mathcal{V}}^2 = \sum_{k=1}^N \int_{(k-1)\tau}^{k\tau} (k\tau - t)^2 \left\| \frac{u_\tau^k - u_\tau^{k-1}}{\tau} \right\|_V^2 = \frac{\tau^2}{3} \|u'_\tau\|_{\mathcal{V}}^2 = \frac{\tau^2}{3} \|\overline{w}_\tau\|_{\mathcal{V}}^2,$$

which means that $\overline{u}_\tau - u_\tau \to 0$ strongly in $\mathcal{V}$ as $\tau \to 0$. On the other hand, from (5.38) and (5.39) we also have $\overline{u}_\tau - u_\tau \to \overline{u} - u$ weakly in $\mathcal{V}$ so from the uniqueness of weak limit we get $\overline{u} = u$. Analogously, we have

$$\|\overline{w}_\tau - w_\tau\|_{\mathcal{V}^*}^2 = \frac{\tau^2}{3} \|w'_\tau\|_{\mathcal{V}^*}^2,$$

which means that $\overline{w}_\tau - w_\tau \to 0$ strongly in $\mathcal{V}^*$. Since $\overline{w}_\tau - w_\tau \to \overline{w} - w$ weakly in $\mathcal{V}$ and the embedding $\mathcal{V} \subset \mathcal{V}^*$ is continuous, we also have $\overline{w}_\tau - w_\tau \to \overline{w} - w$ weakly in $\mathcal{V}^*$. From uniqueness of the weak limit, we have $\overline{w} = w$. Since $\overline{w}_\tau = u'_\tau$ and $u_\tau \to u$ weakly in $\mathcal{V}$, we conclude that

$$w = u'. \tag{5.45}$$

Our goal is to pass to the limit in (5.24) and (5.25). By a standard argument from (5.41) and (5.42) we have $w'_\tau \to w'$ weakly in $\mathcal{V}^*$. Thus, for all $v \in \mathcal{V}$, we obtain

$$(w'_\tau, v)_{\mathcal{H}} = \langle w'_\tau, v \rangle_{\mathcal{V}^* \times \mathcal{V}} \to \langle w', v \rangle_{\mathcal{V}^* \times \mathcal{V}} = (w', v)_{\mathcal{H}}. \tag{5.46}$$

From $H(B)$, it is clear that $\mathcal{B}$ is linear and continuous operator from $\mathcal{V}$ to $\mathcal{V}^*$ and thus also continuous from $w - \mathcal{V}$ to $w - \mathcal{V}^*$. Therefore, since $\overline{u}_\tau \to u$ weakly in $\mathcal{V}$, we get $\mathcal{B}\overline{u}_\tau \to \mathcal{B}u$ weakly in $\mathcal{V}^*$. Thus, we have

$$\langle \mathcal{B}\overline{u}_\tau, v \rangle_{\mathcal{V}^* \times \mathcal{V}} \to \langle \mathcal{B}u, v \rangle_{\mathcal{V}^* \times \mathcal{V}}. \tag{5.47}$$

From (5.44), we get

$$\langle \overline{\eta}_\tau, \bar{\iota} v \rangle_{\mathcal{U}^* \times \mathcal{U}} \to \langle \eta, \bar{\iota} v \rangle_{\mathcal{U}^* \times \mathcal{U}}. \tag{5.48}$$

Since $f_\tau \to f$ strongly in $\mathcal{V}^*$, it is clear that

$$\langle f_\tau, v \rangle_{\mathcal{V}^* \times \mathcal{V}} \to \langle f, v \rangle_{\mathcal{V}^* \times \mathcal{V}}. \tag{5.49}$$

It remains to show that

$$\langle \mathcal{A}\overline{w}_\tau, v \rangle_{\mathcal{V}^* \times \mathcal{V}} \to \langle \mathcal{A}w, v \rangle_{\mathcal{V}^* \times \mathcal{V}}. \tag{5.50}$$

To this end we take any subsequence still denoted by $\{\langle \mathcal{A}\overline{w}_\tau, v \rangle_{\mathcal{V}^* \times \mathcal{V}}\}$. It is enough to show that convergence (5.50) holds for its subsequence. We calculate

$$\begin{aligned} \limsup_{\tau \to 0} \langle \mathcal{A}\overline{w}_\tau, \overline{w}_\tau - w \rangle_{\mathcal{V}^* \times \mathcal{V}} \le & \limsup_{\tau \to 0} \langle f_\tau, \overline{w}_\tau - w \rangle_{\mathcal{V}^* \times \mathcal{V}} \\ & - \liminf_{\tau \to 0} (w'_\tau, \overline{w}_\tau - w)_{\mathcal{H}} \\ & - \liminf_{\tau \to 0} \langle \mathcal{B}\overline{u}_\tau, \overline{w}_\tau - w \rangle_{\mathcal{V}^* \times \mathcal{V}} \\ & - \liminf_{\tau \to 0} \langle \overline{\eta}_\tau, \bar{\iota}\, \overline{w}_\tau - \bar{\iota} w \rangle_{\mathcal{U}^* \times \mathcal{U}}. \end{aligned} \tag{5.51}$$

Since $f_\tau \to f$ strongly in $\mathcal{V}^*$ and $\overline{w}_\tau \to w$ weakly in $\mathcal{V}$, we have

$$\lim_{\tau \to 0} \langle f_\tau, \overline{w}_\tau - w \rangle_{\mathcal{V}^* \times \mathcal{V}} = 0.$$

Next, we observe that

$$\begin{aligned} (w'_\tau, \overline{w}_\tau - w)_{\mathcal{H}} = & (w'_\tau - w', \overline{w}_\tau - w_\tau)_{\mathcal{H}} + \langle w'_\tau, \overline{w}_\tau - w_\tau \rangle_{\mathcal{V}^* \times \mathcal{V}} \\ & + \langle w', \overline{w}_\tau - w \rangle_{\mathcal{V}^* \times \mathcal{V}} \end{aligned}$$

$$= \frac{1}{2}\left(|w_\tau(T) - w(T)|^2 - |u_{1\tau} - u_1|^2\right)$$
$$+ \langle w'_\tau, \overline{w}_\tau - w_\tau\rangle_{\mathcal{V}^*\times\mathcal{V}} + \langle w', \overline{w}_\tau - w\rangle_{\mathcal{V}^*\times\mathcal{V}}.$$

Since $\langle w'_\tau, \overline{w}_\tau - w_\tau\rangle_{\mathcal{V}^*\times\mathcal{V}} \geq 0$, we have $\liminf_{\tau\to 0}(w'_\tau, \overline{w}_\tau - w)_{\mathcal{H}} \geq 0$. Since $\overline{u}_\tau = w'_\tau$ and $w = u'$, it follows that

$$\begin{aligned}\langle \mathcal{B}\overline{u}_\tau, \overline{w}_\tau - w\rangle_{\mathcal{V}^*\times\mathcal{V}} &= \langle \mathcal{B}\overline{u}_\tau, u'_\tau - u'\rangle_{\mathcal{V}^*\times\mathcal{V}} \\ &= \langle \mathcal{B}u_\tau - \mathcal{B}u, u'_\tau - u'\rangle_{\mathcal{V}^*\times\mathcal{V}} + \langle \mathcal{B}u, u'_\tau - u'\rangle_{\mathcal{V}^*\times\mathcal{V}} \\ &\quad + \langle \mathcal{B}\overline{u}_\tau - \mathcal{B}u_\tau, u'_\tau - u'\rangle_{\mathcal{V}^*\times\mathcal{V}}. \end{aligned} \tag{5.52}$$

Since B is linear, symmetric and monotone, it follows that

$$\begin{aligned}\langle \mathcal{B}u_\tau - \mathcal{B}u, u'_\tau - u'\rangle_{\mathcal{V}^*\times\mathcal{V}} &= \frac{1}{2}\langle B(u_\tau(T) - u(T)), u_\tau(T) - u(T)\rangle \\ &\quad - \frac{1}{2}\langle B(u^0_\tau - u^0), u^0_\tau - u_0\rangle \\ &\geq -\frac{1}{2}\|B\|_{\mathcal{L}(V,V^*)}\|u^0_\tau - u^0\|^2. \end{aligned} \tag{5.53}$$

Since $u'_\tau \to u'$ weakly in $\mathcal{V}$, $\overline{u}_\tau \to u_\tau$ strongly in $\mathcal{V}$ and $\mathcal{B}$ is continuous, we get

$$\lim_{\tau\to 0}\langle \mathcal{B}u, u'_\tau - u'\rangle_{\mathcal{V}^*\times\mathcal{V}} + \langle \mathcal{B}\overline{u}_\tau - \mathcal{B}u_\tau, u'_\tau - u'\rangle_{\mathcal{V}^*\times\mathcal{V}} = 0. \tag{5.54}$$

From (5.52)–(5.54), we see that $\liminf_{\tau\to 0}\langle \mathcal{B}\overline{u}_\tau, \overline{w}_\tau - w\rangle_{\mathcal{V}^*\times\mathcal{V}} \geq 0$. Moreover, since $\overline{w}_\tau \to w$ weakly in $\mathcal{V}$ and $\bar{\iota}$ is a compact operator, we have for a subsequence $\bar{\iota}\,\overline{w}_\tau \to \bar{\iota}w$ strongly in $\mathcal{U}$. Since $\overline{\eta}_\tau \to \eta$ weakly in $\mathcal{U}^*$, we have $\lim_{\tau\to 0}\langle \overline{\eta}_\tau, \bar{\iota}\,\overline{w}_\tau - \bar{\iota}w\rangle_{\mathcal{U}^*\times\mathcal{U}} = 0$. Thus, from (5.51) we get

$$\limsup_{\tau\to 0}\langle \mathcal{A}\overline{w}_\tau, \overline{w}_\tau - w\rangle_{\mathcal{V}^*\times\mathcal{V}} \leq 0. \tag{5.55}$$

Since (5.36) holds and $\overline{w}_\tau \to w$ weakly in $\mathcal{V}$, from Lemma 5.13 we deduce (5.50). Using (5.47)–(5.50), we can pass to the limit in (5.24) and obtain

$$\begin{aligned}(w', v)_{\mathcal{H}} + \langle \mathcal{A}w, v\rangle_{\mathcal{V}^*\times\mathcal{V}} + \langle \mathcal{B}u, v\rangle_{\mathcal{V}^*\times\mathcal{V}} + \langle \eta, \bar{\iota}v\rangle_{\mathcal{U}^*\times\mathcal{U}} &= \langle f, v\rangle_{\mathcal{V}^*\times\mathcal{V}} \\ \text{for all } v \in \mathcal{V}. \end{aligned} \tag{5.56}$$

Let $v = h\varphi$ in (5.56) with $h \in V$ and $\varphi \in C_0^\infty(0, T)$. Then, we obtain

$$-\int_0^T \langle w'(t), h\rangle\varphi(t)dt = \int_0^T \langle Aw(t) + Bu(t) + \iota^*\eta(t) - f(t), h\rangle\varphi(t)\,dt. \tag{5.57}$$

From Proposition 23.20 of [16] we deduce

$$-\int_0^T \langle w'(t), h\rangle \varphi(t)dt = \int_0^T (w(t), h)\varphi'(t)\, dt. \tag{5.58}$$

Now we will pass to the limit with (5.23). First we recall that the multifunction $\partial J: U \to 2^{U^*}$ has nonempty, closed and convex values. Furthermore, by Proposition 5.6.10 of [7], it is also upper semicontinuous from U (equipped with the strong topology) into U^* (equipped with the weak topology). We use (5.35) in case $z_\tau = \overline{u}_\tau$ and (5.36) in case $z_\tau = \overline{w}_\tau$ to conclude from $H(\iota)$ that for a subsequence we have $\bar{\iota} z_\tau \to \bar{\iota} z$ strongly in $\mathcal{U}$. Thus, for another subsequence, $\bar{\iota} z_\tau(t) \to (\bar{\iota} z)(t)$ strongly in U for a.e. $t \in (0, T)$, where $z = u$ or $z = w$, depending on which of two problems (P^1) or (P^2) we deal with. Using also (5.44) we are now in a position to apply Proposition 5.12 to find

$$\eta(t) \in \partial J((\bar{\iota} z)(t)) \text{ for a.e. } t \in (0, T). \tag{5.59}$$

Finally, we pass to the limit with the initial conditions on the functions u_τ and w_τ. Since $u\tau \to u$ and $u'_\tau \to u'$ both weakly in $\mathcal{V}$ and the embedding $\{v \in \mathcal{V} |\ v' \in \mathcal{V}\} \subset C(0, T; V)$ is continuous, we have $u_\tau \to u$ weakly in $C(0, T; V)$, so also $u_\tau^0 = u_\tau(0) \to u(0)$ weakly in V. Since by the hypothesis $u_\tau^0 \to u_0$ weakly in V, we conclude that $u(0) = u_0$. Similarly, since $w_\tau \to w$ weakly in $\mathcal{V}$ and $w'_\tau \to w'$ weakly in $\mathcal{V}^*$ and the embedding $\mathcal{W} \subset C(0, T; H)$ is continuous, we have $w_\tau \to w$ weakly in $C(0, T; H)$, so also $w_\tau^0 = w_\tau(0) \to w(0)$ weakly in H. On the other hand, we assume that $w_\tau^0 \to w_0$ strongly in H so we conclude that $w(0) = w_0$. Concerning the above initial conditions together with (5.45), (5.57), (5.58), Remark 5.14 and (5.59) we claim that the triple (u, w, η) solves Problem $(P^m)_*$. Moreover, the convergences required in the thesis, follow directly from (5.38)–(5.44). This completes the proof of the theorem. □

5.4.3 *Uniqueness and Strong Convergence*

In this section we study a problem of uniqueness of solution to the problem (P^2) and a strong convergence of the functions built on the solution of the Rothe problem to the exact one. To this end, we impose more restrictive assumptions on the operator A and the functional J.

$H(A)_1$: $H(A)$ holds and $\langle Av_1 - Av_2, v_1 - v_2\rangle \geq m_1\|v_1 - v_2\|^2 - m_2|v_1 - v_2|^2$ for all $v_1, v_2 \in V$ with $m_1 \geq 0, m_2 \geq 0$,

$H(A)_2$: $H(A)$ holds and the Nemytskii mapping $\mathcal{A}: \mathcal{V} \to \mathcal{V}^*$ is of class (S_+) with respect to the space $M^{2,2}(0, T; V, V^*)$, i.e. if $v_n \to v$ weakly in $\mathcal{V}$, $\limsup_{n\to\infty} \langle \mathcal{A}v_n, v_n - v\rangle_{\mathcal{V}^* \times \mathcal{V}} \leq 0$ and v_n is bounded in $M^{2,2}(0, T; V, V^*)$ then $v_n \to v$ strongly in $\mathcal{V}$,

$H(J)_1$: $H(J)$ holds and $\langle \eta_1 - \eta_2, z_1 - z_2 \rangle_{U^* \times U} \geq -m_3 \|z_1 - z_2\|_U^2$ for all $z_1, z_2 \in U$ and $\eta_i \in \partial J(z_i)$, $i = 1, 2$, with $m_3 \geq 0$,

H_{const}: either H_{aux} holds or $m_1 \geq m_3 \|\iota\|^2$.

Remark 5.20. By Proposition 5.11 and compactness of embedding $V \subset H$, it is easy to observe that $H(A)_1$ is stronger then $H(A)_2$, namely, $H(A)_1$ implies $H(A)_2$.

Now we give a theorem on the uniqueness of solution to the problem $(P^2)_*$.

Theorem 5.21. *Let assumptions $H(A)_1, H(B), H(J)_1, H_0$ and H_{const} hold, and (u_1, w_1, η_1), (u_2, w_2, η_2) be two solutions of problem $(P^2)_*$. Then $u_1 = u_2$ and $w_1 = w_2$.*

Proof. Let (u_1, w_1, η_1), (u_2, w_2, η_2) be two solutions of problem $(P^2)_*$. Then we have

$$\langle (w_1' - w_2')(t), v \rangle + \langle Aw_1(t) - Aw_2(t), v \rangle + \langle Bu_1(t) - Bu_2(t), v \rangle$$
$$+ \langle \eta_1(t) - \eta_2(t), \iota v \rangle_{U^* \times U} = 0 \quad \text{for all } v \in V, \text{ a.e. } t \in (0, T). \tag{5.60}$$

Taking $v = w_1(t) - w_2(t)$ in (5.60), we get

$$\frac{1}{2} \frac{d}{dt} |w_1(t) - w_2(t)|^2 + \langle Aw_1(t) - Aw_2(t), w_1(t) - w_2(t) \rangle$$
$$+ \frac{1}{2} \frac{d}{dt} \langle Bu_1(t) - Bu_2(t), u_1(t) - u_2(t) \rangle$$
$$+ \langle \eta_1(t) - \eta_2(t), \iota w_1(t) - \iota w_2(t) \rangle_{U^* \times U} = 0. \tag{5.61}$$

From $H(A)_1$ and $H(J)_1$, we have

$$\frac{1}{2} \frac{d}{dt} |w_1(t) - w_2(t)|^2 + m_1 \|w_1(t) - w_2(t)\|^2 - m_2 |w_1(t) - w_2(t)|^2$$
$$+ \frac{1}{2} \frac{d}{dt} \langle Bu_1(t) - Bu_2(t), u_1(t) - u_2(t) \rangle - m_3 \|\iota w_1(t) - \iota w_2(t)\|_U \leq 0. \tag{5.62}$$

Integrating (5.62) from 0 to t, for all $t \in (0, T)$ and using $H(B)$, we obtain

$$|w_1(t) - w_2(t)|^2 + 2(m_1 - m_3 \|\iota\|^2) \|w_1 - w_2\|_{\mathcal{V}}^2 \leq 2m_2 \int_0^t |w_1(s) - w_2(s)|^2 \, ds$$

or, if in addition H_{aux} holds,

$$|w_1(t) - w_2(t)|^2 + 2(m_1 - \varepsilon) \|w_1 - w_2\|_{\mathcal{V}}^2$$
$$\leq 2(m_2 + C(\varepsilon)) \int_0^t |w_1(s) - w_2(s)|^2 \, ds.$$

From H_{const} we conclude that there exists a constant $C > 0$ such that

$$|w_1(t) - w_2(t)|^2 \le C \int_0^t |w_1(s) - w_2(s)|^2 \, ds,$$

which by the Gronwall lemma gives $w_1(t) = w_2(t)$ for all $t \in (0, T)$. Since $u_1(t) - u_2(t) = \int_0^t (w_1(s) - w_2(s)) \, ds$, we also get $u_1(t) = u_2(t)$ for all $t \in (0, T)$, which completes the proof. □

Theorem 5.22. *Let u be a solution of Problem $(P^1)_*$ or $(P^2)_*$ and u_τ be its Rothe approximation defined in Sect. 5.4.2. If assumptions $H(A)_2$, $H(B)$, $H(J)$, H_0 and H_{aux} hold, and $u_\tau^0 \to u_0$ strongly in V as $\tau \to 0$, then $u_\tau \to u$ strongly in $C(0, T; V)$ as $\tau \to 0$.*

Proof. Let $u_\tau, \overline{u}_\tau, w_\tau, \overline{w}_\tau$ and $\overline{\eta}_\tau$ be functions built on the solutions of the Rothe problem (P_τ^m) and let (u, w, η) be a solution of the problem (P^m) obtained in Theorem 5.19. Then, for all $v \in V$ and a.e. $t \in (0, T)$, we have

$$\langle w_\tau'(t) - w'(t), v\rangle + \langle A\overline{w}_\tau(t) - Aw(t), v\rangle + \langle B\overline{u}_\tau(t) - Bu(t), v\rangle$$
$$+ \langle \overline{\eta}_\tau(t) - \eta(t), \iota v\rangle_{U^* \times U} = \langle f_\tau(t) - f(t), v\rangle, \tag{5.63}$$

where $\overline{\eta}_\tau(t) \in \partial J(\iota z_\tau(t))$ and $\eta(t) \in \partial J(\iota z(t))$ for a.e. $t \in (0, T)$ and we choose $z_\tau = u_\tau$, $z = u$ in case $m = 1$ and $z_\tau = w_\tau$, $z = w$ in case $m = 2$. Taking $v = \overline{w}_\tau(t) - w(t)$ in (5.63) for a.e. $t \in (0, T)$ we get

$$\frac{1}{2}\frac{d}{dt}|w_\tau(t) - w(t)|^2 + \langle w_\tau'(t) - w'(t), \overline{w}_\tau(t) - w_\tau(t)\rangle$$
$$+ \langle A\overline{w}_\tau(t) - Aw(t), \overline{w}_\tau(t) - w(t)\rangle + \frac{1}{2}\frac{d}{dt}\langle Bu_\tau(t) - Bu(t), u_\tau - u(t)\rangle$$
$$\langle B\overline{u}_\tau(t) - Bu_\tau(t), \overline{w}_\tau(t) - w(t)\rangle$$
$$\le \langle f_\tau(t) - f(t), \overline{w}_\tau(t) - w(t)\rangle$$
$$+ \langle \overline{\eta}_\tau(t) - \eta(t), \iota\overline{w}_\tau(t) - \iota w(t)\rangle_{U^* \times U}. \tag{5.64}$$

Since $\langle w_\tau'(t), \overline{w}_\tau(t) - w_\tau(t)\rangle = |w_\tau'(t)|^2(k\tau - t) \ge 0$ for all $t \in ((k-1)\tau, k\tau)$, we get for a.e. $t \in (0, T)$

$$\frac{1}{2}\frac{d}{dt}|w_\tau(t) - w(t)|^2 + \langle A\overline{w}_\tau(t) - Aw(t), \overline{w}_\tau(t) - w(t)\rangle$$
$$+ \frac{1}{2}\frac{d}{dt}\langle Bu_\tau(t) - Bu(t), u_\tau(t) - u(t)\rangle$$
$$\le \langle f_\tau(t) - f(t), \overline{w}_\tau(t) - w(t)\rangle + \langle \overline{\eta}_\tau(t) - \eta(t), \iota\overline{w}_\tau(t) - \iota w(t)\rangle_{U^* \times U}$$
$$+ \langle w'(t), \overline{w}_\tau(t) - w_\tau(t)\rangle + \langle Bu_\tau(t) - B\overline{u}_\tau(t), \overline{w}_\tau(t) - w(t)\rangle. \tag{5.65}$$

Integrating (5.65) and using $H(B)$, we get

$$
\begin{aligned}
&|w_\tau(t) - w(t)|^2 + \langle \mathcal{A}\overline{w}_\tau, \overline{w}_\tau - w\rangle \\
&\quad \le \langle \mathcal{A}w, \overline{w}_\tau - w\rangle + \|B\|_{\mathcal{L}(V,V^*)}\|u_\tau^0 - u_0\|^2 + 2\langle f_\tau - f, \overline{w}_\tau - w\rangle_{\mathcal{V}^*\times\mathcal{V}} \\
&\qquad + 2\langle \overline{\eta}_\tau - \eta, \overline{\iota}\,\overline{w}_\tau - \overline{\iota}\,w\rangle_{\mathcal{U}^*\times\mathcal{U}} + 2\langle w', \overline{w}_\tau - w_\tau\rangle_{\mathcal{V}^*\times\mathcal{V}} \\
&\qquad + 2\langle \mathcal{B}u_\tau - \mathcal{B}\overline{u}_\tau, \overline{w}_\tau - w\rangle_{\mathcal{V}^*\times\mathcal{V}} + |u_\tau^1 - u_1|^2. \qquad (5.66)
\end{aligned}
$$

We remind that $u_\tau^0 \to u_0$ strongly in V, $f_\tau \to f$ strongly in $\mathcal{V}^*$, $\overline{w}_\tau \to w$ weakly in $\mathcal{V}$ and from $H(\iota)$ also $\overline{\iota}\,\overline{w}_\tau \to \overline{\iota}\,w$ strongly in $\mathcal{U}$. Also $\overline{\eta}_\tau \to \eta$ weakly in $\mathcal{U}^*$ and $\overline{w}_\tau \to w_\tau$ weakly in $\mathcal{V}$. Moreover, $\overline{u}_\tau \to u_\tau$ strongly in $\mathcal{V}$ and, from the continuity of $\mathcal{B}$, also $\mathcal{B}\overline{u}_\tau \to \mathcal{B}u_\tau$ strongly in $\mathcal{V}^*$. Finally $u_\tau^1 \to u_1$ strongly in H. From the above convergences, we conclude that the right hand side of (5.66) converges to 0 as $\tau \to 0$. It follows that $\limsup_{\tau\to 0}\langle \mathcal{A}\overline{w}_\tau, \overline{w}_\tau - w\rangle \le 0$. Thus, since $\overline{w}_\tau \to w$ weakly in $\mathcal{V}$ and (5.36) holds, we conclude from $H(A)_2$, that $\overline{w}_\tau \to w$ strongly in $\mathcal{V}$. Now, for every $t \in (0, T)$, using Hölder inequality we obtain

$$
\begin{aligned}
\|u_\tau(t) - u(t)\| &\le \|u_\tau^0 - u_0\| + \int_0^t \|\overline{w}_\tau(t) - w(t)\|\, ds \\
&\le \|u_\tau^0 - u_0\| + \sqrt{T}\|\overline{w}_\tau - w\|_{\mathcal{V}}^2. \qquad (5.67)
\end{aligned}
$$

From (5.67), we get $u_\tau \to u$ strongly in $C(0, T; V)$, as $\tau \to 0$ which completes the proof of the theorem. □

5.5 Faedo-Galerkin Method

In this section we present the results obtained in [2] on the second order HVIs. In that case only a dependence of subdifferential on the velocity is considered. Moreover, the operator A is allowed to depend on time explicitly. Keeping the notation introduced in Sect. 5.3, we define following problem.

$$
(P^3) \qquad
\begin{cases}
\text{Find } (u, \eta) \in \mathcal{V} \times \mathcal{U} \text{ such that } u' \in \mathcal{W}, \\
u''(t) + A(t, u'(t)) + Bu(t) + \iota^*\eta(t) = f(t) & \text{for a.e. } t \in (0, T), \\
\eta(t) \in \partial J(\iota u'(t)) & \text{for a.e. } t \in (0, T), \\
u(0) = u_0, \quad u'(0) = u_1.
\end{cases}
$$

We impose assumptions on the operator A.

$H(A)_3$: $A : (0, T) \times V \to V^*$ is such that

(i) $A(\cdot, v)$ is measurable on $(0, T)$ for all $v \in V$;
(ii) $A(t, \cdot)$ is pseudomonotone for a.e. $t \in (0, T)$;
(iii) $\|A(t, v)\|_{V^*} \le a(t) + b\|v\|$ for all $v \in V$ with $a \in L^2(0, T), a \ge 0, b > 0$;
(iv) $\langle A(t, v), v\rangle \ge \alpha\|v\|^2$ for all $v \in V$, a.e. $t \in (0, T)$ with $\alpha > 0$;
(v) $\langle A(t, v_1) - A(t, v_2), v_1 - v_2\rangle_{V^*\times V} \ge m_A\|v_1 - v_2\|^2$ for all $v_1, v_2 \in V$ a.e. $t \in (0, T)$ with $m_A > 0$;
(vi) $\|A(t, v_1) - A(t, v_2)\|_{V^*} \le L_A\|v_1 - v_2\|$ for all $v_1, v_2 \in V$ a.e. $t \in (0, T)$ with $L_A > 0$.

Now we introduce the Faedo-Galerkin approximation for Problem (P^3) and formulate a result on the semidiscrete error estimates.

Let V^h be a finite dimensional linear subspace of V equipped with the norm of V, where $h > 0$ denotes the spatial discretization parameter. We use the projection operator $\Pi^h : V \to V^h$ defined by the relation

$$(v - \Pi^h v, v^h)_H = 0 \quad \text{for all } v^h \in V^h. \tag{5.68}$$

The semidiscrete approximation of Problem (P^3) has the following formulation.

$$(P_h^3) \quad \begin{cases} \text{Find } (u^h, \eta^h) \in L^2(0, T, V^h) \times \mathcal{U} \text{ such that} \\ u^{h\,\prime\prime}(t) + A(t, u^{h\,\prime}(t)) + Bu^h(t) + \iota^*\eta^h(t) = f(t) & \text{for a.e. } t \in (0, T), \\ \eta^h(t) \in \partial J(\iota u^{h\,\prime}(t)) & \text{for a.e. } t \in (0, T), \\ u(0) = u_0^h, \quad u'(0) = u_1^h, \end{cases}$$

where $u_0^h = \Pi^h u_0$ and $u_1^h = \Pi^h u_1$.

Now we formulate a theorem on existence and uniqueness of solution for Problems (P^3) and (P_h^3). The proof of the theorem follows the line of the proof of Theorem 5.15 in [13].

Theorem 5.23. *If the assumptions $H(A)_1, H(B), H(J)_1, H(\iota)$ and H_0 hold, and*

$$\alpha > 2c\|\iota\|^2, \tag{5.69}$$

$$m_A > m_3\|\iota\|^2, \tag{5.70}$$

then Problems (P^3) and (P_h^3) have unique solutions.

Next we provide a result on the error estimates between the solutions of Problems (P^3) and (P_h^3).

Theorem 5.24. *Assume that $H(A)_1$, $H(B)$, $H(J)_1$, $H(f)$, H_0 and (5.70) hold. Let u and u^h be solutions of Problems (P^3) and (P_h^3) respectively and $v^h \in L^2(0, T; V^h) \cap \mathcal{W}$. Then there exists a positive constant $\mathcal{M}$ dependent only on the data of the problem, such that*

$$
\|u-u^h\|^2_{C(0,T;V)} + \|u'-u^{h\prime}\|^2_{C(0,T;H)} + \|u'-u^{h\prime}\|^2_{\mathcal{V}}
$$
$$
\leq \mathcal{M}(\|u_0-u_0^h\|^2 + |u_1-u_1^h|\,|u_1-v^h(0)|
$$
$$
+ \|u'-v^h\|^2_{\mathcal{V}} + \|u''-v^{h\prime}\|^2_{\mathcal{H}} + \|\iota u'-\iota v^h\|_{L^2(0,T;U)}). \tag{5.71}
$$

Now we introduce a fully discrete scheme corresponding to Problem (P^3) and analyze the error of fully discrete approximation. First, we impose some additional hypothesis on the data of the problem.

$H(A)_4$: $H(A)_3$ hold and $A(\cdot,v)\in C(0,T;V^*)$ for all $v\in V$.
$H(f)_1$: $f\in C(0,T;V^*)$.

We define a uniform partition of $[0,T]$ denoted by $0=t_0<t_1<\ldots<t_N=T$. Let $k=T/N$ be a time step size and for a continuous function g we denote $g_n=g(t_n)$. Finally for a sequence $\{z_n\}_{n=0}^N$ we denote by $\delta z_n=(z_n-z_{n-1})/k$ for $n=1,\ldots,N$ the divided difference. Thus using the backward Euler scheme the fully discrete approximation of Problem (P^3) is the following.

$$
(P^3_{hk})\quad \begin{cases} \text{Find } \{u_n^{hk}\}_{n=0}^N, \{w_n^{hk}\}_{n=0}^N \subset V^h \text{ and } \{\xi_n^{hk}\}_{n=0}^N\subset U \text{ such that} \\ w_n^{hk}=\delta u_n^{hk} & \text{for } n=1,\ldots,N, \\ \langle \delta w_n^{hk} + A(t_n,w_n^{hk}) + Bu_n^{hk} - f_n, v^h\rangle & \\ = \langle \xi_n^{hk}, \iota v^h\rangle_{U^*\times U} & \text{for all } v^h\in V^h, \\ -\xi_n^{hk}\in \partial J(\iota w_n^{hk}) & \text{for } n=0,\ldots,N, \\ u_0^{hk}=u_0^h,\ \ w_0^{hk}=u_1^h. \end{cases}
$$

Note that from $H(A)_4$ and $H(f)_1$ the values $A(t_n,w_n^{hk})$ and f_n in Problem (P^3_{kh}) are well defined.

Theorem 5.25. *Assume that $H(A)_4$, $H(B)$, $H(J)_1$, $H(f)_1$, H_0 and (5.70) hold. Let (u,w,ξ) be a solution of Problem (P^3), which satisfy the additional regularity assumptions*

$$
u\in C^2(0,T;H)\cap C^1(0,T;V), \tag{5.72}
$$
$$
\xi\in C(0,T;U). \tag{5.73}
$$

Let $\{u_n^{hk}\}_{n=0}^N, \{w_n^{hk}\}_{n=0}^N$ and $\{\xi_n^{hk}\}_{n=0}^N$ be the solution of Problem (P^3_{hk}). Then the following estimate holds, for all $\{v_j^h\}_{j=1}^N\subset V^h$:

$$
\max_{1\leq n\leq N}\{|w_n-w_n^{hk}|^2 + \sum_{j=1}^n k\|w_j-w_j^{hk}\|^2\}
$$
$$
\leq k\sum_{j=1}^N\left(|w_j'-\delta w_j|^2 + \|w_j-v_j^h\|^2\right) + c\max_{1\leq n\leq N}\|\iota w_{\tau n}-\iota v_{\tau n}^h\|_U
$$

$$+\frac{1}{k}\sum_{j=1}^{N-1}|w_j - v_j^h - (w_{j+1} - v_{j+1}^h)|^2 + \max_{1\le n\le N}\|w_n - v_n^h\|_H^2$$

$$+\|u_0 - u_0^h\|^2 + k^2\|u\|_{H^2(0,T;V)}^2 + |w_0 - w_0^h|^2. \tag{5.74}$$

5.6 Applications

In this section we show an application of our results to a mechanical contact problem. We denote by $\mathbb{S}^d$ the space of second order symmetric tensors on $\mathbb{R}^d$ ($d \le 3$), and use "$\cdot$", $\|\cdot\|_{\mathbb{R}^d}$ and $\|\cdot\|_{\mathbb{S}^d}$ for the inner product and the Euclidean norm on $\mathbb{R}^d$ and $\mathbb{S}^d$, respectively, i.e.,

$$u\cdot v = u_i v_i, \quad \|v\|_{\mathbb{R}^d} = (v\cdot v)^{\frac{1}{2}} \quad \text{for all } u, v\in\mathbb{R}^d,$$

$$\sigma\cdot\tau = \sigma_{ij}\tau_{ij}, \quad \|\tau\|_{\mathbb{S}^d} = (\tau\cdot\tau)^{\frac{1}{2}} \quad \text{for all } \sigma,\tau\in\mathbb{S}^d.$$

Here and below the indices i and j run between 1 and d, and the summation convention over repeated indices is adopted.

Let $\Omega\subset\mathbb{R}^d$ be a bounded domain with a Lipschitz boundary Γ and let ν denote the unit outer normal on Γ. We introduce the following function spaces:

$$H = L^2(\Omega)^d = \{\, u = (u_i) \mid u_i\in L^2(\Omega)\,\},\quad H_1 = \{\, u = (u_i) \mid \varepsilon(u)\in Q\,\},$$

$$Q = \{\,\sigma = (\sigma_{ij}) \mid \sigma_{ij} = \sigma_{ji}\in L^2(\Omega)\,\},\qquad Q_1 = \{\,\sigma\in Q \mid \mathrm{Div}\,\sigma\in H\,\}.$$

Here $\varepsilon : H_1 \to Q$ and $\mathrm{Div} : Q_1 \to H$ are the *deformation* and *divergence* operators, respectively, defined by

$$\varepsilon(u) = (\varepsilon_{ij}(u)),\quad \varepsilon_{ij}(u) = \frac{1}{2}(u_{i,j} + u_{j,i}),\quad \mathrm{Div}\,\sigma = (\sigma_{ij,j}),$$

where the index following a comma indicates the partial derivative with respect to the corresponding component of the independent variable. The spaces H, Q, H_1 and Q_1 are real Hilbert spaces endowed with the canonical inner products given by

$$(u,v)_H = \int_\Omega u_i v_i\,dx,\qquad (\sigma,\tau)_Q = \int_\Omega \sigma_{ij}\tau_{ij}\,dx,$$

$$(u,v)_{H_1} = (u,v)_H + (\varepsilon(u),\varepsilon(v))_Q,\quad (\sigma,\tau)_{Q_1} = (\sigma,\tau)_Q + (\mathrm{Div}\sigma,\mathrm{Div}\tau)_H.$$

The associated norms on these spaces are denoted by $\|\cdot\|_H$, $\|\cdot\|_Q$, $\|\cdot\|_{H_1}$ and $\|\cdot\|_{Q_1}$, respectively. Let $H_\Gamma = H^{1/2}(\Gamma)^d$ and let $\overline{\gamma} : H_1 \to H_\Gamma$ be the trace map. For every element $v\in H_1$ we still write v for the trace $\overline{\gamma}v$ of v on Γ and we denote by v_ν and v_τ the *normal* and *tangential* components of v on the boundary Γ given by

$$v_\nu = v \cdot \nu, \quad v_\tau = v - v_\nu \nu. \tag{5.75}$$

Let H_Γ^* be the dual of H_Γ and let $\langle \cdot, \cdot \rangle$ denote the duality pairing between H_Γ^* and H_Γ. For every $\sigma \in Q_1$ there exists an element $\sigma\nu \in H_\Gamma^*$ such that

$$(\sigma, \varepsilon(v))_Q + (\mathrm{Div}\sigma,\ v)_H = \langle \sigma\nu, \overline{\gamma} v \rangle_{H_\Gamma^* \times H_\Gamma} \qquad \text{for all } \ v \in H_1. \tag{5.76}$$

Moreover, if σ is a smooth (say C^1) function, then

$$\langle \sigma\nu, \overline{\gamma} v \rangle_{H_\Gamma^* \times H_\Gamma} = \int_\Gamma \sigma\nu \cdot v \, d\,\Gamma \qquad \text{for all } \ v \in H_1. \tag{5.77}$$

We also denote by σ_ν and σ_τ the *normal* and *tangential* traces of σ and we recall that, when σ is smooth enough, then

$$\sigma_\nu = (\sigma\nu) \cdot \nu, \quad \sigma_\tau = \sigma\nu - \sigma_\nu \nu. \tag{5.78}$$

Now we pass to the description of the mechanical problem. A visco-elastic body occupies an open bounded connected set $\Omega \subset \mathbb{R}^d$ with a Lipschitz boundary Γ that is partitioned into three disjoint parts $\overline{\Gamma}_1$, $\overline{\Gamma}_2$ and $\overline{\Gamma}_3$ with Γ_1, Γ_2 and Γ_3 being relatively open, and meas $(\Gamma_1) > 0$. Let $[0, T]$ be a time interval of interest, $T > 0$. We assume that the body is clamped on Γ_1 and thus the displacement field vanishes there. A volume force of density f_0 acts in Ω and a surface traction of density f_2 acts on Γ_2. The body is in frictional contact with an obstacle on Γ_3. We assume that there is no loss of contact during the process, i.e., the contact is bilateral. Thus, the normal displacement u_ν vanishes on Γ_3. We model the friction by a nonmonotone friction law and the process is assumed to be dynamic.

The classical formulation of the mechanical problem is the following.

Problem P_M**.** *Find a displacement* $u\colon \Omega \times [0,T] \to \mathbb{R}^d$ *and a stress field* $\sigma\colon \Omega \times [0,T] \to \mathbb{S}^d$ *such that*

$$\sigma = \mathcal{C}\varepsilon(u') + \mathcal{G}\varepsilon(u) \qquad \text{in } \Omega \times (0,T), \tag{5.79}$$

$$\rho u'' = Div\, \sigma + f_0 \qquad \text{in } \Omega \times (0,T), \tag{5.80}$$

$$u = 0 \qquad \text{on } \Gamma_1 \times (0,T), \tag{5.81}$$

$$\sigma\nu = f_2 \qquad \text{on } \Gamma_2 \times (0,T), \tag{5.82}$$

$$u_\nu = 0 \qquad \text{on } \Gamma_3 \times (0,T), \tag{5.83}$$

$$\left.\begin{array}{ll}\|\sigma_\tau\|_{\mathbb{R}^d} \le \mu(0)S & \text{if} \quad u'_\tau = 0 \\ -\sigma_\tau = \mu(\|u'_\tau\|_{\mathbb{R}^d})S\, u'_\tau/\|u'_\tau\|_{\mathbb{R}^d} & \text{if} \quad u'_\tau \neq 0\end{array}\right\} \quad \text{on } \Gamma_3 \times (0,T), \tag{5.84}$$

$$u(0) = u_0, \quad u'(0) = u_1 \qquad \text{in } \Omega. \tag{5.85}$$

In (5.84), $\mu(\|u'_\tau\|_{\mathbb{R}^d})S$ represents the magnitude of the limiting friction traction at which slip begins. Here, $S \ge 0$ is a given constant. The friction bound and more precisely, the friction coefficient μ, depends on the magnitude of the slip, $\|u'_\tau\|_{\mathbb{R}^d}$. The strict inequality in (5.84) holds in the stick zone and the equality holds in the slip zone. Due to the basic properties of the Clarke subdifferential, the friction condition (5.84) can be written as a subdifferential inclusion involving a locally Lipshitz, possibly nonconvex superpotential j which depends on the tangential velocity u'_τ. In fact, if the function $j : \mathbb{R}^d \to \mathbb{R}$ is defined by

$$j(\xi) = S \int_0^{\|\xi\|_{\mathbb{R}^d}} \mu(s)\, ds, \quad \text{for all } \xi \in \mathbb{R}^d, \tag{5.86}$$

then we can prove that under assumptions $H(\mu)(a), (b)$, the condition (5.84) is equivalent to the following subdifferential inclusion

$$-\sigma_\tau \in \partial j(u'_\tau) \quad \text{on } \Gamma_3 \times (0,T). \tag{5.87}$$

In the study of the frictional contact problem we need the following assumptions on its data.

$H(\mathcal{C})$: the viscosity operator $\mathcal{C} : \Omega \times [0,T] \times \mathbb{S}^d \to \mathbb{S}^d$ satisfies

$$\left\{\begin{array}{l}
(a)\ \mathcal{C}(\cdot,\cdot,\varepsilon) \text{ is measurable on } \Omega \times [0,T] \text{ for all } \varepsilon \in \mathbb{S}^d; \\
(b)\ \mathcal{C}(x,t,\cdot) \text{ is continuous on } \mathbb{S}^d \text{ for a.e. } (x,t) \in \Omega \times [0,T]; \\
(c)\ \|\mathcal{C}(x,t,\varepsilon)\|_{\mathbb{S}^d} \le a_0(x,t) + a_1\|\varepsilon\|_{\mathbb{S}^d} \text{ for all } \varepsilon \in \mathbb{S}^d, \\
\qquad \text{a.e. } (x,t) \in \Omega \times [0,T] \text{ with } a_0 \in L^2(\Omega \times [0,T]),\ a_0 \ge 0 \text{ and } a_1 > 0; \\
(d)\ \mathcal{C}(x,t,\varepsilon):\varepsilon \ge \alpha\|\varepsilon\|^2_{\mathbb{S}^d} \text{ for all } \varepsilon \in \mathbb{S}^d,\ \text{a.e. } (x,t) \in \Omega\times[0,T] \text{ with } \alpha>0; \\
(e)\ (\mathcal{C}(x,t,\varepsilon_1) - \mathcal{C}(x,t,\varepsilon_2)) : (\varepsilon_1 - \varepsilon_2) \ge m_{\mathcal{C}}\|\varepsilon_1 - \varepsilon_2\|^2_{\mathbb{S}^d} \\
\qquad \text{for all } \varepsilon_1, \varepsilon_2 \in \mathbb{S}^d,\ \text{a.e. } (x,t) \in \Omega \times [0,T] \text{ with } m_{\mathcal{A}} > 0; \\
(f)\ \|\mathcal{C}(x,t,\varepsilon_1) - \mathcal{C}(x,t,\varepsilon_2)\|_{\mathbb{S}^d} \le L_{\mathcal{C}}\|\varepsilon_1 - \varepsilon_2\|_{\mathbb{S}^d} \\
\qquad \text{for all } \varepsilon_1, \varepsilon_2 \in \mathbb{S}^d,\ \text{a.e. } (x,t) \in \Omega \times [0,T] \text{ with } L_{\mathcal{C}} > 0; \\
(g)\ \mathcal{C}(x,\cdot,\varepsilon) \text{ is continuous on } [0,T] \text{ for all } x \in \Omega, \varepsilon \in \mathbb{S}^d.
\end{array}\right.$$

$H(\mathcal{G})$: the elasticity operator $\mathcal{G}: \Omega \times \mathbb{S}^d \to \mathbb{S}^d$ is a bounded symmetric nonnegative definite fourth order tensor, i.e.

$$\begin{cases} (a)\ \mathcal{G}_{ijkl} \in L^\infty(\Omega),\ 1 \le i,j,k,l \le d; \\ (b)\ \mathcal{G}\sigma \cdot \tau = \sigma \cdot \mathcal{G}\tau \ \text{ for all } \sigma, \tau \in \mathbb{S}^d, \ \text{ a.e. in } \Omega; \\ (c)\ \mathcal{G}\tau \cdot \tau \ge 0 \ \text{ for all } \tau \in \mathbb{S}^d, \ \text{ a.e. in } \Omega. \end{cases}$$

$H(f)$: the force and the traction densities satisfy

$$f_0 \in L^2(\Omega)^d, \quad f_2 \in L^2(\Gamma_2)^d.$$

$H(\mu)$: the friction bound $\mu: [0,\infty) \to \mathbb{R}$ satisfies

$$\begin{cases} (a)\ \mu \text{ is continuous}; \\ (b)\ |\mu(s)| \le c_\mu(1+s) \ \text{ for all } s \ge 0, \text{ with } c_\mu > 0; \\ (c)\ \mu(s_1) - \mu(s_2) \ge -\lambda(s_1 - s_2) \ \text{ for all } s_1 > s_2 \ge 0 \text{ with } \lambda > 0. \end{cases}$$

We now turn to the variational formulation for Problem P_M. To this end, we introduce a closed subspace of H_1 defined by

$$V = \{v \in H_1 \mid v = 0 \ \text{ on } \ \Gamma_1,\ v_\nu = 0 \ \text{ on } \ \Gamma_3\}$$

with norm defined by $\|v\|_V = \|\varepsilon(v)\|_Q$ for all $v \in V$. Let $U = L^2(\Gamma_3; \mathbb{R}^d)$ and let $\iota = \overline{\gamma} : V \to U$ be the trace operator. Define the operators $A: (0,T) \times V \to V^*$ and $B: V \to V^*$ by

$$\langle A(t,u), v\rangle = (\mathcal{C}(t, \varepsilon(u)), \varepsilon(v))_Q \quad \text{for } u,\ v \in V \text{ and } t \in (0,T), \tag{5.88}$$

$$\langle Bu, v\rangle = (\mathcal{G}\varepsilon(u), \varepsilon(v))_Q \quad \text{for } u,\ v \in V \tag{5.89}$$

and the functional $J: U \to \mathbb{R}$ by

$$J(v) = \int_{\Gamma_3} j(v(x))\, d\Gamma \quad \text{for } v \in U. \tag{5.90}$$

We also consider the function $f : (0,T) \to V^*$ given by

$$\langle f(t), v\rangle = \int_\Omega f_0(t) \cdot v\, dx + \int_{\Gamma_2} f_2(t) \cdot v\, d\Gamma \tag{5.91}$$

for a.e. $t \in (0,T)$ and for all $v \in V$. Let $\mathcal{V}$, $\mathcal{H}$, $\mathcal{U}$ and $\mathcal{W}$ denote the spaces introduced in Sect. 5.3 and let the Nemytskii operators $\mathcal{A}, \mathcal{B} : \mathcal{V} \to \mathcal{V}^*$ and

$\bar{\iota} : \mathcal{V} \to \mathcal{U}$ be defined by $(\mathcal{A}v)(t) = A(t, v(t))$, $(\mathcal{B}v)(t) = B(v(t))$ and $(\bar{\iota}v)(t) = \iota v(t)$ for all $v \in \mathcal{V}$. Then, we see that as a weak formulation of Problem P_M, we obtain Problem (P^3) formulated in Sect. 5.6. Moreover if the operator A does not depend explicitly on time, namely $(\mathcal{A}v)(t) = A(v(t))$, then Problem (P^3) is equivalent with Problem (P^2) introduced in Sect. 5.3. The following lemmas allow to apply theorems formulated in that chapter to Problem P_M.

Lemma 5.26. *If operator $\mathcal{C}$ satisfies assumptions $H(\mathcal{C})$ (a)–(f) then the operator A satisfies assumptions $H(A)_3$ with the function $a(t) = \sqrt{2}\|a_0(t)\|_{L^2(\Omega)}$ and the constants $b = \sqrt{2a_1}$, $m_A = m_{\mathcal{C}}$ and $L_a = L_{\mathcal{C}}$. If operator $\mathcal{C}$ satisfies assumptions $H(\mathcal{C})$ (a)–(g) then the operator A satisfies assumptions $H(A)_4$-in particular $H(A)_1$ with the function $a(t)$ and the constants defined above. If the operator $\mathcal{C}$ satisfies assumptions $H(\mathcal{C})$ (a)–(e) and moreover it does not depend on time explicitly, namely $\mathcal{C}(x,t,\varepsilon) = \mathcal{C}(x,\varepsilon)$, then the operator A satisfies $H(A)_1$ with the constants $a = \sqrt{2}\|a_0\|_{L^2(\Omega)}$, $b = \sqrt{2a_1}$, $\beta = \gamma = 0$, $m_1 = m_{\mathcal{C}}$ and $m_2 = 0$.*

Lemma 5.27. *If the operator $\mathcal{G}$ satisfies assumptions $H(\mathcal{G})$ then the operator B satisfies assumptions $H(B)$.*

Lemma 5.28. *If the function μ satisfies assumptions $H(\mu)$ then the functional J satisfies assumptions $H(J)_1$ with the constants $c = S\sqrt{2}c_\mu \max\{1, \sqrt{meas_{d-1}(\Gamma_3)}\}$ and $m_3 = S\lambda$.*

Lemma 5.29. *The operator ι satisfies assumption $H(\iota)$.*

Proof. Let $\varepsilon \in (0, \frac{1}{2})$. Then $V \subset H^{1-\varepsilon}(\Omega)$ and the embedding $i : V \to H^{1-\varepsilon}(\Omega)$ is compact. The trace operator $\gamma_1 : H^{1-\varepsilon}(\Omega) \to H^{\frac{1}{2}-\varepsilon}(\Gamma_3)$ is linear and continuous and, finally, the embedding $j : H^{\frac{1}{2}-\varepsilon}(\Gamma_3) \to L^2(\Gamma_3; \mathbb{R}^d) = U$ is also linear and continuous. Thus $\iota = j \circ \gamma_1 \circ i$ is linear, continuous and compact. Moreover the spaces $V \subset H^{1-\varepsilon}(\Omega) \subset V^*$ satisfy assumptions of Proposition 5.11 so the embedding $M^{2,2}(0,T; V, V^*) \subset L^2(0,T; H^{1-\varepsilon}(\Omega))$ is compact. Since the embedding $L^2(0,T; H^{1-\varepsilon}(\Omega)) \subset \mathcal{U}$ is continuous the Nemytskii operator corresponding to ι is compact. □

Lemma 5.30. *The space U satisfies assumption $H(U)$.*

Proof. As in the proof of Lemma 5.29 we take $\varepsilon \in (0, \frac{1}{2})$. Since the embedding $V \subset H^{1-\varepsilon}(\Omega)$ is compact and $H^{1-\varepsilon}(\Omega) \subset H$ is continuous, we can apply the Ehrling Lemma (cf. Lemma 3.1.3 of [18]). Thus for any $\varepsilon > 0$ there is $C(\varepsilon) > 0$ such that for all $v \in V$

$$\|v\|_{H^{1-\varepsilon}(\Omega)} \le \varepsilon\|v\| + C(\varepsilon)|v|. \tag{5.92}$$

Using notation from the proof of Lemma 5.29, we have for all $v \in V$

$$\|\iota v\|_U = \|j \circ \gamma_1 v\|_{H^{1-\varepsilon}(\Omega)} \le c\|v\|_{H^{1-\varepsilon}(\Omega)},$$

with $c > 0$. This together with (5.92) completes the proof. □

From Lemmas 5.26–5.30, we have the following

Corollary 5.31. *If the assumptions $H(\mathcal{C})$, $H(\mathcal{G})$ (a)–(e), $H(\mu)$ and H_0 hold, then the statements of Theorem 5.19 and Theorem 5.21 are valid. Moreover if $H(\mathcal{G})$ (f) holds and*

$$\alpha > 2S\sqrt{2}c_\mu \max\{1, \sqrt{meas_{d-1}(\Gamma_3)}\}\|\iota\|^2$$

and

$$m_{\mathcal{C}} > S\lambda\|\iota\|^2,$$

then so is the statement of Theorem 5.24. Moreover if $H(\mathcal{G})(g)$ and $H(f)_1$ hold then so is the statement of Theorem 5.25.

We note that Theorems 5.24 and 5.25 are the starting points to derive the error rate for the concrete discretization method. In particular let V^h be the space of continuous piecewise affine functions, that is,

$$V^h = \{v^h \in [C(\overline{\Omega})]^d : v^h{}_{|T} \in [P_1(T)]^d \ \forall\, T \in \mathcal{T}^h,\ v^h = 0 \text{ on } \Gamma_1,\ v^h_\nu = 0 \text{ on } \Gamma_3\},$$

where Ω is assumed to be a polygonal domain, $\mathcal{T}^h$ denotes a finite element triangulation of $\overline{\Omega}$, and $P_1(T)$ represents the space of polynomials of total degree less or equal to one in T. Then we have following two corollaries.

Corollary 5.32. *Let assumptions of Theorem 5.24 hold and let u and u^h be solutions of Problems (P^3) and (P^3_h), respectively. Then, under the regularity condition*

$$u, u' \in L^2(0,T;H^2(\Omega;\mathbb{R}^d)), \quad u'_\tau \in L^2(0,T;H^2(\Gamma_3;\mathbb{R}^d)),$$

we have

$$\|u - u^h\|_{C(0,T;V)} + \|\dot{u} - \dot{u}^h\|_{C(0,T;H)} + \|\dot{u} - \dot{u}^h\|_{\mathcal{V}} \le ch, \tag{5.93}$$

with $c > 0$.

Proof. For any $u \in V$ we have the approximation properties

$$\inf_{v^h \in V^h} \|u - v^h\| \le ch\|u\|_{H^2(\Omega;\mathbb{R}^d)},$$

$$\inf_{v^h \in V^h} |u - v^h| \le ch^2\|u\|_{H^2(\Omega;\mathbb{R}^d)},$$

$$\inf_{v^h \in V^h} \|u_\tau - v^h_\tau\|_{L^2(\Gamma_3;\mathbb{R}^d)} \le ch^2\|u_\tau\|_{H^2(\Gamma_3;\mathbb{R}^d)},$$

$$\|\dot{w} - \dot{v}^h\|^2_{\mathcal{V}^*} \le ch^2, \tag{5.94}$$

with $c > 0$. Taking in (5.68) the function v^h defined by $v^h(t) = \Pi^h u(t)$ for all $t \in (0, T)$ and using the inequalities above we obtain (5.93). □

Corollary 5.33. *Let the assumptions of Theorem 5.25 hold and let* (u, w, η) *and* $(\{u_n^{hk}\}_{n=0}^N, \{w_n^{hk}\}_{n=0}^N, \{\xi_n^{hk}\}_{n=0}^N)$ *be solutions of Problem* (P^3) *and* (P^3_{kh}), *respectively. Then, under the regularity conditions*

$$u \in C^1(0, T; H^2(\Omega; \mathbb{R}^d)) \cap H^3(0, T; H), \quad w_\tau \in C(0, T; H^2(\Gamma_3; \mathbb{R}^d)),$$

we have

$$\max_{1 \le n \le N} \{\|u_n - u_n^{hk}\| + |w_n - w_n^{hk}|\} \le c(h + k), \tag{5.95}$$

with $c > 0$.

Proof. We have the following approximation properties of the finite element space V^h:

$$\max_{1 \le n \le N} \inf_{v^h \in V^h} \|u_n - v_n^h\| \le ch\|u\|_{C(0,T;H^2(\Omega;\mathbb{R}^d))},$$

$$\max_{1 \le n \le N} \inf_{v^h \in V^h} |w_n - v_n^h| \le ch^2\|u\|_{C^1(0,T;H^2(\Omega;\mathbb{R}^d))},$$

$$\max_{1 \le n \le N} \inf_{v^h \in V^h} \|w_{n\tau} - v_{n\tau}^h\|_{L^2(\Gamma_3;\mathbb{R}^d)} \le ch^2\|w_\tau\|_{C(0,T;H^2(\Gamma_3;\mathbb{R}^d))},$$

where $c > 0$. Moreover, from the definition of the finite element interpolation operator Π^h, it follows that

$$\|u_0 - u_0^h\| \le ch\|u\|_{C(0,T;H^2(\Omega;\mathbb{R}^d))},$$

$$|w_0 - w_0^h| \le ch^2\|u\|_{C^1(0,T;H^2(\Omega;\mathbb{R}^d))},$$

$$k\sum_{j=1}^N \left(|\dot{w}_j - \delta w_j|^2\right) \le ck^2\|u\|^2_{H^2(0,T;H)}$$

and

$$\frac{1}{k}\sum_{j=1}^{N-1} |w_j - v_j^h - (w_{j+1} - v_{j+1}^h)|^2 \le ch^2\|u\|^2_{H^2(0,T;V)}.$$

Using the above properties we obtain (5.95) from (5.74). □

Acknowledgements This research was supported by the Marie Curie International Research Staff Exchange Scheme Fellowship within the 7th European Community Framework Programme under

Grant Agreement No. 295118, the International Project co-financed by the Ministry of Science and Higher Education of Republic of Poland under grant no. W111/7.PR/2012, the National Science Center of Poland under Maestro Advanced Project no. DEC-2012/06/A/ST1/00262, and the project Polonium "Mathematical and Numerical Analysis for Contact Problems with Friction" 2014/15 between the Jagiellonian University in Krakow and Université de Perpignan Via Domitia and the National Science Center of Poland under Grant no. N N201 604640.

References

1. Aubin, J.P., Frankowska, H.: Set-Valued Analysis. Birkhäuser, Boston/Basel/Berlin (1990)
2. Barboteu, M., Bartosz, K., Han, W., Janiczko, T.: Numerical analysis of a hyperbolic hemivariational inequality arising in dynamic contact. Siam J. Numer. Anal., to be published
3. Barboteu, M., Fernández, J.R., Tarraf, R.: Numerical analysis of a dynamic piezoelectric contact problem arising in viscoelasticity. Comput. Methods Appl. Mech. Eng. **197**, 3724–3732 (2008)
4. Bartosz, K., Cheng, X., Kalita, P., Yu, Y., Zheng, C.: Rothe method for evolution variational-hemivariational inequalities. J. Math. Anal. Appl. **423**(II), 841–862 (2015)
5. Carstensen, C., Gwinner, J.: A theory of discretisation for nonlinear evolution inequalities applied to parabolic Signorini problems. Annali di Matematica (IV) **177**, 363–394 (1999)
6. Clarke, F.H.: Optimization and Nonsmooth Analysis. Wiley-Interscience, New York (1983)
7. Denkowski, Z., Migórski, S., Papageorgiou, N.S.: An Introduction to Nonlinear Analysis, Applications. Kluwer, Dordecht (2003)
8. Haslinger, J., Miettinen, M., Panagiotopoulos, P.D.: Finite Element Method for Hemivariational Inequalities. Theory, Methods and Applications. Kluwer Academic Publishers, Boston/Dordrecht/London (1999)
9. Kalita, P.: Regularity and Rothe method error estimates for parabolic hemivariational inequality. J. Math. Anal. Appl. **389**, 618–631 (2012)
10. Kalita, P.: Convergence of Rothe scheme for hemivariational inequalities of parabolic type. Int. J. Numer. Anal. **10**, 445–465 (2013)
11. Kalita, P.: Semidiscrete variable time-step θ-scheme for nonmonotone evolution inclusion. (see http://arxiv.org/abs/1402.3721).
12. Miettinen, M., Haslinger, J.: Finite element approximation of parabolic hemivariational inequalities. Numer. Funct. Anal. Optim. **19** (5 and 6), 565–585 (1998)
13. Migórski, S., Ochal, A., Sofonea, M.: Nonlinear Inclusions and Hemivariational Inequalities. Models and Analysis of Contact Problems. Advances in Mechanics and Mathematics, vol. 26. Springer, New York (2013)
14. Peng, Z., Liu, Z.: Evolution hemivariational inequality problems with doubly nonlinear operators. J. Global Optim. **51**, 413–427 (2011)
15. Peng, Z., Liu, Z., Liu, X.; Boundary hemivariational inequality problems with doubly nonlinear operators. Math. Ann. **356**, 1339–1358 (2013)
16. Zeidler, E.: Nonlinear Functional Analysis and its Applications, vol. II/A: Linear Monotone Operators. Springer, New York/Berlin/Heidelberg (1990)
17. Zeidler, E.: Nonlinear Functional Analysis and its Applications, vol. II/B: Nonlinear Monotone Operators. Springer, New York/Berlin/Heidelberg (1990)
18. Zheng, S.: Nonlinear Evolution Equations. Chapman & Hall/CRC, London (2004)

Chapter 6
Some Extragradient Algorithms for Variational Inequalities

Changjie Fang and Shenglan Chen

Abstract We present some extragradient algorithms for solving variational inequalities including classical variational inequality, multivalued variational inequality and general variational inequality. The global convergence of the proposed method is established, provided the mapping is continuous and pseudomonotone. Preliminary computational experience is also reported.

Keywords Variational inequality • Single-valued mapping • Multivalued mapping • Extragradient method • Pseudomonotone • Epiconvergence

AMS Classification. 47H04, 47H10, 49J40

6.1 Introduction

It is well known that many problems in nonlinear analysis and optimization can be formulated as the variational inequality problem. Variational inequalities theory has been witnessed to relish an explosive growth in theoretical advances, algorithmic development and applications across all disciplines of pure and applied sciences, see [1, 4, 5, 8–11, 13, 15, 17–20, 22, 23, 25–27, 29, 31] and the references therein.

In 1966, Hartman and Stampacchia [16] introduce the classical variational inequality, denoted by VI(A, K): to find $x^* \in K$ such that

$$\langle A(x^*), y - x^* \rangle \geq 0 \quad \forall\, y \in K, \tag{6.1}$$

where K is a nonempty closed convex set in $\mathbb{R}^d$, $A : K \to \mathbb{R}^d$ is a single-valued mapping, and $\langle \cdot, \cdot \rangle$ and $\|\cdot\|$ denote the inner product and norm in $\mathbb{R}^d$, respectively.

C. Fang (✉) • S. Chen
College of Science, Chongqing University of Posts and Telecommunications, Chongqing 400065, China
e-mail: fangcj@cqupt.edu.cn

W. Han et al. (eds.), *Advances in Variational and Hemivariational Inequalities*, Advances in Mechanics and Mathematics 33, DOI 10.1007/978-3-319-14490-0_6

Projection-type algorithms have been extensively studied in the literature; see [7,8,11,13,18,21,28–30]. An important projection algorithm for solving variational inequality problem (6.1) is the Extragradient Method proposed by Korpelevich [21]; see also [9]. Given the current iterate x_i,calculate

$$y_i = \mathcal{P}_K(x_i - \tau A(x_i)), \tag{6.2}$$

$$x_{i+1} = \mathcal{P}_K(x_i - \tau A(y_i)), \tag{6.3}$$

where τ is a positive parameter. In [21], there is the need to calculate two projections onto K, and convergence is proved under the assumption of Lipschitz continuity and pseudomonotonicity of A. We note that the projection onto a closed convex set K is related to a minimum distance problem. If K is a general closed convex set, this might be computationally expensive. To overcome the difficulty, [27] suggests a projection method for solving the variational inequality problem. In [27], the next iterative point is the projection of the current iteration onto a hyperplane which separates strictly the current iteration from the solution set of the variational inequality problem; see also [17]. Paper [8] proposes the subgradient extragradient algorithm for the single-valued variational inequality in which the Lipschitz continuity assumption is required; see also [7]. Various algorithms for solving the multivalued variational inequality have been extensively studied in the literature [4,5,10–12,14,22–25,30]. The well-known proximal point algorithm [23] requires the multivalued mapping T to be monotone. Li [22] proposes a projection algorithm for solving a multivalued variational inequality with pseudomonotone mapping in which choosing $w_i \in T(x_i)$ requires solving a single-valued variational inequality. Fang [10] presents a double projection algorithm, which is an improvement of the algorithm in [22], so that $w_i \in T(x_i)$ can be taken arbitrarily. In [10], the hyperplane strictly separates the current iterate from the solution set of the variational inequality problem. However, choosing the hyperplane needs computing the supremum in this algorithm, and hence is computationally expensive. To overcome this difficulty, Fang [11] introduces an extragradient algorithm for solving multivalued variational inequality in which computing the supremum is avoided. In this method, the next iterate is a projection onto the feasible set K.

In this chapter, we present some extragradient algorithms for solving variational inequalities. In our method, the current iterate belongs to the set K_i which epiconverge to the feasible set K, and the next iteration is the projection onto the intersection of the hyperplane and the set K_{i+1}. In addition, our Armijo-type line search procedure is also different from those in [10, 11, 14, 17, 22, 27]. We obtain the global convergence of the generalized iteration sequence, assuming that A is pseudomonotone on K with respect to the solution set. We also present numerical results of the proposed method.

This chapter is organized as follows. In Sect. 6.2, we present the algorithm details, some lemmas and preliminary results for convergence analysis for the classical variational inequality. In Sect. 6.3, we suggest the algorithm details for the multivalued variational inequality. The main results for convergence analysis is also provided. Section 6.4, we establish convergence results of algorithms for solving the general variational inequality. Numerical results are reported in the last section.

6.2 The Classical Variational Inequality

In this section, we study the extragradient method for solving the variational inequality problem (6.1). We first recall some basic concepts and lemmas.

A is called pseudomonotone on K in the sense of Karamardian [19], if for any $x, y \in K$,

$$\langle A(y), x - y \rangle \geq 0 \quad \Rightarrow \quad \langle A(x), x - y \rangle \geq 0.$$

Let S_1 be the solution set of (6.1), that is, those points $x^* \in K$ satisfying (6.1). Throughout this chapter, we assume that the solution set S_1 of the problem (6.1) is nonempty and A is pseudomonotone on K with respect to the solution set S_1, i.e.,

$$\langle A(y), y - x \rangle \geq 0 \quad \forall\, y \in K,\ \forall\, x \in S_1. \tag{6.4}$$

The property (6.4) holds if A is pseudomonotone on K.

Let $\mathcal{P}_K$ denote the orthogonal projection onto K and let $\mu > 0$ be a parameter.

Proposition 6.1. *$x \in K$ solves the problem* (6.1) *if and only if*

$$r_\mu(x) := x - \mathcal{P}_K(x - \mu A(x)) = 0.$$

Lemma 6.2 ([32]). *Let K be a closed convex subset of $\mathbb{R}^d$. For any $x, y \in \mathbb{R}^d$ and $z \in K$, the following statements hold:*

(i) $\langle x - \mathcal{P}_K(x), z - \mathcal{P}_K(x) \rangle \leq 0.$
(ii) $\|\mathcal{P}_K(x) - \mathcal{P}_K(y)\|^2 \leq \|x - y\|^2 - \|\mathcal{P}_K(x) - x + y - \mathcal{P}_K(y)\|^2.$

Following [26], we denote by NCCS($\mathbb{R}^d$) the family of all nonempty, closed and convex subsets of $\mathbb{R}^d$. Let $\{K_i\}$ be a sequence of sets in NCCS($\mathbb{R}^d$).

For any $x \in \mathbb{R}^d$, and $\mu > 0$, we denote

$$r^i_\mu(x) := x - \mathcal{P}_{K_i}(x - \mu A(x)).$$

The proof of the following lemma is similar to that of Lemma 3.1 in [6]. For the sake of completeness, we provide the proof.

Lemma 6.3. *For any $x \in \mathbb{R}^d$ and $\mu > 0$,*

$$min\{1, \mu\}\|r^i_1(x)\| \leq \|r^i_\mu(x)\| \leq max\{1, \mu\}\|r^i_1(x)\|. \tag{6.5}$$

Proof. Suppose that $\mu_2 \geq \mu_1 > 0$. We first prove that

$$\frac{\mu_1}{\mu_2}\|r^i_{\mu_2}(x)\| \leq \|r^i_{\mu_1}(x)\| \leq \|r^i_{\mu_2}(x)\|. \tag{6.6}$$

Let $c := \frac{\|r^i_{\mu_1}(x)\|}{\|r^i_{\mu_2}(x)\|}$, then we only need to prove that

$$\frac{\mu_1}{\mu_2} \leq c \leq 1. \tag{6.7}$$

Since $x - r^i_{\mu_1}(x) = P_{K_i}(x - \mu_1 A(x))$, it follows from 6.2(i) that

$$\langle y - (x - r^i_{\mu_1}(x)), x - \mu_1 A(x) - (x - r^i_{\mu_1}(x))\rangle \leq 0,\ \forall\ y \in K_i. \tag{6.8}$$

Since $x - r^i_{\mu_2}(x) \in K_i$, by (6.8), we have

$$\langle (x - r^i_{\mu_2}(x)) - (x - r^i_{\mu_1}(x)), x - \mu_1 A(x) - (x - r^i_{\mu_1}(x))\rangle \leq 0,$$

i.e.,

$$\langle r^i_{\mu_1}(x) - r^i_{\mu_2}(x), r^i_{\mu_1}(x) - \mu_1 A(x)\rangle \leq 0. \tag{6.9}$$

Similarly, we have

$$\langle r^i_{\mu_2}(x) - r^i_{\mu_1}(x), r^i_{\mu_2}(x) - \mu_2 A(x)\rangle \leq 0. \tag{6.10}$$

Multiplying (6.9) and (6.10) by μ_2 and μ_1, respectively, and then adding them, we get

$$\langle r^i_{\mu_1}(x) - r^i_{\mu_2}(x), \mu_2 r^i_{\mu_1}(x) - \mu_1 r^i_{\mu_2}(x)\rangle \leq 0$$

and consequently

$$\mu_1\|r^i_{\mu_2}(x)\|^2 + \mu_2\|r^i_{\mu_1}(x)\|^2 \leq (\mu_1 + \mu_2)\langle r^i_{\mu_1}(x), r^i_{\mu_2}(x)\rangle. \tag{6.11}$$

Using the Cauchy–Schwarz inequality, by (6.11), we obtain

$$\mu_1\|r^i_{\mu_2}(x)\|^2 + \mu_2\|r^i_{\mu_1}(x)\|^2 \leq (\mu_1 + \mu_2)\|r^i_{\mu_2}(x)\|\|r^i_{\mu_1}(x)\|. \tag{6.12}$$

Dividing (6.12) by $\|r^i_{\mu_2}(x)\|^2$, we have

$$\mu_1 + \mu_2 c^2 \leq (\mu_1 + \mu_2)c$$

and thus (6.7) holds. From (6.6), it is easy to see that (6.5) holds. □

Definition 6.4 ([2]). Let K and $\{K_i\}$ be sets in NCCS($\mathbb{R}^d$). The sequence $\{K_i\}$ is said to epiconverge to K (denoted by $K_i \overset{epi}{\to} K$) if

(i) for every $x \in K$, there exists a sequence $\{x_i\}$ such that $x_i \in K_i$ for all $i \geq 0$, and $\lim_{i\to\infty} x_i = x$;
(ii) $x_{i_j} \in K_{i_j}$ for all $j \geq 0$ and $\lim_{j\to\infty} x_{i_j} = x$ imply $x \in K$.

Proposition 6.5 ([26]). *Let K and $\{K_i\}$ be a set and a sequence of sets in NCCS($\mathbb{R}^d$), respectively. If $K_i \overset{epi}{\to} K$ and $\lim_{i\to\infty} x_i = x$, then*

$$\lim_{i\to\infty} \mathcal{P}_{K_i}(x_i) = \mathcal{P}_K(x).$$

Algorithm 6.6. *Let $\{K_i\}$ be a sequence of sets in NCCS($\mathbb{R}^d$) such that $K_i \overset{epi}{\to} K$. Choose $x_1 \in K_1$ and two parameters $\theta, \sigma \in (0, 1)$. Set $i = 1$.*

Step 1. *Let k_i be the smallest nonnegative integer k satisfying*

$$\gamma^k \|A(x_i) - A(\mathcal{P}_{K_i}(x_i - \gamma^k A(x_i)))\| \leq \sigma \|r^i_{\theta^k}(x_i)\|. \tag{6.13}$$

Set $\eta_i = \theta^{k_i}$ and

$$y_i = \mathcal{P}_{K_i}(x_i - \eta_i A(x_i)).$$

Step 2. *Compute $x_{i+1} := \mathcal{P}_{H_i \cap K_{i+1}}(x_i - \eta_i A(y_i))$, where*

$$H_i := \{x \in \mathbb{R}^d : \langle (x_i - \eta_i A(x_i)) - y_i, x - y_i \rangle \leq 0\}.$$

Let $i := i + 1$ and go to Step 1.

Remark 6.7. In view of Lemma 6.2(i) and the definition of H_i, we have

$$\forall\ x \in K_i,\ \langle x_i - \eta_i A(x_i) - y_i, x - y_i \rangle \leq 0.$$

Therefore, $K_i \subseteq H_i$.

First we show that Algorithm 6.6 is well defined and implementable.

Proposition 6.8. *Suppose that $K_i \subseteq K$ for all $i \geq 0$. Then there exists a nonnegative integer k_i satisfying* (6.13).

Proof. If $r^i_{\theta^{n_0}}(x_i) = 0$ for some $n_0 \geq 0$, we take $k_i = n_0$ which satisfies (6.13).

Assume now that $r^i_{\theta^{n_1}}(x_i) \neq 0$ for some $n_1 \geq 0$. Suppose for all k and $y_k = \mathcal{P}_{K_i}(x_i - \theta^k A(x_i))$,

$$\theta^k \|A(x_i) - A(y_k)\| > \sigma \|r^i_{\theta^k}(x_i)\|,$$

i.e.,

$$\begin{aligned}\|A(x_i) - A(y_k)\| &> \frac{\sigma}{\theta^k}\|r^i_{\theta^k}(x_i)\| \\ &\geq \frac{\sigma}{\theta^k}\min\{1, \theta^k\}\|r^i_1(x_i)\| \\ &= \sigma\|r^i_1(x_i)\|, \end{aligned} \tag{6.14}$$

where the second inequality follows from Lemma 6.3 and the equality follows from $\theta \in (0,1)$ and $k \geq 0$. Since $\mathcal{P}_{K_i}(\cdot)$ is continuous and $x_i \in K_i$, $y_k = \mathcal{P}_{K_i}(x_i - \theta^k A(x_i)) \to x_i$ $(k \to \infty)$. Since $r_{\theta^{n_1}}(x_i) \neq 0$, it follows from Lemma 6.3 that

$$0 < \|r^i_{\theta^{n_1}}(x_i)\| \leq \max\{1, \theta^{n_1}\}\|r^i_1(x_i)\| = \|r^i_1(x_i)\|,$$

where the last equality follows from $\theta^{n_1} \leq 1$. Letting $k \to \infty$ in (6.14), we have

$$0 = \|A(x_i) - A(x_i)\| \geq \sigma\|r^i_1(x_i)\| > 0.$$

This contradiction completes the proof. □

Lemma 6.9. *Suppose that $S_1 \subseteq K_i \subseteq K$ for all $i \geq 1$, that $K_i \overset{epi}{\to} K$, and that the assumption* (6.4) *holds. Let $\{x_i\}$ be the sequence generated by Algorithm 6.6 and let $x^* \in S_1$. Then*

$$\|x_{i+1} - x^*\|^2 \leq \|x_i - x^*\|^2 - (1 - \sigma^2)\eta_i^2\|r^i_1(x_i)\|^2. \tag{6.15}$$

Proof. Since $x^* \in S_1$, it follows from assumption (6.4) that

$$\langle A(y_i), y_i - x^*\rangle \geq 0.$$

Thus,

$$\langle A(y_i), x_{i+1} - x^*\rangle \geq \langle A(y_i), x_{i+1} - y_i\rangle. \tag{6.16}$$

By Step 2, we have

$$\langle x_{i+1} - y_i, (x_i - \eta_i A(x_i)) - y_i\rangle \leq 0.$$

Therefore,

$$\begin{aligned}\langle x_{i+1} - y_i, (x_i - \eta_i A(y_i)) - y_i\rangle &= \langle x_{i+1} - y_i, x_i - \eta_i A(x_i) - y_i\rangle \\ &\quad + \eta_i\langle x_{i+1} - y_i, A(x_i) - A(y_i)\rangle \\ &\leq \eta_i\langle x_{i+1} - y_i, A(x_i) - A(y_i)\rangle. \end{aligned} \tag{6.17}$$

Denoting $z_i = x_i - \eta_i A(y_i)$, we obtain

$$\begin{aligned}\|x_{i+1}-x^*\|^2 &= \|\mathcal{P}_{H_i\cap K_{i+1}}(z_i)-x^*\|^2\\ &= \langle \mathcal{P}_{H_i\cap K_{i+1}}(z_i)-z_i+z_i-x^*, \mathcal{P}_{H_i\cap K_{i+1}}(z_i)-z_i+z_i-x^*\rangle\\ &= \|z_i-x^*\|^2+\|z_i-\mathcal{P}_{H_i\cap K_{i+1}}(z_i)\|^2\\ &\quad +2\langle \mathcal{P}_{H_i\cap K_{i+1}}(z_i)-z_i, z_i-x^*\rangle.\end{aligned}$$

Since

$$\begin{aligned}&2\|z_i-\mathcal{P}_{H_i\cap K_{i+1}}(z_i)\|^2+2\langle \mathcal{P}_{H_i\cap K_{i+1}}(z_i)-z_i, z_i-x^*\rangle\\ &\quad = 2\langle z_i-\mathcal{P}_{H_i\cap K_{i+1}}(z_i), x^*-\mathcal{P}_{H_i\cap K_{i+1}}(z_i)\rangle \le 0,\end{aligned}$$

we have

$$\|z_i-\mathcal{P}_{H_i\cap K_{i+1}}(z_i)\|^2+2\langle \mathcal{P}_{H_i\cap K_{i+1}}(z_i)-z_i, z_i-x^*\rangle \le -\|z_i-\mathcal{P}_{H_i\cap K_{i+1}}(z_i)\|^2.$$

Therefore,

$$\begin{aligned}\|x_{i+1}-x^*\|^2 &\le \|z_i-x^*\|^2-\|z_i-\mathcal{P}_{H_i\cap K_{i+1}}(z_i)\|^2\\ &= \|(x_i-\eta_i A(y_i))-x^*\|^2-\|(x_i-\eta_i A(y_i))-\mathcal{P}_{H_i\cap K_{i+1}}(z_i)\|^2\\ &= \|x_i-x^*\|^2-\|x_i-x_{i+1}\|^2+2\eta_i\langle x^*-x_{i+1}, A(y_i)\rangle\\ &\le \|x_i-x^*\|^2-\|x_i-x_{i+1}\|^2+2\eta_i\langle y_i-x_{i+1}, A(y_i)\rangle,\end{aligned}$$

where the last inequality follows from (6.16). Hence,

$$\begin{aligned}\|x_{i+1}-x^*\|^2 &\le \|x_i-x^*\|^2-\|x_i-x_{i+1}\|^2+2\eta_i\langle y_i-x_{i+1}, A(y_i)\rangle\\ &= \|x_i-x^*\|^2-\langle x_i-y_i+y_i-x_{i+1}, x_i-y_i+y_i-x_{i+1}\rangle\\ &\quad +2\eta_i\langle y_i-x_{i+1}, A(y_i)\rangle\\ &= \|x_i-x^*\|^2-\|x_i-y_i\|^2-\|y_i-x_{i+1}\|^2\\ &\quad +2\langle x_{i+1}-y_i, x_i-\eta_i A(y_i)-y_i\rangle\\ &\le \|x_i-x^*\|^2-\|x_i-y_i\|^2-\|y_i-x_{i+1}\|^2\\ &\quad +2\eta_i\langle x_{i+1}-y_i, A(x_i)-A(y_i)\rangle\\ &\le \|x_i-x^*\|^2-\|x_i-y_i\|^2-\|y_i-x_{i+1}\|^2\\ &\quad +2\sigma\|x_{i+1}-y_i\|\|x_i-y_i\|, \qquad (6.18)\end{aligned}$$

where the second inequality follows from (6.17) and the last one follows from the Cauchy–Schwarz inequality and (6.13).

Since

$$
\begin{aligned}
&(\sigma\|x_i - y_i\| - \|x_{i+1} - y_i\|)^2 \\
&= \sigma^2\|x_i - y_i\|^2 - 2\sigma\|x_{i+1} - y_i\|\|x_i - y_i\| + \|y_i - x_{i+1}\|^2 \\
&\geq 0,
\end{aligned}
$$

we obtain

$$
2\sigma\|x_{i+1} - y_i\|\|x_i - y_i\| \leq \sigma^2\|x_i - y_i\|^2 + \|y_i - x_{i+1}\|^2. \tag{6.19}
$$

Combining (6.18) and (6.19), we get

$$
\begin{aligned}
\|x_{i+1} - x^*\|^2 &\leq \|x_i - x^*\|^2 - (1-\sigma^2)\|x_i - y_i\|^2 \\
&= \|x_i - x^*\|^2 - (1-\sigma^2)\|r^i_{\eta_i}(x_i)\|^2.
\end{aligned} \tag{6.20}
$$

By Lemma 6.3,

$$
\|r^i_{\eta_i}(x_i)\| \geq \min\{1, \eta_i\}\|r^i_1(x_i)\| = \eta_i\|r^i_1(x_i)\|. \tag{6.21}
$$

It follows from (6.20) and (6.21) that (6.15) holds. □

Theorem 6.10. *Suppose that $S_1 \subseteq K_i \subseteq K$ for all $i \geq 1$, that $K_i \overset{epi}{\to} K$, and that the assumption* (6.4) *holds. If $A : K \to \mathbb{R}^d$ is continuous on K, then the sequence $\{x_i\}$ generated by Algorithm 6.6 converges to a solution $\overline{x}$ of the problem* (6.1).

Proof. Let $x^* \in S_1$. Since $0 < \sigma < 1$, we have $(1-\sigma^2) \in (0,1)$. It follows from Lemma 6.9 that

$$
(1-\sigma^2)\eta_i^2\|r^i_1(x_i)\|^2 \leq \|x_i - x^*\|^2 - \|x_{i+1} - x^*\|^2.
$$

It follows that the sequence $\{\|x_{i+1} - x^*\|^2\}$ is nonincreasing, and hence is a convergent sequence. Therefore, $\{x_i\}$ is bounded and

$$
0 \leq (1-\sigma^2)\eta_i^2\|r^i_1(x_i)\|^2 \leq \|x_i - x^*\|^2 - \|x_{i+1} - x^*\|^2 \to 0 \ as\ i \to \infty,
$$

which implies that

$$
\lim_{i\to\infty} \eta_i\|r^i_1(x_i)\| = 0. \tag{6.22}
$$

We consider two possible cases. Suppose first that $\limsup_{i\to\infty} \eta_i > 0$. Then, by (6.22), $\liminf_{i\to\infty} \|r_1(x_i)\| = 0$. Since $\{x_i\}$ is bounded, by the continuity of A, there exists an accumulation point $\overline{x}$ of $\{x_i\}$ such that $r_1(\overline{x}) = 0$, i.e., $\overline{x}$ is a solution of the problem (6.1). We show next that the whole sequence $\{x_i\}$ converges to $\overline{x}$. Replacing x^* by $\overline{x}$ in the preceding argument, we obtain that the sequence $\{\|x_i - \overline{x}\|\}$ is nonincreasing and hence converges. Since $\overline{x}$ is an accumulation point of $\{x_i\}$, some subsequence of $\{\|x_i - \overline{x}\|\}$ converges to zero. This shows that the whole sequence $\{\|x_i - \overline{x}\|\}$ converges to zero, hence $\lim_{i\to\infty} x_i = \overline{x}$.

Suppose now that $\lim_{i\to\infty} \eta_i = 0$. By the choice of η_i, we have, for all $k_i \geq 1$,

$$\begin{aligned}\|A(x_i) - A(\mathcal{P}_{K_i}(x_i - \theta^{k_i-1}A(x_i)))\| &> \frac{\sigma}{\theta^{k_i-1}}\|x_i - \mathcal{P}_{K_i}(x_i - \theta^{k_i-1}A(x_i))\| \\ &= \frac{\sigma}{\theta^{k_i-1}}\|r^i_{\theta^{k_i-1}}(x_i)\| \\ &\geq \sigma\|r^i_1(x_i)\|,\end{aligned}$$

where the second inequality follows from Lemma 6.3. Therefore,

$$\|A(x_i) - A(\mathcal{P}_{K_i}(x_i - \theta^{-1}\eta_i A(x_i)))\| > \sigma\|r^i_1(x_i)\|. \tag{6.23}$$

Let $\overline{x}$ be any accumulation point of $\{x_i\}$ and $\{x_{i_j}\}$ be the corresponding subsequence converging to $\overline{x}$. It follows from (6.23) that

$$\|A(x_{i_j}) - A(\mathcal{P}_{K_{i_j}}(x_{i_j} - \theta^{-1}\eta_{i_j} A(x_{i_j})))\| > \sigma\|r^{i_j}_1(x_{i_j})\|. \tag{6.24}$$

Letting $j \to \infty$, by Proposition 6.5 and the continuity of A, we have

$$0 = \|A(\overline{x}) - A(\overline{x})\| \geq \sigma\|r_1(\overline{x})\|.$$

Therefore, $r_1(\overline{x}) = 0$. This implies that $\overline{x}$ solves the variational inequality (6.1). Similar to the preceding proof, we obtain that $\lim_{i\to\infty} x_i = \overline{x}$. □

Algorithm 6.11. *Let $\{K_i\}$ be a sequence of sets in NCCS($\mathbb{R}^d$) such that $K_i \overset{epi}{\to} K$. Choose $x_1 \in K_1$ and two parameters $\theta, \sigma \in (0,1)$. Set $i = 1$.*

Step 1. *Let k_i be the smallest nonnegative integer k satisfying*

$$\theta^k\|A(x_i) - A(\mathcal{P}_{K_i}(x_i - \theta^k A(x_i)))\| \leq \sigma\|r^i_{\theta^k}(x_i)\|. \tag{6.25}$$

Set $\eta_i = \theta^{k_i}$ and

$$y_i = \mathcal{P}_{K_i}(x_i - \eta_i A(x_i)).$$

Step 2. *Compute*

$$x_{i+1} := \mathcal{P}_{K_i}(x_i - \eta_i A(y_i)). \tag{6.26}$$

Let $i := i + 1$ and go to Step 1.

First we show that Algorithm 6.11 is well defined and implementable.

Proposition 6.12. *Suppose that $K_i \subseteq K$ for all $i \geq 0$. Then there exists a nonnegative integer k_i satisfying* (6.25).

The proof is similar to that of Proposition 6.8, and it is omitted.

Lemma 6.13. *Suppose that $S_1 \subseteq K_i \subseteq K$ for all $i \geq 1$, that $K_i \overset{epi}{\to} K$, and that the assumption* (6.4) *holds. Let $\{x_i\}$ be the sequence generated by Algorithm 6.11 and let $x^* \in S_1$. Then*

$$\|x_{i+1} - x^*\|^2 \leq \|x_i - x^*\|^2 - (1-\sigma^2)\eta_i^2 \|r_1^i(x_i)\|^2.$$

Proof. Since $x^* \in S_1$, it follows from assumption (6.4) that

$$\langle A(y_i), y_i - x^* \rangle \geq 0.$$

Therefore,

$$\langle A(y_i), x_{i+1} - x^* \rangle \geq \langle A(y_i), x_{i+1} - y_i \rangle.$$

Since $x_{i+1} \in K_i$, it follows from (6.26) and Lemma 6.2(i) that

$$\langle x_{i+1} - y_i, (x_i - \eta_i A(x_i)) - y_i \rangle \leq 0.$$

Thus,

$$\begin{aligned}\langle x_{i+1} - y_i, (x_i - \eta_i A(y_i)) - y_i \rangle &= \langle x_{i+1} - y_i, x_i - \eta_i A(x_i) - y_i \rangle \\ &\quad + \eta_i \langle x_{i+1} - y_i, A(x_i) - A(y_i) \rangle \\ &\leq \eta_i \langle x_{i+1} - y_i, A(x_i) - A(y_i) \rangle.\end{aligned}$$

The rest of the proof is similar to that of Lemma 6.9 with $\mathcal{P}_{H_i \cap K_{i+1}}$ replaced by $\mathcal{P}_{K_i}$ and we omit the detail. □

Theorem 6.14. *Suppose that $S_1 \subseteq K_i \subseteq K$ for all $i \geq 1$, that $K_i \overset{epi}{\to} K$, and that the assumption* (6.4) *holds. If $A : K \to \mathbb{R}^d$ is continuous on K, then the sequence $\{x_i\}$ generated by Algorithm 6.11 converges to a solution $\overline{x}$ of* (6.1).

The proof is similar to that of Theorem 6.10, and it is omitted.

If $K_i \equiv K$ for all i, then the above two algorithms become the following method for solving the variational inequality problem (6.1).

Algorithm 6.15. *Choose $x_0 \in K$ and two parameters $\theta, \sigma \in (0, 1)$. Set $i = 1$.*

Step 1. *Let k_i be the smallest nonnegative integer k satisfying*

$$\theta^k \|A(x_i) - A(\mathcal{P}_K(x_i - \theta^k A(x_i)))\| \leq \sigma \|r_{\theta^k}(x_i)\|.$$

Set $\eta_i = \theta^{k_i}$ *and*

$$y_i = \mathcal{P}_K(x_i - \eta_i A(x_i)).$$

If $r_{\eta_i}(x_i) = 0$*, stop.*
Step 2. *Compute*

$$x_{i+1} := \mathcal{P}_K(x_i - \eta_i A(y_i)).$$

Let $i := i + 1$ *and go to Step 1.*

As a consequence of Theorem 6.10 or 6.14, we have the following convergence result.

Theorem 6.16. *If* $A : K \to \mathbb{R}^d$ *is continuous on* K *and the assumption* (6.4) *holds, then the sequence* $\{x_i\}$ *generated by Algorithm 6.15 converges to a solution* $\overline{x}$ *of the problem* (6.1).

Remark 6.17. In [21], the mapping A is required to be Lipschitz continuous and pseudomonotone. In Theorem 6.16, the mapping A is assumed to be continuous. Since the assumption (6.4) is weaker than pseudomonotonicity, our assumptions of the mapping A are more general.

6.3 The Multivalued Variational Inequality

We consider the following multivalued variational inequality, denoted by MVI(A, K): find $x^* \in K$ and $w^* \in A(x^*)$ such that

$$\langle w^*, y - x^* \rangle \geq 0 \quad \forall\, y \in K, \tag{6.27}$$

where K is a nonempty closed convex set in $\mathbb{R}^d$, and $A : K \to 2^{\mathbb{R}^d}$ is a multivalued mapping.

Let us recall the definition of a continuous multivalued mapping. A is said to be upper semicontinuous at $x \in K$ if for every open set V containing $A(x)$, there is an open set U containing x such that $A(y) \subset V$ for all $y \in K \cap U$. A is said to be lower semicontinuous at $x \in K$ if given any sequence x_k converging to x and any $y \in A(x)$, there exists a sequence $y_k \in A(x_k)$ that converges to y. A is said to be continuous at $x \in K$ if it is both upper semicontinuous and lower semicontinuous at x. If A is single-valued, then both upper semicontinuity and lower semicontinuity reduce to the continuity of A.

A is called pseudomonotone on K, if for any $x, y \in K$,

$$\langle v, x - y \rangle \geq 0 \text{ for some } v \in A(y) \Rightarrow \langle u, x - y \rangle \geq 0 \text{ for all } u \in A(x).$$

Let S_2 be the solution set of the problem (6.27), that is, those points $x^* \in K$, $w^* \in A(x^*)$ satisfying (6.27). Throughout this chapter, we assume that the solution set S_2 of the problem (6.27) is nonempty and A is continuous on K with nonempty compact convex values satisfying the following property:

$$\langle w, y - x \rangle \geq 0 \quad \forall\, y \in K,\ w \in A(y),\ x \in S_2. \tag{6.28}$$

The property (6.28) holds if A is pseudomonotone on K.

Proposition 6.18. *$x \in K$ and $w \in A(x)$ solve the problem* (6.27) *if and only if*

$$r_\mu(x, w) := x - \mathcal{P}_K(x - \mu w) = 0.$$

For any $x \in \mathbb{R}^d$, $w \in A(x)$ and $\mu > 0$, we denote

$$r_\mu^i(x, w) := x - \mathcal{P}_{K_i}(x - \mu w).$$

Lemma 6.19. *For any $x \in \mathbb{R}^d$, $w \in A(x)$ and $\mu > 0$,*

$$\min\{1, \mu\}\|r_1^i(x, w)\| \leq \|r_\mu^i(x, w)\| \leq \max\{1, \mu\}\|r_1^i(x, w)\|.$$

The proof is similar to that of Lemma 6.3.

Algorithm 6.20. *Let $\{K_i\}$ be a sequence of sets in NCCS($\mathbb{R}^d$) such that $K_i \overset{epi}{\to} K$. Choose $x_1 \in K_1$ and two parameters $\theta, \sigma \in (0, 1)$. Set $i = 1$.*

Step 1. Choose $u_i \in A(x_i)$ and let k_i be the smallest nonnegative integer k satisfying

$$v_i \in A(\mathcal{P}_{K_i}(x_i - \theta^k u_i)), \tag{6.29}$$

$$\theta^k \|u_i - v_i\| \leq \sigma \|r_{\theta^k}^i(x_i, u_i)\|. \tag{6.30}$$

Set $\eta_i = \theta^{k_i}$ and

$$y_i = \mathcal{P}_{K_i}(x_i - \eta_i u_i).$$

Step 2. Compute $x_{i+1} := \mathcal{P}_{H_i \cap K_{i+1}}(x_i - \eta_i v_i)$, where

$$H_i := \{x \in \mathbb{R}^d : \langle x_i - \eta_i u_i - y_i, x - y_i \rangle \leq 0\}.$$

Let $i := i + 1$ and go to Step 1.

The following proposition shows that Algorithm 6.20 is well defined.

Proposition 6.21. *Suppose that $K_i \subseteq K$ for all $i \geq 1$. Then there exists a nonnegative integer k_i satisfying* (6.29) *and* (6.30).

Proof. If $r^i_{\theta^{n_0}}(x_i, u_i) = 0$ for some $n_0 \geq 0$, we take $k_i = n_0$ and $v_i = u_i$ which satisfy (6.29) and (6.30).

Assume now that $r^i_{\theta^{n_1}}(x_i, u_i) \neq 0$ for some $n_1 \geq 0$. Suppose for all k and all $v \in A(\mathcal{P}_{K_i}(x_i - \theta^k u_i))$,

$$\theta^k \|u_i - v\| > \sigma \|r^i_{\theta^k}(x_i, u_i)\|,$$

i.e.,

$$\begin{aligned}
\|u_i - v\| &> \frac{\sigma}{\theta^k}\|r^i_{\theta^k}(x_i, u_i)\| \\
&\geq \frac{\sigma}{\theta^k}\min\{1, \theta^k\}\|r^i_1(x_i, u_i)\| \\
&= \sigma\|r^i_1(x_i, u_i)\|,
\end{aligned}$$

where the second inequality follows from Lemma 6.19 and the equality follows from $\theta \in (0, 1)$ and $k \geq 0$. Since $\mathcal{P}_{K_i}(\cdot)$ is continuous and $x_i \in K_i$, $\mathcal{P}_{K_i}(x_i - \theta^k u_i) \to x_i$ $(k \to \infty)$. Since A is lower semicontinuous, $u_i \in A(x_i)$ and $\mathcal{P}_{K_i}(x_i - \theta^k u_i) \to x_i$ $(k \to \infty)$, there is $v_k \in A(\mathcal{P}_{K_i}(x_i - \theta^k u_i))$ such that $v_k \to u_i (k \to \infty)$. Therefore,

$$\|u_i - v_k\| > \sigma\|r^i_1(x_i, u_i)\| \quad \forall\, k. \tag{6.31}$$

Since $r^i_{\theta^{n_1}}(x_i, u_i) \neq 0$, it follows from Lemma 6.19 that

$$0 < \|r^i_{\theta^{n_1}}(x_i, u_i)\| \leq \max\{1, \theta^{n_1}\}\|r^i_1(x_i, u_i)\| = \|r^i_1(x_i, u_i)\|,$$

where the last equality follows from $\theta^{n_1} \leq 1$. Let $k \to \infty$ in (6.31), we have

$$0 = \|u_i - u_i\| \geq \sigma\|r^i_1(x_i, u_i)\| > 0.$$

This contradiction completes the proof. □

Lemma 6.22. *Suppose that $S_2 \subseteq K_i \subseteq K$ for all $i \geq 1$, that $K_i \stackrel{epi}{\to} K$, and that the assumption* (6.28) *holds. Let $\{x_i\}$ be the sequence generated by Algorithm 6.20 and let $x^* \in S_2$. Then*

$$\|x_{i+1} - x^*\|^2 \leq \|x_i - x^*\|^2 - (1 - \sigma^2)\eta_i^2\|r^i_1(x_i, u_i)\|^2. \tag{6.32}$$

Proof. Since $v_i \in A(y_i)$ and $x^* \in S_2$, it follows from assumption (6.28) that

$$\langle v_i, y_i - x^*\rangle \geq 0.$$

Therefore,

$$\langle v_i, x_{i+1} - x^* \rangle \geq \langle v_i, x_{i+1} - y_i \rangle. \tag{6.33}$$

In view of Step 2, we have

$$\langle x_{i+1} - y_i, x_i - \eta_i u_i - y_i \rangle \leq 0.$$

Thus,

$$\begin{aligned} \langle x_{i+1} - y_i, x_i - \eta_i v_i - y_i \rangle &= \langle x_{i+1} - y_i, x_i - \eta_i u_i - y_i \rangle \\ &\quad + \eta_i \langle x_{i+1} - y_i, u_i - v_i \rangle \\ &\leq \eta_i \langle x_{i+1} - y_i, u_i - v_i \rangle. \end{aligned} \tag{6.34}$$

Denoting $z_i = x_i - \eta_i v_i$, we have

$$\begin{aligned} \|x_{i+1} - x^*\|^2 &= \|\mathcal{P}_{H_i \cap K_{i+1}}(z_i) - x^*\|^2 \\ &= \langle \mathcal{P}_{H_i \cap K_{i+1}}(z_i) - z_i + z_i - x^*, \mathcal{P}_{H_i \cap K_{i+1}}(z_i) - z_i + z_i - x^* \rangle \\ &= \|z_i - x^*\|^2 + \|z_i - \mathcal{P}_{H_i \cap K_{i+1}}(z_i)\|^2 \\ &\quad + 2\langle \mathcal{P}_{H_i \cap K_{i+1}}(z_i) - z_i, z_i - x^* \rangle. \end{aligned}$$

Since

$$\begin{aligned} &2\|z_i - \mathcal{P}_{H_i \cap K_{i+1}}(z_i)\|^2 + 2\langle \mathcal{P}_{H_i \cap K_{i+1}}(z_i) - z_i, z_i - x^* \rangle \\ &\quad = 2\langle z_i - \mathcal{P}_{H_i \cap K_{i+1}}(z_i), x^* - \mathcal{P}_{H_i \cap K_{i+1}}(z_i) \rangle \\ &\quad \leq 0, \end{aligned}$$

we obtain

$$\|z_i - \mathcal{P}_{H_i \cap K_{i+1}}(z_i)\|^2 + 2\langle \mathcal{P}_{H_i \cap K_{i+1}}(z_i) - z_i, z_i - x^* \rangle \leq -\|z_i - \mathcal{P}_{H_i \cap K_{i+1}}(z_i)\|^2.$$

Hence,

$$\begin{aligned} \|x_{i+1} - x^*\|^2 &\leq \|z_i - x^*\|^2 - \|z_i - \mathcal{P}_{H_i \cap K_{i+1}}(z_i)\|^2 \\ &= \|x_i - \eta_i v_i - x^*\|^2 - \|x_i - \eta_i v_i - \mathcal{P}_{H_i \cap K_{i+1}}(z_i)\|^2 \\ &= \|x_i - x^*\|^2 - \|x_i - x_{i+1}\|^2 + 2\eta_i \langle x^* - x_{i+1}, v_i \rangle \\ &\leq \|x_i - x^*\|^2 - \|x_i - x_{i+1}\|^2 + 2\eta_i \langle y_i - x_{i+1}, v_i \rangle, \end{aligned}$$

where the last inequality follows from (6.33). Therefore,

$$\|x_{i+1} - x^*\|^2 \leq \|x_i - x^*\|^2 - \|x_i - x_{i+1}\|^2 + 2\eta_i \langle y_i - x_{i+1}, v_i \rangle$$

$$
\begin{aligned}
&= \|x_i - x^*\|^2 - \langle x_i - y_i + y_i - x_{i+1}, x_i - y_i + y_i - x_{i+1}\rangle \\
&\quad + 2\eta_i \langle y_i - x_{i+1}, v_i\rangle \\
&= \|x_i - x^*\|^2 - \|x_i - y_i\|^2 - \|y_i - x_{i+1}\|^2 \\
&\quad + 2\langle x_{i+1} - y_i, x_i - \eta_i v_i - y_i\rangle \\
&\leq \|x_i - x^*\|^2 - \|x_i - y_i\|^2 - \|y_i - x_{i+1}\|^2 \\
&\quad + 2\eta_i \langle x_{i+1} - y_i, u_i - v_i\rangle \\
&\leq \|x_i - x^*\|^2 - \|r^i_{\eta_i}(x_i, u_i)\|^2 - \|y_i - x_{i+1}\|^2 \\
&\quad + 2\sigma \|x_{i+1} - y_i\| \|r^i_{\eta_i}(x_i, u_i)\|, \qquad (6.35)
\end{aligned}
$$

where the second inequality follows from (6.34) and the last one follows from the Cauchy–Schwarz inequality and (6.30).

Since

$$
2\sigma \|x_{i+1} - y_i\| \|r_{\eta_i}(x_i, u_i)\| \leq \sigma^2 \|r_{\eta_i}(x_i, u_i)\|^2 + \|y_i - x_{i+1}\|^2, \qquad (6.36)
$$

by combining (6.35) and (6.36), we have

$$
\|x_{i+1} - x^*\|^2 \leq \|x_i - x^*\|^2 - (1 - \sigma^2)\|r_{\eta_i}(x_i, u_i)\|^2. \qquad (6.37)
$$

By Lemma 6.19,

$$
\|r^i_{\eta_i}(x_i, u_i)\| \geq \min\{1, \eta_i\}\|r^i_1(x_i, u_i)\| = \eta_i \|r^i_1(x_i, u_i)\|. \qquad (6.38)
$$

It follows from (6.37) and (6.38) that (6.32) holds. □

Theorem 6.23. *Suppose that $S_2 \subseteq K_i \subseteq K$ for all $i \geq 1$, that $K_i \overset{epi}{\to} K$, and that the assumption* (6.28) *holds. If $A : K \to 2^{\mathbb{R}^d}$ is continuous with nonempty compact convex values on K, then the sequence $\{x_i\}$ generated by Algorithm 6.20 converges to a solution $\overline{x}$ of the problem* (6.27).

Proof. Let $x^* \in S_2$. Since $0 < \sigma < 1$, we have $(1 - \sigma^2) \in (0, 1)$. It follows from Lemma 6.22 that

$$
0 \leq (1 - \sigma^2)\eta_i^2 \|r^i_1(x_i, u_i)\|^2 \leq \|x_i - x^*\|^2 - \|x_{i+1} - x^*\|^2. \qquad (6.39)
$$

Thus, the sequence $\{\|x_{i+1} - x^*\|^2\}$ is nonincreasing, and hence, it is a convergent sequence. Therefore, $\{x_i\}$ is bounded. Letting $i \to \infty$ in (6.39), we obtain

$$
\lim_{i \to \infty} \eta_i \|r_1(x_i, u_i)\| = 0. \qquad (6.40)
$$

By the boundedness of $\{x_i\}$, there exists a convergent subsequence $\{x_{i_j}\}$ converging to $\overline{x}$.

If $\overline{x}$ is a solution of the problem (6.27), we show next that the whole sequence $\{x_i\}$ converges to $\overline{x}$. Replacing x^* by $\overline{x}$ in the preceding argument, we obtain that the sequence $\{\|x_i - \overline{x}\|\}$ is nonincreasing and hence converges. Since $\overline{x}$ is an accumulation point of $\{x_i\}$, some subsequence of $\{\|x_i - \overline{x}\|\}$ converges to zero. This shows that the whole sequence $\{\|x_i - \overline{x}\|\}$ converges to zero, hence $\lim_{i\to\infty} x_i = \overline{x}$.

Suppose that $\overline{x}$ is not a solution of the problem (6.27). We show first that k_i in Algorithm 6.20 cannot tend to ∞. Since A is continuous with compact values, Proposition 3.11 in [3] implies that $\{A(x_i) : i \in N\}$ is a bounded set, and so the sequence $\{u_i\}$ is bounded. Therefore, there exists a subsequence $\{u_{i_j}\}$ converging to $\overline{u}$. Since A is upper semicontinuous with compact values, Proposition 3.7 in [3] implies that A is closed, and so $\overline{u} \in A(\overline{x})$. By the definition of k_i, we have

$$\theta^{k_i-1}\|u_i - v\| > \sigma\|r^i_{\theta^{k_i-1}}(x_i, u_i)\| \quad \forall\ v \in A(\mathcal{P}_{K_i}(x_i - \theta^{k_i-1}u_i)),$$

i.e.,

$$\begin{aligned}\|u_i - v\| &> \frac{\sigma}{\theta^{k_i-1}}\|r^i_{\theta^{k_i-1}}(x_i, u_i)\| \\ &\geq \frac{\sigma}{\theta^{k_i-1}}\min\{1, \theta^{k_i-1}\}\|r^i_1(x_i, u_i)\| \\ &= \sigma\|r^i_1(x_i, u_i)\| \quad \forall\ v \in A(\mathcal{P}_{K_i}(x_i - \theta^{k_i-1}u_i)), \forall\ k_i \geq 1,\end{aligned}$$

where the second inequality follows from Lemma 6.19 and the equality follows from $\theta \in (0, 1)$.

If $k_{i_j} \to \infty$, then by Proposition 6.5 and $\overline{x} \in K$, $\mathcal{P}_{K_{i_j}}(x_{i_j} - \theta^{k_{i_j}-1}u_{i_j}) \to \overline{x}$. The lower semicontinuity of A, in turn, implies the existence of $\overline{u}_{i_j} \in A(\mathcal{P}_{K_{i_j}}(x_{i_j} - \theta^{k_{i_j}-1}u_{i_j}))$ such that $\overline{u}_{i_j}$ converges to $\overline{u}$. Therefore

$$\|u_{i_j} - \overline{u}_{i_j}\| > \sigma\|r^i_1(x_{i_j}, u_{i_j})\|.$$

Letting $j \to \infty$, by Proposition 6.5 , we obtain the contradiction

$$0 \geq \sigma\|r_1(\overline{x}, \overline{u})\|^2 > 0.$$

Therefore, $\{k_i\}$ is bounded and so is $\{\eta_i\}$.

By the boundedness of $\{\eta_i\}$ and (6.40), $\lim_{i\to\infty}\|r^i_1(x_i, u_i)\| = 0$. Since the sequences $\{x_i\}$ and $\{u_i\}$ are bounded, it follows from Proposition 6.5 that there exists an accumulation point $(\overline{x}, \overline{u})$ of $\{(x_i, u_i)\}$ such that $r_1(\overline{x}, \overline{u}) = 0$. This implies that $\overline{x}$ solves the variational inequality (6.27). Similar to the preceding proof, we obtain that $\lim_{i\to\infty} x_i = \overline{x}$. □

Algorithm 6.24. *Let $\{K_i\}$ be a sequence of sets in NCCS($\mathbb{R}^d$) such that $K_i \overset{epi}{\to} K$. Choose $x_1 \in K_0$ and two parameters $\theta, \sigma \in (0, 1)$. Set $i = 1$.*

Step 1. *Choose $u_i \in A(x_i)$ and let k_i be the smallest nonnegative integer satisfying*

$$v_i \in A(\mathcal{P}_{K_i}(x_i - \theta^{k_i} u_i)), \tag{6.41}$$

$$\theta^{k_i} \|u_i - v_i\| \le \sigma \|r^i_{\theta^{k_i}}(x_i, u_i)\|. \tag{6.42}$$

Set $\eta_i = \theta^{k_i}$ and

$$y_i = \mathcal{P}_{K_i}(x_i - \eta_i u_i),$$

Step 2. *Compute*

$$x_{i+1} := \mathcal{P}_{K_i}(x_i - \eta_i v_i). \tag{6.43}$$

Let $i := i + 1$ and go to Step 1.

First we show that Algorithm 6.24 is well defined and implementable.

Proposition 6.25. *Suppose that $K_i \subseteq K$ for all $i \ge 1$. Then there exists a nonnegative integer k_i satisfying* (6.41) *and* (6.42).

The proof is similar to that of Proposition 6.21.

Lemma 6.26. *Suppose that $S_2 \subseteq K_i \subseteq K$ for all $i \ge 1$, that $K_i \overset{epi}{\to} K$, and that the assumption* (6.28) *holds. Let $\{x_i\}$ be the sequence generated by Algorithm 6.24 and let $x^* \in S_2$. Then*

$$\|x_{i+1} - x^*\|^2 \le \|x_i - x^*\|^2 - (1 - \sigma^2)\eta_i^2 \|r^i_1(x_i, u_i)\|^2.$$

Proof. Since $v_i \in A(y_i)$ and $x^* \in S_2$, it follows from assumption (6.28) that

$$\langle v_i, y_i - x^* \rangle \ge 0.$$

Thus,

$$\langle v_i, x_{i+1} - x^* \rangle \ge \langle v_i, x_{i+1} - y_i \rangle. \tag{6.44}$$

Since $y_i \in K_i$, it follows from (6.44) and Lemma 6.2(i) that

$$\langle x_{i+1} - y_i, (x_i - \eta_i u_i) - y_i \rangle \le 0.$$

Therefore,

$$\begin{aligned}\langle x_{i+1} - y_i, (x_i - \eta_i v_i) - y_i \rangle &= \langle x_{i+1} - y_i, x_i - \eta_i u_i - y_i \rangle \\ &\quad + \eta_i \langle x_{i+1} - y_i, u_i - v_i \rangle \\ &\le \eta_i \langle x_{i+1} - y_i, u_i - v_i \rangle. \end{aligned} \tag{6.45}$$

The rest of the proof is similar to that of Lemma 6.22 with $\mathcal{P}_{H_i \cap K_{i+1}}$ replaced by $\mathcal{P}_{K_i}$ and we omit it. □

Next we conclude the global convergence of Algorithm 6.24.

Theorem 6.27. *Suppose that* $S_2 \subseteq K_i \subseteq K$ *for all* $i \geq 1$, *that* $K_i \overset{epi}{\to} K$, *and that the assumption* (6.28) *holds. If* $A : K \to 2^{\mathbb{R}^d}$ *is continuous with nonempty compact convex values on* K, *then the sequence* $\{x_i\}$ *generated by Algorithm 6.24 converges to a solution* $\overline{x}$ *of the problem* (6.27).

Proof. We only need to show the result for the case in which $\overline{x}$ is not a solution of the problem (6.27). The rest of the proof is similar to that of Theorem 6.23.

Suppose that $\overline{x}$ is not a solution of the problem (6.27). We show first that k_i in Algorithm 6.24 cannot tend to ∞. Since A is continuous with compact values, Proposition 3.11 in [3] implies that $\{A(x_i) : i \in N\}$ is a bounded set, and so the sequence $\{u_i\}$ is bounded. Therefore, there exists a subsequence $\{u_{i_j}\}$ converging to $\overline{u}$. Since A is upper semicontinuous with compact values, Proposition 3.7 in [3] implies that A is closed, and so $\overline{u} \in A(\overline{x})$. By the definition of k_i, we have

$$\theta^{k_i-1}\|u_i - v\| > \sigma \|r^i_{\theta^{k_i-1}}(x_i, u_i)\| \quad \forall\ v \in A(\mathcal{P}_{K_i}(x_i - \theta^{k_i-1}u_i)),$$

i.e.,

$$\begin{aligned}\|u_i - v\| &> \frac{\sigma}{\theta^{k_i-1}}\|r^i_{\theta^{k_i-1}}(x_i, u_i)\| \\ &\geq \frac{\sigma}{\theta^{k_i-1}}\min\{1, \theta^{k_i-1}\}\|r^i_1(x_i, u_i)\| \\ &= \sigma\|r^i_1(x_i, u_i)\|, \forall\ v \in A(\mathcal{P}_{K_i}(x_i - \theta^{k_i-1}u_i)) \quad \forall\ k_i \geq 1,\end{aligned}$$

where the second inequality follows from Lemma 6.19 and the equality follows from $\theta \in (0, 1)$.

If $k_{i_j} \to \infty$, then by Proposition 6.5 and $\overline{x} \in K$, $\mathcal{P}_{K_{i_j}}(x_{i_j} - \theta^{k_{i_j}-1}u_{i_j}) \to \overline{x}$. The lower semicontinuity of A, in turn, implies the existence of $\overline{u}_{i_j} \in A(\mathcal{P}_{K_{i_j}}(x_{i_j} - \theta^{k_{i_j}-1}u_{i_j}))$ such that $\overline{u}_{i_j}$ converges to $\overline{u}$. Therefore

$$\|u_{i_j} - \overline{u}_{i_j}\| > \sigma\|r^i_1(x_{i_j}, u_{i_j})\|.$$

Letting $j \to \infty$, by Proposition 6.5, we obtain the contradiction

$$0 \geq \sigma\|r_1(\overline{x}, \overline{\xi})\|^2 > 0.$$

Therefore, $\{k_i\}$ is bounded and so is $\{\eta_i\}$. By the boundedness of $\{\eta_i\}$, it follows from (6.39) that

$$\lim_{i\to\infty} \|r_1^i(x_i,\xi_i)\| = 0.$$

Since the sequences $\{x_i\}$ and $\{u_i\}$ are bounded, it follows from Proposition 6.5 that there exists an accumulation point $(\overline{x},\overline{u})$ of $\{(x_i,u_i)\}$ such that $r_1(\overline{x},\overline{u}) = 0$. This implies that $\overline{x}$ solves the variational inequality (6.1). Similar to the proof of Theorem 6.23, we obtain that $\lim_{i\to\infty} x_i = \overline{x}$. □

6.4 The General Variational Inequality

Let K be a nonempty closed convex set in $\mathbb{R}^d$, A and h be single-valued mappings from K to $\mathbb{R}^d$ with nonempty values. We consider the following general variational inequality problem (GVIP): find $x^* \in K$, $h(x^*) \in K$ such that

$$\langle A(x^*), h(y) - h(x^*)\rangle \geq 0 \quad \forall\, y \in K. \tag{6.46}$$

Definition 6.28. Let A and h be single-valued mappings from K to $\mathbb{R}^d$ with nonempty values.

(i) A is called h-pseudomonotone on K, if for any $x, y \in K$,

$$\langle A(y), h(x) - h(y)\rangle \geq 0 \Rightarrow \langle A(x), h(x) - h(y)\rangle \geq 0.$$

(ii) h is α-strongly monotone with respect to *one* solution $\overline{x} \in S$, if for any $x \in K$,

$$\langle h(x) - h(\overline{x}), x - \overline{x}\rangle \geq \alpha\|x - \overline{x}\|^2.$$

(iii) $h^{-1} : \mathbb{R}^d \to 2^{\mathbb{R}^d}$ is locally bounded on K, if it maps bounded subsets of K into bounded sets.

For any $x \in K$, and $\mu > 0$, we denote

$$R_\mu^i(x) := h(x) - \mathcal{P}_{K_i}(h(x) - \mu A(x)),$$

$$R_\mu(x) := h(x) - \mathcal{P}_K(h(x) - \mu A(x)).$$

Lemma 6.29. *$x \in K$ solves the problem* (6.46) *if and only if*

$$R_\mu(x) := h(x) - \mathcal{P}_K(h(x) - \mu A(x)) = 0.$$

Lemma 6.30. *For any $x \in K$ and $\mu > 0$,*

$$min\{1,\mu\}\|R_1^i(x)\| \leq \|R_\mu^i(x)\| \leq max\{1,\mu\}\|R_1^i(x)\|.$$

The proof is similar to the proof of Lemma 6.3.

Algorithm 6.31. *Let $\{K_i\}$ be a sequence of sets in NCCS($\mathbb{R}^d$) such that $K_i \overset{epi}{\to} K$. Choose $x_1 \in K_1$ and two parameters $\theta, \sigma \in (0, 1)$. Set $i = 1$.*

Step 1. *Let k_i be the smallest nonnegative integer k satisfying*

$$\theta^k \| A(h(x_i)) - A(\mathcal{P}_{K_i}(h(x_i) - \theta^k A(x_i)))\| \le \sigma \| R^i_{\theta^k}(x_i)\|. \tag{6.47}$$

Set $\eta_i = \theta^{k_i}$ and

$$y_i = \mathcal{P}_{K_i}(h(x_i) - \eta_i A(x_i)).$$

Step 2. *Find x_{i+1} such that $h(x_{i+1}) := \mathcal{P}_{H_i \cap K_{i+1}}(h(x_i) - \eta_i A(y_i))$, where*

$$H_i := \{x \in \mathbb{R}^d : \langle h(x_i) - \eta_i A(h(x_i)) - h(y_i), x - h(y_i)\rangle \le 0\}.$$

Let $i := i + 1$ and go to Step 1.

Let S_3 be the solution set of the problem (6.46), that is, those points $x^* \in K, h(x^*) \in K$ satisfying the problem (6.46). Throughout this chapter, we assume that the solution set S_3 of the problem (6.46) is nonempty and A is h-pseudomonotone on K with respect to the solution set S_3, i.e.,

$$\langle A(y), h(y) - h(x)\rangle \ge 0 \quad \forall\, y \in K,\ \forall\, x \in S_3. \tag{6.48}$$

The property (6.48) holds if A is h-pseudomonotone on K.

First we show that Algorithm 6.31 is well defined and implementable.

Proposition 6.32. *Suppose that $K_i \subseteq K$ for all $i \ge 0$. Then there exists a nonnegative integer k_i satisfying* (6.47).

Proof. If $R^i_{\theta^{n_0}}(x_i) = 0$ for some $n_0 \ge 0$, we take $k_i = n_0$ which satisfies (6.47).

Assume now that $R^i_{\theta^{n_1}}(x_i) \ne 0$ for some $n_1 \ge 0$. Suppose for all k and $y_k = \mathcal{P}_{K_i}(h(x_i) - \theta^k A(x_i))$,

$$\theta^k \| A(h(x_i)) - A(y_k)\| > \sigma \| R^i_{\theta^k}(x_i)\|,$$

i.e.,

$$\begin{aligned} \| A(h(x_i)) - A(y_k)\| &> \frac{\sigma}{\theta^k} \| R^i_{\theta^k}(x_i)\| \\ &\ge \frac{\sigma}{\theta^k} \min\{1, \theta^k\} \| R^i_1(x_i)\| \\ &= \sigma \| R^i_1(x_i)\|, \end{aligned} \tag{6.49}$$

where the second inequality follows from Lemma 6.30 and the equality follows from $\theta \in (0, 1)$ and $k \geq 0$. Since $\mathcal{P}_{K_i}(\cdot)$ and h are continuous and $h(x_i) \in K_i$, $y_k = \mathcal{P}_{K_i}(h(x_i) - \theta^k A(x_i)) \to h(x_i)$ $(k \to \infty)$. Since $R_{\theta^{n_1}}(x_i) \neq 0$, it follows from Lemma 6.30 that

$$0 < \|R^i_{\theta^{n_1}}(x_i)\| \leq \max\{1, \theta^{n_1}\}\|R^i_1(x_i)\| = \|R^i_1(x_i)\|,$$

where the last equality follows from $\theta^{n_1} \leq 1$. Letting $k \to \infty$ in (6.49), we have

$$0 = \|A(h(x_i)) - A(h(x_i))\| \geq \sigma\|R^i_1(x_i)\| > 0,$$

being A continuous on K. This contradiction completes the proof. □

Theorem 6.33. *Suppose that $S_3 \subseteq K_i \subseteq K$ for all $i \geq 1$, that $K_i \overset{epi}{\to} K$, and that the assumption* (6.48) *holds. Let $\{x_i\}$ be the sequence generated by Algorithm 6.31 and let $x^* \in S_3$. Then*

$$\|h(x_{i+1}) - h(x^*)\|^2 \leq \|h(x_i) - h(x^*)\|^2 - (1 - \sigma^2)\eta_i^2\|R^i_1(x_i)\|^2. \tag{6.50}$$

Proof. Since $x^* \in S_3$, it follows from assumption (6.48) that

$$\langle A(y_i), h(y_i) - h(x^*)\rangle \geq 0.$$

Thus,

$$\langle A(y_i), h(x_{i+1}) - h(x^*)\rangle \geq \langle A(y_i), h(x_{i+1}) - h(y_i)\rangle. \tag{6.51}$$

In view of Step 2 of Algorithm 6.31, we have

$$\langle h(x_{i+1}) - h(y_i), h(x_i) - \eta_i A(h(x_i)) - h(y_i)\rangle \leq 0.$$

Therefore,

$$\begin{aligned}
&\langle h(x_{i+1}) - h(y_i), h(x_i) - \eta_i A(y_i)) - h(y_i)\rangle \\
&\quad = \langle h(x_{i+1}) - h(y_i), h(x_i) - \eta_i A(h(x_i)) - h(y_i)\rangle \\
&\qquad + \eta_i\langle h(x_{i+1}) - h(y_i), A(h(x_i)) - A(y_i)\rangle \\
&\quad \leq \eta_i\langle h(x_{i+1}) - h(y_i), A(h(x_i)) - A(y_i)\rangle.
\end{aligned} \tag{6.52}$$

Denoting $z_i = h(x_i) - \eta_i A(y_i)$, we obtain

$$\begin{aligned}
&\|h(x_{i+1}) - h(x^*)\|^2 \\
&\quad = \|\mathcal{P}_{H_i \cap K_{i+1}}(z_i) - h(x^*)\|^2
\end{aligned}$$

$$
\begin{aligned}
&= \langle \mathcal{P}_{H_i \cap K_{i+1}}(z_i) - z_i + z_i - h(x^*), \mathcal{P}_{H_i \cap K_{i+1}}(z_i) - z_i + z_i - h(x^*) \rangle \\
&= \|z_i - h(x^*)\|^2 + \|z_i - \mathcal{P}_{H_i \cap K_{i+1}}(z_i)\|^2 \\
&\quad + 2\langle \mathcal{P}_{H_i \cap K_{i+1}}(z_i) - z_i, z_i - h(x^*) \rangle.
\end{aligned}
$$

Since

$$
\begin{aligned}
&2\|z_i - \mathcal{P}_{H_i \cap K_{i+1}}(z_i)\|^2 + 2\langle \mathcal{P}_{H_i \cap K_{i+1}}(z_i) - z_i, z_i - h(x^*) \rangle \\
&\quad = 2\langle z_i - \mathcal{P}_{H_i \cap K_{i+1}}(z_i), h(x^*) - \mathcal{P}_{H_i \cap K_{i+1}}(z_i) \rangle \leq 0,
\end{aligned}
$$

we obtain

$$
\begin{aligned}
&\|z_i - \mathcal{P}_{H_i \cap K_{i+1}}(z_i)\|^2 + 2\langle \mathcal{P}_{H_i \cap K_{i+1}}(z_i) - z_i, z_i - h(x^*) \rangle \\
&\quad \leq -\|z_i - \mathcal{P}_{H_i \cap K_{i+1}}(z_i)\|^2.
\end{aligned}
$$

Hence,

$$
\begin{aligned}
&\|h(x_{i+1}) - h(x^*)\|^2 \\
&\quad \leq \|z_i - h(x^*)\|^2 - \|z_i - \mathcal{P}_{H_i \cap K_{i+1}}(z_i)\|^2 \\
&\quad = \|(h(x_i) - \eta_i A(y_i)) - h(x^*)\|^2 - \|(h(x_i) - \eta_i A(y_i)) - \mathcal{P}_{H_i \cap K_{i+1}}(z_i)\|^2 \\
&\quad = \|h(x_i) - h(x^*)\|^2 - \|h(x_i) - h(x_{i+1})\|^2 + 2\eta_i \langle h(x^*) - h(x_{i+1}), A(y_i) \rangle \\
&\quad \leq \|h(x_i) - h(x^*)\|^2 - \|h(x_i) - h(x_{i+1})\|^2 + 2\eta_i \langle h(y_i) - h(x_{i+1}), A(y_i) \rangle,
\end{aligned}
$$

where the last inequality follows from (6.51). Therefore,

$$
\begin{aligned}
&\|h(x_{i+1}) - h(x^*)\|^2 \\
&\quad \leq \|h(x_i) - h(x^*)\|^2 - \|h(x_i) - h(x_{i+1})\|^2 + 2\eta_i \langle h(y_i) - h(x_{i+1}), A(y_i) \rangle \\
&\quad = \|h(x_i) - h(x^*)\|^2 - \langle h(x_i) - h(y_i) + h(y_i) - h(x_{i+1}), h(x_i) - h(y_i) \\
&\qquad + h(y_i) - h(x_{i+1}) \rangle + 2\eta_i \langle h(y_i) - h(x_{i+1}), A(y_i) \rangle \\
&\quad = \|h(x_i) - h(x^*)\|^2 - \|h(x_i) - h(y_i)\|^2 - \|h(y_i) - h(x_{i+1})\|^2 \\
&\qquad + 2\langle h(x_{i+1}) - h(y_i), h(x_i) - \eta_i A(y_i) - h(y_i) \rangle \\
&\quad \leq \|h(x_i) - h(x^*)\|^2 - \|h(x_i) - h(y_i)\|^2 - \|h(y_i) - h(x_{i+1})\|^2 \\
&\qquad + 2\eta_i \langle h(x_{i+1}) - h(y_i), A(h(x_i)) - A(y_i) \rangle \\
&\quad \leq \|h(x_i) - h(x^*)\|^2 - \|h(x_i) - h(y_i)\|^2 - \|h(y_i) - h(x_{i+1})\|^2 \\
&\qquad + 2\sigma \|h(x_{i+1}) - h(y_i)\| \|h(x_i) - h(y_i)\|, \qquad (6.53)
\end{aligned}
$$

where the second inequality follows from (6.52) and the last one follows from the Cauchy–Schwarz inequality and (6.47).

Since

$$0 \le (\sigma\|h(x_i) - h(y_i)\| - \|h(x_{i+1}) - h(y_i)\|)^2 = \sigma^2\|h(x_i) - h(y_i)\|^2$$
$$- 2\sigma\|h(x_{i+1}) - h(y_i)\|\|h(x_i) - h(y_i)\| + \|h(y_i) - h(x_{i+1})\|^2,$$

we have

$$\begin{aligned} &2\sigma\|h(x_{i+1}) - h(y_i)\|\|h(x_i) - h(y_i)\| \\ &\quad \le \sigma^2\|h(x_i) - h(y_i)\|^2 + \|h(y_i) - h(x_{i+1})\|^2. \end{aligned} \tag{6.54}$$

Combining (6.53) and (6.54), we have

$$\begin{aligned} \|h(x_{i+1}) - h(x^*)\|^2 &\le \|h(x_i) - h(x^*)\|^2 - (1-\sigma^2)\|h(x_i) - h(y_i)\|^2 \\ &= \|h(x_i) - h(x^*)\|^2 - (1-\sigma^2)\|R^i_{\eta_i}(x_i)\|^2. \end{aligned} \tag{6.55}$$

By Lemma 6.30,

$$\|R^i_{\eta_i}(x_i)\| \ge \min\{1, \eta_i\}\|R^i_1(x_i)\| = \eta_i\|R^i_1(x_i)\|. \tag{6.56}$$

It follows from (6.55) and (6.56) that (6.50) holds. □

Theorem 6.34. *Suppose that $S_3 \subseteq K_i \subseteq K$ for all $i \ge 1$, that $K_i \overset{epi}{\to} K$, and that the assumption* (6.48) *holds. If $A, h : K \to \mathbb{R}^d$ are continuous on K, h is α-strongly monotone with respect to one solution of GVIP and $h^{-1} : \mathbb{R}^d \to 2^{\mathbb{R}^d}$ is locally bounded on K, then the sequence $\{x_i\}$ generated by Algorithm 6.31 converges to a solution $\overline{x}$ of* (6.46).

Proof. Let $x^* \in S_3$. Since $0 < \sigma < 1$, we have $(1-\sigma^2) \in (0,1)$. It follows from Theorem 6.33 that

$$(1-\sigma^2)\eta_i^2\|R^i_1(x_i)\|^2 \le \|h(x_i) - h(x^*)\|^2 - \|h(x_{i+1}) - h(x^*)\|^2.$$

Thus, the sequence $\{\|h(x_{i+1}) - h(x^*)\|^2\}$ is nonincreasing, and hence is a convergent sequence. Therefore, $\{h(x_i)\}$ is bounded and

$$0 \le (1-\sigma^2)\eta_i^2\|R^i_1(x_i)\|^2 \to 0 \text{ as } i \to \infty,$$

which implies that

$$\lim_{i\to\infty} \eta_i\|R^i_1(x_i)\| = 0. \tag{6.57}$$

We consider two possible cases. Suppose first that $\limsup_{i\to\infty}\eta_i > 0$. Then, by (6.57), $\liminf_{i\to\infty}\|R_1^i(x_i)\| = 0$. Since h^{-1} is locally bounded, it follows that the sequence $\{x_i\}$ is bounded. By Proposition 6.5 and the continuity of A and h, there exists an accumulation point $\overline{x}$ of $\{x_i\}$ such that $R_1(\overline{x}) = 0$, i.e., $\overline{x}$ is a solution of the problem (6.46). We show next that the whole sequence $\{x_i\}$ converges to $\overline{x}$. Replacing x^* by $\overline{x}$ in the preceding argument, we obtain that the sequence $\{\|h(x_i)-h(\overline{x})\|\}$ is nonincreasing and hence converges. Since $\overline{x}$ is an accumulation point of $\{x_i\}$, by the continuity of h, some subsequence of $\{\|h(x_i)-h(\overline{x})\|\}$ converges to zero. This shows that the whole sequence $\{\|h(x_i)-h(\overline{x})\|\}$ converges to zero. By considering the strong monotonicity of h with respect to $\overline{x}\in S$, we have

$$0 \le \alpha\|x_i - x^*\| \le \|h(x_i) - h(x^*)\| \to 0 \text{ as } i \to \infty.$$

This shows that the whole sequence $\{\|x_i - x^*\|\}$ converges to zero, hence $\lim_{i\to\infty} x_i = \overline{x}$.

Suppose now that $\lim_{i\to\infty}\eta_i = 0$. By the choice of η_i, we have, for all $k_i \ge 1$,

$$\begin{aligned}\|A(h(x_i)) - A(\mathcal{P}_{K_i}(h(x_i) - \theta^{k_i-1}A(x_i)))\| &> \frac{\sigma}{\theta^{k_i-1}}\|R^i_{\theta^{k_i-1}}(x_i)\| \\ &\ge \sigma\|R_1^i(x_i)\|, \end{aligned}\tag{6.58}$$

where the second inequality follows from Lemma 6.30. Let $\overline{x}$ be any accumulation point of $\{x_i\}$ and $\{x_{i_j}\}$ is the corresponding subsequence converging to $\overline{x}$. It follows from (6.58) that

$$\|A(h(x_{i_j})) - A(\mathcal{P}_{K_{i_j}}(h(x_{i_j}) - \theta^{-1}\eta_{i_j}A(x_{i_j})))\| > \sigma\|R_1^{i_j}(x_{i_j})\|.$$

Letting $j \to \infty$, by Proposition 6.5 and the continuity of A and h, we have

$$0 = \|A(h(\overline{x})) - A(h(\overline{x}))\| \ge \sigma\|R_1(\overline{x})\|.$$

Therefore, $R_1(\overline{x}) = 0$. This implies that $\overline{x}$ solves the variational inequality (6.46). Similar to the preceding proof, we obtain that $\lim_{i\to\infty} x_i = \overline{x}$. □

Algorithm 6.35. *Let $\{K_i\}$ be a sequence of sets in NCCS($\mathbb{R}^d$) such that $K_i \xrightarrow{epi} K$. Choose $x_1 \in K_1$ and two parameters $\theta, \sigma \in (0,1)$. Set $i = 1$.*

Step 1. *Let k_i be the smallest nonnegative integer k satisfying*

$$\theta^k\|A(h(x_i)) - A(\mathcal{P}_{K_i}(h(x_i) - \theta^k A(x_i)))\| \le \sigma\|R^i_{\theta^k}(x_i)\|.$$

Set $\eta_i = \theta^{k_i}$ and

$$y_i = \mathcal{P}_{K_i}(h(x_i) - \eta_i A(x_i)).$$

Table 6.1 The number of iteration under the different choice of σ

σ	0.1	0.2	0.3	0.5	0.8
$\varepsilon = 10^{-7}$	213	119	82	48	39
$\varepsilon = 10^{-5}$	142	81	56	34	28
$\varepsilon = 10^{-3}$	71	43	31	20	17

Step 2. *Find x_{i+1} such that $h(x_{i+1}) := \mathcal{P}_{K_i}(h(x_i) - \eta_i A(y_i))$.*
Let $i := i + 1$ and go to Step 1.

Remark 6.36. Convergence analysis of Algorithm 6.35 is similar to that of Algorithm 6.31 and we omit it.

6.5 Numerical Experiments

In this section, we present some numerical experiments for Algorithm 6.20. The MATLAB codes are run on a PC (with CPU Intel P-T2390) under MATLAB Version 7.0.1.24704(R14) Service Pack 1. The integers in Table 6.1 denote the number of iterations. The tolerance ε means when $\|r_\mu(x, w)\| \leq \varepsilon$, the procedure stops.

Example 6.37. Let $d = 3$,

$$K := \{x \in \mathbb{R}^d_+ : \sum_{i=1}^{d} x_i = 1\}$$

and $A : K \to 2^{\mathbb{R}^d}$ be defined by

$$A(x) := \{(t, t - x_1, t - x_2) : t \in [0, 1]\}.$$

Then the set K and the mapping A satisfy the assumptions of Theorem 6.23 and $(0, 0, 1)$ is a solution of the multivalued variational inequality. We choose $K_i \equiv K, \theta = 0.8$ for our algorithm. We use $(0, 1, 0) \in K$ as the initial point.

From the above table we observe that the larger σ is, the smaller the number of iteration.

Acknowledgements This research was partially supported by the Natural Science Foundation Project of CQ CSTC of China, No. 2010BB9401, and the Scientific and Technological Research Program of Chongqing Municipal Education Commission of China, No. KJ110509.

References

1. Allevi, E., Gnudi, A., Konnov, I.V.: The proximal point method for nonmonotone variational inequalities. Math. Method Oper. Res. **63**, 553–565 (2006)
2. Attouch, H.: Variational Convergence for Functions and Operators. Pitman, London (1984)
3. Aubin, J.P., Ekeland, I.: Applied Nonlinear Analysis. Wiley, New York (1984)
4. Auslender, A., Teboulle, M.: Lagrangian duality and related multiplier methods for variational inequality problems. SIAM J. Optim. **10**, 1097–1115 (2000)
5. Bao, T.Q., Khanh, P.Q.: A projection-type algorithm for pseudomonotone nonlipschitzian multivalued variational inequalities. In: Eberhard, A., Hadjisavvas, N., Luc, D.T. (eds.) Generalized Convexity, Generalized Monotonicity and Applications, pp. 113–129. Springer, New York (2005)
6. Bnouhachem, A.: A self-adaptive method for solving general mixed variational inequalities. J. Math. Anal. Appl. **309**, 136–150 (2005)
7. Censor, Y., Gibali, A., Reich, S.: The subgradient extragradient method for solving variational inequalities in Hilbert space. J. Optim. Theory Appl. **148**, 318–335 (2011)
8. Censor, Y., Gibali, A., Reich, S.: Extensions of Korpelevich's extragradient method for solving the variational inequality problem in Euclidean space. Optimization **61**, 1119–1132 (2012)
9. Facchinei, F., Pang, J.S.: Finite-Dimensional Variational Inequalities and Complementary Problems. Springer, New York (2003)
10. Fang, C.J., He, Y.R.: A double projection algorithm for multi-valued variational inequalities and a unified framework of the method. Appl. Math. Comput. **217**, 9543–9511 (2011)
11. Fang, C.J., He, Y.R.: An extragradient method for generalized variational inequality. Pac. J. Optim. **9**, 47–59 (2013)
12. Fang, S.C., Peterson, E.L.: Generalized variational inequalities. J. Optim. Theory Appl. **38**, 363–383 (1982)
13. Fang, C.J., Chen, S.L., Yang, C.D.: An algorithm for solving multi-valued variational inequality. J. Inequal. Appl. **2013**, 218 (2013)
14. Fang, C.J., Chen, S.L., Zheng, J.M.: A projection-type method for multivalued variational inequality. Abstr. Appl. Anal. **2013**, 6 (2013). Article ID 836720
15. Fukushima, M.: The primal Douglas-Rachford splitting algorithm for a class of monotone mappings with application to the traffic equilibrium problem. Math. Program. **72**, 1–15 (1996)
16. Hartman, P., Stampacchia, G.: On some nonlinear elliptic differential functional equations. Acta Math. **115**, 271–310 (1966)
17. He, Y.R.: A new double projection algorithm for variational inequalities. J. Comput. Appl. Math. **185**, 166–173 (2006)
18. Iusem, A.N., Svaiter, B.F.: A variant of Korpelevich' method for variational inequalities with a new search strategy. Optimization **42**, 309–321 (1997)
19. Karamardian, S.: Complementarity problems over cones with monotone and pseudomonotone maps. J. Optim. Theory Appl. **18**, 445–454 (1976)
20. Konnov, I.V.: Combined Relaxation Methods for Variational Inequalities. Springer, Berlin (2001)
21. Korpelevich, G.M.: The extragradient method for finding saddle points and other problems. Ekonom. i Mat. Metody **12**, 747–756 (1976)
22. Li, F.L., He, Y.R.: An algorithm for generalized variational inequality with pseudomonotone mapping. J. Comput. Appl. Math. **228**, 212–218 (2009)
23. Rockafellar, R.T.: Monotone operators and the proximal point algorithm. SIAM J. Control Optim. **14**, 877–898 (1976)
24. Saigal, R.: Extension of the generalized complementarity problem. Math. Oper. Res. **1**, 260–266 (1976)
25. Salmon, G., Strodiot, J.J., Nguyen, V.H.: A bundle method for solving variational inequalities. SIAM J. Optim. **14**, 869–893 (2003)

26. Santos, P.S.M., Scheimberg, S.: A projection algorithm for general variational inequalities with perturbed constraint set. Appl. Math. Comput. **181**, 649–661 (2006)
27. Solodov, M.V., Svaiter, B.F.: A new projection method for variational inequality problems. SIAM J. Control Optim. **37**, 765–776 (1999)
28. Sun, D.: A class of iterative methods for solving nonlinear projection equations. J. Optim. Theory Appl. **31**, 123–140 (1996)
29. Wang, Y.J., Xiu, N.H., Zhang, J.Z.: Modified extragradient method for variational inequalities and verification of solution existence. J. Optim. Theory Appl. **119**, 167–183 (2003)
30. Xia, F.Q., Huang, N.J.: A projection-proximal point algorithm for solving generalized variational inequalities. J. Optim. Theory Appl. **150**, 98–117 (2011)
31. Xiu, N., Wang, Y., Zhang, X.: Modified fixed-point equations and related iterative methods for variational inequalities. Comput. Math. Appl. **47**, 913–920 (2004)
32. Zarantonello, E.H.: Projections on convex sets in Hilbert space and spectral theory. In: Zarantonello, E.H. (eds.) Contributions to Nonlinear Functional Analysis. Academic, New York (1971)

Chapter 7
Proximal Methods for the Elastography Inverse Problem of Tumor Identification Using an Equation Error Approach

Mark S. Gockenbach, Baasansuren Jadamba, Akhtar A. Khan, Christiane Tammer, and Brian Winkler

Abstract In this chapter, we study a nonlinear inverse problem in linear elasticity relating to tumor identification by an equation error formulation. This approach leads to a variational inequality as a necessary and sufficient optimality condition. We give complete convergence analysis for the proposed equation error method. Since the considered problem is highly ill-posed, we develop a stable computational framework by employing a variety of proximal point methods and compare their performance with the more commonly used Tikhonov regularization.

Keywords Variational inequality • Elasticity imaging • Elastography inverse problem • Tumor identification • Proximal point method • Regularization

AMS Classification. 35R30, 65N30

M.S. Gockenbach (✉)
Department of Mathematical Sciences, Michigan Technological University, Houghton, MI, USA
e-mail: msgocken@mtu.edu

B. Jadamba • A.A. Khan • B. Winkler
Center for Applied and Computational Mathematics, School of Mathematical Sciences, Rochester Institute of Technology, 85 Lomb Memorial Drive, Rochester, NY 14623, USA
e-mail: bxjsma@rit.edu; aaksma@rit.edu; bcw9368@rit.edu

C. Tammer
Institute of Mathematics, Martin-Luther-University of Halle-Wittenberg, Theodor-Lieser-Str. 5, 06120 Halle-Salle, Germany
e-mail: christiane.tammer@mathematik.uni-halle.de

W. Han et al. (eds.), *Advances in Variational and Hemivariational Inequalities*, Advances in Mechanics and Mathematics 33, DOI 10.1007/978-3-319-14490-0_7

7.1 Introduction

Parameter identification inverse problems in partial differential equations are highly ill-posed and regularization is often needed for an effective solution. Since these problems can be solved most conveniently in an optimization setting, many authors have relied on an optimization framework with variants of Tikhonov regularization among the most widely used. For the tumor identification inverse problem (outlined below), many optimization frameworks have been proposed (output least squares (OLS) [25], modified output least squares (MOLS) [20], energy output least squares (EOLS) [10], and equation error (EE) [8]), where the EOLS and OLS functionals provide nonconvex frameworks and the MOLS and EE functionals provide a convex approach.

For convex frameworks, smooth regularization gives a unique solution to the associated variational inequality, which in turn is a necessary and sufficient optimality condition for solving the inverse problem. For nonconvex frameworks, the variational inequality is only a necessary condition, but a large enough regularization parameter can be chosen to ensure uniqueness. The downside of this approach is that the choice of regularization parameter is largely heuristic and can often introduce error through over-regularization. Thus the proper selection of regularization parameter is of vital interest for practical applications.

Proximal methods are another approach to regularization that seem well-suited for nonlinear inverse problem of parameter identification in partial differential equation. Their general outline consists of the progressive replacement of a single convex optimization by a sequence of strongly convex optimization problems. However, from a theoretical point of view, the proximal approach differs from the Tikhonov approach in one important aspect. For convex problems using Tikhonov regularization, under suitable conditions, the regularized solutions are known to converge to a minimal-norm solution. On the other hand, for proximal point methods, no such characterization concerning the recovered solution is available beforehand. Nonetheless, it is natural to ask whether proximal point methods can be competitive to the more commonly used Tikhonov regularization for nonlinear inverse problems. The use of the proximal methods will allow to put less emphasis on the selection of an optimal regularization parameter. This work addresses this issue for the tumor identification inverse problem for the first time.

We emphasize that besides testing proximal methods for a highly nonlinear inverse problem of tumor identification, we present a new treatment of the equation error approach which is shown to be stable under the H_1-regularization. This is in contrast with the earlier works on the equation error approach (studied only for simpler PDEs) where the H_2 regularization have been used, see Acar [1].

7.1.1 Problem Background

In this work, we will employ the proximal methods for solving a nonlinear inverse problem in linear elasticity relating to tumor identification. As an optimization framework, we use the so-called equation error approach. This inverse problem, known also as the elastography or elasticity imaging inverse problem, arises from a relatively new method for detecting tumors inside the human body using the differing elastic properties between healthy and unhealthy tissue. In elastography, a small external quasi-static compression force is applied to the body and the tissue's axial displacement field or overall motion are measured. A tumor can then be identified from this measurement by recovering the tissue's underlying elasticity. Many researchers have proposed similar elastic imaging methods and some of the details can be found in [2, 4, 5, 9, 11, 24, 25] and the cited references therein.

The underlying mathematical model for the elastography inverse problem is the following system of partial differential equations which describe an isotropic elastic object's response to known body forces and traction applied along its boundary:

$$-\nabla\cdot\sigma = f \quad \text{in } \Omega, \tag{7.1a}$$

$$\sigma = 2\mu\epsilon(u) + \lambda \operatorname{div} u\, I, \tag{7.1b}$$

$$u = f_1 \text{ on } \Gamma_1, \tag{7.1c}$$

$$\sigma n = f_2 \text{ on } \Gamma_2. \tag{7.1d}$$

The domain Ω is a subset of $\mathbb{R}^2$ or $\mathbb{R}^3$ and $\partial\Omega = \Gamma_1 \cup \Gamma_2$ is its Lipschitz boundary. In (7.1), the vector-valued function $u = u(x)$ represents the displacement of the elastic object, f is the body force being applied, n is the unit outward normal, and $\epsilon(u) = \frac{1}{2}(\nabla u + \nabla u^\top)$ is the linearized strain tensor. The stress tensor σ and the stress-strain law (7.1b) hold given the assumption that the elastic object is isotropic and the displacement is small enough to maintain a linear relationship. The Lamé parameters, μ and λ, represent the object's variable elastic properties.

The direct problem in (7.1) is to find the displacement u when the functions f_1, f_2, the coefficients μ and λ, and f are all known. For the elastography inverse problem, we seek to find the parameter μ when a certain measurement z of the displacement u is known. We note that the corresponding inverse problem for many engineering applications is to find both μ and λ, which typically vary in a small range (see [12, 19] and the included citations). However, given that the human body is comprised of mostly water (an incompressible material) it follows that the tissue under consideration is likewise nearly incompressible, i.e. $\lambda \gg \mu$. Thus the tumor identification inverse problem seeks to recover the parameter μ alone.

To introduce the equation error concept, we first consider an exemplar elliptic problem with suitable boundary conditions:

$$-\nabla\cdot(a\nabla u) = f \text{ in } \Omega. \tag{7.2}$$

The output least-squares (OLS) approach for determining the parameter a consists of minimizing the functional

$$J_{\mathrm{OLS}}(a) = \|u(a) - z\|^2, \tag{7.3}$$

with z a measurement of u, $\|\cdot\|$ an appropriate norm, and where $u(a)$ solves the variational problem paired with (7.2). In contrast, the equation error (EE) method for identifying a would be to minimize the residual error between the left and right hand sides of (7.2) given a measurement z of u:

$$J_{\text{EE}}(a) = \frac{1}{2}\|\nabla \cdot (a\nabla z) + f\|^2_{H^{-1}(\Omega)}. \tag{7.4}$$

The OLS approach requires the solution of the weak form of (7.2) to obtain $u(a)$ for any evaluation of J_{OLS}, a significant performance hit for most optimization algorithms. The EE functional does not depend on u and therefore does not require the solution of the underlying variational problem at any point, making it more computationally appealing. We also note that the functional J_{EE} is quadratic in a. Thus the minimization of J_{EE} reduces to the solution of a positive (semi-)definite linear system after discretization, subsequently leading to a convex optimization problem. However, the ∇z term in J_{EE} necessitates the differentiation of the measured data, leaving the EE approach highly susceptible to noise. (See Acar [1], Gockenbach and Khan [15], Al-Jamal and Gockenbach [3] for more on (7.2), and Gockenbach et al. [16] for general elliptic inverse problems.) [7, 13, 14]

Given the advantages of the equation error approach, it is natural to extend it to the tumor identification inverse problem. However, due to the near incompressibility of human tissue as outlined above, standard finite element techniques become ineffective for both the direct and inverse problems.

To better describe the difficulties associated with near incompressibility, we must first introduce some notation. The dot product of two tensors Υ_1 and Υ_2 will be denoted by $\Upsilon_1 \cdot \Upsilon_2$. Given a sufficiently smooth domain $\Omega \subset \mathbb{R}^2$, the L^2-norm of a tensor-valued function $\Upsilon = \Upsilon(x)$ is provided by

$$\|\Upsilon\|^2_{L^2} = \|\Upsilon\|^2_{L^2(\Omega)} = \int_\Omega \Upsilon \cdot \Upsilon = \int_\Omega \left(\Upsilon^2_{11} + \Upsilon^2_{12} + \Upsilon^2_{21} + \Upsilon^2_{22}\right).$$

Alternatively, for a vector-valued function $u(x) = (u_1(x), u_2(x))^\top$, the L^2-norm is given by

$$\|u\|^2_{L^2} = \|u\|^2_{L^2(\Omega)} = \int_\Omega \left(u_1^2 + u_2^2\right),$$

whereas the H^1-norm by

$$\|u\|^2_{H^1} = \|u\|^2_{H^1(\Omega)} = \|u\|^2_{L^2} + \|\nabla u\|^2_{L^2}.$$

For the sake of simplicity, we take $f_1 = 0$ throughout the following. The space of test functions, denoted by $\overline{V}$, is then given by

$$\overline{V} = \{\overline{v} \in H^1(\Omega) \times H^1(\Omega) : \overline{v} = 0 \text{ on } \Gamma_1\}. \tag{7.5}$$

Using Green's identity and applying the boundary conditions from (7.1c) and (7.1d), we get the following weak form of (7.1): Find $\overline{u} \in \overline{V}$ such that

$$\int_{\Omega} 2\mu\epsilon(\overline{u}) \cdot \epsilon(\overline{v}) + \int_{\Omega} \lambda(\operatorname{div} \overline{u})(\operatorname{div} \overline{v}) = \int_{\Omega} f\overline{v} + \int_{\Gamma_2} \overline{v} f_1, \quad \text{for every } \overline{v} \in \bar{V}. \tag{7.6}$$

Continuing, we define $\Psi : \overline{V} \times \overline{V} \to \mathbb{R}$ by

$$\Psi(\overline{u}, \overline{v}) = \int_{\Omega} 2\mu\epsilon(\overline{u}) \cdot \epsilon(\overline{v}) + \int_{\Omega} \lambda(\operatorname{div} \overline{u})(\operatorname{div} \overline{v}).$$

Assuming that both μ and $\mu+\lambda$ are bounded away from zero, it can be shown that there are two positive constants $\psi_1 > 0$ and $\psi_2 > 0$ with $\psi_1 \leq \mu$ and $\psi_2 \geq \lambda + \mu$ such that

$$\psi_1 \|\overline{v}\|_{\overline{V}}^2 \leq \Psi(\overline{v}, \overline{v}) \leq \psi_2 \|\overline{v}\|_{\overline{V}}^2, \quad \text{for every } \overline{v} \in \overline{V}.$$

From the tissue's near incompressibility, we have $\lambda \gg \mu$, and thus the ratio $\psi_3 = \psi_2/\psi_1$ is large. Yet, since the constant Ψ_3 determines the error estimates (as defined by Céa's lemma), it follows that the actual error could easily outweigh the optimal approximation error. This unfortunate situation is well known and has been dubbed the "locking effect" (see Braess [6]).

A variety of approaches have been proposed to overcome the locking effect with one of the most popular being the use of mixed finite elements, an approach which we adopt in this work. By introducing a "pressure" term $p \in Q = L^2(\Omega)$ with

$$p = \lambda \operatorname{div} \overline{u}, \tag{7.7}$$

the weak formulation of (7.7) then becomes:

$$\int_{\Omega} (\operatorname{div} \overline{u})q - \int_{\Omega} \frac{1}{\lambda} pq = 0, \quad \text{for every } q \in Q. \tag{7.8}$$

Using (7.7), the weak form (7.6) then transforms into the following: Find $\overline{u} \in \overline{V}$ such that

$$\int_{\Omega} 2\mu\epsilon(\overline{u}) \cdot \epsilon(\overline{v}) + \int_{\Omega} p(\operatorname{div} \overline{v}) = \int_{\Omega} f\overline{v} + \int_{\Gamma_2} \overline{v} f_1, \quad \text{for every } \overline{v} \in \overline{V}, \tag{7.9}$$

where p is also an unknown.

Thus we have moved from finding $\overline{u} \in \overline{V}$ fulfilling (7.6) to finding $(\overline{u}, p) \in \overline{V} \times Q$ satisfying both (7.8) and (7.9). Throughout the remainder of this chapter, we will study this transformed problem within the ready framework of a saddle point problem.

The remainder of this chapter is organized into four sections. Section 7.2 introduces the equation error optimization formulation for the elastography inverse problem and analyzes the method's convergence. In Sect. 7.3, we present several proximal point approaches for solving the optimization problem arising in Sect. 7.2. Numerical examples and performance analysis are presented in Sect. 7.4, and we conclude with a brief discussion of subsequent directions.

7.2 Equation Error Optimization Framework

In this section we introduce the equation error functional and consider the recovery of the parameter μ within an optimization framework along with an analysis of this method's convergence.

We define the sets

$$\begin{aligned}
\hat{V} &= \left\{u \in H^1(\Omega) \,:\, u = 0 \text{ on } \Gamma_2\right\},\\
\overline{V} &= \hat{V}^2,\\
V &= \overline{V} \times L^2(\Omega),\\
A &= \{\mu \in L^\infty(\Omega) \,:\, \mu \geq \mu_0 \text{ in } \Omega\},
\end{aligned}$$

where $\mu_0 > 0$ is a given constant. For the analysis given below, we will also need the space

$$V^\infty = W^{1,\infty}(\Omega) \times L^\infty(\Omega)$$

with

$$\|u\|_{V^\infty} = \max\left\{\|\overline{u}\|_{W^{1,\infty}}, \|p\|_{L^\infty}\right\},$$

where $u = (\overline{u}, p) \in V^\infty$.

As detailed in Sect. 7.1, the BVP described in Eq. (7.1) is equivalent to the following saddle point problem: Find $u = (\overline{u}, p) \in V$ such that

$$\begin{aligned}
\int_\Omega 2\mu\epsilon(\overline{u}) \cdot \epsilon(\overline{v}) + \int_\Omega p\,(\nabla \cdot \overline{v}) &= \int_\Omega f \cdot \overline{v} + \int_{\Gamma_2} f_2 \cdot \overline{v} \quad \text{for all } \overline{v} \in \overline{V},\\
\int_\Omega (\nabla \cdot \overline{u})\, q - \int_\Omega \frac{1}{\lambda} pq &= 0 \quad \text{for all } q \in L^2(\Omega).
\end{aligned} \tag{7.10}$$

We define $E_1 : L^\infty(\Omega) \times V \to \overline{V}^*$ and $E_2 : V \to L^2(\Omega)^*$ by

$$E_1(\mu, u)v = \int_\Omega 2\mu\epsilon(\overline{u}) \cdot \epsilon(\overline{v}) + \int_\Omega p\,(\nabla \cdot \overline{v}) \quad \text{for all } \overline{v} \in \overline{V},$$

$$E_2(u)v = \int_\Omega (\nabla \cdot \overline{u})\, q - \int_\Omega \frac{1}{\lambda} pq \quad \text{for all } q \in L^2(\Omega),$$

where $u = (\overline{u}, p)$. We also define $m \in \overline{V}^*$ by

$$m(v) = \int_\Omega f \cdot \overline{v} + \int_{\Gamma_2} f_2 \cdot \overline{v} \quad \text{for all } \overline{v} \in \overline{V}.$$

It is important to notice that, while we have defined $E_1(\mu, u)$ for $\mu \in L^\infty(\Omega)$, $u \in V$, the functional is also well-defined for $\mu \in L^2(\Omega)$ and $u \in V^\infty$; that is, we can give up some regularity in μ by requiring more of u. We will take advantage of this below (abusing notation by continuing to write $E_1(\mu, u)$ when μ does not necessarily belong to $L^\infty(\Omega)$).

Here are some preliminary results that we will need below.

Lemma 7.1. *There exist $C_1, C_2 > 0$ such that*

$$\|E_1(\mu, u)\|_{V^*} \leq (C_1\|\mu\|_{L^2} + C_2)\,\|u\|_{V^\infty} \quad \textit{for all } \mu \in L^\infty(\Omega), u \in V^\infty \cap V.$$

Proof. We have

$$\begin{aligned}|E_1(\mu, u)v| &\leq 2\left|\int_\Omega \mu\epsilon(\overline{u}) \cdot \epsilon(\overline{v})\right| + \left|\int_\Omega p\,(\nabla \cdot \overline{v})\right| \\ &\leq 2\|\mu\epsilon(\overline{u})\|_{L^2}\|\epsilon(\overline{v})\|_{L^2} + \|p\|_{L^2}\|\nabla \cdot \overline{v}\|_{L^2}.\end{aligned}$$

We have

$$\begin{aligned}\|\mu\epsilon(\overline{u})\|_{L^2}^2 &= \int_\Omega \mu^2\left(\overline{u}_{1,1}^2 + \frac{1}{2}\left(\overline{u}_{1,2} + \overline{u}_{2,1}\right)^2 + \overline{u}_{2,2}^2\right) \\ &\leq 4\|\overline{u}\|_{W^{1,\infty}}^2 \int_\Omega \mu^2 = 4\|\overline{u}\|_{W^{1,\infty}}^2 \|\mu\|_{L^2}^2,\end{aligned}$$

and hence $\|\mu\epsilon(\overline{u})\|_{L^2} \leq 2\|u\|_{V^\infty}\|\mu\|_{L^2}$. Also,

$$\|p\|_{L^2}^2 = \int_\Omega p^2 \leq |\Omega|\|p\|_{L^\infty}^2 \Rightarrow \|p\|_{L^2} \leq |\Omega|^{1/2}\|p\|_{L^\infty} \leq |\Omega|^{1/2}\|u\|_{V^\infty}.$$

Since

$$\|\epsilon(\overline{v})\|_{L^2} \leq \|\overline{v}\|_{H^1} \leq \|v\|_V \text{ and } \|\nabla \cdot \overline{v}\|_{L^2} \leq \sqrt{2}\|\overline{v}\|_{H^1} \leq \sqrt{2}\|v\|_V,$$

it follows that

$$\begin{aligned}|E_1(\mu,u)v| &\le 4\|u\|_{V^\infty}\|\mu\|_{L^2}\|v\|_V + \sqrt{2}|\Omega|^{1/2}\|u\|_{V^\infty}\|v\|_V \\ &= \left(4\|\mu\|_{L^2} + \sqrt{2}|\Omega|^{1/2}\right)\|u\|_{V^\infty}\|v\|_V,\end{aligned}$$

which proves the desired result (with $C_1 = 4$ and $C_2 = \sqrt{2}|\Omega|^{1/2}$). □

Lemma 7.2. *Assuming λ is bounded away from 0, there exists $C_3 > 0$ with*

$$\|E_2(u)\|_{V^*} \le C_3\|u\|_{V^\infty}.$$

Proof. We have

$$\begin{aligned}E_2(u)v &= \int_\Omega (\nabla\cdot\overline{u})\,q - \int_\Omega \frac{1}{\lambda}pq \\ &\Rightarrow |E_2(u)v| \le \|\nabla\cdot\overline{u}\|_{L^2}\|q\|_{L^2} + \|\lambda^{-1}\|_{L^\infty}\|p\|_{L^2}\|q\|_{L^2},\end{aligned}$$

where $u = (\overline{u}, p)$, $v = (\overline{v}, q)$. Since $\|p\|_{L^2} \le |\Omega|^{1/2}\|u\|_{V^\infty}$, and

$$\begin{aligned}\|\nabla\cdot\overline{u}\|_{L^2}^2 &= \int_\Omega (\overline{u}_{1,1} + \overline{u}_{2,2})^2 \le 2\int_\Omega \left(\overline{u}_{1,1}^2 + \overline{u}_{2,2}^2\right) \\ &\le 2\int_\Omega \left(\|\overline{u}\|_{W^{1,\infty}}^2 + \|\overline{u}\|_{W^{1,\infty}}^2\right) = 4|\Omega|\|\overline{u}\|_{W^{1,\infty}}^2,\end{aligned}$$

and hence $\|\nabla\cdot\overline{u}\|_{L^2} \le 2|\Omega|^{1/2}\|u\|_{V^\infty}$. It follows that

$$|E_2(u)v| \le \left(2|\Omega|^{1/2} + |\Omega|^{1/2}\|\lambda^{-1}\|_{L^\infty}\right)\|u\|_{V^\infty}\|q\|_{L^2} \le C_3\|u\|_{V^\infty}\|v\|_V,$$

with $C_3 = 2|\Omega|^{1/2} + |\Omega|^{1/2}\|\lambda^{-1}\|_{L^\infty}$. This proves the desired result. □

Lemma 7.3. *Suppose $u \in V^\infty$, $\mu \in L^2(\Omega)$, and $\mu_n \in L^2(\Omega)$ for all $n \in \mathbb{Z}^+$. If $\mu_n \to \mu$ in $L^2(\Omega)$, then $E_1(\mu_n, u) \to E_1(\mu, u)$ in V^*.*

Proof. We have

$$\begin{aligned}(E_1(\mu_n,u) - E_1(\mu,u))\,v &= \int_\Omega 2\mu_n\epsilon(\overline{u})\cdot\epsilon(\overline{v}) + \int_\Omega p\,(\nabla\cdot\overline{v}) \\ &\quad - \int_\Omega 2\mu\epsilon(\overline{u})\cdot\epsilon(\overline{v}) - \int_\Omega p\,(\nabla\cdot\overline{v}) \\ &= 2\int_\Omega (\mu_n - \mu)\epsilon(\overline{u})\cdot\epsilon(\overline{v}),\end{aligned}$$

and therefore

$$\begin{aligned}|(E_1(\mu_n,u) - E_1(\mu,u))\,v| &\le 2\|\epsilon(\overline{u})\|_{L^\infty}\|\mu_n - \mu\|_{L^2}\|v\|_V \\ &\le 2\|u\|_{V^\infty}\|\mu_n - \mu\|_{L^2}\|v\|_V.\end{aligned}$$

It follows that $\|E_1(\mu_n, u) - E_1(\mu, u)\|_{V^*} \le 2\|u\|_{V^\infty}\|\mu_n - \mu\|_{L^2}$, and the result follows. □

7.2.1 *The Equation Error Method*

The equation error method aims to estimate μ^* from a measurement z of u^* by minimizing

$$J(\mu; z, \beta) = \|E_1(\mu, z) - m\|_{V^*}^2 + \beta\|\mu\|_{H^1}^2. \tag{7.11}$$

Here we assume that $\mu^* \in A$ and $u^* = (\overline{u}^*, p^*) \in V$ satisfy (7.10).[1]

We first prove that $J(\cdot; z, \beta)$ has a unique minimizer in $H^1(\Omega)$ for each $\beta > 0$, provided z belongs to V^∞.

Theorem 7.4. *Suppose $z \in V^\infty$. Then, for each $\beta > 0$, there exists a unique μ_β satisfying*

$$J(\mu_\beta; z, \beta) \le J(\mu; z, \beta) \quad \text{for all } \mu \in H^1(\Omega).$$

Proof. Since J is bounded below, there exists a minimizing sequence $\{\mu_n\}$ for J. We have

$$\begin{aligned} \beta\|\mu_n\|_{H^1}^2 &\le J(\mu_n; z, \beta) \quad \text{for all } n \\ &\Rightarrow \|\mu_n\|_{H^1}^2 \le \beta^{-1} J(\mu_n; z, \beta) \quad \text{for all } n. \end{aligned}$$

Since the right-hand side is bounded, so is the left. Hence $\{\mu_n\}$ is bounded in $H^1(\Omega)$ and there exists $\mu_\beta \in H^1(\Omega)$ and a subsequence of $\{\mu_n\}$ (still denoted by $\{\mu_n\}$) such that $\mu_n \to \mu_\beta$ weakly in $H^1(\Omega)$ and, by Rellich's theorem, strongly in $L^2(\Omega)$. Since $z \in V^\infty$ and $\mu_n \to \mu_\beta$ in $L^2(\Omega)$ imply, by Lemma 7.3, that $E_1(\mu_n, z) \to E_1(\mu_\beta, z)$ and since the norm is weakly lower semicontinuous, it follows that

$$\begin{aligned} \inf_{\mu \in H^1(\Omega)} J(\mu; z, \beta) &= \lim_{n\to\infty} J(\mu_n; z, \beta) \\ &= \lim_{n\to\infty} \left(\|E_1(\mu_n, z) - m\|_{V^*}^2 + \beta\|\mu_n\|_{H^1}^2\right) \end{aligned}$$

[1]It would be natural to define J by

$$J(\mu; z, \beta) = \|E_1(\mu, z) - m\|_{V^*}^2 + \|E_2(z)\|_{V^*}^2 + \beta\|\mu\|_{H^1}^2.$$

However, $\|E_2(z)\|_{V^*}^2$ is constant with respect to μ and therefore it makes no difference if this term is included or not.

$$\geq \|E_1(\mu_\beta, z) - m\|_{V^*}^2 + \beta\|\mu_\beta\|_{H^1}^2$$
$$= J(\mu_\beta; z, \beta).$$

This shows that μ_β is a minimizer of $J(\cdot; z, \beta)$. The two terms defining this functional are both convex and the second (the regularization term) is strictly convex; hence μ_β is the unique minimizer of $J(\cdot; z, \beta)$. □

The last inequality in the above proof must actually hold as an equality (since $J(\mu_\beta; z, \beta) \geq \inf_{\mu \in H^1(\Omega)} J(\mu; z, \beta)$) and hence $\lim_{n\to\infty} \|\mu_n\|_{H^1} = \|\mu_\beta\|_{H^1}$ must hold. Since $\mu_n \to \mu_\beta$ weakly in $H^1(\Omega)$, this shows that $\{\mu_n\}$ actually converges to μ_β strongly in $H^1(\Omega)$. It follows that any minimizing sequence of $J(\cdot; z, \beta)$ converges in $H^1(\Omega)$ to the unique minimizer μ_β of $J(\cdot; z, \beta)$.

7.2.2 Convergence of the Equation Error Method

Recall that $\mu^* \in A$ and $u^* = (\overline{u}^*, p^*) \in V$ are assumed to satisfy (7.10). We do not assume that μ^* is unique in this regard, so let us define $S = \{\mu \in H^1(\Omega) : E_1(\mu, u^*) = m\}$. Since E_1 is affine in μ, S is convex.

We can now prove the convergence of the equation error method.

Theorem 7.5. *Suppose $u^* \in V^\infty$, $\mu^* \in H^1(\Omega)$ satisfy the saddle point problem* (7.10). *Let $\{z_n\} \subset V^\infty$ be a sequence of observations of u^* that satisfy, with the sequences $\{\epsilon_n\}$, $\{\beta_n\}$, the conditions*

1. $\epsilon_n^2 \leq \beta_n \leq \epsilon_n$ *for all* $n \in \mathbb{Z}^+$;
2. $\epsilon_n^2/\beta_n \to 0$ *as* $n \to \infty$;
3. $\|z_n - u^*\|_{V^\infty} \leq \epsilon_n$ *for all* $n \in \mathbb{Z}^+$;
4. $\epsilon_n \to 0$ *as* $n \to \infty$.

For each $n \in \mathbb{Z}^+$, let μ_n be the unique solution of

$$\min_{\mu \in H^1(\Omega)} J(\mu; z_n, \beta_n).$$

Then there exists $\tilde{\mu} \in S$ such that $\mu_n \to \tilde{\mu}$ in $H^1(\Omega)$. Moreover, $\tilde{\mu}$ satisfies $\|\tilde{\mu}\|_{H^1} \leq \|\mu\|_{H^1}$ for all $\mu \in S$.

Proof. For each $\mu \in S$, we have

$$\beta_n\|\mu_n\|_{H^1}^2 \leq J(\mu; z_n, \beta_n) = \|E_1(\mu, z_n) - m\|_{V^*}^2 + \beta_n\|\mu\|_{H^1}^2$$
$$= \|E_1(\mu, z_n - u^*)\|_{V^*}^2 + \beta_n\|\mu\|_{H^1}^2$$
$$\leq (C_1\|\mu\|_{L^2} + C_2)^2 \|z_n - u^*\|_{V^\infty}^2 + \beta_n\|\mu\|_{H^1}^2.$$

Therefore, for all $\mu \in S$, we have

$$\|\mu_n\|_{H^1}^2 \le (C_1\|\mu\|_{L^2} + C_2)^2 \frac{\epsilon_n^2}{\beta_n} + \|\mu\|_{H^1}^2. \tag{7.12}$$

In particular, we have

$$\|\mu_n\|_{H^1}^2 \le \left(C_1\|\mu^*\|_{L^2} + C_2\right)^2 \frac{\epsilon_n^2}{\beta_n} + \|\mu^*\|_{H^1}^2 \le \left(C_1\|\mu^*\|_{L^2} + C_2\right)^2 + \|\mu^*\|_{H^1}^2,$$

since $\epsilon_n^2 \le \beta_n$ by assumption. This shows that $\{\mu_n\}$ is bounded in $H^1(\Omega)$. Hence, by Rellich's lemma, there exists $\tilde{\mu} \in H^1(\Omega)$ and a subsequence $\{\mu_{n_k}\}$ such that $\mu_{n_k} \to \tilde{\mu}$ weakly in $H^1(\Omega)$ and strongly in $L^2(\Omega)$.

We now show that $E_1(\tilde{\mu}, u^*) = m$, that is, that $\tilde{\mu} \in S$. We have

$$\begin{aligned}
\|E_1(\mu_{n_k}, u^*) - m\|_{V^*}^2 &= \|E_1(\mu_{n_k}, u^*) - E_1(\mu_{n_k}, z_{n_k}) + E_1(\mu_{n_k}, z_{n_k}) - m\|_{V^*}^2 \\
&\le 2\|E_1(\mu_{n_k}, u^* - z_{n_k})\|_{V^*}^2 + 2\|E_1(\mu_{n_k}, z_{n_k}) - m\|_{V^*}^2 \\
&\le 2\left(C_1\|\mu_{n_k}\|_{L^2} + C_2\right)^2 \|z_{n_k} - u^*\|_{V\infty}^2 \\
&\quad + 2\|E_1(\tilde{\mu}, z_{n_k}) - m\|_{V^*}^2 + 2\beta_{n_k}\|\tilde{\mu}\|_{H^1}^2 \\
&\le 2\left(C_1\|\mu_{n_k}\|_{L^2} + C_2\right)^2 \epsilon_{n_k}^2 + 2\left((C_1\|\tilde{\mu}\|_{L^2} + C_2)^2 \epsilon_{n_k}^2 + \beta_{n_k}\|\tilde{\mu}\|_{H^1}^2\right) \\
&\le 2\left(C_1\|\mu_{n_k}\|_{L^2} + C_2\right)^2 \epsilon_{n_k}^2 + 2\left((C_1\|\tilde{\mu}\|_{L^2} + C_2)^2 + \|\tilde{\mu}\|_{H^1}^2\right) \epsilon_{n_k}.
\end{aligned}$$

(Here we used $\epsilon_{n_k}^2 \le \beta_{n_k} \le \epsilon_{n_k}$ and the inequality

$$\|E_1(\tilde{\mu}, z_{n_k}) - m\|_{V^*}^2 + \beta_{n_k}\|\tilde{\mu}\|_{H^1}^2 \le (C_1\|\tilde{\mu}\|_{L^2} + C_2)^2 \epsilon_{n_k}^2 + \beta_{n_k}\|\tilde{\mu}\|_{H^1}^2$$

that was derived above.) Since $\{\|\mu_{n_k}\|_{L^2}\}$ is bounded and $\epsilon_{n_k} \to 0$ as $k \to \infty$, this shows that $\|E_1(\mu_{n_k}, u^*) - m\|_{V^*} \to 0$. Since we also have $E_1(\mu_{n_k}, u^*) \to E_1(\tilde{\mu}, u^*)$ by Lemma 7.3, this shows that $E_1(\tilde{\mu}, u^*) = m$ and hence that $\tilde{\mu} \in S$.

Since $\mu_{n_k} \to \tilde{\mu}$ weakly in $H^1(\Omega)$, we have $\|\tilde{\mu}\|_{H^1} \le \liminf_{k\to\infty} \|\mu_{n_k}\|_{H^1}$. Moreover, by (7.12),

$$\beta_{n_k}\|\mu_{n_k}\|_{H^1}^2 \le (C_1\|\tilde{\mu}\|_{L^2} + C_2)^2 \epsilon_{n_k}^2 + \beta_{n_k}\|\tilde{\mu}\|_{H^1}^2,$$

which implies that

$$\|\mu_{n_k}\|_{H^1}^2 \le (C_1\|\tilde{\mu}\|_{L^2} + C_2)^2 \frac{\epsilon_{n_k}^2}{\beta_{n_k}} + \|\tilde{\mu}\|_{H^1}^2.$$

Since $\epsilon_{n_k}^2/\beta_{n_k} \to \infty$ as $k \to \infty$, this shows that $\limsup_{k\to\infty} \|\mu_{n_k}\|_{H^1} \leq \|\tilde{\mu}\|_{H^1}$. Therefore,

$$\|\tilde{\mu}\|_{H^1} \leq \liminf_{k\to\infty} \|\mu_{n_k}\|_{H^1} \leq \limsup_{k\to\infty} \|\mu_n\|_{H^1} \leq \|\tilde{\mu}\|_{H^1},$$

which proves that $\|\mu_{n_k}\|_{H^1} \to \|\tilde{\mu}\|_{H^1}$, and hence that $\mu_{n_k} \to \tilde{\mu}$ strongly in $H^1(\Omega)$ as $k \to \infty$.

Next, by (7.12),

$$\|\tilde{\mu}\|_{H^1}^2 \leq \lim_{k\to\infty} \|\mu_{n_k}\|_{H^1}^2 \leq \lim_{k\to\infty} \left((C_1\|\mu\|_{L^2} + C_2)^2 \frac{\epsilon_{n_k}^2}{\beta_{n_k}} + \|\mu\|_{H^1}^2 \right) = \|\mu\|_{H^1}^2$$

holds for all $\mu \in S$.

Finally, since S is a convex set, its element of smallest H^1 norm is unique, and we have shown that every convergent subsequence of $\{\mu_n\}$ converges to this unique element $\tilde{\mu}$. Thus $\{\mu_n\}$ itself must converge to $\tilde{\mu}$. This completes the proof. □

7.2.3 *Discrete Formulas*

Before describing the proximal methods which we intend to use, we briefly recall the discretization procedure. We assume that $\mathcal{T}_h$ is a triangulation on Ω, L_h is the space of all piecewise continuous polynomials of degree d_μ relative to $\mathcal{T}_h$, $\overline{U}_h$ is the space of all piecewise continuous polynomials of degree d_u relative to $\mathcal{T}_h$, and Q_h is the space of all piecewise continuous polynomials of degree d_q relative to $\mathcal{T}_h$.

To represent the discrete saddle point problem in a computable form we proceed as follows. We represent bases for L_h, $\overline{U}_h$ and Q_h by $\{\varphi_1, \varphi_2, \ldots, \varphi_m\}$, $\{\psi_1, \psi_2, \ldots, \psi_n\}$, and $\{\chi_1, \chi_2, \ldots, \chi_k\}$, respectively. The space L_h is then isomorphic to $\mathbb{R}^m$ and for any $\mu \in L_h$, we define $\ell \in \mathbb{R}^m$ by $\ell_i = \mu(x_i)$, $i = 1, 2, \ldots, m$, where the nodal basis $\{\varphi_1, \varphi_2, \ldots, \varphi_m\}$ corresponds to the nodes $\{x_1, x_2, \ldots, x_m\}$. Conversely, each $L \in \mathbb{R}^m$ corresponds to $\mu \in L_h$ defined by $\mu = \sum_{i=1}^m \ell_i \varphi_i$. Analogously, $\overline{u} \in \overline{U}_h$ will correspond to $\overline{U} \in \mathbb{R}^n$, where $\overline{U}_i = u(y_i)$, $i = 1, 2, \ldots, n$, and $\overline{u} = \sum_{i=1}^n \overline{U}_i \psi_i$, where $y_1, y_2, \ldots, y_n$ are the nodes of the mesh defining $\overline{U}_h$. Finally, $q \in Q_h$ will correspond to $Q \in \mathbb{R}^k$, where $Q_i = q(z_i)$, $i = 1, 2, \ldots, k$, and $q = \sum_{i=1}^k Q_i \chi_i$, where $z_1, z_2, \ldots, z_k$ are the nodes of the mesh defining Q_h.

We next define $S : R^m \to R^{n+k}$ to be the finite element solution operator that assigns to each coefficient $\mu_h \in A_h$, the unique approximate solution $u_h = (\overline{u}_h, p_h) \in \overline{U}_h \times Q_h$. Then $S(\ell) = U$, where U is defined by

$$K(\ell)U = F, \tag{7.13}$$

where $K(\ell) \in R^{(n+k)\times(n+k)}$ is the stiffness matrix and $F \in R^{n+k}$ is the load vector.

With the above preparation, we have the following discrete version of the equation error functional (7.11) (see Crossen et al. [8] for details):

$$J(\ell) = \frac{1}{2}\left\langle \hat{L}(\overline{Z})\ell + B^T P - F, (K+M)^{-1}\left(\hat{L}(\overline{Z})\ell + B^T P - F\right)\right\rangle,$$

where $\hat{L}$ is the so-called adjoint stiffness matrix, K is the stiffness matrix, M is the mass matrix, and $\overline{Z}$ is the data for $\overline{u}$ (see [8]).

Moreover, for first derivative of $J(\ell)$ is given by:

$$DJ(\ell)(\delta\ell) = \left\langle \hat{L}(\overline{Z})\delta\ell, (K+M)^{-1}(\hat{L}(\overline{Z})\ell + B^T P - F)\right\rangle,$$

implying that the gradient of $J(\ell)$ is given by

$$\nabla J(\ell) = \hat{L}(\overline{Z})^T (K+M)^{-1}(\hat{L}(\overline{Z})\ell + B^T P - F). \tag{7.14}$$

For the second derivative we then have

$$\begin{aligned} D^2 J(\ell)(\delta\ell)(\tilde{\delta\ell}) &= \left\langle \hat{L}(\overline{Z})\delta\ell, (K+M)^{-1}\hat{L}(\overline{Z})\tilde{\delta\ell}\right\rangle \\ &= \left\langle \hat{L}(\overline{Z})^T (K+M)^{-1}\hat{L}(\overline{Z})\delta\ell, \tilde{\delta\ell}\right\rangle, \end{aligned}$$

which implies that the Hessian of $J(\ell)$ is

$$\nabla^2 J(\ell) = \hat{L}(\overline{Z})^T (K+M)^{-1}\hat{L}(\overline{Z}). \tag{7.15}$$

We note that the Hessian does not depend on ℓ, making the coupling of Newton methods with the equation error approach particularly appealing from a computational perspective.

7.3 Proximal Methods

In the previous section, we posed the elastography inverse problem as a convex optimization problem. In this section, our objective is to test the feasibility of the equation error formulation by solving this inverse problem numerically. Evidently, there are numerous methods which can be used to solve the optimization problem at hand. However, as mentioned earlier, due to the stabilizing feature of the proximal methods, we apply several proximal-like optimization algorithms to numerically solve the equation error formulation. We particularly examine several variations on the self-adaptive, inexact Hager and Zhang proximal-point algorithms developed in [17]. We remark that for clarity, the forgoing discussion uses the continuous formulation. However, for computational purposes we use the discrete counterpart given in the previous section. We also note that in our numerical experiments, we often ignore the simple constraints given in the feasible set A and hence solve the unconstrained equation error formulation. This simplification brings us to the framework of [17] where details of the convergence analysis of the used algorithms can be found. We begin with a review of the classical proximal-point algorithm. Drawing on (7.11), we seek the solution to our fundamental constrained convex minimization problem:

$$\min_{\mu \in A} J(\mu) = \|E_1(\mu, z) - m\|_{V*}^2 + \beta \|\mu\|_{H^1}^2 \tag{7.16}$$

where A is a closed and convex set representing our feasible set of parameters.

We now consider the related functional

$$\mathcal{J}_P(\mu) = J(\mu) + \frac{1}{2\lambda^k} \|\mu - \mu^k\|^2, \tag{7.17}$$

where λ_k is a positive number and $\mu^k \in A$. We note that $\mathcal{J}_P(\mu)$ is strictly convex since J and the introduced quadratic term $\frac{1}{2\lambda^k}\|\mu - \mu^k\|_2^2$, known as the *proximal regularization term*, are both also strictly convex. Thus, we have the related, uniquely-solvable subproblem

$$\min_{\mu \in A} \mathcal{J}_P(\mu) \tag{7.18}$$

for which the necessary and sufficient optimality conditions in turn yield the following variational inequality problem: Find $\mu^* \in A$ such that

$$\langle \nabla \mathcal{J}_P(\mu^*), \mu - \mu^* \rangle \geq 0, \quad \forall \mu \in A. \tag{7.19}$$

In this context, the classical proximal-point algorithm generates a sequence $\{\mu^k\}$ such that

$$\mu^{k+1} = \arg\min_{\mu \in A} \left\{ J(\mu) + \frac{1}{2\lambda^k} \|\mu - \mu^k\|^2 \right\} \tag{7.20}$$

where $\{\lambda^k\}$ is a sequence of positive numbers.

Rockafellar [27] showed that if J is strongly convex at a solution of (7.16), then the proximal point method converges linearly when λ^k remains bounded and superlinearly when $\lambda^k \to \infty$. Subsequently, we will consider several variations on the above proximal approach coupled with the method of accelerated convergence outlined by Hager and Zhang [17]. For further details on these methods and their history, we refer the interested reader to [18, 22, 26, 27, 29] and the cited references therein. We note that the so-called auxiliary problem principle which generalizes the proximal point methods have been explored in [23] for elliptic inverse problems.

7.3.1 Hager and Zhang's Proximal Point Method

Hager and Zhang [17] introduced (with $A := \mathbb{R}^n$) two criteria between subsequent iterates of (7.20) for the solution of the subproblem (7.18). One of the criteria that we will be focusing on is that μ^{k+1} is acceptable when

$$\mathcal{J}_P(\mu^{k+1}) \leq J(\mu^k)$$
$$\|\nabla \mathcal{J}_P(\mu^{k+1})\| \leq \theta^k \|\nabla J(\mu^k)\|,$$

where $\theta^k = 1/\lambda^k$.

As they detail, taking the proximal regularization parameter as

$$\theta^k = \tau \|\nabla J(\mu^k)\|^\eta,$$

where $\eta \in [0, 2)$ and $\tau > 0$ are constants, gives quadratic convergence of the iterates to the solution set of (7.18). This gives rise to the following algorithm:

ALGORITHM 1:

Initialization Step: Choose an initial guess μ^0, initialize τ and η, and take $k = 0$. Let $\theta^k = \tau \|\nabla J(\mu^k)\|^\eta$ and let $\gamma = 1$.

Step 1: Find μ^{k+1} satisfying

$$\|\nabla \mathcal{J}_P(\mu^{k+1})\| \leq \theta^k \gamma \|\nabla J(\mu^k)\|. \tag{7.21}$$

Step 2: If μ^{k+1} satisfies

$$\mathcal{J}_P(\mu^{k+1}) \leq J(\mu^k), \tag{7.22}$$

go to Step 3;
else,
set $\gamma = 0.1\gamma$ and go to Step 1.

Step 3: Let

$$\mu^k = \mu^{k+1}.$$

Step 4: Set $k = k + 1$ and go to Step 1.

There are many reasonable stopping criteria to terminate the above algorithms. The particular one that we use is that the gradient norm is bounded by the given tolerance, see Sect. 7.4.1 for details. In Step 1, the subproblem was solved using an unconstrained conjugate-gradient trust-region method to find the subsequent iterate μ^{k+1}. The numerical results for this method are given in Fig. 7.1 and Table 7.1. Convergence analysis of the above scheme can be found in [17, Theorems 3.1 and 4.2].

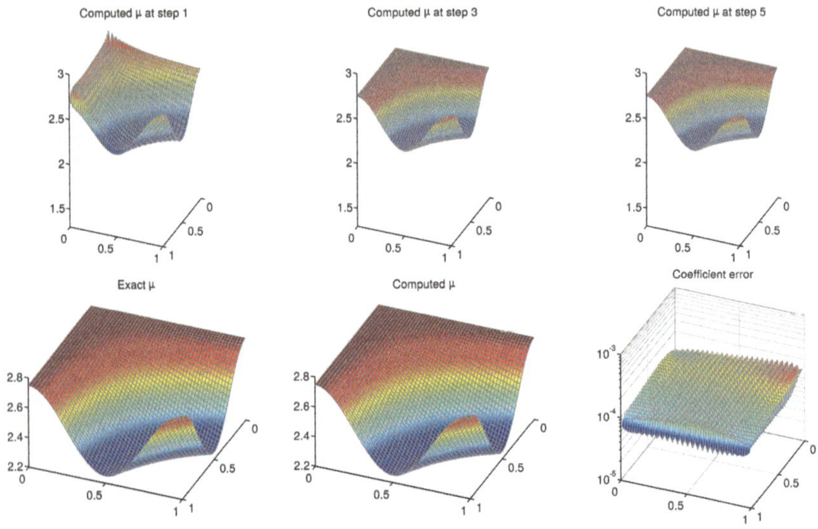

Fig. 7.1 Hager–Zhang method

Table 7.1 Numerical results

Method	J evals	∇J evals	$\nabla^2 J$ evals	Iter.	L^2 error
Hager and Zhang	14, 556	13, 908	–	8	3.556e−05
φ-Divergence	17, 365	16, 593	–	9	2.853e−05
Bregman	24, 292	23, 213	–	12	2.444e−05
Quadratic-φ	**21**	**19**	**19**	**3**	**3.896e−08**
TR using Tikhonov	24,345	24,345	–	9	7.185e−04
TR using Tikhonov (second-order)	9	9	9	8	7.185e−04

7.3.2 Hager and Zhang's Proximal Point Method Using φ-Divergence

The first variant of the classical proximal algorithm we examine replaces the proximal regularization term in (7.17) by what are known as φ-divergences (see Kanzow [21] for further details). For their definition, let Φ denote the class of closed, proper and convex functions $\varphi : \mathbb{R} \to (-\infty, \infty]$ which have domain(φ) $\subset$ $[0, \infty)$ and which possess the following properties:

1. φ is twice continuously differentiable on $\text{int}(\text{domain}(\varphi)) = (0, +\infty)$.
2. φ is strictly convex on its domain.
3. $\lim_{t \to 0+} \varphi'(t) = -\infty$.
4. $\varphi(1) = \varphi'(1) = 0$ and $\varphi''(1) > 0$.
5. There exists $\nu \in \left(\frac{1}{2}\varphi''(1), \varphi''(1)\right)$ such that

$$\left(1-\frac{1}{t}\right)\left(\varphi''(1)+\nu(t-1)\right) \le \varphi'(t) \le \varphi''(1)(t-1) \quad \forall t>0.$$

Then for some $\varphi \in \Phi$, the φ-divergence between two $x, y \in \mathbb{R}^n_+$ is given by

$$d_\varphi(x,y) = \sum_{i=1}^{n} y_i \varphi\left(\frac{x_i}{y_i}\right). \tag{7.23}$$

A few examples of suitable φ functions are

$$\varphi_1(t) = t\log t - t + 1,$$
$$\varphi_2(t) = -\log t + t - 1,$$
$$\varphi_3(t) = \left(\sqrt{t}-1\right)^2.$$

Taking φ_1 above yields the φ-divergence

$$d_{\varphi_1}(x,y) = \sum_{i=1}^{n} x_i \log\frac{x_i}{y_i} + y_i - x_i. \tag{7.24}$$

We now replace the proximal regularization term in (7.17) with (7.24) giving

$$\mathcal{J}_{\varphi_1}(\mu) = J(\mu) + \theta^k d_{\varphi_1}(\mu, \mu^k) \tag{7.25}$$

and subsequently the proximal-like iteration

$$\mu^{k+1} = \arg\min_{\mu\in A} \mathcal{J}_{\varphi_1}(\mu). \tag{7.26}$$

Substituting $\mathcal{J}_{\varphi_1}(\mu)$ for $\mathcal{J}_P(\mu)$ into Algorithm 1 yields the φ-divergence proximal-like algorithm. Numerical results using $\mathcal{J}_{\varphi_1}$ are provided in Fig. 7.2 and in Table 7.1.

7.3.3 Hager and Zhang's Proximal Point Method Using Bregman Functions

Continuing in the manner of φ-divergences above, the next modified algorithm replaces the proximal regularization term with another strictly convex distance function defined by

$$D_\psi(x,y) = \psi(x) - \psi(y) - \nabla\psi(y)^T(x-y), \tag{7.27}$$

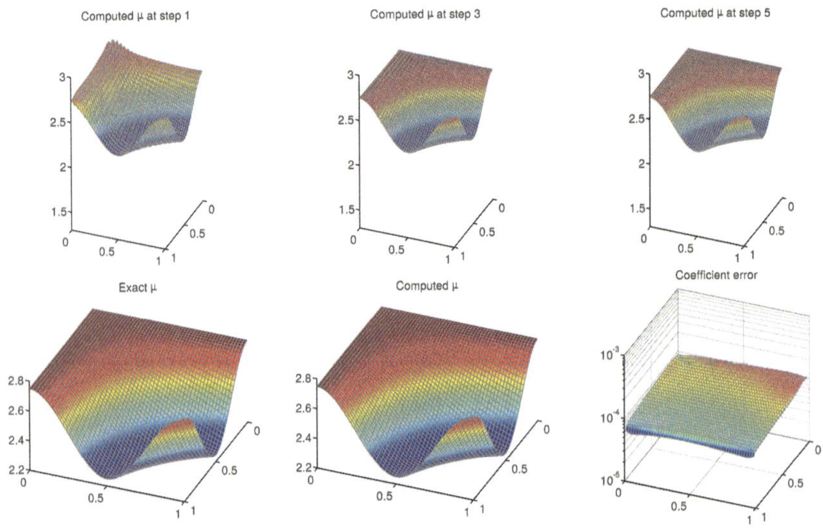

Fig. 7.2 Hager–Zhang method using φ-divergence

where ψ is a so-called Bregman function.

We now review the definition of a Bregman function. Let S be an open and convex set and a let $\psi : \overline{S} \to \mathbb{R}$ be a given mapping. If ψ is a Bregman function, it must satisfy the following criteria:

1. ψ is strictly convex and continuous on $\overline{S}$.
2. ψ is continuously differentiable in S.
3. The partial level set
$$L_\alpha = \{y \in \overline{S} | D_\psi(x, y) \leq \alpha\}$$
is bounded for every $x \in \overline{S}$.
4. If $\{y^k\} \subset S$ converges to x, then $\lim_{k\to\infty} D_\psi(x, y^k) = 0$.

A few examples of Bregman functions are given below:

$$\psi_1(x) = \frac{1}{2}\|x\|^2 \quad \text{with } S = \mathbb{R}^n,$$

$$\psi_2(x) = \sum_{i=1}^{n} x_i \log x_i - x_i \quad \text{with } S = \mathbb{R}^n_+,$$

$$\psi_3(x) = -\sum_{i=1}^{n} \log x_i \quad \text{with } S = \mathbb{R}^n_+.$$

We note that the corresponding distance function for ψ_1 is

$$D_{\psi_1}(x, y) = \frac{1}{2}\|x - y\|^2,$$

which recovers the original proximal regularization term introduced in (7.17). Similarly, for ψ_2 we have

$$D_{\psi_2}(x, y) = \sum_{i=1}^{n} x_i \log \frac{x_i}{y_i} + y_i - x_i$$

which is equivalent to the φ-divergence given by (7.24).

Taking

$$D_{\psi_3}(x, y) = \sum_{i=1}^{n} \frac{x_i}{y_i} - \log \frac{x_i}{y_i} - 1$$

we can again replace the proximal regularization term to get the functional

$$\mathcal{J}_{\psi_3}(\mu) = J(\mu) + \theta^k D_{\psi_3}(\mu, \mu^k) \tag{7.28}$$

and the corresponding subproblem

$$\mu^{k+1} = \arg\min_{\mu \in A} \mathcal{J}_{\psi_3}(\mu). \tag{7.29}$$

We present numerical results using the functional $\mathcal{J}_{\psi_3}$ in Fig. 7.3 and in Table 7.1.

7.3.4 Proximal-Like Methods Using Modified φ-Divergence

To solve the subproblem (7.21) in Algorithm 1 using the φ-divergence method, we made use of a fast conjugate-gradient-based trust-region method.

To avoid possible ill-conditioning of the Hessian (see [21] for a full discussion), we replace the original φ-divergence function with

$$\tilde{d}_\varphi(x, y) = \sum_{i-1}^{n} y_i^2 \varphi\left(\frac{x_i}{y_i}\right) \tag{7.30}$$

giving

$$\nabla^2_{xx} \tilde{d}_\varphi(x, y) = \sum_{i=1}^{n} \varphi''\left(\frac{x_i}{y_i}\right) e_i e_i^T \tag{7.31}$$

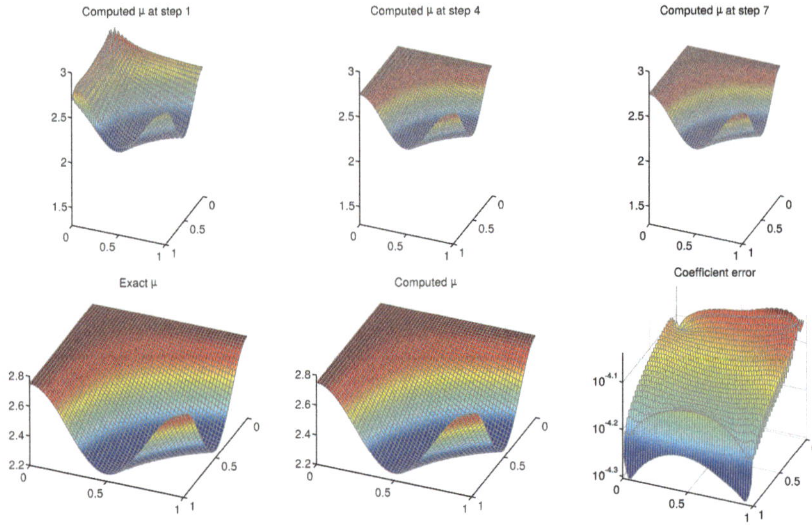

Fig. 7.3 Hager–Zhang method using Bregman function

where e_i is the ith unit basis vector of $\mathbb{R}^n$.

Again taking

$$\mathcal{J}_{\tilde{\varphi}_1}(\mu) = J(\mu; z; \beta) + \tilde{d}_{\varphi_1}(\mu, \mu^k) \tag{7.32}$$

we have the iteration

$$\mu^{k+1} = \arg\min_{\mu \in A} \mathcal{J}_{\tilde{\varphi}_1}(\mu). \tag{7.33}$$

However, now equipped with the Hessian of $\mathcal{J}_{\tilde{\varphi}_1}(\mu)$, we can apply a full Newton-type method to solve the associated subproblem. The numerical performance of this approach is shown in Fig. 7.4.

7.4 Numerical Experiments

In this section we consider a representative example of an elastography inverse problem for the recovery of a variable μ on a two dimensional isotropic domain $\Omega = (0,1) \times (0,1)$ with boundary $\partial\Omega = \Gamma_1 \cup \Gamma_2$. Γ_1, where the Dirichlet boundary conditions hold, is taken as the top boundary of the domain while Γ_2, where the Neumann conditions hold, is taken as the union of the remaining boundary points.

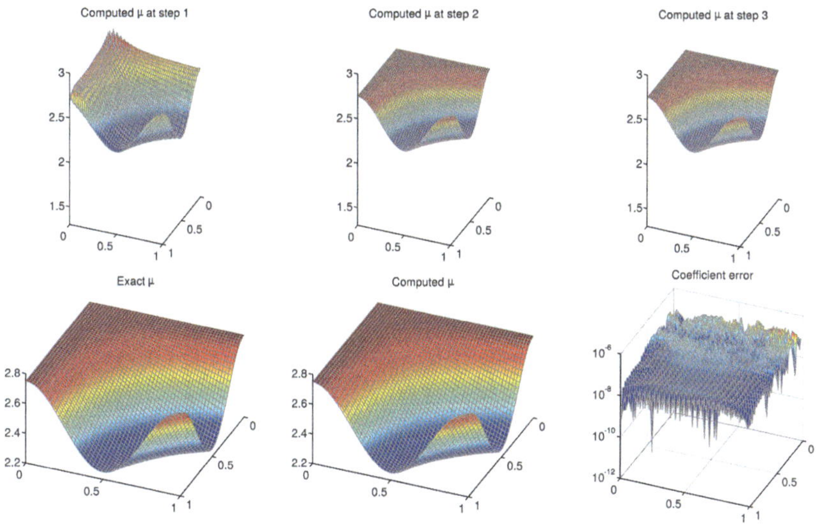

Fig. 7.4 Hager–Zhang using quadratic φ-divergence

The inverse problem is solved on a 50×50 quadrangular mesh with 2,601 quadrangles and 69,254,226 total degrees of freedom (see [8] for a more thorough discussion of the discretization and the use of mixed finite element methods).

In keeping with the near incompressibility inherent in the problem, λ is taken as a large constant, particularly $\lambda = 10^6$.

The functions defining the coefficient, load, and boundary conditions are as follows:

$$\mu(x,y) = 2.5 + \frac{1}{4}\cos(2\pi xy), \qquad f(x,y) = \begin{bmatrix} 1 + \frac{1}{10}x^2 \\ \frac{1}{10}y \end{bmatrix},$$

$$f_1(x,y) = \begin{bmatrix} 0 \\ 0 \end{bmatrix} \text{ on } \Gamma_1, \qquad f_2(x,y) = \begin{bmatrix} \frac{1}{2} + x^2 \\ 0 \end{bmatrix} \text{ on } \Gamma_2.$$

7.4.1 *Comparative Performance Analysis*

For all the numerical results shown in Table 7.1 and in the various figures (Figs. 7.5 and 7.6), we take a constant $\mu^0(x,y) = 1$. For the proximal methods, we take $\tau = 10^{-4}$ and $\eta = 1.99$. For the experiments using Tikhonov regularization, we take the regularization parameter $\beta = 10^{-4}$. Equivalent stopping criteria were applied to both the experiments using proximal methods ($\nabla \mathcal{J} \leq 10^{-10}$) and Tikhonov regularization ($\nabla J \leq 10^{-10}$).

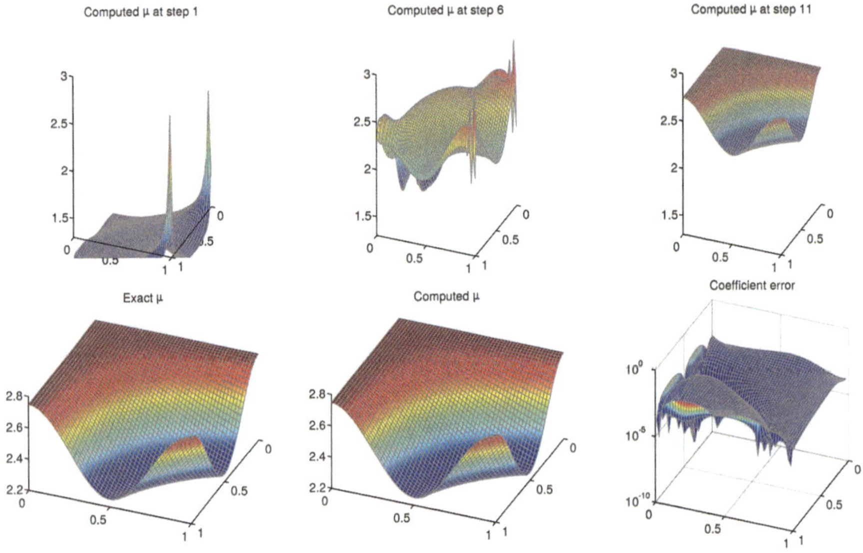

Fig. 7.5 Trust-region using Tikhonov regularization

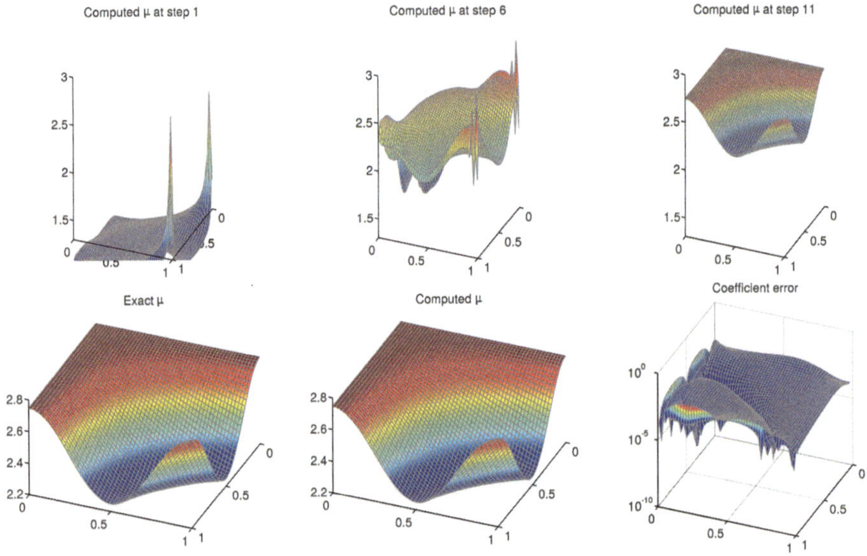

Fig. 7.6 Trust-region using Tikhonov regularization (quadratic)

As the table shows quite dramatically, the second-order Newton method outperforms the other proximal point methods in the solution of the elastography inverse problem. We note, however, that although the cost of calculating the Hessian is small, given that it does not depend on the parameter μ, the inversion of the term $(K + M)$ in (7.15) destroys any sparsity structure inherent in the mass and stiffness matrices and thus places some practical limits on the scale of this method.

Furthermore, the same example was run using Tikhonov regularization and a trust-region dogleg optimization method. For a complete comparison with the other methods, both a quasi-Newton method and a full second-order Newton method were applied to the optimization problem. As can again be seen from Table 7.1, the proximal methods yield results readily comparable to those using Tikhonov regularization. Overall, the proximal methods perform largely better than their Tikhonov counterparts, with fewer function and gradient evaluations and lower error for similar algorithmic stopping criteria. Of particular note is that, although the second-order proximal algorithm has a slightly higher computational cost over its counterpart, the Tikhonov method produces error several orders of magnitude larger.

7.5 Conclusion

In this work, we have examined the equation error approach to the solution of the tumor identification inverse problem and investigated the performance of several proximal-like algorithms for solving the related optimization problem. Several numerical experiments were performed that demonstrate the potential advantages of proximal methods over other commonly-used approaches such as those using Tikhonov regularization, and in particular, where the choice of an optimal regularization parameter is unlikely or impossible.

Given the significant performance advantages afforded by proximal algorithms using full the Newton methods, these preliminary results suggest the exploration of improvements based on these techniques. In future work, the authors seek to apply and perform in-depth analysis of hybrid proximal techniques such as those outlined in [28] for the tumor identification inverse problem, as well as explore second-order methods for solving the proximal subproblem that can overcome the scaling problems inherent in the calculation of the objective function's Hessian.

Acknowledgements The authors are grateful to the referees for their careful reading and suggestions that brought substantial improvements to the work. The work of A.A. Khan is partially supported by RIT's COS D-RIG Acceleration Research Funding Program 2012–2013 and a grant from the Simons Foundation (#210443 to Akhtar Khan).

References

1. Acar, R.: Identification of the coefficient in elliptic equations. SIAM J. Control Optim. **31**, 1221–1244 (1993)
2. Aguilo, M.A., Aquino, W., Brigham, J.C., Fatemi, M.: An inverse problem approach for elasticity imaging through vibroacoustics. IEEE Trans. Med. Imaging **29**, 1012–1021 (2010)

3. Al-Jamal, M.F., Gockenbach, M.S.: Stability and error estimates for an equation error method for elliptic equations. Inverse Prob. **28**, 095006, 15 pp. (2012)
4. Ammari, H., Garapon, P., Jouve, F.: Separation of scales in elasticity imaging: a numerical study. J. Comput. Math. **28**, 354–370 (2010)
5. Arnold, A., Reichling, S., Bruhns, O., Mosler, J.: Efficient computation of the elastography inverse problem by combining variational mesh adaption and clustering technique. Phys. Med. Biol. **55**, 2035–2056 (2010)
6. Braess, D.: Finite Elements: Theory, Fast Solvers, and Applications in Solid Mechanics, 3rd edn. Cambridge University Press, Cambridge (2007)
7. Cahill, N., Jadamba, B., Khan, A.A., Sama, M., Winkler, B.: A first-order adjoint and a second-order hybrid method for an energy output least squares elastography inverse problem of identifying tumor location. Bound. Value Prob. **2013**, 263 (2013). doi:10.1186/1687-2770-2013-263
8. Crossen, E., Gockenbach, M.S., Jadamba, B., Khan, A.A., Winkler, B.: An equation error approach for the elasticity imaging inverse problem for predicting tumor location. Comput. Math. Appl. **67**, 122–135 (2014)
9. Doyley, M.M., Meaney, P.M., Bamber, J.C.: Evaluation of an iterative reconstruction method for quantitative elastography. Phys. Med. Biol. **45**, 1521–1540 (2000)
10. Doyley, M.M., Jadamba, B., Khan, A.A., Sama, M., Winkler, B.: A new energy inversion for parameter identification in saddle point problems with an application to the elasticity imaging inverse problem of predicting tumor location. Numer. Funct. Anal. Optim. **35**, 984–1017 (2014)
11. Fehrenbach, J., Masmoudi, M., Souchon, R., Trompette, P.: Detection of small inclusions by elastography. Inverse Prob. **22**, 1055–1069 (2006)
12. Gockenbach, M.S., Khan, A.A.: Identification of Lamé parameters in linear elasticity: a fixed point approach. J. Ind. Manag. Optim. **1**, 487–497 (2005)
13. Gockenbach, M.S., Khan, A.A.: An abstract framework for elliptic inverse problems. Part 1: an output least-squares approach. Math. Mech. Solids **12**, 259–276 (2007)
14. Gockenbach, M.S., Khan, A.A.: An abstract framework for elliptic inverse problems. Part 2: an augmented Lagrangian approach. Math. Mech. Solids **14**, 517–539 (2009)
15. Gockenbach, M.S., Jadamba, B., Khan, A.A.: Numerical estimation of discontinuous coefficients by the method of equation error. Int. J. Math. Comput. Sci. **1**, 343–359 (2006)
16. Gockenbach, M.S., Jadamba, B., Khan, A.A.: Equation error approach for elliptic inverse problems with an application to the identification of Lamé parameters. Inverse Prob. Sci. Eng. **16**, 349–367 (2008)
17. Hager, W., Zhang, H.: Self-adaptive inexact proximal point methods. Comput. Optim. Appl. **39**, 161–181 (2008)
18. Han, D.: A generalized proximal-point-based prediction-correction method for variational inequality problems. J. Comput. Appl. Math. **221**, 183–193 (2008)
19. Jadamba, B., Khan, A.A., Raciti, F.: On the inverse problem of identifying Lamé coefficients in linear elasticity. Comput. Math. Appl. **56**, 431–443 (2008)
20. Jadamba, B., Khan, A.A., Rus, G., Sama, M., Winkler, B.: A new convex inversion for parameter identification in saddle point problems with an application to the elasticity imaging inverse problem of prediction tumor location. SIAM Appl. Math. **74**, 1486–1510 (2014)
21. Kanzow, C.: Proximal-like methods for convex minimization problems. In: Qi, L., Teo, K., Yang, X. (eds.) Optimization and Control with Applications, pp. 369–392. Springer, New York (2005)
22. Kaplan, A., Tichatschke, R.: Proximal point method and elliptic regularization. Nonlinear Anal. **71**, 4525–4543 (2009)
23. Khan, A.A., Rouhani, B.D.: Iterative regularization for elliptic inverse problems. Comput. Math. Appl. **54**, 850–860 (2007)
24. McLaughlin, J., Yoon, J.R.: Unique identifiability of elastic parameters from time-dependent interior displacement measurement. Inverse Prob. **20**, 25–45 (2004)
25. Oberai, A.A., Gokhale, N.H., Feijóo, G.R.: Solution of inverse problems in elasticity imaging using the adjoint method. Inverse Prob. **19**, 297–313 (2003)

26. Parente, L.A., Lotito, P.A., Solodov, M.V.: A class of inexact variable metric proximal point algorithms. SIAM J. Optim. **19**, 240–260 (2008)
27. Rockafellar, R.T.: Monotone operators and the proximal point algorithm. SIAM J. Control Optim. **14**, 877–898 (1976)
28. Tossings, P.: Mixing proximal regularization, penalization and parallel decomposition in convex programming. Advances in optimization (Lambrecht, 1991), Lecture Notes in Econom. and Math. Systems, Springer, Berlin, **382**, 85–99 (1992)
29. Tossings, P.: The perturbed proximal point algorithm and some of its applications. Appl. Math. Optim. **29**, 125–159 (1994)

Chapter 8
Discontinuous Galerkin Methods for an Elliptic Variational Inequality of Fourth-Order

Fei Wang, Weimin Han, Jianguo Huang, and Tianyi Zhang

Abstract Discontinuous Galerkin (DG) methods are studied for solving an elliptic variational inequality of fourth-order. Numerous discontinuous Galerkin schemes for the Kirchhoff plate bending problem are extended to the variational inequality. Numerical results are presented to illustrate convergence orders of the different methods.

Keywords Variational inequality of fourth-order • Discontinuous Galerkin method • Kirchhoff plates

AMS Classification. 65N30, 49J40

F. Wang (✉)
Department of Mathematics, Pennsylvania State University, University Park, PA 16802, USA

School of Mathematics and Statistics, Huazhong University of Science and Technology, Wuhan 430074, China
e-mail: wangfeitwo@163.com

W. Han
Department of Mathematics & Program in Applied Mathematical and Computational Sciences, University of Iowa, Iowa City, IA 52242, USA

School of Mathematics and Statistics, Xi'an Jiaotong University, Xi'an, Shaanxi 710049, China
e-mail: weimin-han@uiowa.edu

J. Huang
Department of Mathematics, and MOE–LSC, Shanghai Jiao Tong University, Shanghai 200240, China

Division of Computational Science, E-Institute of Shanghai Universities, Shanghai Normal University, Shanghai, China
e-mail: jghuang@sjtu.edu.cn

T. Zhang
Program in Applied Mathematical and Computational Sciences, University of Iowa, Iowa City, IA 52242, USA
e-mail: tianyi-zhang@uiowa.edu

W. Han et al. (eds.), *Advances in Variational and Hemivariational Inequalities*,
Advances in Mechanics and Mathematics 33, DOI 10.1007/978-3-319-14490-0_8

8.1 Introduction

In this chapter, we introduce and study several discontinuous Galerkin (DG) methods for solving an elliptic variational inequality of fourth-order.

8.1.1 Discontinuous Galerkin Methods

Discontinuous Galerkin methods form an important family of nonconforming finite element methods for boundary value or initial-boundary value problems of hyperbolic, parabolic and elliptic partial differential equations. We refer to [10] for a historical account about DG methods. Discontinuous Galerkin methods use piecewise smooth functions without much global smoothness to approximate the problem solution, and connect the information between two neighboring elements by the so-called numerical traces. The practical interest in DG methods is due to their flexibility in mesh design and adaptivity, in that they allow elements of arbitrary shapes, irregular meshes with hanging nodes, and the discretionary local shape function spaces. In addition, the increase of the locality in discretization enhances the degree of parallelizability.

There are two basic approaches to construct DG methods for linear elliptic boundary value problems. The first approach is to choose an appropriate bilinear form that contains penalty terms to penalize jumps across neighboring elements to make the scheme stable. The second approach is to choose appropriate numerical fluxes to make the method consistent, conservative and stable. In [1, 2], Arnold, Brezzi, Cockburn, and Marini provided a unified error analysis of DG methods for linear elliptic boundary value problems of second-order and succeeded in building a bridge between these two families of DG methods, establishing a framework to understand their properties, differences and the connections between them. In [21], numerous DG methods were extended for solving elliptic variational inequalities of second-order, and a priori error estimates were established, which are of optimal order for linear elements. DG methods for the Signorini problem and a quasistatic contact problem were also studied in [22, 23], respectively. In this chapter, we study DG methods to solve an elliptic variational inequality of fourth-order for the Kirchhoff plates. The novel difficulty in constructing stable DG methods for such problems is caused by their high order of four. The major known DG methods for the biharmonic equation in the literature are primal DG methods, namely variations of interior penalty (IP) methods [4, 5, 7, 12, 16–18, 20]. Fully discontinuous IP methods, which cover meshes with hanging nodes and locally varying polynomial degrees, ideally suited for hp-adaptivity, were investigated systematically in [16–18, 20] for biharmonic problems. In [12], a C^0 IP formulation was introduced for Kirchhoff plates and quasi-optimal error estimates were obtained for smooth solutions. Unlike fully discontinuous Galerkin methods, C^0 type DG methods do not "double" the degrees of freedom at element boundaries. To consider the C^0 IP method under a weak regularity assumption on the solution, a rigorous error analysis was presented in [7]. A weakness of this method is that the penalty parameter can not be precisely quantified a priori, and the penalty parameter must be chosen suitably large to guarantee stability. However, a large penalty parameter has a negative impact on accuracy. Based on this observation, a C^0 DG (CDG) method was introduced in [24], where the stability condition can be precisely quantified. In [15], a consistent and stable CDG method, called the LCDG method, was derived for the Kirchhoff plate bending problem, which can be viewed as an extension of the LDG method studied in [8, 9]. We will extend these three methods and propose two other new CDG methods to solve the elliptic variational inequality of fourth-order.

8.1.2 Kirchhoff Plate Bending Problem

We now describe a Kirchhoff plate bending problem. Let $\Omega \subset \mathbb{R}^2$ be a bounded polygonal domain with boundary Γ. The boundary value problem of a clamped Kirchhoff plate under a given scaled vertical load $f \in L^2(\Omega)$ is (cf. [19])

$$\begin{cases} \sum_{i,j=1}^{2} \mathcal{M}_{ij,ij}(u) + f = 0 \ \text{ in } \Omega, \\ u = \partial_\nu u = 0 \ \text{ on } \Gamma, \end{cases} \tag{8.1}$$

where

$$\mathcal{M}_{ij}(u) := -(1-\kappa)\partial_{ij}u - \kappa \sum_{k=1}^{2} \partial_{kk}u\delta_{ij}, \quad 1 \le i, j \le 2,$$

δ_{ij} is the usual Kronecker delta, $\kappa \in (0, 0.5)$ denotes the Poisson ratio of an elastic thin plate occupying the region Ω and $\boldsymbol{\nu}$ stands for the unit outward normal vector on Γ. As in [15], we introduce an auxiliary matrix-valued function by

$$\boldsymbol{\sigma} := -(1-\kappa)\nabla^2 u - \kappa\, \mathrm{tr}(\nabla^2 u)\boldsymbol{I}, \tag{8.2}$$

where $\boldsymbol{I}$ is the identity matrix of order 2 and $\mathrm{tr}(\cdot)$ is the trace operation on matrices. Here, we denote the gradient of v by ∇v and the Hessian of v by $\nabla^2 v$, i.e.,

$$\nabla^2 v := \nabla(\nabla v) = \nabla((\partial_1 v, \partial_2 v)^t) = \begin{pmatrix} \partial_{11}v \ \partial_{12}v \\ \partial_{21}v \ \partial_{22}v \end{pmatrix}.$$

Then, the problem (8.1) can be rewritten as

$$\begin{cases} \dfrac{1}{1-\kappa}\boldsymbol{\sigma} - \dfrac{\kappa}{1-\kappa^2}(\mathrm{tr}\boldsymbol{\sigma})\boldsymbol{I} = -\nabla^2 u \ \text{ in } \Omega, \\ -\nabla \cdot (\nabla \cdot \boldsymbol{\sigma}) = f \ \text{ in } \Omega, \\ u = \partial_\nu u = 0 \ \text{ on } \Gamma. \end{cases} \tag{8.3}$$

For a vector-valued function $\boldsymbol{v} = (v_1, v_2)^t$ and a matrix-valued function $\boldsymbol{\sigma} = (\sigma_{ij})_{2\times 2}$, we define their divergences by

$$\nabla \cdot \boldsymbol{v} := v_{1,1} + v_{2,2}, \quad \nabla \cdot \boldsymbol{\sigma} := (\sigma_{11,1} + \sigma_{21,2}, \sigma_{12,1} + \sigma_{22,2})^t.$$

We denote the normal and tangential components of a vector $\boldsymbol{v}$ on the boundary by $v_\nu = \boldsymbol{v} \cdot \boldsymbol{\nu}$ and $\boldsymbol{v}_\tau = \boldsymbol{v} - v_\nu \boldsymbol{\nu}$. Similarly, for a tensor $\boldsymbol{\sigma}$, we define its normal component $\sigma_\nu = \boldsymbol{\sigma}\boldsymbol{\nu} \cdot \boldsymbol{\nu}$ and tangential component $\boldsymbol{\sigma}_\tau = \boldsymbol{\sigma}\boldsymbol{\nu} - \sigma_\nu \boldsymbol{\nu}$. We have the decomposition formula

$$(\boldsymbol{\sigma}\boldsymbol{\nu}) \cdot \boldsymbol{v} = (\sigma_\nu \boldsymbol{\nu} + \boldsymbol{\sigma}_\tau) \cdot (v_\nu \boldsymbol{\nu} + \boldsymbol{v}_\tau) = \sigma_\nu v_\nu + \boldsymbol{\sigma}_\tau \cdot \boldsymbol{v}_\tau.$$

For two matrices $\boldsymbol{\tau}$ and $\boldsymbol{\sigma}$, their double dot inner product and corresponding norm are $\boldsymbol{\sigma} : \boldsymbol{\tau} = \sum_{i,j=1}^{2} \sigma_{ij}\tau_{ij}$ and $|\boldsymbol{\tau}| = (\boldsymbol{\tau} : \boldsymbol{\tau})^{1/2}$.

The following result is very useful for the analysis of DG methods, which can be verified directly by integration by part.

Lemma 8.1. *Let D be a bounded domain with a Lipschitz boundary ∂D. For a symmetric matrix-valued function $\boldsymbol{\tau}$ and a scalar function v, the following two identities hold*

$$\int_D v\,\nabla \cdot (\nabla \cdot \boldsymbol{\tau})\,dx = \int_D \nabla^2 v : \boldsymbol{\tau}\,dx - \int_{\partial D} \nabla v \cdot (\boldsymbol{\tau}\boldsymbol{n})\,ds + \int_{\partial D} v\,\boldsymbol{n} \cdot (\nabla \cdot \boldsymbol{\tau})\,ds,$$

$$\int_D \nabla^2 v : \boldsymbol{\tau}\,dx = -\int_D \nabla v \cdot (\nabla \cdot \boldsymbol{\tau})\,dx + \int_{\partial D} \nabla v \cdot (\boldsymbol{\tau}\boldsymbol{n})\,ds,$$

whenever the terms appearing on both sides of the above identities make sense. Here $\boldsymbol{n}$ is the unit outward normal to ∂D.

Multiplying the second equation in (8.3) by a test function $v \in H_0^2(\Omega)$ and noticing $v = \partial_\nu v = 0$, we get the following equation by applying Lemma 8.1,

$$-\int_\Omega \boldsymbol{\sigma} : \nabla^2 v\,dx = \int_\Omega f v\,dx. \tag{8.4}$$

With the definition of $\boldsymbol{\sigma}$, the weak formulation of the problem (8.3) can be derived from (8.4) as follows:

$$\text{Find } u \in H_0^2(\Omega): \quad a(u,v) = (f,v) \quad \forall\, v \in H_0^2(\Omega), \tag{8.5}$$

where the bilinear form is

$$a(u,v) = \int_\Omega \left[\Delta u\,\Delta v + (1-\kappa)\,(2\,\partial_{12}u\,\partial_{12}v - \partial_{11}u\,\partial_{22}v - \partial_{22}u\,\partial_{11}v)\right] dx, \tag{8.6}$$

and the linear form is

$$(f, v) = \int_{\Omega} f\, v\, dx.$$

In this chapter, we consider an elliptic variational inequality (EVI) of the fourth-order for Kirchhoff plates [11]:

$$\text{Find } u \in K: \quad a(u, v-u) \geq (f, v-u) \quad \forall\, v \in K. \tag{8.7}$$

Here,

$$K = \left\{ v \in H^2(\Omega) \cap H_0^1(\Omega) : \partial_\nu v \geq 0 \text{ on } \Gamma \right\}. \tag{8.8}$$

Applying the standard theory on elliptic variational inequalities (e.g., [3, 13]), we know the problem (8.7) has a unique solution $u \in K$. This variational inequality describes a simply supported plate. The displacement u is held fixed on the boundary, and the rotation of the plate is unilateral on the boundary.

In error analysis of numerical solutions for the problem (8.7), we need to take advantage of pointwise relations satisfied by the solution u.

Proposition 8.2. *Assume the solution of the problem* (8.7) *has the regularity* $u \in H^3(\Omega)$. *Then,*

$$\begin{aligned} &-\nabla \cdot (\nabla \cdot \boldsymbol{\sigma}) = f \quad \text{a.e. in } \Omega, \\ &\boldsymbol{\sigma}_\tau = \boldsymbol{0},\ \sigma_\nu \leq 0,\ \partial_\nu u \geq 0,\ \sigma_\nu \partial_\nu u = 0 \quad \text{a.e. on } \Gamma. \end{aligned} \tag{8.9}$$

Proof. Note that $\boldsymbol{\sigma}$ is defined by (8.2). Then $\boldsymbol{\sigma} \in H^1(\Omega)^{2\times 2}$. We rewrite (8.7) as

$$\int_{\Omega} \left[-\boldsymbol{\sigma} : \nabla^2(v-u) - f\,(v-u) \right] dx \geq 0 \quad \forall\, v \in K.$$

Take $v = u \pm \varphi$ for any $\varphi \in C_0^\infty(\Omega)$ to obtain

$$-\int_{\Omega} \boldsymbol{\sigma} : \nabla^2 \varphi\, dx = \int_{\Omega} f\, \varphi\, dx \quad \forall\, \varphi \in C_0^\infty(\Omega).$$

Thus,

$$-\nabla \cdot (\nabla \cdot \boldsymbol{\sigma}) = f \quad \text{in the sense of distribution.}$$

Since $f \in L^2(\Omega)$, we deduce that $\nabla \cdot (\nabla \cdot \boldsymbol{\sigma}) \in L^2(\Omega)$ and

$$-\nabla \cdot (\nabla \cdot \boldsymbol{\sigma}) = f \quad \text{a.e. in } \Omega.$$

This is the first relation in (8.9).

Since $\nabla \cdot \boldsymbol{\sigma} \in L^2(\Omega)^2$ and $\nabla \cdot (\nabla \cdot \boldsymbol{\sigma}) \in L^2(\Omega)$, we have

$$-\int_\Omega \nabla \cdot (\nabla \cdot \boldsymbol{\sigma})\, v\, dx = \int_\Omega (\nabla \cdot \boldsymbol{\sigma}) \cdot \nabla v\, dx \quad \forall\, v \in C_0^\infty(\Omega).$$

From this relation, we obtain

$$-\int_\Omega \nabla \cdot (\nabla \cdot \boldsymbol{\sigma})\, v\, dx = \int_\Omega (\nabla \cdot \boldsymbol{\sigma}) \cdot \nabla v\, dx \quad \forall\, v \in H_0^1(\Omega).$$

Therefore, for any $v \in H_0^1(\Omega) \cap H^2(\Omega)$,

$$\begin{aligned}-\int_\Omega \nabla \cdot (\nabla \cdot \boldsymbol{\sigma})\, v\, dx &= \int_\Omega (\nabla \cdot \boldsymbol{\sigma}) \cdot \nabla v\, dx\\ &= \int_\Gamma (\boldsymbol{\sigma} \boldsymbol{\nu}) \cdot \nabla v\, ds - \int_\Omega \boldsymbol{\sigma} : \nabla^2 v\, dx,\end{aligned}$$

i.e.,

$$a(u, v) = \int_\Omega f\, v\, dx - \int_\Gamma (\boldsymbol{\sigma} \boldsymbol{\nu}) \cdot \nabla v\, ds \quad \forall\, v \in H_0^1(\Omega) \cap H^2(\Omega).$$

Recalling the inequality (8.7), we then have

$$-\int_\Gamma \nabla(v - u) \cdot (\boldsymbol{\sigma} \boldsymbol{\nu})\, ds \geq 0 \quad \forall\, v \in K. \tag{8.10}$$

In (8.10), we choose $v = 0$ and $2u$ in turn to obtain

$$\int_\Gamma \nabla u \cdot (\boldsymbol{\sigma} \boldsymbol{\nu})\, ds = 0. \tag{8.11}$$

Hence,

$$\int_\Gamma \nabla v \cdot (\boldsymbol{\sigma} \boldsymbol{\nu})\, ds \leq 0 \quad \forall\, v \in K. \tag{8.12}$$

By the arbitrariness of $v \in K$, it follows that $\boldsymbol{\sigma}_\tau = \mathbf{0}$. Then we get

$$\int_\Gamma \sigma_\nu \partial_{\boldsymbol{\nu}} v\, ds \leq 0 \quad \forall\, v \in K. \tag{8.13}$$

By the arbitrariness of $\partial_{\boldsymbol{\nu}} v \geq 0$ on Γ for v in K, we have $\sigma_\nu \leq 0$ a.e. on Γ. Back to (8.11), we further deduce $\sigma_\nu \partial_{\boldsymbol{\nu}} u = 0$ a.e. on Γ. So the relations on the boundary Γ in (8.9) hold. □

Throughout the chapter, we assume the solution of the problem (8.7) has the regularity $u \in H^3(\Omega)$. In [14, pp. 323–327], one can find regularity results $u \in H^3(\Omega)$ for solutions of some variational inequalities of fourth-order.

The rest of the chapter is as follows. In Sect. 8.2, we introduce some notations and derive some C^0 discontinuous Galerkin methods for solving the Kirchhoff plate bending problem, and extend them to solve the elliptic variational inequality of fourth-order. In Sect. 8.3, consistency of the CDG methods, boundedness and stability of the bilinear forms are presented. In the final section, we report results from a numerical example.

8.2 DG Methods for Kirchhoff Plate Problem

8.2.1 Notations

We introduce some notations frequently used later on. For a given function space B, let $(B)_s^{2\times 2} := \left\{\boldsymbol{\tau} \in (B)^{2\times 2} : \boldsymbol{\tau}^t = \boldsymbol{\tau}\right\}$. Given a bounded set $D \subset \mathbb{R}^2$ and a positive integer m, $H^m(D)$ is the usual Sobolev space with the corresponding norm $\|\cdot\|_{m,D}$ and semi-norm $|\cdot|_{m,D}$, which are abbreviated by $\|\cdot\|_m$ and $|\cdot|_m$, respectively, when D is chosen as Ω. $\|\cdot\|_D$ is the norm of the Lebesgue space $L^2(D)$. We assume Ω is a polygonal domain and denote by $\{\mathcal{T}_h\}_h$ a family of triangulations of $\overline{\Omega}$, with the minimal angle condition satisfied. Let $h_T = \mathrm{diam}(T)$ and $h = \max\{h_T : T \in \mathcal{T}_h\}$. For a triangulation $\mathcal{T}_h$, let $\mathcal{E}_h$ be the union of all edges. We have the non-overlapping decomposition $\mathcal{E}_h = \mathcal{E}_h^i \cup \mathcal{E}_h^b$, where $\mathcal{E}_h^i \subset \mathcal{E}_h$ is the union of all interior edges, i.e., the union of all edges in $\mathcal{E}_h$ that do not lie on Γ, and similarly, $\mathcal{E}_h^b \subset \mathcal{E}_h$ is the union of the edges on Γ. For any $e \in \mathcal{E}_h$, denote by h_e its length. Related to the triangulation $\mathcal{T}_h$, let

$$
\begin{aligned}
\boldsymbol{\Sigma} &:= \left\{\boldsymbol{\tau} \in \left(L^2(\Omega)\right)_s^{2\times 2} : \tau_{ij}|_T \in H^1(T)\ \forall\, T \in \mathcal{T}_h,\ i,j = 1,2\right\},\\
V &:= \left\{v \in H_0^1(\Omega) : v|_T \in H^2(T)\ \forall\, T \in \mathcal{T}_h\right\}.
\end{aligned}
$$

The corresponding finite element spaces are

$$
\begin{aligned}
\boldsymbol{\Sigma}_h &:= \left\{\boldsymbol{\tau}_h \in \left(L^2(\Omega)\right)_s^{2\times 2} : \tau_{hij}|_T \in P_l(T)\ \forall\, T \in \mathcal{T}_h, i,j = 1,2\right\},\\
V_h &:= \left\{v_h \in H_0^1(\Omega) : v_h|_T \in P_2(T)\ \forall\, T \in \mathcal{T}_h\right\}.
\end{aligned}
$$

Here, for a triangle $T \in \mathcal{T}_h$, $P_l(T)$ and $P_2(T)$ are the polynomial spaces of degrees l and 2, respectively, with $l = 0, 1$. Note that we have the following property

$$
\nabla_h^2 V_h \subset \boldsymbol{\Sigma}_h, \quad \frac{1}{1-\kappa}\boldsymbol{\Sigma}_h - \frac{\kappa}{1-\kappa^2}(\mathrm{tr}\boldsymbol{\Sigma}_h)\,\boldsymbol{I} \subset \boldsymbol{\Sigma}_h, \tag{8.14}
$$

where $\nabla_h^2 V_h|_T := \nabla^2(V_h|_T)$ for any $T \in \mathcal{T}_h$. $\Delta_h v$ is defined by the relation $\Delta_h v = \Delta v$ on any element $T \in \mathcal{T}_h$. Considering the CDG methods for the variational inequality (8.7), we introduce the corresponding finite element set

$$K_h := \{v_h \in V_h : \partial_\nu v_h \geq 0 \text{ at all vertex nodes on } \Gamma\}.$$

On each element T, $v_h|_T$ is a quadratic polynomial function, so $\partial_\nu v_h$ is a linear polynomial on each edge. With the constraint that $\partial_\nu v_h \geq 0$ at all vertex nodes on Γ, we know

$$\partial_\nu v_h \geq 0 \quad \text{on } \Gamma. \tag{8.15}$$

For a function $v \in L^2(\Omega)$ with $v|_T \in H^m(T)$ for all $T \in \mathcal{T}_h$, define broken norm and seminorm by

$$\|v\|_{m,h} = \left(\sum_{T\in\mathcal{T}_h} \|v\|_{m,T}^2\right)^{1/2}, \quad |v|_{m,h} = \left(\sum_{T\in\mathcal{T}_h} |v|_{m,T}^2\right)^{1/2}.$$

The above symbols are used in a similar manner when v is a vector or matrix-valued function. Throughout this chapter, C denotes a generic positive constant independent of h and other parameters, which may take different values at different occurrences. To avoid writing these constants repeatedly, we use "$x \lesssim y$" to mean that "$x \leq Cy$". For two vectors $\boldsymbol{u}$ and $\boldsymbol{v}$, $\boldsymbol{u} \otimes \boldsymbol{v}$ is a matrix with $u_i v_j$ as its (i, j)th component.

Consider two elements T^+ and T^- with a common edge $e \in \mathcal{E}_h^i$ and let $\boldsymbol{n}^+$ and $\boldsymbol{n}^-$ be their outward unit normals on e. For a scalar-valued function v, set its restriction on $T^\pm$ by $v^\pm = v|_{T^\pm}$. Similarly, for a matrix-valued function $\boldsymbol{\tau}$, write $\boldsymbol{\tau}^\pm = \boldsymbol{\tau}|_{T^\pm}$. Then define averages and jumps on $e \in \mathcal{E}_h^i$ as follows:

$$\begin{aligned}
\{v\} &= \frac{1}{2}(v^+ + v^-), \quad [v] = v^+\boldsymbol{n}^+ + v^-\boldsymbol{n}^-,\\
\{\nabla v\} &= \frac{1}{2}(\nabla v^+ + \nabla v^-), \quad [\nabla v] = \nabla v^+ \cdot \boldsymbol{n}^+ + \nabla v^- \cdot \boldsymbol{n}^-,\\
\{\boldsymbol{\tau}\} &= \frac{1}{2}(\boldsymbol{\tau}^+ + \boldsymbol{\tau}^-), \quad [\boldsymbol{\tau}] = \boldsymbol{\tau}^+\boldsymbol{n}^+ + \boldsymbol{\tau}^-\boldsymbol{n}^-.
\end{aligned}$$

For $e \in \mathcal{E}_h^b$, the above definitions are modified:

$$\begin{aligned}
\{v\} &= v, \quad [v] = v\boldsymbol{\nu},\\
\{\nabla v\} &= \nabla v, \quad [\nabla v] = \nabla v \cdot \boldsymbol{\nu},\\
\{\boldsymbol{\tau}\} &= \boldsymbol{\tau}, \quad [\boldsymbol{\tau}] = \boldsymbol{\tau}\boldsymbol{\nu}.
\end{aligned}$$

The jump $[\![\cdot]\!]$ of the vector ∇v is

$$[\![\nabla v]\!] = \frac{1}{2}(\nabla v^+ \otimes \boldsymbol{n}^+ + \boldsymbol{n}^+ \otimes \nabla v^+ + \nabla v^- \otimes \boldsymbol{n}^- + \boldsymbol{n}^- \otimes \nabla v^-) \quad \text{on } e \in \mathcal{E}_h^i,$$

$$[\![\nabla v]\!] = \frac{1}{2}(\nabla v \otimes \boldsymbol{\nu} + \boldsymbol{\nu} \otimes \nabla v) \quad \text{on } e \in \mathcal{E}_h^b.$$

Define a global lifting operator $\boldsymbol{r}_i : \left(L^2(\mathcal{E}_h^i)\right)_s^{2\times 2} \to \boldsymbol{\Sigma}_h$ by

$$\int_\Omega \boldsymbol{r}_i(\boldsymbol{\phi}) : \boldsymbol{\tau}\, dx = -\int_{\mathcal{E}_h^i} \boldsymbol{\phi} : \{\boldsymbol{\tau}\}\, ds \quad \forall\, \boldsymbol{\tau} \in \boldsymbol{\Sigma}_h,\ \boldsymbol{\phi} \in \left(L^2(\mathcal{E}_h^i)\right)_s^{2\times 2}. \tag{8.16}$$

Moreover, for each $e \in \mathcal{E}_h$, introduce a local lifting operator $\boldsymbol{r}_e : \left(L^2(e)\right)_s^{2\times 2} \to \boldsymbol{\Sigma}_h$ by

$$\int_\Omega \boldsymbol{r}_e(\boldsymbol{\phi}) : \boldsymbol{\tau}\, dx = -\int_e \boldsymbol{\phi} : \{\boldsymbol{\tau}\}\, ds \quad \forall\, \boldsymbol{\tau} \in \boldsymbol{\Sigma}_h,\ \boldsymbol{\phi} \in \left(L^2(e)\right)_s^{2\times 2}. \tag{8.17}$$

It is easy to check that the following identity holds

$$\boldsymbol{r}_i(\boldsymbol{\phi}) = \sum_{e\in\mathcal{E}_h^i} \boldsymbol{r}_e(\boldsymbol{\phi}|_e) \quad \forall\, \boldsymbol{\phi} \in \left(L^2(\mathcal{E}_h^i)\right)_s^{2\times 2},$$

so we have

$$\|\boldsymbol{r}_i(\boldsymbol{\phi})\|^2 = \|\sum_{e\in\mathcal{E}_h^i} \boldsymbol{r}_e(\boldsymbol{\phi}|_e)\|^2 \le 3 \sum_{e\in\mathcal{E}_h^i} \|\boldsymbol{r}_e(\boldsymbol{\phi}|_e)\|^2. \tag{8.18}$$

8.2.2 Discontinuous Galerkin Formulations

We first present the derivation of a general primal formulation of CDG methods for the problem (8.5). By the first equation in (8.3), the first relation in (8.9) and Lemma 8.1, we have

$$\int_T \left(\frac{1}{1-\kappa}\boldsymbol{\sigma} : \boldsymbol{\tau} - \frac{\kappa}{1-\kappa^2}\operatorname{tr}\boldsymbol{\sigma}\ \operatorname{tr}\boldsymbol{\tau}\right) dx = \int_T \nabla u \cdot (\nabla \cdot \boldsymbol{\tau})\, dx - \int_{\partial T} \nabla u \cdot (\boldsymbol{\tau}\boldsymbol{n}_T)\, ds$$

for any smooth second-order tensor-valued function $\boldsymbol{\tau}$, and

$$-\int_T f v\, dx = \int_T \nabla^2 v : \boldsymbol{\sigma}\, dx - \int_{\partial T} \nabla v \cdot (\boldsymbol{\sigma}\boldsymbol{n}_T)\, ds + \int_{\partial T} \boldsymbol{n}_T \cdot (\nabla \cdot \boldsymbol{\sigma}) v\, ds$$

for any smooth scalar-valued function v. Thus, consider a CDG approximate solution $(\boldsymbol{\sigma}_h, u_h) \in \boldsymbol{\Sigma}_h \times V_h$ governed by

$$\int_T \left(\frac{1}{1-\kappa} \boldsymbol{\sigma}_h : \boldsymbol{\tau}_h - \frac{\kappa}{1-\kappa^2} \mathrm{tr}\boldsymbol{\sigma}_h \mathrm{tr}\boldsymbol{\tau}_h \right) dx$$

$$= \int_T \nabla u_h \cdot (\nabla \cdot \boldsymbol{\tau}_h)\, dx - \int_{\partial T} \widehat{\nabla u_h} \cdot (\boldsymbol{\tau}_h \boldsymbol{n}_T)\, ds, \tag{8.19}$$

$$-\int_T f v_h\, dx = \int_T \nabla_h^2 v_h : \boldsymbol{\sigma}_h\, dx - \int_{\partial T} \nabla v_h \cdot (\hat{\boldsymbol{\sigma}}_h \boldsymbol{n}_T)\, ds, \tag{8.20}$$

for all $(\boldsymbol{\tau}_h, v_h) \in \boldsymbol{\Sigma}_h \times V_h$ and all $T \in \mathcal{T}_h$. Here we take $\widehat{\nabla \cdot \boldsymbol{\sigma}_h} = \mathbf{0}$ in the last equation for sake of simplicity. To derive numerous CDG methods, we first introduce an identity. For a scalar function v and a symmetric matrix-valued function $\boldsymbol{\tau}$, smooth on each element of the partition $\mathcal{T}_h$, after a direct manipulation, we have

$$\sum_{T \in \mathcal{T}_h} \int_{\partial T} \nabla v \cdot (\boldsymbol{\tau} \boldsymbol{n}_T)\, ds = \sum_{e \in \mathcal{E}_h^i} \int_e [\boldsymbol{\tau}] \cdot \{\nabla v\}\, ds + \sum_{e \in \mathcal{E}_h} \int_e \{\boldsymbol{\tau}\} : [\![\nabla v]\!]\, ds. \tag{8.21}$$

We now sum Eqs. (8.19) and (8.20) over all $T \in \mathcal{T}_h$ and apply Lemma 8.1 and (8.21) to obtain

$$\int_\Omega \left(\frac{1}{1-\kappa} \boldsymbol{\sigma}_h : \boldsymbol{\tau}_h - \frac{\kappa}{1-\kappa^2} \mathrm{tr}\boldsymbol{\sigma}_h \mathrm{tr}\boldsymbol{\tau}_h \right) dx$$

$$= -\int_\Omega \nabla_h^2 u_h : \boldsymbol{\tau}_h\, dx + \int_{\mathcal{E}_h^i} \{\nabla u_h - \widehat{\nabla u_h}\} \cdot [\boldsymbol{\tau}_h]\, ds$$

$$+ \int_{\mathcal{E}_h} [\![\nabla u_h - \widehat{\nabla u_h}]\!] : \{\boldsymbol{\tau}_h\}\, ds, \tag{8.22}$$

$$-\int_\Omega f v_h\, dx = \int_\Omega \nabla_h^2 v_h : \boldsymbol{\sigma}_h\, dx - \int_{\mathcal{E}_h^i} \{\nabla v_h\} \cdot [\hat{\boldsymbol{\sigma}}_h]\, ds - \int_{\mathcal{E}_h} [\![\nabla v_h]\!] : \{\hat{\boldsymbol{\sigma}}_h\}\, ds. \tag{8.23}$$

Taking $\boldsymbol{\tau}_h = (1-\kappa)\nabla_h^2 v_h + \kappa\, \mathrm{tr}\left(\nabla_h^2 v_h\right) \boldsymbol{I}$ in (8.22), we have

$$\int_\Omega \nabla_h^2 v_h : \boldsymbol{\sigma}_h\, dx = -\int_\Omega (1-\kappa)\nabla_h^2 u_h : \nabla_h^2 v_h\, dx - \int_\Omega \kappa\, \mathrm{tr}\left(\nabla_h^2 u_h\right) \mathrm{tr}\left(\nabla_h^2 v_h\right) dx$$

$$+ \int_{\mathcal{E}_h^i} \{\nabla u_h - \widehat{\nabla u_h}\} \cdot \left((1-\kappa)[\nabla_h^2 v_h] + \kappa[\mathrm{tr}\left(\nabla_h^2 v_h\right)]\right) ds$$

$$+ \int_{\mathcal{E}_h} [\![\nabla u_h - \widehat{\nabla u_h}]\!] : \left((1-\kappa)\{\nabla_h^2 v_h\} + \kappa\, \mathrm{tr}\left(\{\nabla_h^2 v_h\}\right) \boldsymbol{I}\right) ds.$$

Combining the last equation and (8.23), we obtain an equation which does not rely on $\boldsymbol{\sigma}_h$ explicitly:

$$B_h(u_h, v_h) = \int_\Omega f v_h \, dx \quad \forall\, v_h \in V_h, \tag{8.24}$$

where

$$\begin{aligned} B_h(u_h, v_h) :=& \int_\Omega (1-\kappa)\nabla_h^2 u_h : \nabla_h^2 v_h \, dx + \int_\Omega \kappa \operatorname{tr}\left(\nabla_h^2 u_h\right) \operatorname{tr}\left(\nabla_h^2 v_h\right) dx \\ &+ \int_{\mathcal{E}_h^i} \{\widehat{\nabla u_h} - \nabla u_h\} \cdot \left((1-\kappa)[\nabla_h^2 v_h] + \kappa[\operatorname{tr}\left(\nabla_h^2 v_h\right)]\right) ds \\ &+ \int_{\mathcal{E}_h} [\![\widehat{\nabla u_h} - \nabla u_h]\!] : \left((1-\kappa)\{\nabla_h^2 v_h\} + \kappa \operatorname{tr}\left(\{\nabla_h^2 v_h\}\right) \boldsymbol{I}\right) ds \\ &+ \int_{\mathcal{E}_h^i} \{\nabla v_h\} \cdot [\hat{\boldsymbol{\sigma}}_h] \, ds + \int_{\mathcal{E}_h} [\![\nabla v_h]\!] : \{\hat{\boldsymbol{\sigma}}_h\} \, ds. \end{aligned} \tag{8.25}$$

CDG methods for the problem (8.5) can be obtained from (8.24)–(8.25) by proper choices of numerical traces $\hat{\boldsymbol{\sigma}}_h$ and $\widehat{\nabla u_h}$.

The relations (8.24) and (8.25) are also the starting point for designing CDG methods for solving the variational inequality (8.7) through choice of suitable numerical traces to guarantee consistency and stability. For example, taking

$$\begin{cases} \widehat{\nabla u_h} = \{\nabla u_h\} & \text{on } e \in \mathcal{E}_h, \\ \hat{\boldsymbol{\sigma}}_h = -(1-\kappa)\{\nabla_h^2 u_h\} - \kappa \operatorname{tr}\left(\{\nabla_h^2 u_h\}\right) \boldsymbol{I} + \dfrac{\eta}{h_e}[\![\nabla u_h]\!] & \text{on } e \in \mathcal{E}_h^i, \\ \hat{\boldsymbol{\sigma}}_\tau = \boldsymbol{0},\ \hat{\sigma}_{h\nu} \le 0,\ \hat{\sigma}_{h\nu}\partial_{\boldsymbol{\nu}} u_h = 0 & \text{on } e \in \mathcal{E}_h^b, \end{cases}$$

we obtain from (8.24) and (8.25) that

$$B_{1,h}^{(1)}(u_h, v_h) = \int_\Omega f v_h \, dx - \int_\Gamma \hat{\sigma}_{h\nu} \partial_{\boldsymbol{\nu}} v_h \, ds, \tag{8.26}$$

where

$$\begin{aligned} B_{1,h}^{(1)}(u_h, v_h) =& \int_\Omega (1-\kappa)\nabla_h^2 u_h : \nabla_h^2 v_h \, dx + \int_\Omega \kappa \operatorname{tr}\left(\nabla_h^2 u_h\right) \operatorname{tr}\left(\nabla_h^2 v_h\right) dx \\ &- \int_{\mathcal{E}_h^i} [\![\nabla u_h]\!] : \left((1-\kappa)\{\nabla_h^2 v_h\} + \kappa \operatorname{tr}\left(\{\nabla_h^2 v_h\}\right) \boldsymbol{I}\right) ds \\ &- \int_{\mathcal{E}_h^i} [\![\nabla v_h]\!] : \left((1-\kappa)\{\nabla_h^2 u_h\} + \kappa \operatorname{tr}\left(\{\nabla_h^2 u_h\}\right) \boldsymbol{I}\right) ds \\ &+ \int_{\mathcal{E}_h^i} \eta h_e^{-1} [\![\nabla u_h]\!] : [\![\nabla v_h]\!] \, ds. \end{aligned} \tag{8.27}$$

Here η is a function, which is equal to a constant η_e on each $e \in \mathcal{E}_h^i$, with $\{\eta_e\}_{e\in\mathcal{E}_h^i}$ having a uniform positive bound from above and below. In (8.26), let $v_h = w_h - u_h$ with $w_h \in K_h$. Since $\hat{\sigma}_{h\nu} \le 0$, $\hat{\sigma}_{h\nu}\partial_\nu u_h = 0$ on $e \in \mathcal{E}_h^b$, we obtain

$$B_{1,h}^{(1)}(u_h, w_h - u_h) \ge \int_\Omega f(w_h - u_h)\,dx \quad \forall\, w_h \in K_h. \tag{8.28}$$

For a compact formulation, we can use lifting operator $\boldsymbol{r}_i$ to get

$$\begin{aligned}
B_{2,h}^{(1)}(u_h, v_h) &= \int_\Omega (1-\kappa)\nabla_h^2 u_h : \left(\nabla_h^2 v_h + \boldsymbol{r}_i([\![\nabla v_h]\!])\right) dx \\
&\quad + \int_\Omega \kappa \,\mathrm{tr}\left(\nabla_h^2 u_h\right) \mathrm{tr}\left(\nabla_h^2 v_h + \boldsymbol{r}_i([\![\nabla v_h]\!])\right) dx \\
&\quad + \int_\Omega \boldsymbol{r}_i\left([\![\nabla u_h]\!]\right) : \left((1-\kappa)\nabla_h^2 v_h + \kappa\,\mathrm{tr}\left(\nabla_h^2 v_h\right) \boldsymbol{I}\right) dx \\
&\quad + \int_{\mathcal{E}_h^i} \eta h_e^{-1} [\![\nabla u_h]\!] : [\![\nabla v_h]\!]\, ds.
\end{aligned} \tag{8.29}$$

This is the C^0 interior penalty (IP) formulation. A similar C^0 IP method was studied in [7].

The two formulas (8.27) and (8.29) are equivalent on the finite element spaces V_h, so either form can be used to compute the finite element solution u_h. In this chapter, we give a priori error estimates strictly based on the first formula $B_{1,h}^{(1)}$. Because of the equivalence of these two formulations on V_h, we will prove the stability for the second formula $B_{2,h}^{(1)}$ on V_h, which ensures the stability for the first formulation $B_{1,h}^{(1)}$ on V_h. This comment is valid for the rest of the CDG methods.

We now introduce four more CDG methods for the variational inequality (8.7). The methods are all of the form (8.28), and so we will only list the corresponding bilinear form.

Comparing with the DG methods for the second order elliptic problem, we can give the C^0 non-symmetric interior penalty (NIPG) formulations,

$$\begin{aligned}
B_{1,h}^{(2)}(u_h, v_h) &= \int_\Omega (1-\kappa)\nabla_h^2 u_h : \nabla_h^2 v_h\, dx + \int_\Omega \kappa\,\mathrm{tr}\left(\nabla_h^2 u_h\right) \mathrm{tr}\left(\nabla_h^2 v_h\right) dx \\
&\quad + \int_{\mathcal{E}_h^i} [\![\nabla u_h]\!] : \left((1-\kappa)\{\nabla_h^2 v_h\} + \kappa\,\mathrm{tr}\left(\{\nabla_h^2 v_h\}\right) \boldsymbol{I}\right) ds \\
&\quad - \int_{\mathcal{E}_h^i} [\![\nabla v_h]\!] : \left((1-\kappa)\{\nabla_h^2 u_h\} + \kappa\,\mathrm{tr}\left(\{\nabla_h^2 u_h\}\right) \boldsymbol{I}\right) ds \\
&\quad + \int_{\mathcal{E}_h^i} \eta h_e^{-1} [\![\nabla u_h]\!] : [\![\nabla v_h]\!]\, ds,
\end{aligned}$$

or equivalently,

$$
\begin{aligned}
B_{2,h}^{(2)}(u_h, v_h) = & \int_\Omega (1-\kappa)\nabla_h^2 u_h : \left(\nabla_h^2 v_h + \boldsymbol{r}_i([\![\nabla v_h]\!])\right) dx \\
& + \int_\Omega \kappa \operatorname{tr}\left(\nabla_h^2 u_h\right) \operatorname{tr}\left(\nabla_h^2 v_h + \boldsymbol{r}_i([\![\nabla v_h]\!])\right) dx \\
& - \int_\Omega \boldsymbol{r}_i\left([\![\nabla u_h]\!]\right) : \left((1-\kappa)\nabla_h^2 v_h + \kappa \operatorname{tr}\left(\nabla_h^2 v_h\right) \boldsymbol{I}\right) dx \\
& + \int_{\mathcal{E}_h^i} \eta h_e^{-1} [\![\nabla u_h]\!] : [\![\nabla v_h]\!] \, ds.
\end{aligned}
$$

That is, we solve the variational inequality

$$
B_{1,h}^{(2)}(u_h, w_h - u_h) \geq \int_\Omega f(w_h - u_h)\, dx \quad \forall\, w_h \in K_h. \tag{8.30}
$$

Using the local lifting operator $\boldsymbol{r}_e$, we can give the third example. Taking

$$
\begin{cases}
\widehat{\nabla u_h} = \{\nabla u_h\} \quad \text{on } e \in \mathcal{E}_h, \\
\hat{\boldsymbol{\sigma}}_h = -(1-\kappa)\{\nabla_h^2 u_h\} - \kappa \operatorname{tr}(\{\nabla_h^2 u_h\})\boldsymbol{I} - (1-\kappa)\{\boldsymbol{r}_i([\![\nabla u_h]\!])\} \\
\qquad - \kappa\{\operatorname{tr}(\boldsymbol{r}_i([\![\nabla u_h]\!]))\boldsymbol{I}\} \\
\qquad - (1-\kappa)\{\eta\, \boldsymbol{r}_e([\![\nabla u_h]\!])\} - \kappa\{\eta \operatorname{tr}(\boldsymbol{r}_e([\![\nabla u_h]\!]))\}\boldsymbol{I} \quad \text{on } e \in \mathcal{E}_h^i, \\
\hat{\boldsymbol{\sigma}}_\tau = 0,\ \hat{\sigma}_{h\nu} \leq 0,\ \hat{\sigma}_{h\nu}\partial_{\boldsymbol{\nu}} u_h = 0 \quad \text{on } e \in \mathcal{E}_h^b,
\end{cases}
$$

we get from (8.25) that

$$
\begin{aligned}
B_{1,h}^{(3)}(u_h, v_h) = & \int_\Omega (1-\kappa)\nabla_h^2 u_h : \nabla_h^2 v_h\, dx + \int_\Omega \kappa \operatorname{tr}\left(\nabla_h^2 u_h\right) \operatorname{tr}\left(\nabla_h^2 v_h\right) dx \\
& - \int_{\mathcal{E}_h^i} [\![\nabla u_h]\!] : \left((1-\kappa)\{\nabla_h^2 v_h\} + \kappa \operatorname{tr}\left(\{\nabla_h^2 v_h\}\right) \boldsymbol{I}\right) ds \\
& - \int_{\mathcal{E}_h^i} [\![\nabla v_h]\!] : \left((1-\kappa)\{\nabla_h^2 u_h\} + \kappa \operatorname{tr}\left(\{\nabla_h^2 u_h\}\right) \boldsymbol{I}\right) ds \\
& + \int_\Omega \boldsymbol{r}_i([\![\nabla v_h]\!]) : \left((1-\kappa)\boldsymbol{r}_i([\![\nabla u_h]\!]) + \kappa \operatorname{tr}(\boldsymbol{r}_i([\![\nabla u_h]\!]))\boldsymbol{I}\right) dx \\
& + \sum_{e \in \mathcal{E}_h^i} \int_\Omega \eta\big((1-\kappa)\boldsymbol{r}_e([\![\nabla u_h]\!]) : \boldsymbol{r}_e([\![\nabla v_h]\!]) \\
& \qquad + \kappa \operatorname{tr}(\boldsymbol{r}_e([\![\nabla u_h]\!]))\operatorname{tr}(\boldsymbol{r}_e([\![\nabla v_h]\!]))\big)\, dx,
\end{aligned}
$$

or equivalently,

$$B_{2,h}^{(3)}(u_h, v_h) = \int_\Omega (1-\kappa)\left(\nabla_h^2 u_h + \boldsymbol{r}_i([\![\nabla u_h]\!])\right) : \left(\nabla_h^2 v_h + \boldsymbol{r}_i([\![\nabla v_h]\!])\right) dx$$
$$+ \int_\Omega \kappa \operatorname{tr}\left(\nabla_h^2 u_h + \boldsymbol{r}_i([\![\nabla u_h]\!])\right) \operatorname{tr}\left(\nabla_h^2 v_h + \boldsymbol{r}_i([\![\nabla v_h]\!])\right) dx$$
$$+ \sum_{e\in\mathcal{E}_h^i} \int_\Omega \eta(\,(1-\kappa)\boldsymbol{r}_e([\![\nabla u_h]\!]) : \boldsymbol{r}_e([\![\nabla v_h]\!])$$
$$+ \kappa \operatorname{tr}(\boldsymbol{r}_e([\![\nabla u_h]\!]))\operatorname{tr}(\boldsymbol{r}_e([\![\nabla v_h]\!]))\,)\, dx,$$

which is the CDG formulation proposed in [24]. That is, we solve the variational inequality

$$B_{1,h}^{(3)}(u_h, w_h - u_h) \geq \int_\Omega f(w_h - u_h)\, dx \quad \forall\, w_h \in K_h. \tag{8.31}$$

With the choice of

$$\begin{cases} \widehat{\nabla u}_h = \{\nabla u_h\} \quad \text{on } e \in \mathcal{E}_h, \\ \hat{\boldsymbol{\sigma}}_h = -(1-\kappa)\{\nabla_h^2 u_h\} - \kappa \operatorname{tr}(\{\nabla_h^2 u_h\})\boldsymbol{I} - (1-\kappa)\{\eta\, \boldsymbol{r}_e([\![\nabla u_h]\!])\} \\ \qquad - \kappa\{\eta \operatorname{tr}(\boldsymbol{r}_e([\![\nabla u_h]\!]))\}\boldsymbol{I} \quad \text{on } e \in \mathcal{E}_h^i, \\ \hat{\boldsymbol{\sigma}}_\tau = \boldsymbol{0},\ \hat{\sigma}_{h\nu} \leq 0,\ \hat{\sigma}_{h\nu}\partial_{\boldsymbol{\nu}} u_h = 0 \quad \text{on } e \in \mathcal{E}_h^b, \end{cases}$$

we obtain

$$B_{1,h}^{(4)}(u_h, v_h) = \int_\Omega (1-\kappa)\nabla_h^2 u_h : \nabla_h^2 v_h\, dx + \int_\Omega \kappa \operatorname{tr}\left(\nabla_h^2 u_h\right) \operatorname{tr}\left(\nabla_h^2 v_h\right) dx$$
$$- \int_{\mathcal{E}_h^i} [\![\nabla u_h]\!] : \left((1-\kappa)\{\nabla_h^2 v_h\} + \kappa \operatorname{tr}\left(\{\nabla_h^2 v_h\}\right) \boldsymbol{I}\right) ds$$
$$- \int_{\mathcal{E}_h^i} [\![\nabla v_h]\!] : \left((1-\kappa)\{\nabla_h^2 u_h\} + \kappa \operatorname{tr}\left(\{\nabla_h^2 u_h\}\right) \boldsymbol{I}\right) ds$$
$$+ \sum_{e\in\mathcal{E}_h^i} \int_\Omega \eta(\,(1-\kappa)\boldsymbol{r}_e([\![\nabla u_h]\!]) : \boldsymbol{r}_e([\![\nabla v_h]\!])$$
$$+ \kappa \operatorname{tr}(\boldsymbol{r}_e([\![\nabla u_h]\!]))\operatorname{tr}(\boldsymbol{r}_e([\![\nabla v_h]\!]))\,)\, dx,$$

or equivalently,

$$
\begin{aligned}
B_{2,h}^{(4)}(u_h, v_h) = &\int_{\Omega} (1-\kappa)\nabla_h^2 u_h : \left(\nabla_h^2 v_h + \boldsymbol{r}_i([\![\nabla v_h]\!])\right) dx \\
&+ \int_{\Omega} \kappa \operatorname{tr}\left(\nabla_h^2 u_h\right) \operatorname{tr}\left(\nabla_h^2 v_h + \boldsymbol{r}_i([\![\nabla v_h]\!])\right) dx \\
&+ \int_{\Omega} \boldsymbol{r}_i\left([\![\nabla u_h]\!]\right) : \left((1-\kappa)\nabla_h^2 v_h + \kappa \operatorname{tr}\left(\nabla_h^2 v_h\right) \boldsymbol{I}\right) dx \\
&+ \sum_{e\in\mathcal{E}_h^i} \int_{\Omega} \eta(\,(1-\kappa)\boldsymbol{r}_e([\![\nabla u_h]\!]) : \boldsymbol{r}_e([\![\nabla v_h]\!]) \\
&\qquad + \kappa \operatorname{tr}(\boldsymbol{r}_e([\![\nabla u_h]\!]))\operatorname{tr}(\boldsymbol{r}_e([\![\nabla v_h]\!]))\,)\, dx,
\end{aligned}
$$

which is the CDG formulation extended from the DG method of [6] for elliptic problem of second order. That is, we solve the variational inequality

$$
B_{1,h}^{(4)}(u_h, w_h - u_h) \geq \int_{\Omega} f(w_h - u_h)\, dx \quad \forall\, w_h \in K_h. \tag{8.32}
$$

Choosing

$$
\begin{cases}
\widehat{\nabla u}_h = \{\nabla u_h\} \quad \text{on } e \in \mathcal{E}_h, \\
\hat{\boldsymbol{\sigma}}_h = -(1-\kappa)\{\nabla_h^2 u_h\} - \kappa \operatorname{tr}(\{\nabla_h^2 u_h\})\boldsymbol{I} - (1-\kappa)\{\boldsymbol{r}_i([\![\nabla u_h]\!])\} \\
\qquad - \kappa\{\operatorname{tr}(\boldsymbol{r}_i([\![\nabla u_h]\!]))\boldsymbol{I}\} + \eta h_e^{-1}[\![\nabla u_h]\!] \quad \text{on } e \in \mathcal{E}_h^i, \\
\hat{\boldsymbol{\sigma}}_\tau = \boldsymbol{0},\ \hat{\sigma}_{h\nu} \leq 0,\ \hat{\sigma}_{h\nu}\partial_{\boldsymbol{\nu}} u_h = 0 \quad \text{on } e \in \mathcal{E}_h^b,
\end{cases}
$$

we get the LCDG method [15],

$$
\begin{aligned}
B_{1,h}^{(5)}(u_h, v_h) := &\int_{\Omega} (1-\kappa)\nabla_h^2 u_h : \nabla_h^2 v_h\, dx + \int_{\Omega} \kappa \operatorname{tr}\left(\nabla_h^2 u_h\right) \operatorname{tr}\left(\nabla_h^2 v_h\right) dx \\
&- \int_{\mathcal{E}_h^i} [\![\nabla u_h]\!] : \left((1-\kappa)\{\nabla_h^2 v_h\} + \kappa \operatorname{tr}\left(\{\nabla_h^2 v_h\}\right) \boldsymbol{I}\right) ds \\
&- \int_{\mathcal{E}_h^i} [\![\nabla v_h]\!] : \left((1-\kappa)\{\nabla_h^2 u_h\} + \kappa \operatorname{tr}\left(\{\nabla_h^2 u_h\}\right) \boldsymbol{I}\right) ds \\
&+ \int_{\Omega} \boldsymbol{r}_i([\![\nabla v_h]\!]) : \left((1-\kappa)\boldsymbol{r}_i([\![\nabla u_h]\!]) + \kappa \operatorname{tr}(\boldsymbol{r}_i([\![\nabla u_h]\!]))\boldsymbol{I}\right) dx \\
&+ \int_{\mathcal{E}_h^i} \eta h_e^{-1} [\![\nabla u_h]\!] : [\![\nabla v_h]\!]\, ds,
\end{aligned}
$$

or equivalently,

$$\begin{aligned} B_{2,h}^{(5)}(u_h, v_h) := & \int_\Omega (1-\kappa) \left(\nabla_h^2 u_h + \boldsymbol{r}_i([\![\nabla u_h]\!])\right) : \left(\nabla_h^2 v_h + \boldsymbol{r}_i([\![\nabla v_h]\!])\right) dx \\ & + \int_\Omega \kappa \operatorname{tr}\left(\nabla_h^2 u_h + \boldsymbol{r}_i([\![\nabla u_h]\!])\right) \operatorname{tr}\left(\nabla_h^2 v_h + \boldsymbol{r}_i([\![\nabla v_h]\!])\right) dx \\ & + \int_{\mathcal{E}_h^i} \eta h_e^{-1} [\![\nabla u_h]\!] : [\![\nabla v_h]\!] \, ds. \end{aligned}$$

That is, we solve the variational inequality

$$B_{1,h}^{(5)}(u_h, w_h - u_h) \geq \int_\Omega f(w_h - u_h)\, dx \quad \forall\, w_h \in K_h. \tag{8.33}$$

In the following sections, we will study CDG methods for the EVI (8.7) defined as follows: Find $u_h \in K_h$ such that

$$B_h(u_h, v_h - u_h) \geq (f, v_h - u_h) \quad \forall\, v_h \in K_h, \tag{8.34}$$

where the bilinear form $B_h(w, v) = B_{1,h}^{(j)}(w, v)$ with $j = 1, \cdots, 5$.

8.3 Consistency, Boundedness and Stability

We present some properties of the five DG methods introduced in Sect. 8.2. First, we address the consistency of the methods (8.34).

Lemma 8.3. *Assume $u \in H^3(\Omega)$ is the solution of the problem* (8.7). *For all the five CDG methods $B_h(w, v) = B_{1,h}^{(j)}(w, v)$ with $j = 1, \cdots, 5$, we have*

$$B_h(u, v_h - u) \geq (f, v_h - u) \quad \forall\, v_h \in K_h.$$

Proof. Noting $[\![\nabla u]\!] = 0$ on each edge $e \in \mathcal{E}_h^i$, we use (8.2) to get

$$\begin{aligned} B_h(u, v_h - u) = & \int_\Omega (1-\kappa) \nabla^2 u : \nabla_h^2 (v_h - u)\, dx \\ & + \int_\Omega \kappa \operatorname{tr}\left(\nabla^2 u\right) \operatorname{tr}\left(\nabla_h^2 (v_h - u)\right) dx \\ & - \int_{\mathcal{E}_h^i} [\![\nabla (v_h - u)]\!] : \left((1-\kappa)\nabla^2 u + \kappa \operatorname{tr}\left(\nabla^2 u\right) \boldsymbol{I}\right) ds \\ = & - \sum_{T \in \mathcal{T}_h} \int_T \boldsymbol{\sigma} : \nabla_h^2 (v_h - u)\, dx + \int_{\mathcal{E}_h^i} [\![\nabla (v_h - u)]\!] : \boldsymbol{\sigma}\, ds. \end{aligned}$$

Using Lemma 8.1 and noticing $[\boldsymbol{\sigma}] = \mathbf{0}$ on each edge $e \in \mathcal{E}_h^i$, we have

$$\begin{aligned}\sum_{T\in\mathcal{T}_h}\int_T \boldsymbol{\sigma} : \nabla_h^2(v_h - u)\,dx &= -\sum_{T\in\mathcal{T}_h}\int_T \nabla(v_h - u)\cdot(\nabla\cdot\boldsymbol{\sigma})\,dx \\ &\quad + \sum_{T\in\mathcal{T}_h}\int_{\partial T} \nabla(v_h - u)\cdot(\boldsymbol{\sigma}\boldsymbol{n}_T)\,ds \\ &= -\int_\Omega \nabla(v_h - u)\cdot(\nabla\cdot\boldsymbol{\sigma})\,dx \\ &\quad + \int_{\mathcal{E}_h} [\![\nabla(v_h - u)]\!] : \boldsymbol{\sigma}\,ds.\end{aligned}$$

Combining the above two equations,

$$B_h(u, v_h - u) = \int_\Omega \nabla(v_h - u)\cdot(\nabla\cdot\boldsymbol{\sigma})\,dx - \int_\Gamma [\![\nabla(v_h - u)]\!] : \boldsymbol{\sigma}\,ds.$$

Since $v_h - u = 0$ on Γ and $\partial_{\boldsymbol{\nu}} v_h \geq 0$ on Γ for all $v_h \in K_h$, we use Lemma 8.1 and (8.9) to obtain

$$\begin{aligned}B_h(u, v_h - u) &= -\int_\Omega (v_h - u)\nabla\cdot(\nabla\cdot\boldsymbol{\sigma})\,dx - \int_\Gamma \nabla(v_h - u)\cdot(\boldsymbol{\sigma}\boldsymbol{\nu})\,ds \\ &= \int_\Omega f(v_h - u)\,dx - \int_\Gamma \sigma_\nu \partial_{\boldsymbol{\nu}} v_h\,ds \\ &\geq \int_\Omega f(v_h - u)\,dx.\end{aligned}$$

So the stated result holds. □

Let $V(h) := V_h + H_0^1(\Omega) \cap H^3(\Omega)$ and define two mesh-dependent energy norms by

$$|v|_*^2 = |v|_{2,h}^2 + \sum_{e\in\mathcal{E}_h^i} h_e^{-1}\|[\![\nabla v]\!]\|_{0,e}^2, \quad |||v|||^2 = |v|_*^2 + \sum_{T\in\mathcal{T}_h} h_T^2 |v|_{3,T}^2, \quad v \in V(h).$$

To show these formulas define norms, we only need prove that $|v|_* = 0$ and $v \in V(h)$ imply $v = 0$. From $|v|_{2,h} = 0$, we have $v|_T \in P_1(T)$ and so ∇v is piecewise constant. Let e be a common edge of two elements T^+ and T^-. From $\|[\![\nabla v]\!]\|_{0,e} = 0$, we obtain $(\nabla v)^+ = (\nabla v)^-$. Thus, ∇v is constant in Ω and so $v \in P_1(\Omega)$. Since $v = 0$ on Γ, we conclude $v = 0$ in Ω.

Before presenting boundedness and stability results of the bilinear forms, we give a useful estimate for the lifting operator $\boldsymbol{r}_e$.

Lemma 8.4. *For any $v \in V(h)$ and $e \in \mathcal{E}_h^i$,*

$$C_1 h_e^{-1} \|[\![\nabla v]\!]\|_{0,e}^2 \leq \|\boldsymbol{r}_e([\![\nabla v]\!])\|_{0,h}^2 \leq C_2 h_e^{-1} \|[\![\nabla v]\!]\|_{0,e}^2. \tag{8.35}$$

Proof. The second inequality was proved in [15]. For $v \in H^3(\Omega)$, $[\![\nabla v]\!] = 0$ on $e \in \mathcal{E}_h^i$. So we only need to consider the case $v \in V_h$. By the formula between (4.4) and (4.5) in [2], we know

$$h_e^{-1} \|\boldsymbol{\varphi}\|_{0,e}^2 \lesssim \|\boldsymbol{r}_e^*(\boldsymbol{\varphi})\|_{0,\Omega}^2 \lesssim h_e^{-1} \|\boldsymbol{\varphi}\|_{0,e}^2 \quad \forall \boldsymbol{\varphi} \in [P_1(e)]^2, \tag{8.36}$$

where the lifting operator $\boldsymbol{r}_e^* : (L^2(e))^2 \to W_h$ is defined by

$$\int_\Omega \boldsymbol{r}_e^*(\boldsymbol{v}) \cdot \boldsymbol{w}_h \, dx = -\int_e \boldsymbol{v} \cdot \{\boldsymbol{w}_h\} \, ds \quad \forall \boldsymbol{w}_h \in W_h.$$

Here, $W_h := \left\{\boldsymbol{w}_h \in \left(L^2(\Omega)\right)^2 : \boldsymbol{w}_h|_K \in [P_l(K)]^2 \ \forall K \in \mathcal{T}_h\right\}$.

For two matrix-valued functions $\boldsymbol{\phi} = (\phi_{ij})_{2\times 2}$ and $\boldsymbol{\tau} = (\tau_{ij})_{2\times 2}$, let $\boldsymbol{\phi}_1 = (\phi_{11}, \phi_{21})^t$, $\boldsymbol{\phi}_2 = (\phi_{12}, \phi_{22})^t$, $\boldsymbol{\tau}_1 = (\tau_{11}, \tau_{21})^t$, $\boldsymbol{\tau}_2 = (\tau_{12}, \tau_{22})^t$, so that $\boldsymbol{\phi} = (\boldsymbol{\phi}_1, \boldsymbol{\phi}_2)$, $\boldsymbol{\tau} = (\boldsymbol{\tau}_1, \boldsymbol{\tau}_2)$. Then

$$\begin{aligned}
\int_\Omega \boldsymbol{r}_e(\boldsymbol{\phi}) : \boldsymbol{\tau} \, dx &= -\int_e \boldsymbol{\phi} : \{\boldsymbol{\tau}\} \, ds = -\int_e \boldsymbol{\phi}_1 \cdot \{\boldsymbol{\tau}_1\} \, ds - \int_e \boldsymbol{\phi}_2 \cdot \{\boldsymbol{\tau}_2\} \, ds \\
&= \int_\Omega \boldsymbol{r}_e^*(\boldsymbol{\phi}_1) \cdot \boldsymbol{\tau}_1 \, dx + \int_\Omega \boldsymbol{r}_e^*(\boldsymbol{\phi}_2) \cdot \boldsymbol{\tau}_2 \, dx \\
&= \int_\Omega (\boldsymbol{r}_e^*(\boldsymbol{\phi}_1), \boldsymbol{r}_e^*(\boldsymbol{\phi}_2)) : \boldsymbol{\tau} \, dx,
\end{aligned}$$

for all $\boldsymbol{\tau} \in \boldsymbol{\sigma}_h$. So $\boldsymbol{r}_e(\boldsymbol{\phi}) = (\boldsymbol{r}_e^*(\boldsymbol{\phi}_1), \boldsymbol{r}_e^*(\boldsymbol{\phi}_2))$, $\|\boldsymbol{r}_e(\boldsymbol{\phi})\|_{0,\Omega}^2 = \|\boldsymbol{r}_e^*(\boldsymbol{\phi}_1)\|_{0,\Omega}^2 + \|\boldsymbol{r}_e^*(\boldsymbol{\phi}_2)\|_{0,\Omega}^2$, and

$$\begin{aligned}
h_e^{-1} \|\boldsymbol{\phi}\|_{0,e}^2 &= h_e^{-1}(\|\boldsymbol{\phi}_1\|_{0,e}^2 + \|\boldsymbol{\phi}_2\|_{0,e}^2) \\
&\lesssim \|\boldsymbol{r}_e^*(\boldsymbol{\phi}_1)\|_{0,\Omega}^2 + \|\boldsymbol{r}_e^*(\boldsymbol{\phi}_2)\|_{0,\Omega}^2 = \|\boldsymbol{r}_e(\boldsymbol{\phi})\|_{0,\Omega}^2.
\end{aligned}$$

Let $\boldsymbol{\phi} = [\![\nabla v]\!]$, the first inequality is proved. □

From (8.35) and (8.18), we have

$$\|\boldsymbol{r}_i([\![\nabla v]\!])\|_{0,h}^2 = \|\sum_{e \in \mathcal{E}_h^i} \boldsymbol{r}_e([\![\nabla v]\!])\|_{0,h}^2 \leq 3C_2 \sum_{e \in \mathcal{E}_h^i} h_e^{-1} \|[\![\nabla v]\!]\|_{0,e}^2.$$

For the boundedness of the primal forms $B_{1,h}^{(j)}$ with $j = 1, \cdots, 5$, first notice that $\|\mathrm{tr}(\boldsymbol{\tau})\|_{0,h} \lesssim \|\boldsymbol{\tau}\|_{0,h}$. By the Cauchy–Schwarz inequality and Lemma 8.4, we get the following inequalities:

$$\int_{\Omega} \nabla_h^2 w : \nabla_h^2 v \, dx \leq |w|_{2,h} |v|_{2,h}, \tag{8.37}$$

$$\int_{\Omega} \boldsymbol{r}_i([\![\nabla w]\!]) : \boldsymbol{r}_i([\![\nabla v]\!]) \, dx \lesssim \left(\sum_{e \in \mathcal{E}_h^i} h_e^{-1} \|[\![\nabla w]\!]\|_{0,e}^2 \right)^{1/2} \left(\sum_{e \in \mathcal{E}_h^i} h_e^{-1} \|[\![\nabla v]\!]\|_{0,e}^2 \right)^{1/2}, \tag{8.38}$$

$$\int_{\mathcal{E}_h^i} \eta h_e^{-1} [\![\nabla w]\!] : [\![\nabla v]\!] \, ds \leq \sup_{e \in \mathcal{E}_h^i} \eta_e \left(\sum_{e \in \mathcal{E}_h^i} h_e^{-1} \|[\![\nabla w]\!]\|_{0,e}^2 \right)^{1/2} \left(\sum_{e \in \mathcal{E}_h^i} h_e^{-1} \|[\![\nabla v]\!]\|_{0,e}^2 \right)^{1/2}, \tag{8.39}$$

$$\sum_{e \in \mathcal{E}_h^i} \int_{\Omega} \eta \boldsymbol{r}_e([\![\nabla w]\!]) : \boldsymbol{r}_e([\![\nabla v]\!]) \, dx \lesssim \sup_{e \in \mathcal{E}_h^i} \eta_e \left(\sum_{e \in \mathcal{E}_h^i} h_e^{-1} \|[\![\nabla w]\!]\|_{0,e}^2 \right)^{1/2} \left(\sum_{e \in \mathcal{E}_h^i} h_e^{-1} \|[\![\nabla v]\!]\|_{0,e}^2 \right)^{1/2}. \tag{8.40}$$

Using the trace inequality $\|\nabla^2 v\|_{0,e} \lesssim h_e^{-1} |v|_{2,K}^2 + h_e |v|_{3,K}^2$ with e an edge of K, we have

$$\begin{aligned} \int_{\mathcal{E}_h^i} [\![\nabla w]\!] : \{\nabla_h^2 v\} \, ds &= \sum_{e \in \mathcal{E}_h^i} \int_e [\![\nabla w]\!] : \{\nabla_h^2 v\} \, ds \\ &\leq \left(\sum_{e \in \mathcal{E}_h^i} h_e^{-1} \|[\![\nabla w]\!]\|_{0,e}^2 \right)^{1/2} \left(\sum_{e \in \mathcal{E}_h^i} h_e \|\{\nabla_h^2 v\}\|_{0,e}^2 \right)^{1/2} \\ &\lesssim \left(\sum_{e \in \mathcal{E}_h^i} h_e^{-1} \|[\![\nabla w]\!]\|_{0,e}^2 \right)^{1/2} \left(\sum_{T \in \mathcal{T}_h} (|v|_{2,T}^2 + h_T^2 |v|_{3,T}^2) \right)^{1/2}. \end{aligned} \tag{8.41}$$

The inequalities (8.37) and (8.41) are needed by all bilinear forms. For the CDG methods with the bilinear form $B_{1,h}^{(j)}$, $j = 1, 2, 5$, the inequality (8.39) is needed.

The inequality (8.38) is needed by the formulas $B_{1,h}^{(j)}$ with $j = 3, 5$. The methods with the bilinear forms $B_{1,h}^{(j)}$, $j = 3, 4$, need the inequality (8.40). Then we have the following result.

Lemma 8.5 (Boundedness). *Let $B_h = B_{1,h}^{(j)}$ with $j = 1, \cdots, 5$. Then*

$$B_h(w, v) \lesssim |||w||| \; |||v||| \quad \forall (w, v) \in V(h) \times V(h). \tag{8.42}$$

For stability over V_h, note that $|||v||| = |v|_*$ for any $v \in V_h$. Formulations $B_{1,h}^{(j)}$ and $B_{2,h}^{(j)}$ are equivalent on V_h, so we just need to prove the stability for $B_{2,h}^{(j)}$ based on $|\cdot|_*$. We use the Cauchy–Schwarz inequality and Lemma 8.4 to get

$$\begin{aligned}
B_{2,h}^{(1)}(v,v) &= (1-\kappa)\int_\Omega \nabla_h^2 v : \nabla_h^2 v \, dx + \kappa \int_\Omega \left(\mathrm{tr}(\nabla_h^2 v)\right)^2 dx \\
&\quad + 2(1-\kappa)\int_\Omega \nabla_h^2 v : \boldsymbol{r}_i([\![\nabla v]\!])\, dx \\
&\quad + 2\kappa \int_\Omega \mathrm{tr}\left(\nabla_h^2 v\right) \mathrm{tr}\left(\boldsymbol{r}_i([\![\nabla v]\!])\right) dx + \int_{\mathcal{E}_h^i} \eta h_e^{-1} |[\![\nabla v]\!]|^2 ds \\
&\geq (1-\kappa)|v|_{2,h}^2 + \kappa \|\Delta_h v\|_{0,h}^2 - (1-\kappa)\left(\epsilon |v|_{2,h}^2 + \frac{1}{\epsilon}\|\boldsymbol{r}_i([\![\nabla v]\!])\|_{0,h}^2\right) \\
&\quad - \kappa\left(\|\Delta_h v\|_{0,h}^2 + \|\mathrm{tr}\left(\boldsymbol{r}_i([\![\nabla v]\!])\right)\|_{0,h}^2\right) + \eta_0 \sum_{e \in \mathcal{E}_h^i} h_e^{-1} \|[\![\nabla v]\!]\|_{0,e}^2 \\
&\geq (1-\epsilon)(1-\kappa)|v|_{2,h}^2 \\
&\quad + \left(\eta_0 - \frac{3(1-\kappa)C_2}{\epsilon} - 6C_2\kappa\right) \sum_{e \in \mathcal{E}_h^i} h_e^{-1} \|[\![\nabla v]\!]\|_{0,e}^2,
\end{aligned}$$

where $0 < \epsilon < 1$ is a constant and C_2 is the generic positive constant in (8.35). Therefore, stability is valid for C^0 IP method when $\min_{e \in \mathcal{E}_h^i} \eta_e = \eta_0 > 3(1-\kappa)C_2 + 6C_2\kappa = 3(1+\kappa)C_2$.

$$\begin{aligned}
B_{2,h}^{(2)}(v,v) &= \int_\Omega (1-\kappa)\nabla_h^2 v : \nabla_h^2 v \, dx + \int_\Omega \kappa \left(\mathrm{tr}(\nabla_h^2 v)\right)^2 dx \\
&\quad + \int_{\mathcal{E}_h^i} \eta h_e^{-1} \left([\![\nabla v]\!]\right)^2 ds \\
&\geq (1-\kappa)|v|_{2,h}^2 + \eta_0 \sum_{e \in \mathcal{E}_h^i} h_e^{-1} \|[\![\nabla v]\!]\|_{0,e}^2.
\end{aligned}$$

So stability is valid for the C^0 NIPG method for any $\eta_0 > 0$. This property is the reason why the method with the bilinear form $B_{2,h}^{(2)}$ is useful even though $B_{2,h}^{(2)}$ is not symmetric.

$$\begin{aligned}
B_{2,h}^{(4)}(v,v) \geq{}& (1-\kappa)|v|_{2,h}^2 + \kappa\|\Delta_h v\|_{0,h}^2 + 2(1-\kappa)\int_\Omega \nabla_h^2 v : \boldsymbol{r}_i([\![\nabla v]\!])\,dx \\
&+ 2\kappa \int_\Omega \Delta_h v\,\mathrm{tr}\,(\boldsymbol{r}_i([\![\nabla v]\!]))\,dx \\
&+ \eta_0 \sum_{e\in\mathcal{E}_h^i} \left((1-\kappa)\|\boldsymbol{r}_e([\![\nabla v]\!])\|_{0,h}^2 + \kappa\|\mathrm{tr}(\boldsymbol{r}_e([\![\nabla v]\!]))\|_{0,h}^2\right) \\
\geq{}& (1-\kappa)|v|_{2,h}^2 + \kappa\|\Delta_h v\|_{0,h}^2 - (1-\kappa)\left(\epsilon|v|_{2,h}^2 + \frac{1}{\epsilon}\|\boldsymbol{r}_i([\![\nabla v]\!])\|_{0,h}^2\right) \\
&- \kappa\|\Delta_h v\|_{0,h}^2 - \kappa\|\mathrm{tr}\,(\boldsymbol{r}_i([\![\nabla v]\!]))\,\|_{0,h}^2 \\
&+ \eta_0 C_1(1-\kappa)\sum_{e\in\mathcal{E}_h^i} h_e^{-1}\|[\![\nabla v]\!]\|_{0,e}^2 + \eta_0\kappa \sum_{e\in\mathcal{E}_h^i}\|\mathrm{tr}(\boldsymbol{r}_e([\![\nabla v]\!]))\|_{0,h}^2 \\
\geq{}& (1-\epsilon)(1-\kappa)|v|_{2,h}^2 + (1-\kappa)\left(\eta_0 C_1 - \frac{3C_2}{\epsilon}\right)\sum_{e\in\mathcal{E}_h^i} h_e^{-1}\|[\![\nabla v]\!]\|_{0,e}^2 \\
&+ (\eta_0\kappa - 3\kappa)\sum_{e\in\mathcal{E}_h^i}\|\mathrm{tr}(\boldsymbol{r}_e([\![\nabla v]\!]))\|_{0,h}^2.
\end{aligned}$$

Since $C_2 > C_1$, $\eta_0 > 3$ is guaranteed from $\eta_0 > 3C_2/C_1$. Thus, stability is valid for this CDG formulation when $\eta_0 > 3C_2/C_1$. For the method of Wells–Dung corresponding to the form $B_{2,h}^{(3)}$ and the LCDG method corresponding to the form $B_{2,h}^{(5)}$, stability can be analyzed by a similar argument (cf. [15,24], respectively), with $\eta_0 > 0$.

Lemma 8.6 (Stability). *Let $B_h = B_{2,h}^{(j)}$ with $j = 1, \cdots, 5$. Assume*

$$\min_{e\in\mathcal{E}_h^i} \eta_e > 3\,(1+\kappa)\,C_2 \text{ for } j = 1$$

and

$$\min_{e\in\mathcal{E}_h^i} \eta_e > 3\,C_2/C_1 \text{ for } j = 4,$$

with C_1 and C_2 the constants in the inequality (8.35). *Then,*

$$|\!|\!|\, v|\!|\!|^2 \lesssim B_h(v,v) \qquad \forall\, v \in V_h. \tag{8.43}$$

It follows from Lemma 8.6 that under the stated conditions, the stability property is also valid for $B_{1,h}^{(j)}$ with $j = 1, \cdots, 5$.

8.4 Numerical Results

In this section, we present a numerical example on the five methods studied in solving the elliptic variational inequality (8.7). To solve the discretized variational inequality, we use the optimization toolbox in MATLAB for the associated quadratic optimization problem. Let $\Omega = (-1, 1) \times (-1, 1)$, $\kappa = 0.3$. Choose the right hand side function to be

$$f(x, y) = 24(1 - x^2)^2 + 24(1 - y^2)^2 + 32(3x^2 - 1)(3y^2 - 1).$$

We use uniform triangulations $\mathcal{T}_h$ of the region $\overline{\Omega}$ and piecewise quadratic polynomials, i.e.,

$$V_h = \{v_h \in H_0^1(\Omega) : v_h|_T \in P_2(T)\ \forall\, T \in \mathcal{T}_h\}.$$

Since the domain is rectangular, the outward normal is not defined at the four corner points, $(-1, -1)$, $(-1, 1)$, $(1, -1)$, and $(1, 1)$. So in terms of the restriction in the optimization problem, at each of the four corners, we specify the constraint $\partial_\nu v_h \geq 0$ twice (corresponding to the two sides intersecting at the corner).

Tables 8.1, 8.2, 8.3, 8.4 and 8.5 provide numerical solution errors in the energy norm $|||\cdot|||$ for the five DG methods discussed in this chapter. Since the true solution of the variational inequality (8.7) is not known, we use the numerical solution corresponding to the meshsize $h = 0.0625$ as the true solution in computing the errors.

We observe that all the five DG methods perform well, and the methods 1, 2, and 5 converge faster than the methods 3 and 4. For the methods 1, 2, and 5,

Table 8.1 Error $|||u - u_h|||$ for C^0 IP method (8.28)

h	1	0.5	0.25	0.125
$\eta = 1$	6.1327	3.6713	1.7921	0.8921
$\eta = 10$	4.4264	2.0188	1.0202	0.5076
$\eta = 100$	2.7621	1.1607	0.5767	0.2860
$\eta = 1{,}000$	1.5772	0.6544	0.3245	0.1628

Table 8.2 Error $|||u - u_h|||$ for NIPG method (8.30)

h	1	0.5	0.25	0.125
$\eta = 1$	5.5212	3.2523	1.7682	0.8987
$\eta = 10$	4.3318	2.0225	1.0215	0.5082
$\eta = 100$	2.7521	1.1610	0.5767	0.2860
$\eta = 1{,}000$	1.5766	0.6545	0.3245	0.1628

Table 8.3 Error $|||u - u_h|||$ for Wells–Dung DG formulation (8.31) as in [24]

h	1	0.5	0.25	0.125
$\eta = 1$	5.1659	3.1617	2.3018	1.6206
$\eta = 10$	4.2043	2.3618	1.6532	1.1411
$\eta = 100$	2.9413	1.4600	0.9954	0.7116
$\eta = 1{,}000$	1.7224	0.9311	0.5570	0.3836

Table 8.4 Error $|||u - u_h|||$ for Baker-DG formulation (8.32) as in [5]

h	1	0.5	0.25	0.125
$\eta = 1$	5.1509	3.1566	2.3016	1.6204
$\eta = 10$	4.1924	2.3418	1.6425	1.1408
$\eta = 100$	2.8889	1.4448	0.9818	0.6774
$\eta = 1{,}000$	1.7181	0.9300	0.5569	0.3837

Table 8.5 Error $|||u - u_h|||$ for LCDG method (8.33)

h	1	0.5	0.25	0.125
$\eta = 1$	5.4865	3.0717	1.7165	0.8831
$\eta = 10$	4.2776	1.9847	1.0158	0.5071
$\eta = 100$	2.7463	1.1584	0.5764	0.2859
$\eta = 1{,}000$	1.5762	0.6543	0.3244	0.1608

the numerical convergence order in the norm $|||u - u_h|||$ is around 1, whereas for the methods 3 and 4, the numerical convergence order in the norm $|||u - u_h|||$ is around 1/2.

Acknowledgements The work was partially supported by the Chinese National Science Foundation (Grant No. 11101168), the Marie Curie International Research Staff Exchange Scheme Fellowship within the seventh European Community Framework Programme under Grant Agreement No. 295118, the National Science Center of Poland under Maestro Advanced Project no. DEC-2012/06/A/ST1/00262, grants from Simons Foundation (No. 207052, No. 228187), the NSFC (Grant No. 11171219), and E-Institutes of Shanghai Municipal Education Commission (E03004).

References

1. Arnold, D.N., Brezzi, F., Cockburn, B., Marini, L.D.: Discontinuous Galerkin methods for elliptic problems. In: Cockburn, B., Karniadakis, G.E., Shu, C.-W. (eds.) Discontinuous Galerkin Methods: Theory, Computation and Applications. Lecture Notes in Computer Science and Engineering, vol. 11, pp. 89–101. Springer, New York (2000)
2. Arnold, D.N., Brezzi, F., Cockburn, B., Marini, L.D.: Unified analysis of discontinuous Galerkin methods for elliptic problems. SIAM J. Numer. Anal. **39**, 1749–1779 (2002)
3. Atkinson, K., Han, W.: Theoretical Numerical Analysis: A Functional Analysis Framework. Texts in Applied Mathematics, vol. 39, 3rd edn. Springer, New York (2009)
4. Babuška, I., Zlámal, M.: Nonconforming elements in the finite element method with penalty. SIAM J. Numer. Anal. **10**, 863–875 (1973)
5. Baker, G.A.: Finite element methods for elliptic equations using nonconforming elements. Math. Comput. **31**, 44–59 (1977)

6. Bassi, F., Rebay, S., Mariotti, G., Pedinotti, S., Savini, M.: A high-order accurate discontinuous finite element method for inviscid and viscous turbomachinery flows. In: Decuypere, R., Dibelius, G. (eds.) Proceedings of 2nd European Conference on Turbomachinery, Fluid Dynamics and Thermodynamics, pp. 99–108. Technologisch Instituut, Antwerpen (1997)
7. Brenner, S.C., Sung, L.: C^0 interior penalty methods for fourth order elliptic boundary value problems on polygonal domains. J. Sci. Comput. **22/23**, 83–118 (2005)
8. Castillo, P., Cockburn, B., Perugia, I., Schötzau, D.: An a priori error analysis of the local discontinuous Galerkin method for elliptic problems. SIAM J. Numer. Anal. **18**, 1676–1706 (2000)
9. Cockburn, B.: Discontinuous Galerkin methods. Z. Angew. Math. Mech. **83**, 731–754 (2003)
10. Cockburn, B., Karniadakis, G.E., Shu, C.-W. (eds.): Discontinuous Galerkin Methods. Theory, Computation and Applications. Lecture Notes in Computer Science and Engineering, vol. 11. Springer, New York (2000)
11. Duvaut, G., Lions, J.-L.: Inequalities in Mechanics and Physics. Springer, Berlin (1976)
12. Engel, G., Garikipati, K., Hughes, T., Larson, M., Mazzei, L., Taylor, R.: Continuous/discontinuous finite element approximations of fourth-order elliptic problems in structural and continuum mechanics with applications to thin beams and plates, and strain gradient elasticity. Comput. Methods Appl. Mech. Eng. **191**, 3669–3750 (2002)
13. Glowinski, R.: Numerical Methods for Nonlinear Variational Problems. Springer, New York (1984)
14. Glowinski, R., Lions, J.-L., Trémolières, R.: Numerical Analysis of Variational Inequalities. North-Holland, New York (1981)
15. Huang, J., Huang, X., Han, W.: A new C^0 discontinuous Galerkin method for Kirchhoff plates. Comput. Methods Appl. Mech. Eng. **199**, 1446–1454 (2010)
16. Mozolevski, I., Bösing, P.R.: Sharp expressions for the stabilization parameters in symmetric interior-penalty discontinuous Galerkin finite element approximations of fourth-order elliptic problems. Comput. Methods Appl. Math. **7**, 365–375 (2007)
17. Mozolevski, I., Süli, E.: A priori error analysis for the hp-version of the discontinuous Galerkin finite element method for the biharmonic equation. Comput. Methods Appl. Math. **3**, 596–607 (2003)
18. Mozolevski, I., Süli, E., Bösing, P.R.: hp-version a priori error analysis of interior penalty discontinuous Galerkin finite element approximations to the biharmonic equation. J. Sci. Comput. **30**, 465–491 (2007)
19. Reddy, J.N.: Theory and Analysis of Elastic Plates and Shells, 2nd edn. CRC Press, New York (2007)
20. Süli, E., Mozolevski, I.: hp-version interior penalty DGFEMs for the biharmonic equation. Comput. Methods Appl. Mech. Eng. **196**, 1851–1863 (2007)
21. Wang, F., Han, W., Cheng, X.: Discontinuous Galerkin methods for solving elliptic variational inequalities. SIAM J. Numer. Anal. **48**, 708–733 (2010)
22. Wang, F., Han, W., Cheng, X.: Discontinuous Galerkin methods for solving Signorini problem. IMA J. Numer. Anal. **31**, 1754–1772 (2011)
23. Wang, F., Han, W., Cheng, X.: Discontinuous Galerkin methods for solving the quasistatic contact problem. Numer. Math. **126**, 771–800 (2014)
24. Wells, G.N., Dung, N.T.: A C^0 discontinuous Galerkin formulation for Kirchhoff plates. Comput. Methods Appl. Mech. Eng. **196**, 3370–3380 (2007)

Part III
Applications

Chapter 9
Dynamic Gao Beam in Contact with a Reactive or Rigid Foundation

Kevin T. Andrews, Kenneth L. Kuttler, and Meir Shillor

Abstract This chapter constructs and analyzes a model for the dynamic behavior of nonlinear viscoelastic beam, which is acted upon by a horizontal traction, that may come in contact with a rigid or reactive foundation underneath it. We use a model, first developed and studied by D.Y. Gao, that allows for the buckling of the beam when the horizontal traction is sufficiently large. In contrast with the behavior of the standard Euler–Bernoulli linear beam, it can have three steady states, two of which are buckled. Moreover, the Gao beam can vibrate about such buckled states, which makes it important in engineering applications. We describe the contact process with either the normal compliance condition when the foundation is reactive, or with the Signorini condition when the foundation is perfectly rigid. We use various tools from the theory of pseudomonotone operators and variational inequalities to establish the existence and uniqueness of the weak or variational solution to the dynamic problem with the normal compliance contact condition. The main step is in the truncation of the nonlinear term and then establishing the necessary a priori estimates. Then, we show that when the viscosity of the material approaches zero and the stiffness of the foundation approaches infinity, making it perfectly rigid, the associated solutions of the problem with normal compliance converge to a solution of the elastic problem with the Signorini condition.

Keywords Gao beam • Dynamic contact • Normal compliance • The Signorini contact condition • Existence and uniqueness

AMS Classification. 35L86; 74H20, 74H25, 74K10, 74M15

K.T. Andrews (✉) • M. Shillor
Department of Mathematics and Statistics, Oakland University, Rochester, MI 48309, USA
e-mail: andrews@oakland.edu; shillor@oakland.edu

K.L. Kuttler
Department of Mathematics, Brigham Young University, Provo, UT 84602, USA
e-mail: klkuttle@math.byu.edu

W. Han et al. (eds.), *Advances in Variational and Hemivariational Inequalities*,
Advances in Mechanics and Mathematics 33, DOI 10.1007/978-3-319-14490-0_9

9.1 Introduction

In this chapter we analyze a model for the dynamic evolution of nonlinear viscoelastic beam that may contact a rigid or reactive foundation that is situated below it. The beam is subject to a general horizontal traction at one end. We use a model for the motion of the beam that was first developed and studied by D.Y. Gao in a series of papers [5, 7, 8]. This model exhibits several interesting and novel features that contrast with the behavior of the standard Euler–Bernoulli linear beam. Among other things, there are nonzero steady states present when the lateral traction exerted on one end of the beam is sufficiently compressive. These so-called buckled steady states are stable, being the minima of the stored elastic energy, while the unbuckled zero steady state is unstable, being a local maximum of the stored energy. Consequently, the Gao beam may vibrate about its buckled states. By contrast, the Euler–Bernoulli beam only vibrates around its zero steady state.

The process of contact between deformable bodies is usually modeled with either the *normal compliance condition* that assumes reactive contact surfaces or with the *Signorini condition* or *unilateral contact condition* which assumes that one of the contacting surfaces is perfectly rigid (see explanations in [9, 20] and the numerous references therein). The second condition may be viewed as an idealization of the first, obtained in the limit as the stiffness of the foundation increases to infinity. In this chapter, we prove that the dynamic contact problem involving a viscoelastic Gao beam and a reactive foundation has a unique weak solution. We do so by writing the evolution equation in abstract variational form and using truncation, approximations, a priori estimates and results for evolution problems; in particular, the basic existence result for pseudomonotone operators [12] underlies the proof. With viscosity present we have sufficient regularity and compactness to obtain the necessary a priori estimates. We obtain two other existence results by passing to the limit in two different ways. In the first case, we let the viscosity tend to zero so that the beam is now purely elastic. In this process, as expected, we lose some regularity but still obtain an existence result, now without the assertion of uniqueness. In the second, we let the stiffness tend to infinity and so obtain an existence result for the dynamic contact problem of a viscoelastic Gao beam and a rigid foundation. Once again, there is a loss of regularity which prevents us from establishing uniqueness and which is consistent with the expectation that contact with a rigid foundation may cause discontinuity in the velocity. It is very likely that this possible nonuniqueness is related to the need to include a coefficient of restitution that measures the energy loss during the contact process.

Other recent work involving the viscoelastic Gao beam and dynamic contact with a reactive foundation may be found in [14], which considers a setting where the beam is subject to damage caused by the growth of a crack, leading to eventual breaking of the beam. In [15] the setting includes a stochastic force input and a random gap. Here, we complement that study with the full analysis that, in particular, includes the limiting cases of a purely elastic beam, and the completely rigid obstacle. All of this work has been done with longer term aim of investigating computationally how contact with the obstacle affects the various vibrational modes of the buckled beam. Indeed, some preliminary results show that very complicated types of behavior may take place. However, there is intrinsic mathematical interest in the analysis of the model presented here. Additional analysis and simulations of dynamic problems involving the Gao beam can be found in [1–4, 16, 19] and the references therein.

We note that when this work was essentially complete, the article [17] came to our attention, where a draft version of one of our results can be found.

The rest of this chapter is organized as follows. The model is presented in Sect. 9.2 where the classical formulations of the problems with the normal compliance (reactive obstacle) and the Signorini (perfectly rigid obstacle) contact conditions are given. The variational formulations of the two models are described in Sect. 9.3 where the assumptions on the problem data and the statements of our main results on the existence of the unique weak solution to the problem with viscosity and normal compliance, Theorem 9.5, and the existence of a weak solution to the problem without viscosity and the Signorini condition, Theorem 9.6, are presented. Section 9.4 deals with the truncated problem, where the cubic nonlinearity present in the Gao beam is modified. Proposition 9.8 guarantees that, for each truncation ceiling R, there exists a unique weak solution. The proof is based on the existence results for pseudomonotone operators to be found in [12]. Then, in Sect. 9.5 the necessary a priori estimates are derived that allow us to pass to the limit $R \to \infty$ and obtain Theorem 9.5. In Sect. 9.6 we obtain the estimates and pass to the inviscid limit, as the viscosity coefficient approaches zero and obtain the existence of a weak solution to the purely elastic problem with the normal compliance condition, Theorem 9.14. In the final section, Sect. 9.7, we pass to the limit in which the obstacle becomes rigid, when the normal compliance constant tends to infinity, and obtain the existence of a weak solution to the problem without viscosity, purely elastic beam and perfectly rigid obstacle.

9.2 The Model

We derive a model for the vibrations of the dynamic Gao beam in contact with a rigid or reactive foundation. The beam is assumed to be viscoelastic and of moderate thickness, and its motion is restricted by an *obstacle*, the so-called *foundation*, that is situated below it. This is the setting depicted in Fig. 9.1. We denote by $w = w(x,t)$ the displacement of the beam's central axis at location x and time t and the model is scaled so that the reference configuration occupies $0 \leq x \leq 1$ and the density (per unit length) is 1. The beam is subjected to a distributed transverse load f, and a horizontal traction $p = p(t)$ that is acting at the end $x = 1$ that may cause buckling. We say that when $p < 0$ the end $x = 1$ is being compressed and when $p > 0$ it is under tension. We let $g = g(x) \leq 0$ denote the *gap* between the central axis of the beam (in its reference configuration $[0, 1]$) and the foundation. The beam is in

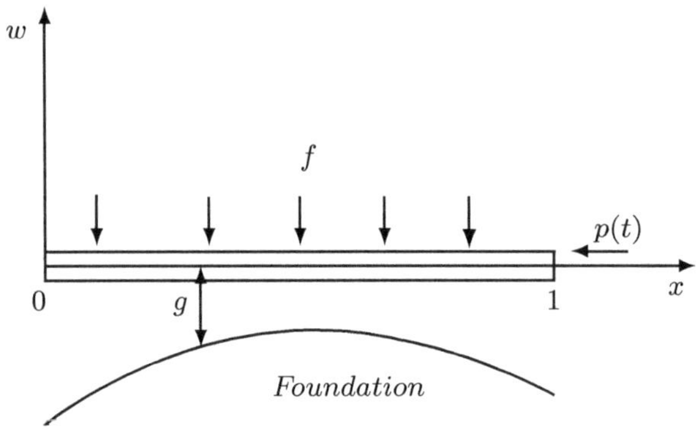

Fig. 9.1 The beam, foundation, gap g, and traction p

contact with the foundation when the vertical displacement satisfies $w \leq g$. In the case when the foundation is reactive, it is possible for $w < g$.

When the beam is not in contact, the motion of the beam is governed by

$$w_{tt} - \sigma_x = f, \tag{9.1}$$

where the viscoelastic shear stress σ [5, 8], is given by

$$\sigma = -kw_{xxx} - \gamma w_{txxx} + \frac{1}{3}aw_x^3 - \nu p w_x. \tag{9.2}$$

Here k is the scaled elasticity modulus, γ the viscosity coefficient, a the Gao coefficient that allows for buckling, and ν is the scaled stiffness coefficient for lateral compression or tension. All four of these coefficients are assumed to be positive constants. The cubic term in w_x is the novelty in this model since it creates a double-well elastic potential energy with one or three steady states. When the traction p is sufficiently large there are three steady states: the steady state $w = 0$ is unstable and the other two are stable buckled states. (See [4, 6, 8] for further details.)

The initial displacement and velocity of the beam are given by

$$w(x,0) = w_0(x), \qquad w_t(x,0) = v_0(x), \tag{9.3}$$

and the beam is assumed to be clamped at both ends, so

$$w(0,t) = w_x(0,t) = 0, \quad w(1,t) = w_x(1,t) = 0. \tag{9.4}$$

We now describe the contact between the beam and the reactive foundation. When $w > g$ the beam is not in contact, thus,

$$w > g \implies w_{tt} - \sigma_x = f. \tag{9.5}$$

When $w \leq g$ the beam is in contact with the foundation, and the interpenetration of the surface asperities makes $w < g$ possible. However, the foundation reacts to the interpenetration with a force ζ that is directed upward, i.e., $\zeta \geq 0$. The force balance equation in this case is

$$w_{tt} - \sigma_x = f + \zeta.$$

We assume that the reaction force ζ depends on the interpenetration, i.e., $\zeta = q(g - w)$, where $q(\cdot)$ is the *normal compliance function*, to be described shortly. Therefore,

$$w \leq g \implies w_{tt} - \sigma_x = f + q(g - w). \tag{9.6}$$

We assume that the normal compliance function is such that $q(r) = 0$ when $r \leq 0$, with a typical choice in the literature [11, 13, 18, 20] being

$$q(g - w) = \kappa(g - w)_+,$$

where $(r)_+ = \max\{0, r\}$ is the positive part function and κ is the normal compliance stiffness constant, a measure of the stiffness of the foundation. The use of $(g - w)_+$ guarantees that the reaction vanishes when there is no contact, i.e., when $w \geq g$, and that the reaction is proportional to the interpenetration when in contact, i.e., when $w < g$. A perfectly rigid obstacle is obtained in the limit $\kappa \to \infty$, as we describe below.

The classical formulation of the problem of the *dynamic frictionless vibrations of the Gao beam in contact with a reactive foundation* is thus as follows.

Problem 9.1. Find the displacement field $w = w(x,t)$, for $x \in (0,1)$ and $t \in (0,T)$, such that

$$w_{tt} - \sigma_x = f + q(g - w), \tag{9.7}$$

$$\sigma = -k w_{xxx} - \gamma w_{txxx} + \frac{1}{3} a w_x^3 - \nu p w_x, \tag{9.8}$$

$$w(0,t) = w_x(0,t) = 0, \tag{9.9}$$

$$w(1,t) = w_x(1,t) = 0, \tag{9.10}$$

$$w(x,0) = w_0(x), \quad w_t(x,0) = v_0(x). \tag{9.11}$$

The nonlinear term $\frac{1}{3} a w_x^3$ in (9.7) makes buckling possible but introduces mathematical difficulties since it prevents us from obtaining the necessary estimates in the usual way, and so it adds a layer of complexity to the problem. We will deal with this issue by truncating the term, and this truncated variational problem is considered in Sect. 9.4. We then derive estimates needed to pass to the limit in the truncated problem so as to establish a unique solution to the above problem.

When the foundation is assumed to be perfectly rigid, then it is always true that $g \leq w$ and the reaction force ζ acts only when $w = g$, at which time we must have $\zeta \geq 0$. This leads to the so-called *Signorini nonpenetration condition* or *unilateral contact condition*. The classical formulation of the problem of the *dynamic frictionless vibrations of the Gao beam in contact with a perfectly rigid foundation* is as follows.

Problem 9.2. Find the displacement field $w = w(x,t)$, for $x \in (0,1)$ and $t \in (0,T)$, such that

$$w_{tt} - \sigma_x = f + \zeta, \tag{9.12}$$

$$\sigma = -kw_{xxx} - \gamma w_{txxx} + \frac{1}{3}aw_x^3 - \nu p w_x, \tag{9.13}$$

$$w \geq g, \quad \zeta \geq 0, \quad \zeta(g - w) = 0, \tag{9.14}$$

$$w(0,t) = w_x(0,t) = 0, \tag{9.15}$$

$$w(1,t) = w_x(1,t) = 0, \tag{9.16}$$

$$w(x,0) = w_0(x), \quad w_t(x,0) = v_0(x). \tag{9.17}$$

Formally, if we replace q with $\frac{1}{\epsilon}q$ in (9.7), then one obtains the Signorini condition (9.14) in the limit $\epsilon \to 0$. We show below that we actually can pass to the limit and obtain the variational solution to Problem 9.2 in the limit $\epsilon \to 0$ of solutions to Problem 9.1, so the statement is not only formal.

9.3 Variational Formulations

In this section we present the variational formulations of Problems 9.1 and 9.2, list the assumptions on the problem data and state our existence results. We denote by $(\cdot,\cdot)$ the inner product on $L^2(0,1)$. We use standard notation for Sobolev spaces, in particular, $H^2(0,1)$ is the space of functions that are square integrable and have first and second square integrable distributional derivatives.

Let V be a closed subspace of $H^2(0,1)$, given by

$$V = \{u \in H^2(0,1) : u = u_x = 0 \text{ at } x = 0, 1\}.$$

We seek solutions w such that $w, w_t \in \mathcal{V} \equiv L^2(0,T;V)$ and $w_{tt} \in \mathcal{V}^* \equiv L^2(0,T;V^*))$, where V^* denotes the dual of V. Let $H = L^2(0,1)$, then, since $C_c^\infty(0,1)$ is dense in V and in H, we have that (V, H, V^*) is a Gelfand or evolution triple, and if we let $\mathcal{H} = L^2(0,T;H)$, then $(\mathcal{V}, \mathcal{H}, \mathcal{V}^*)$ is also a Gelfand triple. We denote the respective duality pairings by $\langle\cdot,\cdot\rangle_V$ and $\langle\cdot,\cdot\rangle_{\mathcal{V}}$. On V we use the norm

$\|u\|_V^2 = (u_{xx}, u_{xx})$, which is equivalent to the usual $H^2(0,1)$ norm. We will also denote by W the closure of V in $H^1(0,1)$ and let $\mathcal{W} = L^2(0,T;W)$. We can then consider the spaces

$$\mathcal{V} \subseteq \mathcal{W} \subseteq \mathcal{H} = \mathcal{H}^* \subseteq \mathcal{W}^* \subseteq \mathcal{V}^*,$$

where each space is densely embedded in the following one. Below we will use the prime to denote the (distributional) time derivative, and let $v = w_t = w'$.

We consider Problem 9.1 first. Applying (9.7), with (9.8), to a test function $u \in V$ gives, for $t \in (0,T)$,

$$\langle v'(t), u\rangle_V + \langle \sigma(t), u_x\rangle_V = \langle f(t), u\rangle_V + \langle q(g-w), u\rangle_V.$$

Using the usual manipulations and the boundary conditions, we obtain the following *variational formulation of the problem of dynamic contact of a Gao beam with a reactive foundation.*

Problem 9.3. Find a displacement field $w : [0,T] \to V$ and a velocity field $v : [0,T] \to V$, with $v' \in \mathcal{V}^*$, such that for a.a. $t \in [0,T]$ and all $u \in V$,

$$\begin{aligned}\langle v'(t), u\rangle_V &+ k(w_{xx}(t), u_{xx}) + \gamma(v_{xx}(t), u_{xx}) + \frac{1}{3}a(w_x^3(t), u_x)\\ &- \nu p(t)(w_x(t), u_x) = \langle f(t), u\rangle_V + (q(g-w), u),\end{aligned} \tag{9.18}$$

$$w(t) = w_0 + \int_0^t v(\tau)\,d\tau, \quad v(0) = v_0. \tag{9.19}$$

To obtain the variational formulation of the problem with a rigid foundation, we let K be the following closed and convex set in V where we seek $w(t)$,

$$K = \{\psi \in V : \psi \geq g\}.$$

We select $u \in K$ and apply (9.12), together with (9.13), to the test function $u - w(t)$. We use integration by parts and the boundary conditions, and obtain

$$\begin{aligned}\langle v'(t), u - w(t)\rangle_V &+ k(w_{xx}(t), u_{xx} - w_{xx}(t)) + \gamma(v_{xx}(t), u_{xx} - w_{xx}(t))\\ &+ \frac{1}{3}a(w_x^3(t), u_x - w_x(t)) - \nu p(t)(w_x(t), u_x - w_x(t))\\ &= \langle f(t), u - w(t)\rangle_V + \langle \zeta, u - w(t)\rangle_V.\end{aligned} \tag{9.20}$$

Next, we consider the last term on the right-hand side. If $w(x,t) > g(x)$ there is no contact between the beam and the foundation at x, therefore, $\zeta = 0$, which implies that $\langle \zeta, u - w(t)\rangle_V = 0$. If $w(x,t) = g(x)$, the beam is in contact with the foundation, which reacts with a reactive force $\zeta \geq 0$, and since $u \in K$, we have $u \geq g$, therefore

$$\langle \zeta, u - w(t) \rangle_V \geq 0.$$

The last observation allows us to drop this term and upon integration on $[0, t]$, and dropping the viscosity term, i.e., setting $\gamma = 0$, we have the following *variational inequality formulation of the problem of contact between an elastic Gao beam and a rigid foundation.*

Problem 9.4. Find the displacement field $w : [0, T] \to K$ and the velocity field $v : [0, T] \to V$, such that for all $u \in \mathcal{V}$ such that $u' \in L^\infty(0, T; H)$ and $u(t) \in K$ for each t,

$$\begin{aligned}
&(v(t), w(t) - u(t)) \\
&- \int_0^t (v(s), w'(s) - u'(s))\, ds + k \int_0^t (w_{xx}(s), w_{xx}(s) - u_{xx}(s))\, ds \\
&+ \frac{1}{3} a \int_0^t (w_x^3(s), w(s) - u(s))\, ds - \nu \int_0^t p(s)(w_x(s), w_x(s) - u_x(s))\, ds \\
&\qquad \leq \int_0^t \langle f(s), w(s) - u(s) \rangle_V\, ds + (v_0, w(0) - u(0)), \qquad (9.21)
\end{aligned}$$

together with (9.19).

To state our main existence results we make the following assumptions on the data:

$$f \in \mathcal{V}^*, \qquad (9.22)$$

$$p \in W^{1,\infty}(0, T), \quad |p|, |p'| \leq p_0, \quad p_0 > 0, \qquad (9.23)$$

$$w_0, \; v_0 \in V. \qquad (9.24)$$

We also assume that the normal compliance function $q : \mathbb{R} \to \mathbb{R}$ satisfies the following conditions:

$$\begin{cases} |q(r_1) - q(r_2)| \leq k_q |r_1 - r_2|, \quad k_q > 0, \\ (q(r_1) - q(r_2))(r_1 - r_2) \geq 0, \\ q(r) = 0, \quad r \leq 0. \end{cases} \qquad (9.25)$$

The main results of this work are the following two theorems.

Theorem 9.5. *Assume that conditions* (9.22)–(9.25) *hold. Then, there exists a unique solution* (w, v) *of Problem 9.3 such that*

$$w \in L^\infty(0,T;V), \qquad v \in L^\infty(0,T;H)\cap \mathcal{V}, \qquad v' \in \mathcal{V}^*. \tag{9.26}$$

The proof of the theorem is given in Sect. 9.5, and is based on the truncated problem that is studied in Sect. 9.4.

Next, we have the following result for the rigid foundation problem without viscosity.

Theorem 9.6. *Assume that the conditions* (9.22)–(9.24) *hold and that additionally* $w_0 \in K$ *and* $f \in \mathcal{H}$. *Then, there exists a solution* (w, v) *of Problem 9.4 such that,*

$$\begin{aligned} w \in & L^\infty(0,T;V), \quad v \in L^\infty(0,T;H)\cap C\left(0,T;V'\right), \\ & w(t) \in K \quad for \ a.a. \ t \in [0,T]. \end{aligned} \tag{9.27}$$

The proof of the theorem is given in Sect. 9.7, and is based on passing to the limits in the normal compliance and the viscosity terms in Problem 9.3. We note that we do not assert that the solution is unique. This question is unresolved, and uniqueness seems to be unlikely without additional assumptions, such as the addition of a coefficient of restitution that measures the energy loss during the contact process. We also note that we do not claim any additional regularity for v', except that it is a distribution.

9.4 The Truncated Problem

The two nonlinearities present in the model are the cubic term in the stress function and the contact condition. These do not fit into any of the usual nonlinear formats and so we must use methods adapted to this particular situation to treat them. To deal mathematically with the cubic term, we introduce the truncation operator Ψ_R,

$$\Psi_R(r) = \begin{cases} R & \text{for } R \le r, \\ r & \text{for } |r| \le R, \\ -R & \text{for } r \le -R, \end{cases} \tag{9.28}$$

where R is a large number. Then, we replace the term w_x^3 with $\Psi_R^2(w_x)w_x$. This yields the following truncated variational problem.

Problem 9.7. Find the displacement field $w : [0,T] \to V$ and the velocity field $v : [0,T] \to V$, with $v' \in \mathcal{V}^*$, such that for a.a. $t \in [0,T]$ and all $\varphi \in V$,

$$\begin{aligned} \langle v^*(t), \varphi\rangle_V + k(w_{xx}(t), \varphi_{xx}) + \gamma(v_{xx}(t), \varphi_{xx}) + \frac{1}{3}a(\Psi_R^2(w_x(t))w_x(t), \varphi_x) \\ - \nu p(t)(w_x(t), \varphi_x) = \langle f(t), \varphi\rangle_V + (q(g-w), \varphi), \end{aligned} \tag{9.29}$$

$$w(t) = w_0 + \int_0^t v(\tau)\,d\tau, \quad v(0) = v_0. \tag{9.30}$$

We have the following existence and uniqueness result for this truncated problem.

Proposition 9.8. *Assume that* (9.22)–(9.25) *hold. Then, for each* $R > 0$ *and* $T > 0$ *there exists a unique solution* $(w, v) = (w_R, v_R)$ *to Problem 9.7 on* $[0, T]$ *such that*

$$w \in C([0,T]; V), \qquad v \in \mathcal{V}, \qquad v' \in \mathcal{V}^*. \tag{9.31}$$

To prove this result, we rewrite Problem 9.7 in an abstract form and use the existence results in [10, 12]. To that end, we define the operators B, K, K_R, J : $\mathcal{V} \to \mathcal{V}^*$, for $w, \phi \in \mathcal{V}$, by

$$\langle B(w), \phi\rangle_{\mathcal{V}} = \int_0^T \int_0^1 p(t) w_x \phi_x \,dxdt, \tag{9.32}$$

$$\langle K(w), \phi\rangle_{\mathcal{V}} = \int_0^T \int_0^1 w_{xx} \phi_{xx} \,dxdt, \tag{9.33}$$

$$\langle K_R(w), \phi\rangle_{\mathcal{V}} = \int_0^T \int_0^1 \Psi_R^2(w_x) w_x \phi_x \,dxdt, \tag{9.34}$$

$$\langle J(w), \phi\rangle_{\mathcal{V}} = \int_0^T \int_0^1 q(g - w)\phi \,dx\,dt, \tag{9.35}$$

The abstract formulation of (9.29) and (9.30) now can be given as:

Problem 9.9. Find a pair $(w, v) \in \mathcal{V} \times \mathcal{V}$ such that

$$v' + kK(w) + \gamma K(v) + \frac{1}{3} a K_R(w) - \nu B(w) - J(w) = f \quad \text{in } \mathcal{V}^*, \tag{9.36}$$

together with (9.30).

We now rewrite this problem as a first order abstract system. To that end, we let $Y = V \times V$ and $\mathcal{Y} = \mathcal{V} \times \mathcal{V}$, and use the product norm $\|y\|_Y = \|\phi\|_V + \|\psi\|_V$, for $y = (\phi, \psi) \in Y$, and similarly for $\mathcal{Y}$. We define the operator $A : \mathcal{Y} \to \mathcal{Y}^*$ by

$$A \begin{pmatrix} w \\ v \end{pmatrix} = \begin{pmatrix} -kK(v) \\ kK(w) + \gamma K(v) + \frac{1}{3} a K_R(w) - \nu B(w) - J(w) \end{pmatrix}, \tag{9.37}$$

and $D : \mathcal{Y} \to \mathcal{Y}^*$ by

$$D \begin{pmatrix} w \\ v \end{pmatrix} = \begin{pmatrix} kK(w) \\ v \end{pmatrix}. \tag{9.38}$$

Let $F = (0, f)$ and $u_0 = (w_0, v_0)$, and let $\mathcal{Z} = \{u \in \mathcal{Y} : (Du)' \in \mathcal{Y}^*\}$ with norm given by $\|z\|_{\mathcal{Z}} = \|z\|_{\mathcal{Y}} + \|(Dz)'\|_{\mathcal{Y}^*}$. We may now rewrite Problem 9.9 as follows.

Problem 9.10. Find $u = (w, v) \in \mathcal{Z}$ such that

$$(Du)' + Au = F, \qquad \text{in } \mathcal{Y}^* \tag{9.39}$$

$$Du(0) = Du_0, \qquad \text{in } Y^*. \tag{9.40}$$

The first order system above is an implicit evolution problem of the type considered in [10, 12]. To use the existence results of these papers it suffices to show that, for all sufficiently large λ , the following three conditions hold true:

(i) There exist constants C_0 and C_1, which depend on the data but are independent of $u \in \mathcal{Y}$, such that $||Au||_{\mathcal{Y}'} \leq C_0 + C_1||u||_{\mathcal{Y}}$ for all $u \in \mathcal{Y}$.

(ii) $\displaystyle\lim_{||u||_{\mathcal{Y}}\to\infty} \frac{\langle \lambda Du, u\rangle_{\mathcal{Y}} + \langle Au, u\rangle_{\mathcal{Y}}}{||u||_{\mathcal{Y}}} = \infty.$

(iii) $u \to Au$ is a pseudomonotone map from $\mathcal{Z}$ to $\mathcal{Z}'$.

We now derive the necessary estimates needed to establish these conditions in the following three lemmas. We set $u = (w, v) \in \mathcal{Y}$ and denote by C, C_0 and C_1 generic positive constants whose values may change from line to line but which are independent of u.

Lemma 9.11. *The operator A is bounded, i.e., it satisfies condition (i).*

Proof. We consider in turn the various terms which appear in $\langle A(u), y\rangle_{\mathcal{Y}}$ where $y = (\phi, \psi) \in \mathcal{Y}$. First, we have that

$$|\langle K(v), \phi\rangle_{\mathcal{V}}| \leq \int_0^T \int_0^1 |v_{xx}||\phi_{xx}|\, dxdt \leq \|v\|_{\mathcal{V}}\|\phi\|_{\mathcal{V}} \leq \|u\|_{\mathcal{Y}}\|y\|_{\mathcal{Y}}.$$

Similarly, $|\langle K(w), \psi\rangle_{\mathcal{V}}| \leq \|u\|_{\mathcal{Y}}\|y\|_{\mathcal{Y}}$. Next,

$$|\langle K_R(w), \psi\rangle_{\mathcal{V}}| \leq R^2 \int_0^T \int_0^1 |w_x||\psi_x|\, dxdt \leq R^2\|w\|_{\mathcal{V}}\|\psi\|_{\mathcal{V}} \leq R^2\|u\|_{\mathcal{Y}}\|y\|_{\mathcal{Y}}.$$

This is where the truncation is used. Next, using $|p(t)| \leq p^*$, we obtain

$$|\langle B(w), \psi\rangle_{\mathcal{V}}| \leq \int_0^T \int_0^1 |p(t)||w_x||\psi_x|\, dxdt \leq p^*\|w\|_{\mathcal{V}}\|\psi\|_{\mathcal{V}} \leq p^*\|u\|_{\mathcal{Y}}\|y\|_{\mathcal{Y}}.$$

Finally,

$$|\langle J(w), \psi\rangle_{\mathcal{V}}| =| \int_0^T \int_0^1 q(g - w(x,t))\psi(x,t)\, dx\, dt|$$

$$\leq (C_0 + C_1\|w\|_{\mathcal{V}})\|\psi\|_{\mathcal{V}} \leq (C_0 + C_1\|u\|_{\mathcal{Y}})\|\psi\|_{\mathcal{Y}}.$$

Collecting these estimates shows that

$$|\langle A(u), y\rangle_{\mathcal{Y}}| \leq (C_0 + C_1\|u\|_{\mathcal{Y}})\|\psi\|_{\mathcal{Y}},$$

as desired. □

We now show that condition (ii) holds:

Lemma 9.12. *The operator $(\lambda D + A)$ is coercive for all sufficiently large λ, i.e.,*

$$\lim_{||u||_{\mathcal{Y}}\to\infty} \frac{\langle \lambda Du, u\rangle_{\mathcal{Y}} + \langle Au, u\rangle_{\mathcal{Y}}}{||u||_{\mathcal{Y}}} = \infty. \tag{9.41}$$

Proof. We have,

$$\langle \lambda Du, u\rangle_{\mathcal{Y}} = \lambda\langle K(w), w\rangle_{\mathcal{V}} + \lambda\langle v, v\rangle_{\mathcal{V}} = \lambda\|w\|^2_{\mathcal{V}} + \lambda\|v\|^2_{\mathcal{H}},$$

and

$$\langle Au, u\rangle_{\mathcal{Y}} = \langle -K(v), w\rangle_{\mathcal{V}} + k\langle K(w), v\rangle_{\mathcal{V}} + \gamma\langle K(v), v\rangle_{\mathcal{V}}$$
$$+ \frac{1}{3}a\langle K_R(w), v\rangle_{\mathcal{V}} - \nu\langle pB(w), v\rangle_{\mathcal{V}} + \langle J(w), v\rangle_{\mathcal{V}}.$$

We deal with each term which appears on the right in turn. We use the Hölder inequality as we did in the previous lemma and the Cauchy inequality with ϵ in most of the estimates. First, we note that

$$\langle -K(v), w\rangle_{\mathcal{V}} \geq -\|v\|_{\mathcal{V}}\|w\|_{\mathcal{V}} \geq -\frac{\gamma}{8}\|v\|^2_{\mathcal{V}} - \frac{2}{\gamma}\|w\|^2_{\mathcal{V}}.$$

Similarly,

$$k\langle K(w), v\rangle_{\mathcal{V}} \geq -k\|v\|_{\mathcal{V}}\|w\|_{\mathcal{V}} \geq -\frac{\gamma}{8}\|v\|^2_{\mathcal{V}} - \frac{2k^2}{\gamma}\|w\|^2_{\mathcal{V}}.$$

We also have that

$$\gamma\langle K(v), v\rangle_{\mathcal{V}} = \gamma\|v\|^2_{\mathcal{V}}.$$

Next,

$$\frac{1}{3}a\langle K_R(w), v\rangle_{\mathcal{V}} \geq -\frac{1}{3}aR^2\|w_x\|_{\mathcal{H}}\|v_x\|_{\mathcal{H}}$$

$$\geq -\frac{\gamma}{8}\|v\|_{\mathcal{V}}^2 - \frac{2a^2R^4}{9\gamma}\|w\|_{\mathcal{V}}^2.$$

Also,

$$\begin{aligned} -\nu\langle B(w), v\rangle_{\mathcal{V}} &\geq -\nu p^*\|w_x\|_{\mathcal{H}}\|v_x\|_{\mathcal{H}} \\ &\geq -\frac{\gamma}{8}\|v\|_{\mathcal{V}}^2 - \frac{2\nu^2 p^{*2}}{\gamma}\|w\|_{\mathcal{V}}^2. \end{aligned}$$

Finally, we have that

$$\begin{aligned} \langle J(w), v\rangle_{\mathcal{V}} &\geq (-C_1 - C_0\|w\|_{\mathcal{V}})\|v\|_{\mathcal{V}} \\ &\geq -\frac{\gamma}{8}\|v\|_{\mathcal{V}}^2 - \frac{2C_0^2}{\gamma}\|w\|_{\mathcal{V}}^2 - C. \end{aligned}$$

Collecting these estimates and rearranging the constants yields

$$\langle\lambda Du, u\rangle_{\mathcal{Y}} + \langle Au, u\rangle_{\mathcal{Y}} \geq (\lambda - C)\|w\|_{\mathcal{V}}^2 + \lambda\|v\|_{\mathcal{H}}^2 + \frac{1}{4}\gamma\|v\|_{\mathcal{V}}^2 - C.$$

Dividing this estimate by $||u||_{\mathcal{Y}} = ||w||_{\mathcal{V}} + ||v||_{\mathcal{V}}$ and letting $||u||_{\mathcal{Y}} \to \infty$ leads to (9.41), for each sufficiently large λ. □

We note that the presence of the viscosity term is essential to obtaining the previous result. We now show that condition (iii) holds.

Lemma 9.13. *The operator $A : \mathcal{Z} \to \mathcal{Z}'$ is pseudomonotone.*

Proof. We note that we may write $A = A_1 + A_2$ where

$$A_1\begin{pmatrix} w \\ v \end{pmatrix} = \begin{pmatrix} -kK(v) \\ kK(w) + \gamma K(v) \end{pmatrix}, \tag{9.42}$$

and

$$A_2\begin{pmatrix} w \\ v \end{pmatrix} = \begin{pmatrix} 0 \\ \frac{1}{3}aK_R(w) - \nu B(w) + J(w) \end{pmatrix}. \tag{9.43}$$

Since it is easy to check that the operator A_1 is linear, bounded and monotone, we only need to prove that the operator A_2 is completely continuous to conclude that the sum $A_1 + A_2$ is pseudomonotone (see, e.g., [22, Proposition 27.6]). This, in turn, follows from examining the operators K_R, B and J. To that end, let $\{u_m\} = \{(w_m, v_m)\}$ be a sequence which converges weakly to $u = (w, v)$ in $\mathcal{Z}$. Then, $w_m \to w$ and $v_m \to v$ weakly in $L^2(0, T; V)$, and also $w'_m \to w'$ and $v'_m \to v'$ weakly in $L^2(0, T; V^*)$. It follows from Corollary 4 of [21], that $\{w_m\}$ and $\{v_m\}$ are relatively compact in $L^2(0,T;H^1(0,1))$. By passing to subsequences we may assume that $\{w_m\}$ and $\{v_m\}$ converge strongly in $L^2(0,T;H^1(0,1))$ and pointwise a.a. to w and v, respectively. Now we have, for $\phi \in \mathcal{V}$, that

$$|\langle B(w_m) - B(w), \phi\rangle_{\mathcal{V}}| \leq p^* \|w_{mx} - w_x\|_{\mathcal{H}} \|\phi_x\|_{\mathcal{H}}.$$

Since $\|\phi_x\|_{L^2(0,T;H)} \leq \|\phi\|_{\mathcal{V}}$, we obtain

$$\|B(w_m) - B(w)\|_{\mathcal{V}^*} \leq C\|w_{mx} - w_x\|_{\mathcal{H}} \to 0 \quad \text{as} \quad m \to \infty.$$

Next,

$$|\langle K_R(w_m) - K_R(w), \phi\rangle_{\mathcal{V}}| \leq R^2 \|w_{mx} - w_x\|_{L^2(0,T;H)} \|\phi_x\|_{\mathcal{H}}$$
$$+ \|(\Psi_R^2(w_{mx}) - \Psi_R^2(w_x))w_x\|_{\mathcal{H}} \|\phi_x\|_{\mathcal{H}}.$$

Now, since the truncation function Ψ_R is Lipschitz continuous and bounded it follows that $\|(\Psi_R^2(w_{mx}) - \Psi_R^2(w_x))w_x\|_{L^2(0,T;H)} \to 0$ by the dominated convergence theorem. It follows that

$$\|K_R(w_m) - K_R(w)\|_{\mathcal{V}^*} \to 0 \quad \text{as} \quad m \to \infty.$$

Finally, using the assumptions on the operator q and straightforward manipulations, we obtain

$$|\langle J(w_m) - J(w), \phi\rangle_{\mathcal{V}}| \leq C\|w_m - w\|_{L^2(0,T;H^1(0,1))} \|\phi\|_{L^2(0,T;H^1(0,1))}.$$

Then, dividing both sides by $\|\phi\|_{\mathcal{V}}$, we obtain

$$\|J(w_m) - J(w)\|_{\mathcal{V}^*} \leq C\|w_m - w\|_{L^2(0,T;H^1(0,1))} \to 0$$

as $m \to \infty$. We conclude that the operator $A_2 : \mathcal{Z} \to \mathcal{Z}^*$ is completely continuous. □

We now have all the ingredients to provide the proof of Proposition 9.8.

Proof. Since the conditions (i)–(iii) on page 235 have now been established we may apply the abstract existence theorem in [10] to conclude that a time shifted version of the above system has a solution. Specifically, in the shifted system the new dependent variable $\tilde{u}$ is given by $\tilde{u} = ue^{-\lambda t}$. The above conditions are sufficient to conclude existence for the shifted system and thus we have existence of a solution $u = (w, v) \in \mathcal{Z}$ for the original system. This implies that w, $w' = v \in \mathcal{V}$ and so $w \in C([0,T]; V)$, which completes the proof of the existence part in Proposition 9.8. Thus, for each R sufficiently large, there exists a solution (w_R, v_R) to Problem 9.7.

We now consider the question of uniqueness. The function $x \to \Psi_R^2(x)x$ is clearly Lipschitz continuous. Let C_R be the Lipschitz constant. Then,

$$\left| \int_0^t \int_0^1 \left(\Psi_R^2(w_{1x}) w_{1x} - \Psi_R^2(w_{1x}) w_{1x} \right) (v_{1x} - v_{2x})\, dx dt \right|$$
$$\leq C_R \int_0^t \int_0^1 |w_{1x} - w_{2x}|\, |v_{1x} - v_{2x}|\, dx ds.$$

Thus,

$$\left(\int_0^t \left\langle \frac{1}{3} a K_R(w_1) - \frac{1}{3} a K_R(w_2), v_1 - v_2 \right\rangle ds \right)$$
$$\leq \frac{1}{3} a C_R \int_0^t \|w_1 - w_2\|_W \, \|v_1 - v_2\|_W \, ds.$$

Similarly, but with less trouble,

$$\int_0^t \langle (-\nu B(w_1) + J(w_1)) - (-\nu B(w_2) + J(w_2)), v_1 - v_2 \rangle \, ds$$
$$\leq C \int_0^t \|w_1 - w_2\|_W \, \|v_1 - v_2\|_W \, ds.$$

Here, and to the end of this section, C depends on R. Hence,

$$\int_0^t \left\langle A_2 \begin{pmatrix} w_1 \\ v_1 \end{pmatrix} - A_2 \begin{pmatrix} w_2 \\ v_2 \end{pmatrix}, \begin{pmatrix} w_1 \\ v_1 \end{pmatrix} - \begin{pmatrix} w_2 \\ v_2 \end{pmatrix} \right\rangle ds$$
$$\leq C \int_0^t \|w_1 - w_2\|_W \, \|v_1 - v_2\|_W \, ds$$

and so

$$\lambda \left\langle D \begin{pmatrix} w_1 \\ v_1 \end{pmatrix} - D \begin{pmatrix} w_2 \\ v_2 \end{pmatrix}, \begin{pmatrix} w_1 \\ v_1 \end{pmatrix} - \begin{pmatrix} w_2 \\ v_2 \end{pmatrix} \right\rangle$$
$$+ \left\langle A \begin{pmatrix} w_1 \\ v_1 \end{pmatrix} - A \begin{pmatrix} w_2 \\ v_2 \end{pmatrix}, \begin{pmatrix} w_1 \\ v_1 \end{pmatrix} - \begin{pmatrix} w_2 \\ v_2 \end{pmatrix} \right\rangle$$
$$\geq \lambda k \|w_1 - w_2\|_V^2 + \lambda |v_1 - v_2|_H^2 + \gamma \|v_1 - v_2\|_V^2$$
$$- C \|w_1 - w_2\|_W \, \|v_1 - v_2\|_W$$
$$\geq \lambda k \|w_1 - w_2\|_V^2 + \lambda |v_1 - v_2|_H^2 + \gamma \|v_1 - v_2\|_V^2$$
$$- C \left(\|w_1 - w_2\|_W^2 + \|v_1 - v_2\|_W^2 \right)$$

then it follows from compactness of the embedding of V into W, that

$$\geq \lambda k \|w_1 - w_2\|_V^2 + \lambda |v_1 - v_2|_H^2 + \gamma \|v_1 - v_2\|_V^2$$
$$- C_R \left(\|w_1 - w_2\|_W^2 + \frac{\gamma}{2C_R} \|v_1 - v_2\|_V^2 + C(\gamma, C_R) |v_1 - v_2|_H^2 \right)$$
$$> 0,$$

if λ is chosen large enough. Therefore, the operator $\lambda D + A$ is strictly monotone for such λs and this proves that the solution in Proposition 9.8 is unique, which concludes the proof. □

9.5 Proof of Theorem 9.5

We have obtained in Proposition 9.8 the solution of the abstract equation

$$v' + kK(w) + \gamma K(v) + \frac{1}{3} a K_R(w) - \nu B(w) - J(w) = f \quad \text{in } \mathcal{V}^*, \tag{9.44}$$

where the operators are given in (9.32)–(9.35), the truncation operator Ψ_R in (9.28), and v and w are related by

$$w(t) = w_0 + \int_0^t v(s)\, ds \text{ in } V.$$

Consider now the function $\Psi_R(x)^2 x$, whose graph is a monotone function of the form depicted in Fig. 9.2 below. We now define

$$\Phi_R(w(t)) = \int_0^t \Psi_R(w(s))^2 w(s) w'(s)\, ds.$$

We note that Φ_R is a convex and positive function. Furthermore,

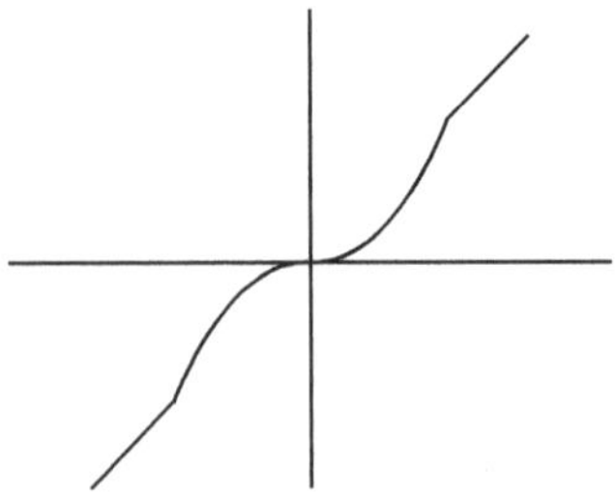

Fig. 9.2 The function $\Psi_R(x)^2 x$

$$\Phi_R(x) = \frac{x^4}{4} \text{ if } |x| < R, \quad \text{and} \quad \lim_{R\to\infty} \Phi_R(x) = \frac{x^4}{4}.$$

for each x. Moreover, Φ_R is increasing in R. We now observe that

$$\begin{aligned}
\left|\int_0^t \int_0^1 p(t) w_x v_x \, dx dt\right| &= \left|\frac{1}{2}\int_0^1 \int_0^t p(t) \frac{d}{ds}\left(w_x^2\right) ds dx\right| \\
&= \left|\frac{1}{2}\int_0^1 \left[w_x^2 p|_0^t - \int_0^t w_x^2(s)\, p'(s)\, ds\right]\right| \\
&\leq \frac{1}{2}\int_0^1 \left(w_x^2(t)\, p_0 + w_{0x}^2 p(0)\right) \\
&\quad + \frac{1}{2} p_0 \int_0^t |w_x(s)|_H^2 \, ds.
\end{aligned}$$

Moreover,

$$\begin{aligned}
\int_0^t -J(w)\, v ds &= \int_0^t \int_0^1 q(g-w)(-v)\, dx\, dt \\
&= \int_0^1 \int_0^t \frac{d}{dt} Q(g-w)\, ds dx \\
&= \int_0^1 Q(g-w)\, dx,
\end{aligned}$$

where $Q' = q$. Since q is given to be monotone (see assumption (9.25)), we can take $Q \geq 0$.

Now we apply both sides of (9.44) to v and integrate from 0 to t. Then, using the results just derived, we obtain an expression of the form

$$\begin{aligned}
&\frac{1}{2}|v(t)|_H^2 - \frac{1}{2}|v_0|^2 + \frac{k}{2}\|w(t)\|_V^2 - \frac{k}{2}\|w_0\|_H^2 \\
&+ \gamma \int_0^t \|v\|_V^2 \, ds + \frac{1}{3} a \int_0^1 \Phi_R(w_x(t))\, dx \\
&- \nu\left(\frac{1}{2}\int_0^1 \left(w_x^2(t)\, p_0 + w_{0x}^2 p(0)\right) + \frac{1}{2} p_0 \int_0^t |w_x(s)|_H^2 \, ds\right) \\
&+ \int_0^1 Q(q-w)\, dx \leq \int_0^t |f|_H |v|_H \, ds.
\end{aligned}$$

Ignoring the two nonnegative terms with Q and a, we obtain,

$$|v(t)|_H^2 + k\,\|w(t)\|_V^2 - p_0\nu\,\|w(t)\|_W^2 + \gamma \int_0^t \|v\|_V^2\,ds$$
$$\le C(w_0, p_0, v_0, f) + \int_0^t |v|_H^2\,ds + p_0\nu \int_0^t \|w(s)\|_W^2\,ds.$$

We note that the estimate depends on the problem data but does not depend on R. Now, we use the compactness of the embedding of V into W to obtain

$$p_0\nu\,\|w\|_W^2 \le \frac{k}{2}\|w\|_V^2 + C_k\,|w|_H^2 = \frac{k}{2}\|w\|_V^2 + 2C_k\,|w_0|_H^2 + 2C_kT\int_0^t |v|_H^2\,.$$

This leads to the inequality

$$|v(t)|_H^2 + \frac{k}{2}\|w(t)\|_V^2 + \gamma\int_0^t \|v\|_V^2\,ds$$
$$\le C(w_0, p_0, v_0, f, k) + C(k,T)\int_0^t |v|_H^2\,ds + p_0\nu\int_0^t \|w(s)\|_W^2\,ds.$$

Then, Gronwall's inequality implies that

$$|v(t)|_H^2 + \frac{k}{2}\|w(t)\|_V^2 + \gamma\int_0^t \|v\|_V^2\,ds \le C(w_0, p_0, v_0, f, k, T)\,. \tag{9.45}$$

The constant on the right-hand side is independent of γ and R. Now, by the continuous embedding of $H^1(0,1)$ into $C([0,1])$, it follows that, for large enough R, the solution of the truncated problem satisfies $|w_x(t,x)| < R$ and so the constraint imposed by the truncation Ψ_R is never in operation. Thus, the solution to Problem 9.7, for R sufficiently large, is also a solution to Problem 9.3. Since the solution to Problem 9.7 is unique, so is the solution to Problem 9.3. This completes the proof of Theorem 9.5.

9.6 The Elastic Beam with the Normal Compliance Condition

Next we consider obtaining the existence of a solution in the case when the beam is elastic, so that $\gamma = 0$. In this case, uniqueness is lost. From the estimate (9.45), it follows from the equation solved and the boundedness of the operators that there also exists an estimate of the form

$$\|v'\|_{V^*}^2 + |v(t)|_H^2 + \frac{k}{2}\|w(t)\|_V^2 + \gamma\int_0^t \|v\|_V^2\,ds \le C(w_0, p_0, v_0, f, k, T)\,.$$

Now let the solution in the above viscoelastic problem be denoted by (w_γ, v_γ). By the above estimate, there exists a subsequence, still denoted with the subscript γ, such that, as $\gamma \to 0$,

$$\begin{aligned}
v_\gamma &\to v \text{ weakly in } L^\infty(0,T;H),\\
w_\gamma &\to w \text{ strongly in } C([0,T];Z),\\
\gamma K v_\gamma &\to 0 \text{ weakly in } \mathcal{V}^*,\\
v'_\gamma &\to v' \text{ in } \mathcal{V}^*,\\
K w_\gamma &\to K w \text{ weakly in } \mathcal{V}^*,\\
B w_\gamma &\to B w \text{ weakly in } \mathcal{V}^*.
\end{aligned}$$

Here Z is an intermediate space satisfying the conditions that the embedding of V into Z is compact and that the embedding of Z into $C([0,1])$ is continuous. For example, one could take for Z the closure of V in $H^{\frac{11}{6}}([0,1])$.

These convergences are easily enough to pass to a limit in the equation solved by (w_γ, v_γ), including the nonlinear term involving w_x^3. This yields a solution to the equation

$$\begin{aligned}
\langle v'(t), u\rangle_V + k(w_{xx}(t), u_{xx}) + \frac{1}{3}a(w_x^3(t), u_x) - \nu p(t)(w_x(t), u_x)&\\
= \langle f(t), u\rangle_V + (q(g-w), u) \qquad \forall\, u \in V.&
\end{aligned} \tag{9.46}$$

Because of the manner of obtaining this solution, it is not at all clear that it is unique. All that is clear is that it exists and satisfies the regularity implied by the above convergences. This yields the following existence theorem for the inviscid problem.

Theorem 9.14. *There exists a solution (v, w) to the inviscid problem (9.46) along with the initial condition $v(0) = v_0$ and the relation between v and w given by*

$$w(t) = w_0 + \int_0^t v(s)\,ds, \quad w_0 \in H^2([0,1]).$$

This solution satisfies $v' \in \mathcal{V}^*$, $v \in \mathcal{V}$.

9.7 The Elastic Beam with the Signorini Condition

Now we will pass to a limit in the elastic beam as the normal compliance condition becomes increasingly stiff. Thus, we replace q with $\frac{1}{\varepsilon}q$ and consider the solution to the abstract beam equation given below which we denote with a subscript of ε:

$$v'_\varepsilon + kK(w_\varepsilon) + \frac{1}{3}aw^3_{\varepsilon x} - \nu B(w_\varepsilon) - \frac{1}{\varepsilon}J(w_\varepsilon) = f. \tag{9.47}$$

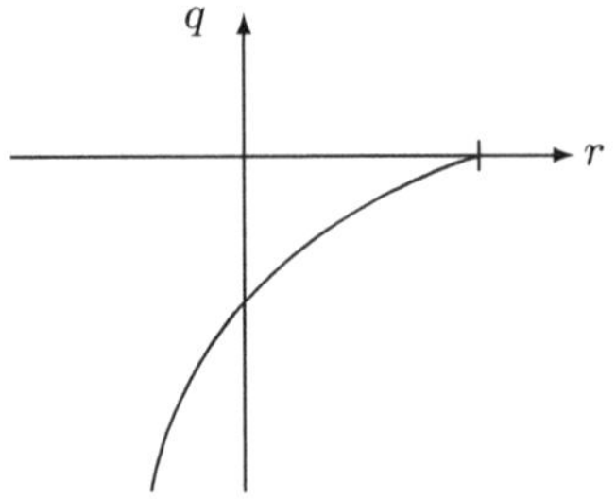

Fig. 9.3 The graph of the function $-q(g-r)$

We let this equation act on v_ε, recall that we now assume that $f \in \mathcal{H}$, and integrate from 0 to t to obtain

$$\frac{1}{2}|v_\varepsilon(t)|_H^2 - \frac{1}{2}|v_0|^2 + \frac{k}{2}\|w_\varepsilon(t)\|_V^2 - \frac{k}{2}\|w_0\|_H^2$$
$$+ \frac{a}{12}\int_0^1 w_{\varepsilon x}^4(t)dx - \frac{a}{12}\int_0^1 w_{0x}^4 dx - \nu\int_0^t \langle B(w_\varepsilon(s)), v_\varepsilon(s)\rangle ds$$
$$+ \int_0^t\int_0^1\left(-\frac{1}{\varepsilon}q(g-w_\varepsilon(s))\right)v_\varepsilon(s)\,dxds = \int_0^t (f(s), v_\varepsilon(s))ds. \quad (9.48)$$

We now focus our attention on the term containing q. From the above assumptions, the graph of $r \to -q(g-r)$ is of the general form depicted above (Fig. 9.3) for each $x \in [0,1]$. We define $Q(x,w) = \int_{q(x)}^{w}(-q(g-\alpha))\,dr$. Then, $Q(x,w) \geq 0$ and

$$\frac{\partial Q(x,\omega)}{\partial w} = -q(g(x)-w), \qquad Q(x, w_0(x)) = 0,$$

the last equality holds since $w_0 \geq g(x)$. Then, using these facts in the term containing q and rearranging terms, we obtain the inequality

$$\frac{1}{2}|v_\varepsilon(t)|_H^2 + \frac{k}{2}\|w_\varepsilon(t)\|_V^2 + \frac{a}{12}\int_0^1 w_{\varepsilon x}^4 dx - \nu\int_0^1\int_0^t p(t)\,w_{\varepsilon x}v_{\varepsilon x}dsdx$$
$$+\frac{1}{\varepsilon}\int_0^1 Q(x, w_\varepsilon(t,x))\,dx \leq C(f, v_0, w_0) + \int_0^t |v_\varepsilon(s)|_H^2\,ds. \quad (9.49)$$

We now consider the term containing p. It can be written as

$$\int_0^1\int_0^t p(s)\frac{d}{dt}|w_{\varepsilon x}|^2\,dsdx = \int_0^1\left[p(s)|w_{\varepsilon x}|^2|_0^t - \int_0^t |w_{\varepsilon x}(s)|^2 p'(s)\,ds\right]dx,$$

and so it is dominated by an expression of the form

$$p_0 \|w_\varepsilon(t)\|_W^2 + p_0 \int_0^t \|w_\varepsilon(s)\|_W^2 \, ds + C(w_0).$$

Using this result in (9.49), we obtain the estimate

$$\frac{1}{2}|v_\varepsilon(t)|_H^2 + \frac{k}{2}\|w_\varepsilon(t)\|_V^2 + \frac{a}{12}\int_0^1 w_{\varepsilon x}^4 dx + \frac{1}{\varepsilon}\int_0^1 Q(x, w_\varepsilon(t,x))\, dx$$
$$\leq C(f, v_0, w_0) + p_0\|w_\varepsilon(t)\|_W^2 + p_0\int_0^t \|w_\varepsilon(s)\|_W^2 \, ds + \int_0^t |v_\varepsilon(s)|_H^2 \, ds.$$

Using the compactness of the embedding of V into W and the relationship between w_ε and v_ε, we have that

$$\frac{1}{2}|v_\varepsilon(t)|_H^2 + \frac{k}{4}\|w_\varepsilon(t)\|_V^2 + \frac{a}{12}\int_0^1 w_{\varepsilon x}^4 dx + \frac{1}{\varepsilon}\int_0^1 Q(x, w_\varepsilon(t,x))\, dx$$
$$\leq C(f, v_0, w_0) + C\int_0^t |v_\varepsilon(s)|_H^2 \, ds + p_0\int_0^t \|w_\varepsilon(s)\|_W^2 \, ds.$$

Therefore, by Gronwall's inequality, there is a constant $C(f, v_0, w_0)$, independent of ε, such that

$$|v_\varepsilon(t)|_H^2 + \|w_\varepsilon(t)\|_V^2 + \int_0^1 w_{\varepsilon x}^4 dx + \frac{C}{\varepsilon}\int_0^1 Q(x, w_\varepsilon(t,x))\, dx \leq C(f, v_0, w_0). \tag{9.50}$$

It follows that there exists a subsequence, still denoted with ε, such that, as $\varepsilon \to 0$,

$$v_\varepsilon \to v \text{ weak* in } L^\infty(0,T;H),$$
$$w_\varepsilon \to w \text{ strongly in } C([0,T];Z),$$
$$w_\varepsilon \to w \text{ weak * in } L^\infty(0,T;V),$$
$$Kw_\varepsilon \to Kw \text{ weakly in } \mathcal{V}^*,$$
$$Bw_\varepsilon \to Bw \text{ weakly in } \mathcal{V}^*,$$
$$-\frac{1}{\varepsilon}q(g - w_\varepsilon) \to \xi \text{ weakly in } \mathcal{V}^*.$$

Here, Z is the intermediate space between V and $C[0,1]$ described in the previous section. We also note that multiplication of (9.50) by ε and then passing to a limit as $\varepsilon \to 0$ implies that

$$\int_0^1 Q(x, w(t,x))\, dx = 0,$$

and so, for each t,

$$g(x) \leq w(t, x).$$

We now return to (9.47). We let this equation act on $u \in \mathcal{V}$ with $u' \in L^{\infty}(0, T; H)$. The above convergences are enough to pass to a limit in this equation and obtain the following

$$\begin{aligned}(v(t), u(t)) &- (v_0, u(0)) - \int_0^t \big(v(s), u'(s)\big)\, ds + k \int_0^t \langle Kw, u(s)\rangle\, ds \\ &+ \frac{1}{3} a \int_0^t (w_x^3(s), u(s)) ds - \nu \int_0^t \langle Bw(s), u(s)\rangle\, ds + \int_0^t \langle \xi, u\rangle_V\, ds \\ &= \int_0^t (f(s), u(s)) ds. \end{aligned} \tag{9.51}$$

It only remains to consider the term containing ξ. Since the graph of $r \mapsto -q(g - r)$ shows that it is monotone and since we have now shown that $w(t, x) \geq g(x)$ for each x, it follows that, for each t, and any $u(t) \in V$ with $u(t, x) \geq g(x)$,

$$-\frac{1}{\varepsilon} q(g - w_\varepsilon(t, x))(w_\varepsilon(t, x) - u(t, x)) \geq 0.$$

It follows that for such u,

$$\begin{aligned}\int_0^t \int_0^1 -\frac{1}{\varepsilon} q(g - w_\varepsilon(t, x))(w_\varepsilon(t, x) - u(t, x))\, dx ds \\ = \int_0^t \left\langle -\frac{1}{\varepsilon} q(g - w_\varepsilon), w_\varepsilon - u\right\rangle_V ds \geq 0.\end{aligned}$$

Then, passing to the limit, we obtain

$$\int_0^t \langle \xi, w - u\rangle_V\, ds \geq 0. \tag{9.52}$$

In (9.51), we now replace u with $w - u$ where w is as described in the above limit and $u(t, x) \geq g(x), u \in \mathcal{V}$, and $u' \in L^{\infty}(0, T; H)$, thus,

$$\begin{aligned}&(v(t), w(t) - u(t)) - (v_0, w(0) - u(0)) \\ &- \int_0^t \big(v(s), w'(s) - u'(s)\big)\, ds + k \int_0^t \langle Kw, w(s) - u(s)\rangle\, ds\end{aligned}$$

$$+\frac{1}{3}a\int_0^t (w_x^3(s), w(s) - u(s))ds - \nu\int_0^t \langle Bw(s), w(s) - u(s)\rangle\, ds$$

$$+\int_0^t \langle \xi, w - u\rangle\, dt = \int_0^t (f(s), w(s) - u(s))ds.$$

Thus, replacing the various operators with their definitions and using (9.52), we obtain the variational inequality (9.21) that is valid for all $u \in \mathcal{V}$ such that $u' \in L^\infty(0, T, H)$ and $u(t,x) \geq g(x)$ for each t. This concludes the proof of Theorem 9.6.

References

1. Ahn, J., Kuttler, K.L., Shillor, M.: Dynamic contact of two Gao beams. Electron. J. Differ. Equ. **2012**(194), 1–42 (2012)
2. Andrews, K.T., M'Bengue, M.F., Shillor, M.: Vibrations of a nonlinear dynamic beam between two stops. Discrete Continuous Dyn. Syst. Ser. B **12**, 23–38 (2009)
3. Andrews, K.T., Dumont, Y., M'Bengue, M.F., Purcell, J., Shillor, M.: Analysis and simulations of a nonlinear dynamic beam. J. Appl. Math. Phys. (ZAMP) **63**, 1005–1019 (2012)
4. Cai, K., Gao, D.Y., Qin, Q.H.: Post-buckling solutions of hyper-elastic beam by canonical dual finite element method. Math. Mech. Solids (MMS) (2013). doi:10.1177/1081286513482483
5. Gao, D.Y.: Nonlinear elastic beam theory with application in contact problems and variational approaches. Mech. Res. Commun. **23**, 11–17 (1996)
6. Gao, D.Y.: Bi-complementarity and duality: a framework in nonlinear equilibria with applications to the contact problems of elastoplastic beam theory. J. Appl. Math. Anal. **221**, 672–697 (1998)
7. Gao, D.Y.: Finite deformation beam models and triality theory in dynamical post-buckling analysis. Int. J. Non-Linear Mech. **35**, 103–131 (2000)
8. Gao, D.Y.: Duality Principles in Nonconvex Systems: Theory, Methods and Applications. Kluwer Academic Publishers, Dordrecht (2000)
9. Han, W., Sofonea, M.: Quasistatic Contact Problems in Viscoelasticity and Viscoplasticity. Studies in Advanced Mathematics. American Mathematical Society/RI and International Press, Providence/Somerville, MA (2002)
10. Kuttler, K.L.: Time dependent implicit evolution equations. Nonlinear Anal. **10**, 447–463 (1986)
11. Kuttler, K.L.: Dynamic friction contact problems for general normal and friction laws. Nonlinear Anal. **28**, 559–575 (1997)
12. Kuttler, K.L., Shillor, M.: Set-valued pseudomonotone maps and degenerate evolution equations. Commun. Contemp. Math. **1**, 87–123 (1999)
13. Kuttler, K.L., Shillor, M.: Dynamic contact with normal compliance wear and discontinuous friction coefficient. SIAM J. Math. Anal. **34**, 1–27 (2002)
14. Kuttler, K.L., Purcell, J., Shillor, M.: Analysis and simulations of a contact problem for a nonlinear dynamic beam with a crack. Q. J. Mech. Appl. Math. **65**, 1–25 (2012)
15. Kuttler, K.L., Li, J., Shillor, M.: Existence for dynamic contact of a stochastic viscoelastic Gao beam. Nonlinear Analysis Real World Appl. **22**, 568–580 (2015)
16. M'Bengue, M.F.: Analysis of models for nonlinear dynamic beams with or without damage or frictionless contact. Oakland University Doctoral Dissertation (2008)
17. M'Bengue, M.F.: Analysis of a nonlinear beam in contact with a foundation. J. Appl. Math. Phys. (ZAMP) (2013). doi:10.1007/s00033-013-0348-7

18. Rochdi, M., Shillor, M., Sofonea, M.: Quasistatic viscoelastic contact with normal compliance and friction. J. Elast. **51**, 105–126 (1998)
19. Russell, D.L., White, L.W.: A nonlinear elastic beam system with inelastic contact constraints. Appl. Math. Optim. **46**, 291–312 (2002)
20. Shillor, M., Sofonea, M., Telega, J.J.: Models and Analysis of Quasistatic Contact. Lecture Notes in Physics. Springer, Berlin (2004)
21. Simon, J.: Compact sets in the space $L^p(0,T;B)$. Ann. Mat. Pura. Appl. **146**, 65–96 (1987)
22. Zeidler, E.: Nonlinear Functional Analysis and its Applications II/B. Springer, New York (1990)

Chapter 10
A Hyperelastic Dynamic Frictional Contact Model with Energy-Consistent Properties

Mikael Barboteu, David Danan, and Mircea Sofonea

Abstract In this chapter we present an energy-consistent numerical model for the dynamic frictional contact between a hyperlastic body and a foundation. Our contribution has two traits of novelty. The first one arises from the specific frictional contact model we consider, which provides intrinsic energy-consistent properties. The contact is modeled with a normal compliance condition of such a type that the penetration is limited with unilateral constraint and, the friction is described with a version of Coulomb's law of dry friction. The second trait of novelty consists in the construction and the analysis of an energy-consistent scheme, based on recent energy-controlling time integration methods for nonlinear elastodynamics. Some numerical results for representative impact problems are provided. They illustrate both the specific properties of the contact model and the energy-consistent properties of the numerical scheme.

Keywords Contact and friction • Normal compliance • Unilateral constraint • Coulomb friction • Nonlinear elastodynamics • Hyperelasticity • Time integration schemes • Energy-conserving algorithms

AMS Classification. 74M15, 74M20, 74M10, 74B20, 74H15, 74S30, 49M15, 90C53

10.1 Introduction

An important topic concerning the modelling of dynamic frictional contact problems is the enforcement of suitable contact interface conditions with energy-consistent properties. This leads to consideration of acceptable physical models and of numerical schemes with long term time integration accuracy and stability properties. During the last 20 years, many works have been devoted to the construction of

M. Barboteu (✉) • D. Danan • M. Sofonea
Laboratoire de Mathématiques et Physique (LAMPS), Université de Perpignan Via Domitia, 52 Avenue Paul Alduy, 66860 Perpignan, France
e-mail: barboteu@univ-perp.fr; david.danan@univ-perp.fr; sofonea@univ-perp.fr

W. Han et al. (eds.), *Advances in Variational and Hemivariational Inequalities*, Advances in Mechanics and Mathematics 33, DOI 10.1007/978-3-319-14490-0_10

energy-conserving methods for elastodynamic contact problems, see [2, 4, 6, 18–20,26,29,30], for instance. Consistent energy dissipation extensions can be found in [4,6,20,30] and [5], in the study of models with friction and viscosity, respectively.

In this current work we consider new frictional contact conditions between a hyperelastic body and a foundation, which take into account the asperities of the contact surface and some energy-consistent properties, as well. More precisely, we consider a specific frictional contact model which provides intrinsic energy-consistent properties, characterized by a conserving behaviour for frictionless impacts and admissible dissipation for friction phenomena. In our model the contact is described with a normal compliance condition of such a type that the penetration is limited with unilateral constraint. This penetration can be assimilated to the flattening of the asperities on the contact surface of the foundation. The associated model of friction, which represents a generalization of the models introduced in [10,11], is based on a version of Coulomb's law in which we assume that the friction bound depends on both the depth of the penetration and the slip rate.

In order to provide numerical energy-consistent properties related to the specific frictional contact, a crucial issue is to focus on computational aspects with long term time integration accuracy and stability properties. To this end, we consider an energy-consistent scheme based on recent energy-controlling time integration methods for nonlinear elastodynamics, developed in [2,4,6,18–20,26,30]. In particular, we use a Newton continuation method and augmented Lagrangian arguments, already used in [4]. Next, to obtain additional energy conserving properties, we combine the specific penalized methods considered in [4, 17, 19], with the procedure of equivalent mass matrix introduced in [26]. We also provide some numerical experiments of representative impact problems which illustrate, in terms of conservation of energy, the good properties of both the frictional contact model and the numerical scheme. More precisely, we compare our numerical results with results obtained by using time integration schemes used in [4,19,26] and we discuss issues related to energy conservation properties.

The rest of the chapter is structured as follows. In Sect. 10.2 we introduce the notation as well as some preliminary material used for the physical setting of hyperelastic contact problems. In Sect. 10.3 we describe the specific contact model, including the associated Coulomb's law of friction. In Sect. 10.4, we present the classical formulation of the dynamic frictional contact problem and derive its variational formulation. Then, in Sect. 10.5 we introduce the fully discrete approximation of the problem. Section 10.6 is devoted to the analysis of the energy-consistent approach used to solve nonlinear elastodynamic frictional contact problems. Thus, after presenting the usual energy-conserving frameworks used, we focus on the analysis of the discrete energy evolution of the method. Finally, in Sect. 10.7 we present numerical results in the study of two representative two-dimensional examples with linear elastic and hyperelastic materials, respectively.

10.2 Notation and Physical Setting

In this section we present the notation we shall use and some preliminary material. Everywhere in this chapter we use $\mathbb{N}$ for the set of positive integers and $\mathbb{R}_-$ will represent the set of non positive real numbers, i.e. $\mathbb{R}_- = (-\infty, 0]$. We denote by $\mathbb{M}^d$ the space of second order tensors on $\mathbb{R}^d$ or, equivalently, the space of square matrices of order d. The inner product and norm on $\mathbb{R}^d$ and $\mathbb{M}^d$ are defined by

$$\boldsymbol{u} \cdot \boldsymbol{v} = u_i v_i \,, \qquad \|\boldsymbol{v}\| = (\boldsymbol{v} \cdot \boldsymbol{v})^{\frac{1}{2}} \qquad \forall\, \boldsymbol{u}, \boldsymbol{v} \in \mathbb{R}^d,$$

$$\boldsymbol{\Pi} : \boldsymbol{\tau} = \Pi_{ji} \tau_{ij} \,, \qquad \|\boldsymbol{\tau}\| = (\boldsymbol{\tau} : \boldsymbol{\tau})^{\frac{1}{2}} \qquad \forall\, \boldsymbol{\Pi}, \boldsymbol{\tau} \in \mathbb{M}^d.$$

Let $\Omega \subset \mathbb{R}^d$ $(d = 1, 2, 3)$ be an open bounded connected set with a Lipschitz boundary Γ. We use the notation $\boldsymbol{x} = (x_i)$ for a typical point in $\Omega \cup \Gamma$ and we denote by $\boldsymbol{\nu} = (\nu_i)$ the outward unit normal at Γ. Here and below the indices i, j, k, l run between 1 and d and, unless stated otherwise, the summation convention over repeated indices is used. An index that follows a comma represents the partial derivative with respect to the corresponding component of the spatial variable, e.g. $u_{i,j} = \partial u_i / \partial x_j$. We consider the spaces

$$V = \{\boldsymbol{v} \in H^1(\Omega; \mathbb{R}^d) \ : \ \boldsymbol{v} = \boldsymbol{0} \ \text{on} \ \Gamma_1\}, \quad H = L^2(\Omega; \mathbb{R}^d).$$

These are real Hilbert spaces endowed with their standard inner products $(\boldsymbol{u}, \boldsymbol{v})_V$ and $(\boldsymbol{\Pi}, \boldsymbol{\tau})_H$ and their associated norms $\|\cdot\|_V$ and $\|\cdot\|_H$, respectively. Note that $V \subset H \subset V^*$ is an evolution triple, with all embeddings being continuous, compact and dense. The duality pairing between V^* and V will be denoted by $\langle \boldsymbol{u}, \boldsymbol{v} \rangle_{V^* \times V}$.

For an element $\boldsymbol{v} \in V$ we still write $\boldsymbol{v}$ for its trace and we denote by v_ν and $\boldsymbol{v}_\tau$ the normal and tangential components of $\boldsymbol{v}$ on Γ, given by $v_\nu = \boldsymbol{v} \cdot \boldsymbol{\nu}$ and $\boldsymbol{v}_\tau = \boldsymbol{v} - v_\nu \boldsymbol{\nu}$, respectively. Also, for a regular stress function $\boldsymbol{\Pi}$ we use the notation Π_ν and $\boldsymbol{\Pi}_\tau$ for its normal and tangential components, i.e. $\Pi_\nu = (\boldsymbol{\Pi} \boldsymbol{\nu}) \cdot \boldsymbol{\nu}$ and $\boldsymbol{\Pi}_\tau = \boldsymbol{\Pi} \boldsymbol{\nu} - \Pi_\nu \boldsymbol{\nu}$. Moreover, we recall that the divergence operator is defined by the equality $\operatorname{Div} \boldsymbol{\Pi} = (\Pi_{ij,j})$ and, finally, the following Green's formula holds:

$$\int_\Omega \boldsymbol{\Pi} : \nabla \boldsymbol{v} \, dx + \int_\Omega \operatorname{Div} \boldsymbol{\Pi} \cdot \boldsymbol{v} \, dx = \int_\Gamma \boldsymbol{\Pi} \boldsymbol{\nu} \cdot \boldsymbol{v} \, d\Gamma \quad \forall\, \boldsymbol{v} \in V. \tag{10.1}$$

In the rest of the chapter we consider the time interval of interest $[0, T]$ with $T > 0$. We denote by $t \in [0, T]$ the time variable and, as already mentioned, $\boldsymbol{x} \in \Omega \cup \Gamma$ will represent the spatial variable. In order to simplify the notation, we do not indicate the dependence of the functions on $\boldsymbol{x}$ and t. Moreover, we use the dots above to represent the derivatives with respect to the time. We also use $\boldsymbol{u}$ for the displacement field and $\boldsymbol{\Pi}$ for the first Piola-Kirchoff stress tensor.

We consider a dynamic contact problem between a deformable body and a foundation in the framework of finite deformations theory. The material's behavior

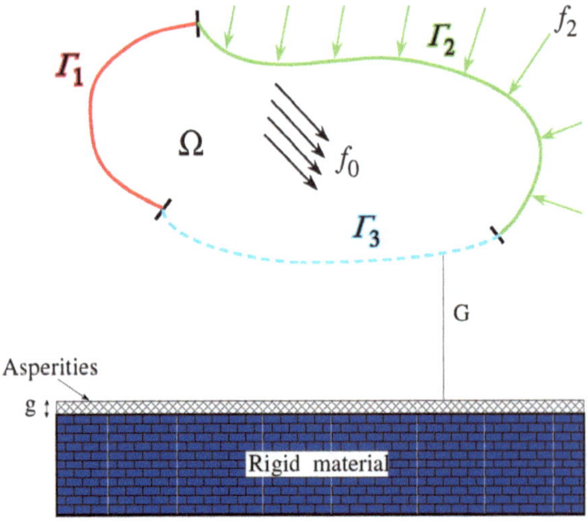

Fig. 10.1 A deformable body in dynamic contact with a foundation

is described with a hyperelastic constitutive law. We recall that hyperelastic constitutive laws are characterized by the first Piola-Kirchoff tensor $\boldsymbol{\Pi}$ which derives from an internal hyperelastic energy density $W(\mathbf{F})$, i.e. $\boldsymbol{\Pi} = \partial_{\mathbf{F}} W(\mathbf{F})$. Here $\mathbf{F}$ is the deformation gradient defined by $\mathbf{F} = \mathbf{I} + \nabla \boldsymbol{u}$ and $\partial_{\mathbf{F}}$ represents the differential with respect to the variable $\mathbf{F}$, see [12] for details. The deformable body occupies the domain Ω with the boundary partitioned into three disjoint parts $\overline{\Gamma}_1$, $\overline{\Gamma}_2$ and $\overline{\Gamma}_3$ with Γ_1, Γ_2 and Γ_3 being relatively open. The part Γ_1 is the part in which the displacement field is prescribed. A volume force of density $\boldsymbol{f}_0$ acts in $\Omega \times (0, T)$, and we assume that a density of traction forces, denoted by $\boldsymbol{f}_2$, acts on the part Γ_2, i.e.

$$\boldsymbol{\Pi} \boldsymbol{\nu} = \boldsymbol{f}_2 \quad \text{on} \quad \Gamma_2 \times (0, T).$$

On the part Γ_3 the body can arrive in frictional contact with an obstacle, the so-called foundation, as shown in Fig. 10.1.

Various contact boundary conditions have been used to model contact phenomena, both in engineering and mathematical literature, see for instance [1, 14, 16, 27, 28, 33, 34, 36, 37, 39–41] and the references therein. One of the most popular is the Signorini condition, introduced in [38], which describes the contact with a perfectly rigid foundation. Expressed in terms of unilateral constraints for the displacement field, this condition leads to highly nonlinear and nonsmooth mathematical problems. The unilateral contact conditions with a gap between a deformable body and a rigid foundation are given by

$$u_\nu - G \leq 0, \quad \Pi_\nu \leq 0, \quad \Pi_\nu (u_\nu - G) = 0 \quad \text{on} \quad \Gamma_3 \times (0, T), \tag{10.2}$$

where the gap function G measures the distance between a point on Γ_3 and its projection onto the rigid obstacle. In the following, for simplicity, we consider that the gap function vanishes, i.e. we use condition (10.2) with $G = 0$. The Signorini contact condition is idealistic since a foundation is never perfectly rigid, due to the presence of microscopic asperities on its surface. Furthermore, it induces non negligible dissipation of energy during impacts, as explained in [2, 29]. For this reason, an unilateral contact condition expressed in terms of velocity field has been considered in the literature. Its form is given by

$$\dot{u}_\nu \leq 0, \quad \Pi_\nu \leq 0, \quad \Pi_\nu \dot{u}_\nu = 0 \quad \text{on} \quad \Gamma_3 \times (0, T). \tag{10.3}$$

Even if it induces good properties of conservation of energy, condition (10.3) is not realistic, since it could lead to non controlled penetrations.

The contact with a deformable foundation is modelled by the so-called normal compliance contact condition. It assigns a reactive normal pressure that depends on the interpenetration of the asperities on the body's surface and those on the foundation. The normal compliance contact condition was first introduced in [33, 36] in the study of dynamic contact problems with elastic and viscoelastic materials. A general expression for the normal compliance condition with a zero gap function is

$$-\Pi_\nu = p(u_\nu) \quad \text{on} \quad \Gamma_3 \times (0, T), \tag{10.4}$$

where $p(\cdot)$ is a nonnegative prescribed function which vanishes for negative argument. This condition can be viewed as a regularization of the Sigorini unilateral condition. It is obvious to see that the normal compliance condition is characterized by a non limited penetration.

In the case of the frictional contact, the contact condition is usually associated to Coulomb's law of dry friction, given by

$$\begin{cases} \|\boldsymbol{\Pi}_\tau\| \leq -\mu \Pi_\nu & \text{if} \quad \dot{\boldsymbol{u}}_\tau = \boldsymbol{0}, \\ -\boldsymbol{\Pi}_\tau = -\mu \Pi_\nu \frac{\dot{\boldsymbol{u}}_\tau}{\|\dot{\boldsymbol{u}}_\tau\|} & \text{if} \quad \dot{\boldsymbol{u}}_\tau \neq \boldsymbol{0}, \end{cases} \quad \text{on } \Gamma_3 \times (0, T). \tag{10.5}$$

Here μ is a positive variable, the coefficient of friction.

As noted in [35], the tribological laws (10.2) and (10.5) can be written in the form of subdifferential inclusions which derive from non-differentiable convex potentials:

$$\Pi_\nu \in \partial I_{\mathbb{R}^-}(u_\nu) \quad \text{and} \quad -\boldsymbol{\Pi}_\tau \in -\mu \Pi_\nu \partial \|\dot{\boldsymbol{u}}_\tau\| \quad \text{on} \quad \Gamma_3 \times (0, T).$$

Here $\partial I_{\mathbb{R}^-}$ and $\partial \|\dot{\boldsymbol{u}}_\tau\|$ denote the subdifferential of the indicator function $I_{\mathbb{R}^-}$ of the negative half-line of $\mathbb{R}$ and the subdifferential of the norm of the slip rate, $\|\dot{\boldsymbol{u}}_\tau\|$, respectively. Furthermore, in the case of large deformations, we can refer to [14] for the definition of objective quantities related to contact mechanics.

10.3 A Specific Frictional Contact Model

The purpose of this section is to present a specific frictional contact model which provides intrinsic energy-consistent properties. The contact is modeled with a normal compliance condition of such a type that the penetration is limited with unilateral constraint and the friction is modeled with a specific version of Coulomb's law.

In order to obtain energy conservation properties, the work of normal contact reactions, denoted by $\mathcal{W}_c = \int_{\Gamma_3} \Pi_\nu \dot{u}_\nu \, d\Gamma$, has to vanish. Therefore, for energy conservation purposes, as explained in [2, 29], the following persistency condition has to be added:

$$\dot{u}_\nu \Pi_\nu = 0 \quad \text{on} \quad \Gamma_3 \times (0, T). \tag{10.6}$$

This condition means that normal contact reactions can appear only during persistent contact. Note that the unilateral contact condition (10.3), expressed in terms of velocity, could lead to displacements which do not satisfy the non penetration condition. Furthermore, as explained in [4, 6], it is impossible to enforce, at a given moment, both the complementarity condition in displacement and in velocity.

To overcome this drawback, in order to take into account the deformability of the foundation (arising from the existence of micro asperities on its surface), we consider the following normal compliance condition restricted by unilateral constraint:

$$\begin{cases} \Pi_\nu + p(u_\nu) \le 0, \\ u_\nu - g \le 0, \\ (\Pi_\nu + p(u_\nu))(u_\nu - g) = 0, \end{cases} \qquad \text{on} \ \ \Gamma_3 \times (0, T). \tag{10.7}$$

This condition was introduced for an elastic-visco-plastic problem in [23]. In this model, the contact follows a normal compliance condition with penetration but up to the limit g and then, when this limit is reached, the contact follows a Signorini-type unilateral condition with the gap g and without any additional penetration in the foundation. We conclude from above that condition (10.7) models the contact with a foundation which is composed by a thin deformable layer of asperities of thickness g which covers a perfect rigid material. This contact model has two intrinsic advantages: the adequation with energy conservation properties during penetrations for the impact phase ($-\Pi_\nu = p(u_\nu)$ for $0 \le u_\nu < g$), on one hand, and the limitation of penetration into the foundation ($\Pi_\nu \le 0,\ u_\nu - g \le 0,\ \Pi_\nu(u_\nu - g) = 0$), on the other hand. Note that the energy conservation property during penetrations comes from the specific form of the normal compliance function p. Indeed, let us consider a normal compliance function p defined by

$$p(u_\nu) = r u_\nu^+ \qquad \text{with} \qquad u_\nu^+ = \max(0, u_\nu) = \text{dist}_{\mathbb{R}^-}(u_\nu), \tag{10.8}$$

where r is a positive constant which represents the deformability of the foundation. Then, with (10.8), the persistency condition is no longer necessary to obtain energy conservation properties and the work $\mathcal{W}_c$ of contact reactions takes the form

$$\mathcal{W}_c = \int_{\Gamma_3} \Pi_\nu \dot{u}_\nu \, d\Gamma = -\int_{\Gamma_3} r u_\nu^+ \dot{u}_\nu \, d\Gamma = -\int_{\Gamma_3} \frac{r}{2} \frac{d}{dt} \{u_\nu^+\}^2 \, d\Gamma. \tag{10.9}$$

If we consider the Gonzalez approach, [15], then the conservation of the energy for a contact system with a normal compliance condition of the form (10.8) is provided by the following energy assessment on $[0, t]$:

$$\begin{aligned} E(t) - E(0) = & \int_0^t \int_\Omega \mathbf{f} \cdot \dot{\boldsymbol{u}} \, d\Omega + \int_0^t \int_{\Gamma_2} \mathbf{g} \cdot \dot{\boldsymbol{u}} \, d\Gamma \\ & - \frac{r}{2} \int_{\Gamma_3} \left[(u_\nu^+(t))^2 - (u_\nu^+(0))^2 \right] d\Gamma. \end{aligned} \tag{10.10}$$

Here $E(t)$ represents the internal energy of the body Ω at time t and is defined by

$$E(t) = \frac{1}{2} \int_\Omega \rho \dot{u}^2 d\Omega + \int_\Omega W(\mathbf{F}) \, d\Omega. \tag{10.11}$$

We refer to Sect. 10.6.2 or [19] for more details about this energy assessment properties for a normal compliance condition of the form (10.8). The same statement can be established for the angular and the linear momentum, as shown in [39].

We turn now to the friction conditions. Our goal is to consider a friction model suited to the previous contact conditions. To this end, we introduce a specific version of Coulomb's law of dry friction in which the friction bound depends both on the depth of the penetration for $0 \le u_\nu \le g$ and on the normal contact stress Π_ν. Therefore, we consider the following friction condition:

$$\begin{cases} \|\boldsymbol{\Pi}_\tau\| \le -\mu(u_\nu)\Pi_\nu & \text{if} \quad \dot{\boldsymbol{u}}_\tau = \mathbf{0}, \\ -\boldsymbol{\Pi}_\tau = -\mu(u_\nu)\Pi_\nu \frac{\dot{\boldsymbol{u}}_\tau}{\|\dot{\boldsymbol{u}}_\tau\|} & \text{if} \quad \dot{\boldsymbol{u}}_\tau \neq \mathbf{0}, \end{cases} \quad \text{on } \Gamma_3 \times (0, T). \tag{10.12}$$

Here μ denotes the coefficient of friction and is assumed to depend on the penetration u_ν as long as $u_\nu < g$. When there is penetration, as far as the normal displacement does not reach the bound g (i.e. $0 \le u_\nu < g$), the contact is described with a normal compliance condition associated to the classical Coulomb's law of dry friction with the friction bound $\mu(u_\nu)p(u_\nu)$. Details on the normal compliance contact condition associated to Coulomb's law of dry friction can be found in [16, 37, 40], for instance. After the complete flattening of the asperities, i.e. when the normal displacement reaches the bound g, the magnitude of the normal stress is larger than $p(g)$ and, moreover, friction follows a Coulomb's law associated to unilateral contact, with the friction bound $-\mu(g)\Pi_\nu$. Note that the friction bound is characterized by the friction coefficient μ which depends on the depth of the penetration u_ν and on the size of the asperities g. In what follows, we consider two examples, namely

$$\mu_1(\eta) = \begin{cases} 0 \text{ for } \eta \leq 0, \\ \frac{\eta}{g}\mu_0 \text{ for } \eta \in (0, g), \\ \mu_0 \text{ for } \eta \geq g. \end{cases} \quad \mu_2(\eta) = \begin{cases} 0 \text{ for } \eta \leq 0, \\ (2 - \frac{\eta}{g})\mu_0 \text{ for } \eta \in (0, g), \\ \mu_0 \text{ for } \eta \geq g. \end{cases} \quad (10.13)$$

Here $\mu_0 = \mu(g)$ denotes a given coefficient of friction associated to the unilateral contact ($u_\nu = g$). In the case of function μ_1, we remark that the friction bound increases with respect to the flattening of the asperities. In contrast, in the case of the function μ_2, the friction bound decreases with respect to the flattening of the asperities. Several experimental studies have demonstrated the dependence of the friction coefficient with respect to the normal compression load and the flattening of asperities. This behavior is generated by the wear of asperities on the contact surfaces. References on this matter include [22,31,32], among others.

10.4 Mechanical Problem and Variational Formulation

10.4.1 Mechanical Problem

With these preliminaries the formulation of hyperelastodynamic frictional contact problem is the following.

Problem $\mathcal{P}_M$. *Find a displacement field $\boldsymbol{u} : \Omega \times [0,T] \to \mathbb{R}^d$ and a stress field $\boldsymbol{\Pi} : \Omega \times [0,T] \to \mathbb{M}^d$ such that*

$$\boldsymbol{\Pi} = \partial_{\mathbf{F}} W(\mathbf{F}) \quad \text{in } \Omega \times (0,T), \qquad (10.14)$$

$$\rho \ddot{\boldsymbol{u}} - \operatorname{Div} \boldsymbol{\Pi} - \boldsymbol{f}_0 = \boldsymbol{0} \quad \text{in } \Omega \times (0,T), \qquad (10.15)$$

$$\boldsymbol{u} = \boldsymbol{0} \quad \text{on } \Gamma_1 \times (0,T), \qquad (10.16)$$

$$\boldsymbol{\Pi}\boldsymbol{\nu} = \boldsymbol{f}_2 \quad \text{on } \Gamma_2 \times (0,T), \qquad (10.17)$$

$$\begin{cases} u_\nu \leq g, \ \Pi_\nu + p(u_\nu) \leq 0, \\ (u_\nu - g)(\Pi_\nu + p(u_\nu)) = 0 \end{cases} \quad \text{on } \Gamma_3 \times (0,T), \qquad (10.18)$$

$$\begin{cases} \|\boldsymbol{\Pi}_\tau\| \leq -\mu(u_\nu)\Pi_\nu & \text{if } \dot{\boldsymbol{u}}_\tau = \boldsymbol{0}, \\ -\boldsymbol{\Pi}_\tau = -\mu(u_\nu)\Pi_\nu \frac{\dot{\boldsymbol{u}}_\tau}{\|\dot{\boldsymbol{u}}_\tau\|} & \text{if } \dot{\boldsymbol{u}}_\tau \neq \boldsymbol{0}, \end{cases} \quad \text{on } \Gamma_3 \times (0,T), \qquad (10.19)$$

$$\boldsymbol{u}(0) = \boldsymbol{u}_0, \ \dot{\boldsymbol{u}}(0) = \boldsymbol{u}_1 \quad \text{in } \Omega. \qquad (10.20)$$

We recall that Eq. (10.14) is the hyperelastic constitutive law. Equation (10.15) represents the equation of motion in which ρ denotes the density of the material and is assumed to be constant, for the sake of simplicity. Conditions (10.16) and (10.17) are the displacement and traction boundary conditions, respectively. Finally, (10.20) represent the initial conditions in which $\boldsymbol{u}_0$ and $\boldsymbol{u}_1$ are the initial displacement and velocity, respectively.

Next, conditions (10.18) and (10.19) represent the frictional contact conditions, already introduced in Sect. 10.3. We recall that (10.18) represents a contact condition with normal compliance contact and unilateral constraint, in which the penetration is limited to the value g. Condition (10.19) represents a version of Coulomb's law of dry friction in relation with the contact conditions (10.18). Note that the condition (10.18) is equivalent to

$$-\Pi_\nu \in p(u_\nu) + \partial I_{(-\infty,g]}(u_\nu) \quad \text{on } \Gamma_3 \times (0,T), \tag{10.21}$$

where ∂ represents the subdifferential operator in the sense of the convex analysis and I_A denotes the indicator function of the set $A \subset \mathbb{R}$. In the same way, we observe that the condition (10.19) is equivalent to

$$-\boldsymbol{\Pi}_\tau \in -\mu(u_\nu)\Pi_\nu \partial\|\dot{\boldsymbol{u}}_\tau\| \quad \text{on } \Gamma_3 \times (0,T). \tag{10.22}$$

In the rest of the chapter, we will consider the frictional contact conditions in their subdifferential form (10.21), (10.22).

10.4.2 *Variational Formulation*

We now introduce a hybrid variational formulation of Problem $\mathcal{P}_M$ in which the dual variables corresponding to Lagrange multipliers are related to the contact stress and the friction force. In this case, the Lagrange multipliers verify extended subdifferential inclusions derived from the pointwise subdifferential inclusions defined in (10.21) and (10.22). To this end we consider the trace spaces $X_\nu = \{\, v_\nu|_{\Gamma_3} : \boldsymbol{v} \in V \,\}$ and $X_\tau = \{\, v_\tau|_{\Gamma_3} : \boldsymbol{v} \in V \,\}$, equipped with their usual norms. We denote by X_ν' and X_τ' the duals of the spaces X_ν and X_τ, respectively. Moreover, we denote by $\langle\cdot,\cdot\rangle_{X_\nu',X_\nu}$ and $\langle\cdot,\cdot\rangle_{X_\tau',X_\tau}$ the corresponding duality pairing mappings.

To establish the variational formulation, we need additional notations. Thus, we consider the function $\boldsymbol{f} : (0,T) \to V^*$ and the operator $B : V \to V^*$ defined by

$$\langle \boldsymbol{f}(t), \boldsymbol{v}\rangle_{V^*\times V} = (\boldsymbol{f}_0(t), \boldsymbol{v})_H + (\boldsymbol{f}_2(t), \boldsymbol{v})_{L^2(\Gamma_2;\mathbb{R}^d)}, \tag{10.23}$$

$$\langle B\boldsymbol{u}, \boldsymbol{v}\rangle_{V^*\times V} = \int_\Omega \boldsymbol{\Pi}(\boldsymbol{u}) : \nabla \boldsymbol{v}\, dx \tag{10.24}$$

for all $t \in (0,T)$, $\boldsymbol{u}, \boldsymbol{v} \in V$. For the contact conditions, we introduce a function $\varphi_\nu : X_\nu \to (-\infty,+\infty]$ and an operator $L : X_\nu \to X_\nu'$ defined by

$$\varphi_\nu(u_\nu) = \int_{\Gamma_3} I_{(-\infty,g]}(u_\nu)\, d\Gamma \quad \forall\, u_\nu \in X_\nu,$$

$$L : X_\nu \to X_\nu', \quad \langle L u_\nu, v_\nu\rangle_{X_\nu',X_\nu} = \int_{\Gamma_3} p(u_\nu) v_\nu\, d\Gamma \quad \forall\, u_\nu,\, v_\nu \in X_\nu.$$

We note that, for all $t \in (0,T)$, condition (10.21) leads to the subdifferential inclusion

$$-\Pi_\nu(t) \in \partial\varphi_\nu(u_\nu(t)) + L u_\nu(t) \quad \text{in } X_\nu'. \tag{10.25}$$

To reformulate the friction law, we introduce the function $\varphi_\tau : L^2(\Gamma_3;\mathbb{R}^d) \to (-\infty,+\infty]$ defined by

$$\varphi_\tau(\boldsymbol{u}_\tau) = \int_{\Gamma_3} \|\boldsymbol{u}_\tau\|\, d\Gamma \quad \forall\, \boldsymbol{u}_\tau \in L^2(\Gamma_3;\mathbb{R}^d).$$

We note that for all $t \in (0,T)$, condition (10.22) leads to the subdifferential inclusion

$$-\boldsymbol{\Pi}_\tau(t) \in -\mu(u_\nu(t))\Pi_\nu \partial\varphi_\tau(\dot{\boldsymbol{u}}_\tau(t)) \quad \text{in } X_\tau'. \tag{10.26}$$

Inclusions (10.25) and (10.26) suggest to introduce two new unknowns, the Lagrange multipliers, which represent the normal and tangential stresses on the contact surface, and which will be denoted in what follows by ξ_ν and $\boldsymbol{\xi}_\tau$, respectively. Therefore, multiplying the equation of motion (10.15) by the test function $\boldsymbol{v}$, integrating the result over $\Omega \times (0,T)$ and using the Green formula (10.1) and the inclusions (10.25)–(10.26), we obtain the following hybrid variational formulation of Problem $\mathcal{P}_M$, in terms of three unknown fields.

Problem $\mathcal{P}_V$**.** *Find a displacement field $\boldsymbol{u} : [0,T] \to V$, a normal stress field $\xi_\nu : [0,T] \to X_\nu'$ and a tangential stress field $\boldsymbol{\xi}_\tau : [0,T] \to X_\tau'$ such that*

$$\langle \rho\ddot{\boldsymbol{u}}(t) + \boldsymbol{B}\boldsymbol{u}(t), \boldsymbol{v}\rangle_{V^*\times V} = \langle \boldsymbol{f}(t), \boldsymbol{v}\rangle_{V^*\times V} + \langle \xi_\nu(t), v_\nu\rangle_{X_\nu',X_\nu} + \langle \boldsymbol{\xi}_\tau(t), v_\tau\rangle_{X_\tau',X_\tau} \quad \forall\, \boldsymbol{v} \in V, \tag{10.27}$$

$$-\xi_\nu(t) \in \partial\varphi_\nu(u_\nu(t)) + L u_\nu(t), \tag{10.28}$$

$$-\boldsymbol{\xi}_\tau(t) \in -\mu(u_\nu(t))\xi_\nu \partial\varphi_\tau((\dot{\boldsymbol{u}}_\tau(t)), \tag{10.29}$$

for all $t \in [0,T]$ and, moreover,

$$\boldsymbol{u}(0) = \boldsymbol{u}_0, \qquad \dot{\boldsymbol{u}}(0) = \boldsymbol{u}_1. \tag{10.30}$$

We note that the existence and uniqueness of weak solution of Problem $\mathcal{P}_V$ represents, at the best of our knowledge, an open mathematical problem. Nevertheless, the solvability of a dynamic viscoelastic frictional contact problem with a regularized version of the frictional contact conditions (10.28)–(10.29) has been established in [9], in the case of small deformations theory. We also recall that the question of weak solvability for several contact problems is discussed in detail in the books [16, 34, 40].

10.5 Discretization and Variational Approximation

This section is devoted to the discretization of the variational problem $\mathcal{P}_V$, based on arguments similar to those used in [4, 5, 7–11].

First, we recall some preliminary material concerning the time discretization step. Let N be an integer, let $k = \frac{T}{N}$ be the time step and define

$$t_n = n\,k, \quad 0 \leq n \leq N.$$

Below, for a continuous function $f(t)$ with values in a function space, we use the notation $f_j = f(t_j)$, for $0 \leq j \leq N$. In what follows, we consider a collection of discrete times $\{t_n\}_{n=0}^{N}$ which define a uniform partition of the time interval $[0, T] = \bigcup_{n=1}^{N} [t_{n-1}, t_n]$ with $t_0 = 0$, $t_n = t_{n-1} + k$ and $t_N = T$. Finally, for a sequence $\{w_n\}_{n=1}^{N}$, we denote the midpoint divided differences by

$$\delta w_{n-\frac{1}{2}} = (w_n - w_{n-1})/k = \frac{1}{2}(\delta w_n + \delta w_{n-1}), \tag{10.31}$$

and, equivalently, we have $\delta w_n = -\delta w_{n-1} + \frac{2}{k}(w_n - w_{n-1})$. In the rest of the paper, we use the notation $\square_{n-\frac{1}{2}} = \frac{1}{2}(\square_n + \square_{n-1})$, where $\square_n$ represents the approximation of $\square(t_n)$. Note that the time integration scheme we use is based on the implicit second order midpoint rule given in (10.31).

We now present some material concerning the spatial discretization step. Let Ω be a polyhedral domain. Moreover, let $\{\mathcal{T}^h\}$ be a regular family of triangular finite element partitions of $\overline{\Omega}$ that are compatible with the boundary decomposition $\Gamma = \overline{\Gamma_1} \cup \overline{\Gamma_2} \cup \overline{\Gamma_3}$, i.e., if one side of an element $Tr \in \mathcal{T}^h$ has more than one point on Γ, then the side lies entirely on $\overline{\Gamma_1}$, $\overline{\Gamma_2}$ or $\overline{\Gamma_3}$. The space V is approximated by the finite dimensional space $V^h \subset V$ of continuous and piecewise affine functions, that is,

$$V^h = \{\, \boldsymbol{v}^h \in [C(\overline{\Omega})]^d \;:\; \boldsymbol{v}^h|_T \in [P_1(Tr)]^d \quad \forall\, Tr \in \mathcal{T}^h,$$
$$\boldsymbol{v}^h = \boldsymbol{0} \text{ at the nodes on } \Gamma_1 \,\},$$

where $P_1(Tr)$ represents the space of polynomials of degree less or equal to one in Tr. For the discretization of the normal contact terms, we consider the space

$$X_\nu^h = \{ v_{\nu|\Gamma_3}^h : \boldsymbol{v}^h \in V^h \}$$

equipped with its usual norm. Let us consider the discrete space of piecewise constant functions $Y_\nu^h \subset L^2(\Gamma_3)$ related to the discretization of the normal stress ξ_ν. Then, we note that the contact condition (10.28) leads to the following discrete subdifferential inclusion at time $t_{n-\frac{1}{2}}$:

$$-\xi_{\nu}{}_{n-\frac{1}{2}}^{hk} \in \partial\varphi_\nu(u_{\nu}{}_{n-\frac{1}{2}}^{hk}) + L u_{\nu}{}_{n-\frac{1}{2}}^{hk} \quad \text{in } Y_\nu^h.$$

For the discretization of the tangential friction terms, let us consider the space

$$X_\tau^h = \{ \boldsymbol{v}_{\tau|\Gamma_3}^h : \boldsymbol{v}^h \in V^h \}$$

equipped with its usual norm. We also consider the discrete space of piecewise constants $Y_\tau^h \subset L^2(\Gamma_3)^d$ related to the discretization of the friction density $\boldsymbol{\xi}_\tau$. In a similar way, we note that the friction condition (10.29) leads to the following discrete subdifferential inclusion at time $t_{n-\frac{1}{2}}$

$$-\boldsymbol{\xi}_{\tau}{}_{n-\frac{1}{2}}^{hk} \in -\mu(u_\nu^h)\xi_{\nu}{}_{n-\frac{1}{2}}^{hk}\,\partial\varphi_\tau(\delta \boldsymbol{u}_{\tau}{}_{n-\frac{1}{2}}^{hk}) \quad \text{in} \quad Y_\tau^h.$$

More details about the discretization step can be found in [24, 25, 41].

Let $\boldsymbol{u}_0^h \in V^h$ and $\boldsymbol{u}_1^h \in V^h$ be finite element approximations of $\boldsymbol{u}_0$ and $\boldsymbol{u}_1$, respectively. Then, using the previous notations and the midpoint scheme (10.31), the fully discrete approximation of the Problem $\mathcal{P}_V$ at the time $t_{n-\frac{1}{2}}$ is the following.

Problem $\mathcal{P}_V^{hk}$. *Find a discrete displacement field $\boldsymbol{u}^{hk} = \{\boldsymbol{u}_n^{hk}\}_{n=0}^N \subset V^h$, a discrete normal stress field $\xi_\nu^{hk} = \{\xi_{\nu n}^{hk}\}_{n=0}^N \subset Y_\nu^h$ and a discrete tangential stress field $\boldsymbol{\xi}_\tau^{hk} = \{\boldsymbol{\xi}_{\tau n}^{hk}\}_{n=0}^N \subset Y_\tau^h$ such that, for all $n = 1, \ldots, N$,*

$$\langle \frac{\rho}{k}\left(\delta \boldsymbol{u}_n^{hk} - \delta \boldsymbol{u}_{n-1}^{hk}\right) + B\boldsymbol{u}_{n-\frac{1}{2}}^{hk}, \boldsymbol{v}^h\rangle_{V^*\times V} = \langle \boldsymbol{f}_{n-\frac{1}{2}}^{hk}, \boldsymbol{v}^h\rangle_{V^*\times V}$$

$$+ \langle \xi_{\nu}{}_{n-\frac{1}{2}}^{hk}, v_\nu^h\rangle_{Y_\nu^h, X_\nu^h} + \langle \boldsymbol{\xi}_{\tau}{}_{n-\frac{1}{2}}^{hk}, \boldsymbol{v}_\tau^h\rangle_{Y_\tau^h, X_\tau^h} \quad \forall\, \boldsymbol{v} \in V^h, \tag{10.32}$$

$$-\xi_{\nu}{}_{n-\frac{1}{2}}^{hk} \in \partial\varphi_\nu(u_{\nu}{}_{n-\frac{1}{2}}^{hk}) + L u_{\nu}{}_{n-\frac{1}{2}}^{hk}, \tag{10.33}$$

$$-\boldsymbol{\xi}_{\tau}{}_{n-\frac{1}{2}}^{hk} \in -\mu(u_{\nu}{}_{n-\frac{1}{2}}^{hk})\xi_{\nu}{}_{n-\frac{1}{2}}^{hk}\,\partial\varphi_\tau(\delta \boldsymbol{u}_{\tau}{}_{n-\frac{1}{2}}^{hk}), \tag{10.34}$$

$$\boldsymbol{u}_0^{hk} = \boldsymbol{u}_0^h, \qquad \delta \boldsymbol{u}_0^{hk} = \boldsymbol{u}_1^h. \tag{10.35}$$

Note that the discrete frictional contact conditions (10.33) and (10.34) are considered at the time $t_{n-\frac{1}{2}}$. As a consequence the solution $\boldsymbol{u}_n^{hk}$ does not verify the contact conditions at the desired time t_n and the scheme does not control penetration. In order to overcome this drawback, several authors impose the contact conditions at the time t_n, see [4, 5, 15, 19, 24, 30], for instance. Here we use an implicit backward divided difference for the discretization of the tangential velocity $\dot{\boldsymbol{u}}_\tau(t)$ given by $\delta\boldsymbol{u}_{\tau n} = (\boldsymbol{u}_{\tau n} - \boldsymbol{u}_{\tau n-1})/k$, which leads to the following discrete problem.

Problem $\overline{\mathcal{P}}_V^{hk}$**.** *Find a discrete displacement field* $\boldsymbol{u}^{hk} = \{\boldsymbol{u}_n^{hk}\}_{n=0}^N \subset V^h$, *a discrete normal stress field* $\xi_\nu^{hk} = \{\xi_{\nu n}^{hk}\}_{n=0}^N \subset Y_\nu^h$ *and a discrete tangential stress field* $\boldsymbol{\xi}_\tau^{hk} = \{\boldsymbol{\xi}_{\tau n}^{hk}\}_{n=0}^N \subset Y_\tau^h$ *such that, for all* $n = 1, \ldots, N$,

$$\langle \frac{\rho}{k}\left(\delta\boldsymbol{u}_n^{hk} - \delta\boldsymbol{u}_{n-1}^{hk}\right) + B\boldsymbol{u}_{n-\frac{1}{2}}^{hk}, \boldsymbol{v}^h\rangle_{V^*\times V} = \langle \boldsymbol{f}_{n-\frac{1}{2}}^{hk}, \boldsymbol{v}^h\rangle_{V^*\times V}$$

$$+ \langle \xi_{\nu n}^{hk}, v_\nu^h\rangle_{Y_\nu^h, X_\nu^h} + \langle \boldsymbol{\xi}_{\tau n}^{hk}, \boldsymbol{v}_\tau^h\rangle_{Y_\tau^h, X_\tau^h} \quad \forall\, \boldsymbol{v} \in V^h, \tag{10.36}$$

$$-\xi_{\nu n}^{hk} \in \partial\varphi_\nu(u_{\nu n}^{hk}) + L u_{\nu n}^{hk}, \tag{10.37}$$

$$-\boldsymbol{\xi}_{\tau n}^{hk} \in -\mu(u_{\nu n}^{hk})\xi_{\nu n}^{hk}\,\partial\varphi_\tau(\delta\boldsymbol{u}_{\tau n}^{hk}), \tag{10.38}$$

$$\boldsymbol{u}_0^{hk} = \boldsymbol{u}_0^h \quad \text{in} \quad \delta\boldsymbol{u}_0^{hk} = \boldsymbol{u}_1^h. \tag{10.39}$$

Note that the specific discretization used in Problem $\overline{\mathcal{P}}_V^{hk}$ represents the starting point to develop improved energy-conserving algorithms for the solution of elastodynamic contact problems with long term time integration accuracy and stability. Some details on the energy-conserving framework can be found in the next section.

10.6 Numerical Solution with Energy-Consistent Properties

10.6.1 Usual Discrete Energy-Conserving Framework

We start by recalling some preliminaries concerning the usual discrete energy-conserving framework in the case without contact. In the rest of the section, to simplify the notation and the readability, we do not indicate the dependence of various variables with respect to the discretization parameters k and h, i.e., for example, we write $\boldsymbol{u}$ instead of $\boldsymbol{u}^{hk}$.

In order to solve a nonlinear elastodynamic problem, we have to use adapted time integration schemes. When nonlinear dynamic problems are considered, the standard implicit schemes (θ-method, Newmark schemes, midpoint or HHT methods) lose their unconditional stability, as explained in [21, 28]. Therefore, there is a need to use implicit energy conservative schemes as those used in

[3, 15, 17, 28, 39], which are appropriate, due to their long term time integration accuracy and stability. In all these methods the corresponding discrete mechanical conservation properties are satisfied. To establish these discrete energy conservative properties, one of the most used implicit time integration scheme is the second order midpoint scheme given by

$$\delta \boldsymbol{u}_n = -\delta \boldsymbol{u}_{n-1} + \frac{2}{k}(\boldsymbol{u}_n - \boldsymbol{u}_{n-1}). \tag{10.40}$$

Moreover, according the time integration scheme of Gonzalez [15], the variational inequality (10.36) is characterized by the operator B defined by

$$\langle B\boldsymbol{u}_{n-\frac{1}{2}}, \boldsymbol{v}\rangle_{V^*\times V} = \int_\Omega \boldsymbol{\Pi}^{algo} : \nabla \boldsymbol{v}\, dx \ \text{ for } \boldsymbol{v} \in V^h, \tag{10.41}$$

in which the discrete tensor $\boldsymbol{\Pi}^{algo}$ is introduced in order to satisfied exact discrete energy properties. This tensor defined by Gonzalez in [15] takes the form

$$\begin{cases} \boldsymbol{\Pi}^{algo} = \mathbf{F}_{n-\frac{1}{2}}\boldsymbol{\Sigma}^{algo}, \\ \boldsymbol{\Sigma}^{algo} = 2\frac{\partial \tilde{W}}{\partial \mathbf{C}}(\mathbf{C}_{n-\frac{1}{2}}) + 2[\tilde{W}(\mathbf{C}_n) - \tilde{W}(\mathbf{C}_{n-1}) \\ \qquad -\frac{\partial \tilde{W}}{\partial \mathbf{C}}(\mathbf{C}_{n-\frac{1}{2}}) : \Delta\mathbf{C}_{n-1}]\frac{\Delta\mathbf{C}_{n-1}}{\Delta\mathbf{C}_{n-1}:\Delta\mathbf{C}_{n-1}}, \end{cases} \tag{10.42}$$

where $\Delta\mathbf{C}_{n-1} = \mathbf{C}_n - \mathbf{C}_{n-1}$ and $\mathbf{C}_{n-1} = {}^t\mathbf{F}_{n-1}\mathbf{F}_{n-1}$. As shown in [12], the axiom of frame indifference implies that $W(\mathbf{F}) = \tilde{W}(\mathbf{C})$. Then, using the arguments in [15], it follows that (10.42) verifies exactly the energy conservation condition characterized by

$$\boldsymbol{\Pi}^{algo} : (\nabla \boldsymbol{u}_n - \nabla \boldsymbol{u}_{n-1}) = \tilde{W}(\mathbf{C}_n) - \tilde{W}(\mathbf{C}_{n-1}). \tag{10.43}$$

For more details on standard energy-conserving framework, we refer the reader to [3, 15, 17, 28, 39].

10.6.2 An Improved Energy-Consistent Approach

Many works have been devoted to extend the previous conservative properties to frictionless impact; more precisely, Laursen and Chawla [29] and Armero and Petocz [2] have shown the benefit of the persistency condition to conserve the energy in the discrete framework. Nevertheless, in all these works the numerical method shows that the interpenetration vanishes only when the time step tends towards zero. In order to overcome this drawback, Laursen and Love [30] have developed an efficient method, by introducing a discrete jump in velocity; however, this method requires the solution of an auxiliary system in order to compute the velocity update results. Furthermore, Hauret and Le Tallec [19] have considered a specific penalized enforcement of the contact conditions which allows to provide energy conservation properties. Then, Khenous, Laborde and Renard [26] have introduced the Equivalent Mass Matrix method (EMM), based on a procedure of redistribution of the mass matrix. Interpretations and extensions of this method can be found in [18]. The resulting problem exhibits Lipschitz regularity in time and achieves good energy evolution properties, due to the fact that the persistency condition is automatically satisfied. This equivalent mass matrix approach was studied and used in many works; for instance, theoretical and computational aspects related to this model can be found in [17, 20, 26].

In what follows, we present an improved energy-conserving method for hyper-elastodynamic contact problems with its extension to frictional dissipation phenomena. This method permits to enforce the normal compliance with unilateral constraint during each time step with controlled contact penetrations and with energy-consistent properties. The strategy developed is based on the solution of the system (10.36) by taking into account only the normal compliance condition with friction, in the first step, then the normal compliance restricted by unilateral constraint with friction, in the second step. This strategy is employed successively when passing from the time moment t_{n-1} to the time moment t_n. To this end, we developed an adapted continuation Newton method, decomposed in two steps, which could be summarized as follows:

step (a): Newton scheme to solve the nonlinear system (10.36)

$$\text{with} \quad \begin{cases} -\xi_{\nu n} = L u_{\nu n} & \text{on } \Gamma_3, \\ -\boldsymbol{\xi}_{\tau n} \in -\mu(u_{\nu n})\xi_{\nu n}\varphi_\tau(\delta \boldsymbol{u}_{\tau n}) & \text{on } \Gamma_3. \end{cases} \tag{10.44}$$

step (b): Continuation of the Newton scheme to solve (10.36)

$$\text{with} \quad \begin{cases} \text{if } u_{\nu n}^{(a)} < g & -\xi_{\nu n} = L u_{\nu n} \quad \text{on } \Gamma_3, \\ \text{if } u_{\nu n}^{(a)} \geq g & -\xi_{\nu n} \in \partial\varphi_\nu(u_{\nu n}) + Lg \quad \text{on } \Gamma_3, \\ & -\boldsymbol{\xi}_{\tau n} \in -\mu_0 \xi_{\nu n}\varphi_\tau(\delta \boldsymbol{u}_{\tau n}) \quad \text{on } \Gamma_3. \end{cases} \tag{10.45}$$

According the work of Hauret and Le Tallec [19], we reproduce in the discrete framework the conservation properties described in (10.10) by taking into account a specific form of the normal contact reaction $\xi_{\nu n}$ defined by

$$-\xi_{\nu n} = L u_{\nu n} = r p(u_{\nu_n}) \text{ with } p(u_{\nu_n}) = \frac{[(u_{\nu n})_+]^2 - [(u_{\nu n-1})_+]^2}{2(u_{\nu n} - u_{\nu n-1})}. \tag{10.46}$$

Note that $p(u_{\nu_n})$ represents a specific form of the normal compliance function at t_n. Here r is a penalization parameter interpreted as the stiffness coefficient of the asperities of the foundation. In the following, the continuation Newton method with the normal compliance form (10.46) will be called the Improved Penalized Method (IPM). To keep this paper in reasonable length, we skip the details of the solution of the nonlinear system (10.36) with conditions (10.44) and (10.45), and we restrict ourselves to recall that the presentation of the algorithms together with their numerical implementation can be found in [4, 9]. Details on the discretization step and Computational Contact Mechanics, including algorithms similar to that used here, can be found in [1, 4, 5, 24, 25, 28, 41].

10.6.3 Analysis of the Discrete Energy Evolution

This section is devoted to establish energy-consistent properties induced by the improved penalized method described in the previous Sect. 10.6.2. Below we use the notation E_n and E_{n-1} for the energy E of the hyperelastic frictional contact system at times t_n and t_{n-1}, respectively. For instance, the discrete energy E_n can be written as follows:

$$E_n = \frac{1}{2}\int_\Omega \rho\,[\delta \boldsymbol{u}_n]^2\,dx + \int_\Omega \tilde{W}(\mathbf{C}_n)\,d\Gamma. \tag{10.47}$$

The notation $E_n^{(a)}$, $E_n^{(b)}$, $E_{n-1}^{(a)}$ and $E_{n-1}^{(b)}$ have similar significance, being related to the steps (a) and (b) of the numerical method introduced in Sect. 10.6.2.

The general assessment of the discrete energy of the frictional contact Problem $\overline{\mathcal{P}}_V^{hk}$ between times t_{n-1} and t_n is based on the following proposition.

Proposition 10.1. *The following equality holds:*

$$E_n - E_{n-1} = k\langle \boldsymbol{f}_{n-\frac{1}{2}}, \boldsymbol{u}_{n-\frac{1}{2}}\rangle_{V^*\times V}$$

$$+\,k\int_{\Gamma_3}\left(\xi_{\nu_n}\delta u_{\nu n} + \boldsymbol{\xi}_{\tau_n}\cdot\delta \boldsymbol{u}_{\tau n}\right)d\Gamma. \tag{10.48}$$

Here E_n and E_{n-1} denote the internal energy E of the hyperelastic frictional contact system at times t_n and t_{n-1}, respectively.

Proof. We use the variational formulation (10.36) with

$$\boldsymbol{v} = \delta\boldsymbol{u}_{n-\frac{1}{2}} = \frac{\boldsymbol{u}_n - \boldsymbol{u}_{n-1}}{k} = \frac{\delta\boldsymbol{u}_n + \delta\boldsymbol{u}_{n-1}}{2}.$$

Then, we use (10.42) to get the hyperelastic energy conservation. As a consequence we obtain the equality

$$\frac{1}{2k}\int_\Omega \rho(\delta\boldsymbol{u}_n - \delta\boldsymbol{u}_{n-1}).(\delta\boldsymbol{u}_n + \delta\boldsymbol{u}_{n-1})dx + \frac{1}{k}\int_\Omega \boldsymbol{\Pi}^{algo} : \nabla(\boldsymbol{u}_n - \boldsymbol{u}_{n-1})dx$$

$$= \langle \boldsymbol{f}_{n-\frac{1}{2}}, \boldsymbol{u}_{n-\frac{1}{2}}\rangle_{V^*\times V} + \int_{\Gamma_3}[\xi_{\nu_n}\delta u_{\nu_n} + \boldsymbol{\xi}_{\tau_n}\cdot\delta\boldsymbol{u}_{\tau_n}]d\Gamma.$$

Furthermore, using the identity

$$(\delta \boldsymbol{u}_n - \delta \boldsymbol{u}_{n-1})(\delta \boldsymbol{u}_n + \delta \boldsymbol{u}_{n-1}) = [\delta \boldsymbol{u}_n]^2 - [\delta \boldsymbol{u}_{n-1}]^2$$

and the conservation property of the Gonzalez scheme given in (10.43), we obtain that

$$\frac{1}{2k}\int_\Omega \rho([\delta \boldsymbol{u}_n]^2 - [\delta \boldsymbol{u}_{n-1}]^2)\,dx + \frac{1}{k}\int_\Omega \tilde{W}(\mathbf{C}_n) - \tilde{W}(\mathbf{C}_{n-1})\,dx$$
$$= \langle \boldsymbol{f}_{n-\frac{1}{2}}, \boldsymbol{u}_{n-\frac{1}{2}}\rangle_{V^*\times V} + \int_{\Gamma_3} [\xi_{\nu_n}\delta u_{\nu_n} + \boldsymbol{\xi}_{\tau_n}.\delta \boldsymbol{u}_{\tau_n}]\,d\Gamma.$$

Finally, using the definition (10.47) of the discrete energy we obtain the identity (10.48). □

Remark 10.2. Similar results for the discrete angular and linear momenta can also be established, see for instance [17, 19].

Based on the Proposition 10.1, we can state, at the end of the step (a), the assessment of the discrete energy for the specific normal compliance contact.

Proposition 10.3. *The following equality holds:*

$$E_n^{(a)} - E_{n-1}^{(a)} = \langle \boldsymbol{f}_{n-\frac{1}{2}}, \boldsymbol{u}_{n-\frac{1}{2}}\rangle_{V^*\times V} - \int_{\Gamma_3} \frac{r}{2}\left([(u_{\nu n})_+]^2 - [(u_{\nu n-1})_+]^2\right) d\Gamma$$
$$+ k\int_{\Gamma_3} \boldsymbol{\xi}_{\tau n}\cdot \delta \boldsymbol{u}_{\tau n} d\Gamma. \qquad (10.49)$$

Proof. We use similar arguments as those used in the proof of Proposition 10.1, in particular equality (10.46) combined with equality $\delta u_{\nu n} = (u_{\nu n} - u_{\nu n-1})/k$. □

We remark that the form (10.46) of the normal contact reaction allows, in the frictionless case, an energy assessment which is agreement with the formula (10.10). In addition, when the external forces vanishes, the energy statements in Propositions 10.1 and 10.3 during the steps (a) and (b), respectively, allow us to obtain the following situations.

Case without friction

step (a): $\xi_{\nu_n}\delta u_{\nu n} \approx 0 \Rightarrow E_n^{(a)} \approx E_{n-1}^{(a)}$,

step (b): $\xi_{\nu_n}\delta u_{\nu n} \leq 0 \Rightarrow E_n^{(b)} \leq E_{n-1}^{(b)}$.

Case with friction

step (a): $\xi_{\nu_n}\delta u_{\nu n} \approx 0,\ \boldsymbol{\xi}_{\tau n}\cdot\delta \boldsymbol{u}_{\tau n} \leq 0 \Rightarrow E_n^{(a)} \leq E_{n-1}^{(a)}$,

step (b): $\xi_{\nu_n}\delta u_{\nu n} \leq 0,\ \boldsymbol{\xi}_{\tau n}\cdot\delta \boldsymbol{u}_{\tau n} \leq 0 \Rightarrow E_n^{(b)} \leq E_{n-1}^{(b)}$.

In the case without friction ($\boldsymbol{\xi}_{\tau n} = 0$), we remark that during step (a), the energy of the system is almost conserved. Indeed, the difference $[(u_{\nu n-1})_+]^2 - [(u_{\nu n})_+]^2$ is very small since the penetrations $(u_{\nu n-1})_+$ and $(u_{\nu n})_+$ are small. During the step (b), the enforcement of the unilateral constraint allows to limit the penetrations obtained in step (a) by the value g, which represents a small given value. On the other hand, we can easily prove that the product $\xi_{\nu_n} \delta u_{\nu_n}$ is always negative and this represents an unacceptable physical behavior, since it generates energy dissipation. However, this energy dissipation is low because the impact has occurred during the step (a). Furthermore, when the friction case is considered, in both steps (a) and (b) we observe an admissible dissipation of energy. Indeed, in this case the inner product $\boldsymbol{\xi}_{\tau_n} \cdot \delta \boldsymbol{u}_{\tau_n}$ is always negative. In other words, this strategy limits the energy dissipation between times t_n and t_{n-1}, in the frictionless case, and allows the energy dissipation, in the frictional one. To resume, the advantages of the method arise in the fact that both the dissipation of energy and the penetrations are limited during the impact.

10.7 Numerical Experiments

In order to recover the theoretical numerical behaviour of the fully discrete scheme discussed in Sect. 10.6.3, we carried out some numerical simulations based on two representative dynamic contact problems: the impact without friction of a linearly elastic ball against a foundation (Sect. 10.7.1) and the impact with friction of a hyperelastic ring against a foundation (Sect. 10.7.2).

10.7.1 Impact of a Linearly Elastic Ball Against a Foundation

This representative benchmark problem describes the frictionless impact of a linearly elastic ball against a foundation (see [24]). The elastic ball is thrown with an initial velocity ($\boldsymbol{u}_1 = (0, -10)$ m/s) toward the foundation $\{(x_1, x_2) \in \mathbb{R}^2 : \ x_2 \leq 0\}$. The material's behavior is described by an elastic linear constitutive law defined by the energy function

$$W(\boldsymbol{\epsilon}) = \frac{E\kappa}{2(1+\kappa)(1-2\kappa)}(\mathrm{tr}\boldsymbol{\epsilon})^2 + \frac{E}{2(1+\kappa)}\mathrm{tr}(\boldsymbol{\epsilon}^2) \quad \forall\, \boldsymbol{\epsilon} \in \mathbb{M}^n.$$

Here E and κ are Young's modulus and Poisson's ratio of the material and $\mathrm{tr}(\cdot)$ denotes the trace function, respectively. Note that $\boldsymbol{\epsilon} = \frac{1}{2}(\nabla \boldsymbol{u}^T + \nabla \boldsymbol{u})$ represents the linearized strain tensor in the framework of the small deformations theory ($\|\boldsymbol{u}\| << 1$ and $\|\nabla \boldsymbol{u}\| << 1$ in Ω).

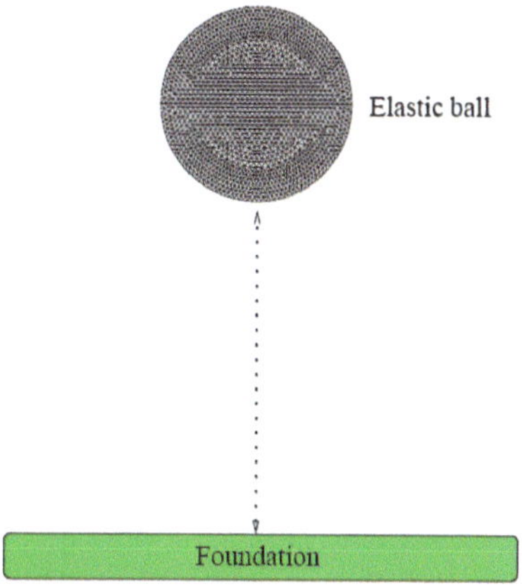

Fig. 10.2 Discretization of the elastic ball in contact with a foundation

The physical setting is depicted in Fig. 10.2. There,

$$\begin{aligned}
\Omega &= \left\{(x_1, x_2) \in \mathbb{R}^2 : \ (x_1 - 100)^2 + (x_2 - 100)^2 \leq 100\right\}, \\
\Gamma_1 &= \emptyset, \qquad \Gamma_2 = \emptyset, \\
\Gamma_3 &= \left\{(x_1, x_2) \in \mathbb{R}^2 : \ (x_1 - 100)^2 + (x_2 - 100)^2 = 100\right\}.
\end{aligned}$$

The domain Ω represents the cross section of the ball, under the assumption of the plane stress. No volume forces are assumed to act on the body during the process. For the discretization of the contact problem depicted we use 7,820 elastic nodes and 128 Lagrange multiplier nodes. For the numerical experiments, we use the following data:

$$\begin{aligned}
&\rho = 1{,}000\,\mathrm{kg/m^3}, \quad T = 2\,\mathrm{s}, \quad k = 0.001, \\
&\boldsymbol{u}_0 = (0,0)\,\mathrm{m}, \quad \boldsymbol{u}_1 = (0,-10)\,\mathrm{m/s}, \\
&E = 100\,\mathrm{GPa}, \quad \kappa = 0.35, \quad \boldsymbol{f}_0 = (0,0)\,\mathrm{Pa}, \\
&g = 5.10^{-4}\,\mathrm{m}, \quad r = 1{,}000\,\mathrm{Pa}, \quad \mu = 0.
\end{aligned}$$

Note that we consider a very small value for g in order to limit (or to neglect) the penetration. This value represents 0.005 % of the radius of the ball.

In Fig. 10.3, the sequence of the deformed ball together with contact forces are presented before, during and after the impact. The interest of this representative example is to compare the numerical results obtained by using the continuation method (presented in Sect. 10.6.2) with numerical results obtained by using some classical methods. To this end, we consider five existing methods:

- The solution of the problem with $g = 0\,\mathrm{m}$ which corresponds to a classical method with Signorini contact condition.

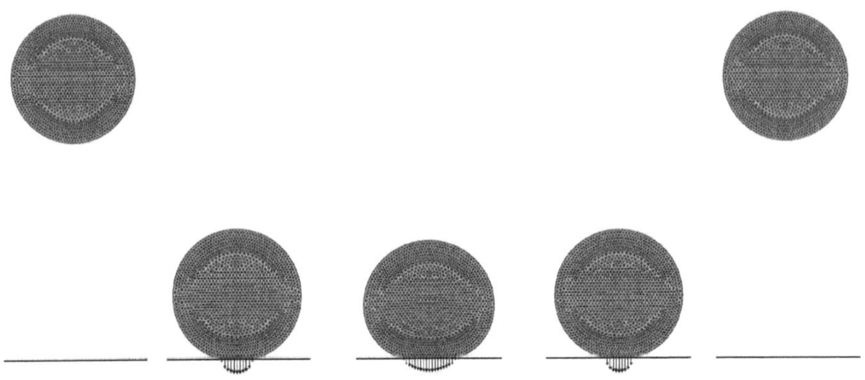

Fig. 10.3 Sequence of the deformed ball and contact forces before, during and after the impact

- The solution with a large value for g ($g = 1{,}000\,\text{m}$) which corresponds to a classical method with a normal compliance condition of the form $\xi_{\nu n} = -r(u_{\nu_n})_+$.
- The specific penalization method developed by Hauret and Le Tallec [19] in which the normal compliance condition is given by (10.46).
- The Equivalent Mass Matrix (EMM) method proposed by Khenous et al. [24], which represents a specific distribution of the mass matrix with no inertia of the contact nodes. This method is characterized by relevant stability properties of the contact stress.
- The adapted Newton continuation method developed by Ayyad and Barboteu [4], which is characterized by the enforcement following two steps of the unilateral contact law and the persistency condition (10.6) during each time increment.

In what follows we analyze the methods in terms of discrete energy evolution. To this end, the total discrete energy at time t_n is defined by the following formula:

$$E_n = \frac{1}{2}\int_\Omega \rho \dot{\boldsymbol{u}}_n^2 dx + \int_\Omega \boldsymbol{\sigma}_n : \boldsymbol{\varepsilon}(\boldsymbol{u}_n)\, dx,$$

where $\boldsymbol{\sigma} = \frac{\partial W(\epsilon)}{\partial \epsilon}$ denotes the stress tensor for infinitesimal deformations.

Figure 10.4 represents the evolution of the total discrete energy of the dynamic system. According to it, we note that after the impact (i.e. for $t \geq 1.52$) and for the considered time step $k = 0.001$, the classicals method with Signorini law (curve (⊖)) as well as the method with standard normal compliance condition (curve (⊟)) are characterized by a non conservation of the energy, which is not realistic from the physical point of view. We also remark that the EMM method (curve ▼) strongly reduces the dissipation of the energy, without obtaining the exact conservation. Furthermore, the scheme developed by Ayyad and Barboteu [4] (curve (•)) and the specific penalized method (curve (■)) conserve the energy after the

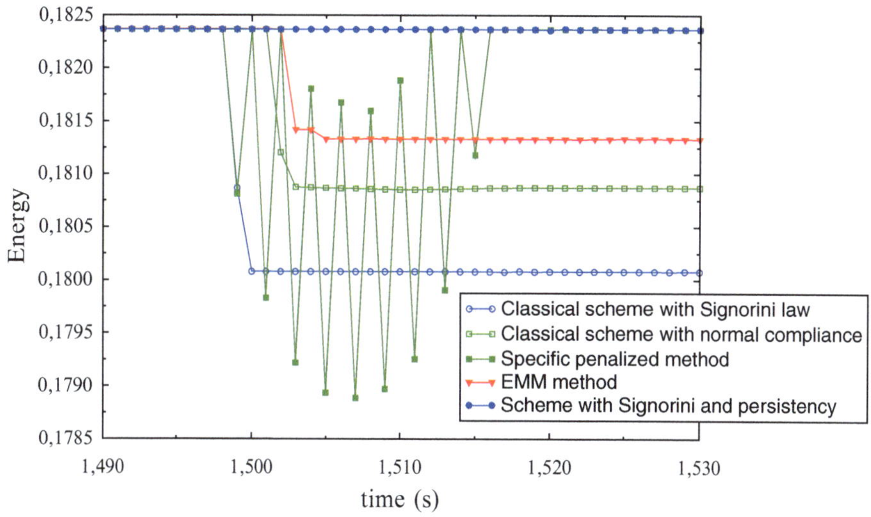

Fig. 10.4 Discrete energy behaviour of selected time integration schemes during the impact

impact. However, for the penalized method we find some energy fluctuations which disappear after the impact. Furthermore, both for this method and the method used in [4], the unilateral contact is not exactly satisfied. Indeed, the specific penalized method generates a maximal error on the normal contact displacement of 0.0034 m, and 0.0051 m for the method of Ayyad and Barboteu [4].

In order to correct these drawbacks, we considered the Improved Penalized Method (curve (▲)) based on the combination of the specific penalized method with the normal compliance law with finite penetration introduced in Sect. 10.6.2. According to Fig. 10.5, we can see that the Improved Penalized Method (IPM) enables to obtain a better conservation of the energy and, in addition, it limits the penetration. Nevertheless, this method generates some fluctuations of the discrete energy during the impact. For this reason, we considered an improvement of the IPM method, obtained by adding the EMM procedure (curve (♦)). This last strategy (IPM + EMM) enables to conserve almost the discrete energy and to limit the penetration. Indeed, we obtain: 0.1 % of dissipation and 0.005 % of penetration. In Fig. 10.6, we analyze the discrete energy behaviour of the Improved Penalized Method with EMM (IPM + EMM) according to the depth g of the penetration. For $g = 0.001$, the IPM + EMM method recovers the same behaviour as the Specific Penalized Method with EMM procedure. We can also note that the numerical results obtained by using the IPM + EMM method approach the numerical results obtained by using the EMM method, as g tends to zero.

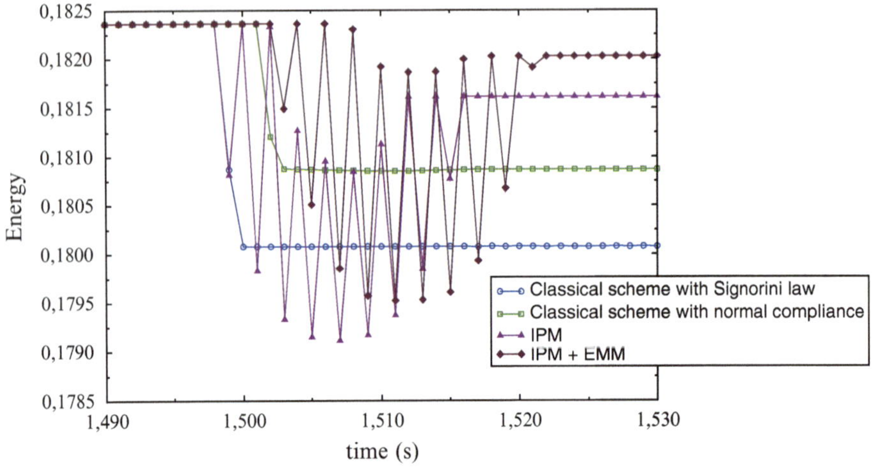

Fig. 10.5 Discrete energy behaviour of variants of the improved penalized method (IPM) during the impact

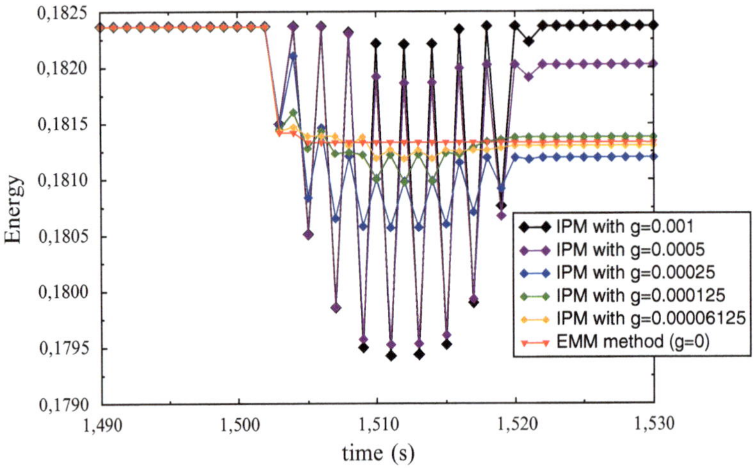

Fig. 10.6 Discrete energy behaviour of the improved penalized method with EMM with respect to the depth g of the penetration

10.7.2 Impact of a Hyperelastic Ring Against a Foundation

In order to highlight the conservative or the dissipative behaviour of the method in the hyperelastic case we consider a second representative application, introduced by Laursen in [28]. This application concerns an academic frictional impact problem with a hyperelastic constitutive behavior of the material: the impact with friction of

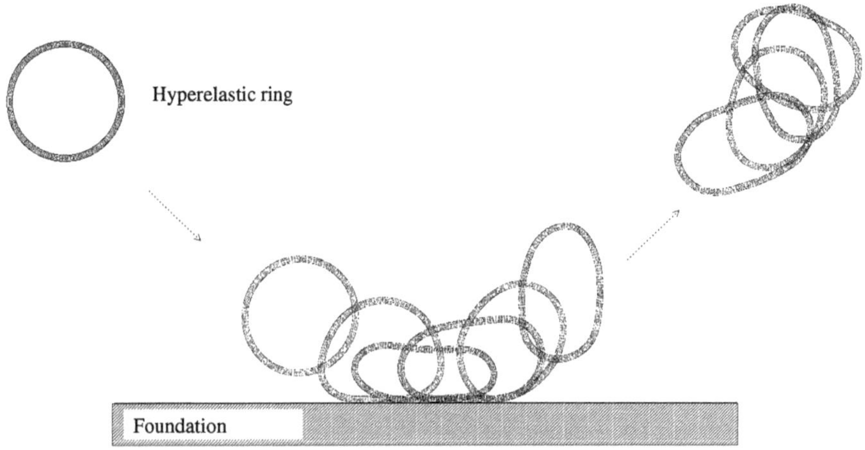

Fig. 10.7 Sequence of the deformed hyperelastic ring during and after impact

the ring against a foundation. The compressible material response, considered here, is governed by an Ogden constitutive law (see [13]) defined by the energy density

$$W(\mathbf{C}) = c_1(I_1 - 3) + c_2(I_2 - 3) + a(I_3 - 1) - (c_1 + 2c_2 + a)\ln I_3.$$

Here I_1, I_2 and I_3 represent the three invariants of the tensor $\mathbf{C}$. This example allows us to assess the performance and to check the conservative or dissipative behavior of the methods. We implemented the IPM method and we compared it to various time integration schemes. Details on the physical setting of the problem are given below:

$$\begin{aligned}
\Omega &= \left\{(x_1, x_2) \in \mathbb{R}^2 : \ 81 \le (x_1 - 100)^2 + (x_2 - 100)^2 \le 100\right\}, \\
\Gamma_1 &= \emptyset, \qquad \Gamma_2 = \emptyset, \\
\Gamma_3 &= \left\{(x_1, x_2) \in \mathbb{R}^2 : \ (x_1 - 100)^2 + (x_2 - 100)^2 = 100\right\}.
\end{aligned}$$

As for the first numerical example, the domain Ω represents the cross section of a three-dimensional deformable body under the plane stress hypothesis. The elastic ring is thrown with an initial velocity at 45° angle toward a foundation as depicted in Fig. 10.7. The foundation is given by $\left\{(x_1, x_2) \in \mathbb{R}^2 : \ x_2 \le 0\right\}$. For the discretization, we use 1,664 elastic nodes and 128 Lagrange multiplier nodes. For the numerical experiments, the data are:

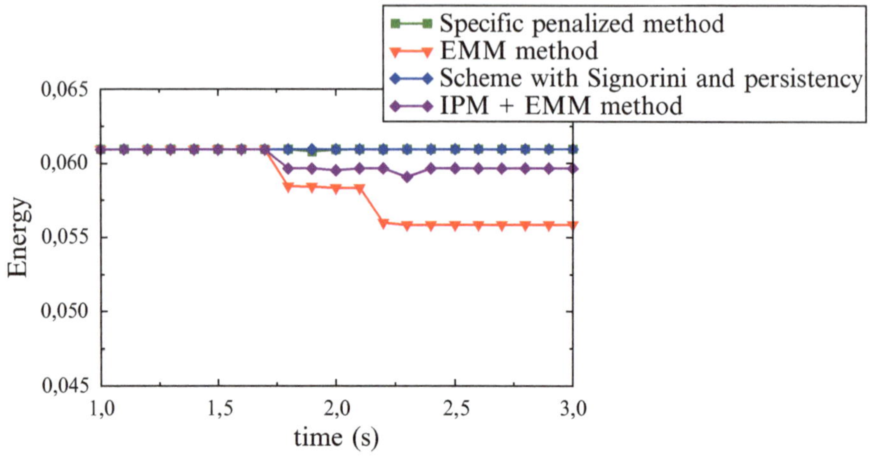

Fig. 10.8 Discrete energy behavior of selected time integration schemes during the impact (frictionless case)

$$\rho = 1{,}000\,\text{kg/m}^3, \quad T = 10\,\text{s}, \quad k = \tfrac{1}{300},$$
$$\boldsymbol{u}_0 = (0,0)\,\text{m}, \quad \boldsymbol{u}_1 = (10,-10)\,\text{m/s},$$
$$c_1 = 0.5\,\text{MPa}, \quad c_2 = 0.5 * 10^{-2}\,\text{MPa}, \quad a = 0.35\,\text{MPa},$$
$$g = 0.002\,\text{m}, \quad r = 1{,}000\,\text{Pa}, \quad \mu_0 = 0.2$$

In Fig. 10.8 we present the evolution of the total discrete energy of the dynamic system without friction for various time integration schemes considered in the previous numerical example, i.e. the specific penalized method (curve (■)), the EMM method (curve ▼), the scheme developed by Ayyad and Barboteu [4] (curve (•)) and the Improved Penalized Method with the EMM procedure (IPM + EMM illustrated by curve (♦)). Let us consider the discrete energy at time t_n defined as follows,

$$E_n = \frac{1}{2}\int_\Omega \rho \dot{\boldsymbol{u}}_n^2 d\Omega + \int_\Omega \tilde{W}(\mathbf{C}_n) d\Omega.$$

In the frictionless case (see Fig. 10.8), we observe that the IPM+EMM method reduces dissipation, when compared with the EMM procedure. Furthermore, this method is characterized by a small penetration ($g = 0.002$ m). In contrast, the specific penalized method and the method presented in [4] generate a maximal error on the normal contact displacement (0.058 and 0.071 m, respectively).

In the frictional case, we consider two cases for the friction function μ which defines the friction bound $-\mu(u_\nu)\Pi_\nu$ of the friction law (10.12). Note that during the flattening of the asperities, i.e. $0 \le u_\nu < g$, the friction follows a Coulomb's law associated to normal compliance contact and, therefore, the friction bound is

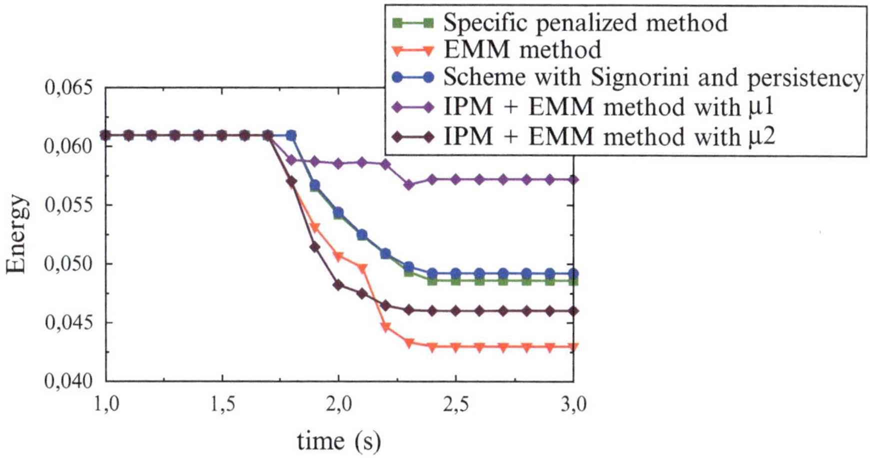

Fig. 10.9 Discrete energy behavior of selected time integration schemes during the impact with friction

equal to $\mu(u_\nu)p(u_\nu)$. In the numerical experiments, we consider the two examples of function μ given in (10.13). Note that the values of the decreasing function $\mu_2(\cdot)$ are larger than the values of the increasing function $\mu_1(\cdot)$ for a penetration u_ν such that $0 < u_\nu < g$. According to Fig. 10.9, we observe that the use of the function μ_1 permits to limit the energy dissipation induced by the friction while the use of the function μ_2 is characterized by a strong energy dissipation. In conclusion, the IPM+EMM method allows the energy dissipation. This seems to be reasonable from a physical point of view, due to the complex phenomena which appear during the flattening and wearing of the asperities. Recall that the numerical modelling of contact surfaces with asperities was one of the main objectives of the present work.

Acknowledgements This research was supported by the Marie Curie International Research Staff Exchange Scheme Fellowship within the 7th European Community Framework Programme under Grant Agreement No. 295118 and the project Polonium "Mathematical and Numerical Analysis for Contact Problems with Friction" 2014/15.

References

1. Alart, P., Curnier, A.: A mixed formulation for frictional contact problems prone to Newton like solution methods. Comput. Methods Appl. Mech. Eng. **92**, 353–375 (1991)
2. Armero, F., Petocz, E.: Formulation and analysis of conserving algorithms for frictionless dynamic contact/impact problems. Comput. Methods Appl. Mech. Eng. **158**, 269–300 (1998)
3. Armero, F., Romero, I.: On the formulation of high-frequency dissipative time-stepping algorithms for nonlinear dynamics. Part II: second-order methods. Comput. Methods Appl. Mech. Eng. **190**, 6783–6824 (2001)

4. Ayyad, Y., Barboteu, M.: Formulation and analysis of two energy-consistent methods for nonlinear elastodynamic frictional contact problems. J. Comput. Appl. Math. **228**, 254–269 (2009)
5. Ayyad, Y., Barboteu, M., Fernández, J.R.: A frictionless viscoelastodynamic contact problem with energy-consistent properties: numerical analysis and computational aspects. Comput. Methods Appl. Mech. Eng. **198**, 669–679 (2009)
6. Barboteu, M.: An energy-conserving algorithm for non linear elastodynamic contact problems - extension to a frictional dissipation phenomenon. Analysis and Simulation of Contact Problems. In: Lecture Notes in Applied Computational Mechanics, vol. 27, pp. 71–78. Springer, Berlin/Heidelberg (2005)
7. Barboteu, M., Matei, A., Sofonea, M.: Analysis of quasistatic viscoplastic contact problems with normal compliance. Q. J. Mech. Appl. Math. **65**, 555–579 (2012)
8. Barboteu, M., Bartosz, K., Kalita, P.: An analytical and numerical approach to a bilateral contact problem with nonmonotone friction. Int. J. Appl. Math. Comput. Sci. **23**, 263–276 (2013)
9. Barboteu, M., Bartosz, K., Kalita, P.: A dynamic viscoelastic contact problem with normal compliance, finite penetration and nonmonotone slip rate dependent friction. Nonlinear Anal. Real World Appl. **22**, 452–472 (2015)
10. Barboteu, M., Bartosz, K., Kalita, P., Ramadan, A.: Analysis of a contact problem with normal compliance, finite penetration and nonmonotone slip dependent friction. Commun. Contemp. Math. **16**(1), 29 (2014). doi:10.1142/S0219199713500168
11. Barboteu, M., Cheng, X., Sofonea, M.: Analysis of a contact problem with unilateral constraint and slip-dependent friction. Math. Mech. Solids. (2014) doi:10.1177/1081286514537289
12. Ciarlet, P.G.: Mathematical Elasticity: Three-Dimensional Elasticity. In: Studies in Mathematics and its Applications, vol. 1. North-Holland, Amsterdam (1988)
13. Ciarlet, P.G., Geymonat, G.: Sur les lois de comportement en élasticité non-linéaire compressible. C. R. Acad. Sci. **295(I)**, 423–426 (1982)
14. Curnier, A., He, Q.-C., Klarbring, A.: Continuum mechanics modelling of large deformation contact with friction. In: Raus, M., et al. (eds.) Contact Mechanics, pp. 145–158. Plenum Press, New York (1995)
15. Gonzalez, O.: Exact energy and momentum conserving algorithms for general models in non linear elasticity. Comput. Methods Appl. Mech. Eng. **190**, 1763–1783 (2000)
16. Han, W., Sofonea, M.: Quasistatic Contact Problems in Viscoelasticity and Viscoplasticity. Studies in Advanced Mathematics, vol. 30. American Mathematical Society/International Press, Providence, RI/Somerville, MA (2002)
17. Hauret, P.: Numerical methods for the dynamic analysis of two-scale incompressible nonlinear structures. Ph.D. thesis, Ecole Polytechnique, France (2004)
18. Hauret, P.: Mixed interpretation and extensions of the equivalent mass matrix approach for elastodynamics with contact. Comput. Methods Appl. Mech. Eng. **199**(45–48), 2941–2957 (2010)
19. Hauret, P., Le Tallec, P.: Energy-controlling time integration methods for nonlinear elastodynamics and low-velocity impact. Comput. Methods Appl. Mech. Eng. **195**, 4890–4916 (2006)
20. Hauret, P., Salomon, J., Weiss, A., Wohlmuth, B.: Energy consistent co-rotational schemes for frictional contact problems. SIAM J. Sci. Comput. **30**, 2488–2511 (2008)
21. Hilber, H., Hughes, T., Taylor, R.: Improved numerical dissipation for time integration algorithms in structural dynamics. Earthq. Eng. Struct. Dyn. **5**, 283–292 (1977)
22. Hwang, I.-H., Lee, Y.-G., Lee, J.-H.: A micromachined friction meter for silicon sidewalls with consideration of contact surface shape. J. Micromech. Microeng. **16**, 2475–2481 (2006)
23. Jarušek, J., Sofonea, M.: On the solvability of dynamic elastic-visco-plastic contact problems. Z. Angew. Mat. Mech. **88**, 3–22 (2008)
24. Khenous, H.B., Laborde, P., Renard, Y.: On the discretization of contact problems in elastodynamics. Analysis and simulation of Contact Problems, In: Lecture Notes in Applied Computational Mechanics, vol. 27, pp. 31–38. Springer, Berlin/Heidelberg (2006)

25. Khenous, H.B., Pommier, J., Renard, Y.: Hybrid discretization of the Signorini problem with Coulomb friction. Theoretical aspects and comparison of some numerical solvers. Appl. Numer. Math. **56**, 163–192 (2006)
26. Khenous, H.B., Laborde, P., Renard, Y.: Mass redistribution method for finite element contact problems in elastodynamics. Eur. J. Mech. A Solids **27**, 918–932 (2008)
27. Kikuchi, N., Oden, J.T.: Contact Problems in Elasticity: A Study of Variational Inequalities and Finite Element Methods. SIAM, Philadelphia (1988)
28. Laursen, T.: Computational Contact and Impact Mechanics. Springer, Berlin (2002)
29. Laursen, T., Chawla, V.: Design of energy-conserving algorithms for frictionless dynamic contact problems. Int. J. Numer. Methods Eng. **40**, 863–886 (1997)
30. Laursen, T., Love, G.: Improved implicit integrators for transient impact problems: dynamic frictional dissipation within an admissible conserving framework. Comput. Methods Appl. Mech. Eng. **192**, 2223–2248 (2003)
31. Lemu, H., Trzepiecinski, T.: Numerical and experimental study of frictional behavior in bending under tension test. J. Mech. Eng. **59**, 41–49 (2013)
32. Len, B., Ara, P., Yun, Z.: Physics of the coefficient of friction in CMP. Trans. Electr. Electron. Mater. **8**, 79–83 (2007)
33. Martins, J.A.C., Oden, J.T.: Existence and uniqueness results for dynamic contact problems with nonlinear normal and friction interface laws. Nonlinear Anal. TMA **11**, 407–428 (1987)
34. Migórski, S., Ochal, A., Sofonea, M.: Nonlinear Inclusions and Hemivariational Inequalities. Models and Analysis of Contact Problems. Advances in Mechanics and Mathematics, vol. 26. Springer, New York (2013)
35. Moreau, J.-J.: On unilateral constraints, friction and plasticity. In: Capriz, G., Stampacchia, G. (eds.) New Variational Techniques in Mathematical Physics. CIME II, pp. 175–322. Springer, Berlin/Heidelberg (1973)
36. Oden, J.T., Martins, J.A.C.: Models and computational methods for dynamic friction phenomena. Comput. Methods Appl. Mech. Eng. **52**, 527–634 (1985)
37. Shillor, M., Sofonea, M., Telega, J.: Models and Variational Analysis of Quasistatic Contact. Lecture Notes in Physics, vol. 655. Springer, Berlin/Heidelberg (2004)
38. Signorini, A.: Sopra alcune questioni di elastostatica. Atti della Società Italiana per il Progresso delle Scienze (1933)
39. Simo, J., Tarnow, N.: The discrete energy-momentum method. part I: conserving algorithms for nonlinear elastodynamics. Z. Angew. Mat. Phys. **43**, 757–793 (1992)
40. Sofonea, M., Matei, A.: Mathematical Models in Contact Mechanics.London Mathematical Society Lecture Note Series, vol. 398. Cambridge University Press, Cambridge (2012)
41. Wriggers, P.: Computational Contact Mechanics. Wiley, Chichester (2002)

Chapter 11
A Non-clamped Frictional Contact Problem with Normal Compliance

Oanh Chau, Daniel Goeleven, and Rachid Oujja

Abstract In this chapter we study a dynamic frictional contact problem with normal compliance and non-clamped contact conditions, for thermo-viscoelastic materials. The weak formulation of the problem leads to a general system defined by a second order quasivariational evolution inequality coupled with a first order evolution equation. We state and prove an existence and uniqueness result, by using arguments on parabolic variational inequalities, monotone operators and fixed point. Then, we provide a numerical scheme of approximations and various numerical computations.

Keywords Thermo-viscoelasticity • Dynamic frictional process • Non-clamped condition • Normal compliance • Evolution inequality • Fixed point • Weak solution • Numerical simulations

AMS Classification. 74M15, 74M10, 74F05; 74S05, 74S20, 74H20, 74H25, 47J22

11.1 Introduction

Contact problems are omnipresent in mechanics, civil engineering, industry and everyday life, and represent a challenging topic, due to their important applications and various open questions they involve. In order to describe the behavior of deformable bodies subjected to various nonlinear and non-smooth solicitations such as contact, friction and thermal effects, mathematical models are necessary. They are useful in the study of a large number of problems related to impacts, cracks, packing, transport, process engineering and heat transfer. For this reason, the engineering and mathematical literature devoted to dynamic and quasistatic frictional contact

O. Chau (✉) • D. Goeleven • R. Oujja
University of La Réunion, PIMENT EA4518, 97715 Saint-Denis Messag cedex 9 La Réunion, France
e-mail: oanh.chau@univ-reunion.fr; daniel.goeleven@univ-reunion.fr; rachid.oujja@univ-reunion.fr

W. Han et al. (eds.), *Advances in Variational and Hemivariational Inequalities*, Advances in Mechanics and Mathematics 33, DOI 10.1007/978-3-319-14490-0_11

problems, including their mathematical modeling, mathematical analysis, numerical analysis and numerical simulations, is continuously increasing.

An early study of contact problems within the mathematical analysis framework was done in the pioneering references [6, 11, 15]. For the error estimates analysis and numerical approximation, we refer the reader to [5, 7, 9, 19]. Functional nonlinear analysis results useful in the study of contact problems could be found in [2–4, 12, 20]. Mathematical models of frictional contact with viscoelastic and viscoplastic plastic materials have been studied in [10, 17, 18]. One of the purpose of these works was to show the cross-fertilization between various new and nonstandard models arising in contact mechanics and the abstract theory of variational inequalities. Further extensions to nonconvex contact conditions with nonmonotone and possible multivalued constitutive laws led to the recent domain of non-smooth mechanics within the framework of the so-called hemivariational inequalities. References in the field include [8, 13, 16].

This chapter is a companion work of our previous paper [1]. There, we studied a dynamic contact problem with friction, for thermo-viscoelastic materials with long memory and subdifferential boundary conditions. The model led to a system defined by a second order evolution inequality coupled with a first order evolution equation. An existence and uniqueness result for the displacement and the temperature fields has been established. Finally, a fully discrete scheme for numerical approximations was introduced and various numerical computations in dimension two have been provided.

In contrast, in this current work we investigate a dynamic contact problem with normal compliance and friction for thermo-viscoelastic materials with short memory. As in [1], the usual clamped condition has been deleted. This leads to a new and non-standard model of system defined by a second order quasi-variational inequality, coupled with a first order evolution equation. The main difficulties in the analysis of this model arise from the fact that Korn's inequality cannot be applied any more. Moreover, the model presents a strong nonlinearity due to the fact that the process is assumed to be frictional. Such kind of semi-coercive problems were first studied in [6] for Coulomb's friction models where the inertial term of the dynamic process has been used in order to compensate the loss of coerciveness in the a priori estimates. By a change of variable, we bring the coupled second order evolution inequality into a classical first order evolution inequality. Then, using a fixed point method frequently used in [10, 17], combined with monotonicity and convexity arguments, we prove the existence and uniqueness of the displacement and the temperature fields. Finally, to complete our study, we introduce a numerical scheme for the approximation of the solution and we perform numerical computations.

The chapter is organized as follows. In Sect. 11.2 we describe the mechanical problem, list the assumptions on the data, derive the variational formulation and then we state our main existence and uniqueness result, Theorem 11.1. In Sect. 11.3 we give the proof of the claimed result. In Sect. 11.4 we present several numerical simulations in the study of a two dimensional problem, which illustrate the evolution of the displacement and temperature fields.

11.2 The Contact Problem

The physical setting is as follows. A thermo-viscoelastic body occupies a bounded domain $\Omega \subset \mathbb{R}^d$ $(d = 2, 3)$ with a Lipschitz boundary Γ that is partitioned into two disjoint measurable parts, Γ_F and Γ_C. Let $[0, T]$ be the time interval of interest, where $T > 0$. We assume that a volume force of density $\boldsymbol{f}_0$ acts in $\Omega \times (0, T)$ and that surface tractions of density $\boldsymbol{f}_F$ apply on $\Gamma_F \times (0, T)$. The body may come in contact with an obstacle, the foundation, over the potential contact surface Γ_C. The contact is described with a normal compliance condition, with friction and heat exchange. Our aim is to study the dynamic evolution of the body, by using an appropriate mathematical model.

To this end, let us recall some classical notations, see e.g. [6, 10, 14] for further details. We denote by $\mathbb{S}^d$ the space of second order symmetric tensors on $\mathbb{R}^d$, while "$\cdot$" and $\|\cdot\|$ will represent the inner product and the Euclidean norm on $\mathbb{S}^d$ and $\mathbb{R}^d$. Let $\boldsymbol{\nu}$ denote the unit outer normal on Γ. Everywhere in the sequel, the indices i, j, k, h run from 1 to d, summation over repeated indices is implied and the index that follows a comma represents the partial derivative with respect to the corresponding component of the independent variable. We use standard notation for continuous, L^p and Sobolev spaces of functions defined on Ω and Γ. In addition, we use the following notation:

$$H = L^2(\Omega)^d, \qquad \mathcal{H} = \{\boldsymbol{\sigma} = (\sigma_{ij}) \mid \sigma_{ij} = \sigma_{ji} \in L^2(\Omega),\ 1 \le i, j \le d\},$$
$$H_1 = \{\boldsymbol{u} \in H \mid \boldsymbol{\varepsilon}(\boldsymbol{u}) \in \mathcal{H}\}, \qquad \mathcal{H}_1 = \{\boldsymbol{\sigma} \in \mathcal{H} \mid \operatorname{Div} \boldsymbol{\sigma} \in H\}.$$

Here $\boldsymbol{\varepsilon} : H_1 \to \mathcal{H}$ and $\operatorname{Div} : \mathcal{H}_1 \to H$ are the deformation and the divergence operators, respectively, defined by

$$\boldsymbol{\varepsilon}(\boldsymbol{u}) = (\varepsilon_{ij}(\boldsymbol{u})), \quad \varepsilon_{ij}(\boldsymbol{u}) = \frac{1}{2}(u_{i,j} + u_{j,i}), \quad \operatorname{Div} \boldsymbol{\sigma} = (\sigma_{ij,j}).$$

The spaces H, $\mathcal{H}$, H_1 and $\mathcal{H}_1$ are real Hilbert spaces endowed with the canonical inner products given by

$$(\boldsymbol{u}, \boldsymbol{v})_H = \int_\Omega u_i v_i \, dx, \qquad (\boldsymbol{\sigma}, \boldsymbol{\tau})_{\mathcal{H}} = \int_\Omega \sigma_{ij} \tau_{ij} \, dx,$$
$$(\boldsymbol{u}, \boldsymbol{v})_{H_1} = (\boldsymbol{u}, \boldsymbol{v})_H + (\boldsymbol{\varepsilon}(\boldsymbol{u}), \boldsymbol{\varepsilon}(\boldsymbol{v}))_{\mathcal{H}}, \quad (\boldsymbol{\sigma}, \boldsymbol{\tau})_{\mathcal{H}_1} = (\boldsymbol{\sigma}, \boldsymbol{\tau})_{\mathcal{H}} + (\operatorname{Div} \boldsymbol{\sigma}, \operatorname{Div} \boldsymbol{\tau})_H.$$

The mechanical problem is then formulated as follows.

Problem $\mathcal{P}$. *Find a displacement field $\boldsymbol{u} : \Omega \times [0, T] \longrightarrow \mathbb{R}^d$, a stress field $\boldsymbol{\sigma} : \Omega \times [0, T] \longrightarrow \mathbb{S}^d$ and a temperature field $\theta : \Omega \times [0, T] \longrightarrow \mathbb{R}_+$ such that*

$$\boldsymbol{\sigma}(t) = \boldsymbol{\varepsilon}(\boldsymbol{u}'(t)) + \mathcal{B}\boldsymbol{\varepsilon}(\boldsymbol{u}(t)) - \theta(t)\, C_e \quad \text{in } \Omega, \tag{11.1}$$

$$\boldsymbol{u}''(t) = \operatorname{Div} \boldsymbol{\sigma}(t) + \boldsymbol{f}_0(t) \quad \text{in } \Omega, \tag{11.2}$$

$$\boldsymbol{\sigma}(t)\nu = \boldsymbol{f}_F(t) \quad \text{on } \Gamma_F, \tag{11.3}$$

$$-\sigma_\nu(t) = c_\nu \, (u_\nu(t) - g)_+ \quad \text{on } \Gamma_C, \tag{11.4}$$

$$\left.\begin{array}{l} \|\boldsymbol{\sigma}_\tau(t)\| \le \mu_\tau \, c_\nu \, (u_\nu(t) - g)_+, \\ \|\boldsymbol{\sigma}_\tau(t)\| < \mu_\tau \, c_\nu \, (u_\nu(t) - g)_+ \implies \boldsymbol{u}'_\tau(t) = \boldsymbol{0}, \\ \|\boldsymbol{\sigma}_\tau(t)\| = \mu_\tau \, c_\nu \, (u_\nu(t) - g)_+ \\ \quad \implies \boldsymbol{u}'_\tau(t) = -\lambda \, \boldsymbol{\sigma}_\tau(t) \text{ for some } \lambda \ge 0, \end{array}\right\} \text{ on } \Gamma_C, \tag{11.5}$$

$$\theta'(t) - \operatorname{div}(K_c \, \nabla\theta(t)) = -c_{ij} \, \frac{\partial \, u'_i}{\partial \, x_j}(t) + q(t) \quad \text{on } \Omega, \tag{11.6}$$

$$-k_{ij} \, \frac{\partial \, \theta}{\partial \, x_j}(t) \, n_i = k_e \, (\theta(t) - \theta_R) \quad \text{on } \Gamma_C, \tag{11.7}$$

$$\theta(t) = 0 \quad \text{on } \Gamma_F, \tag{11.8}$$

for all $t \in [0, T]$ *and, moreover,*

$$\boldsymbol{u}(0) = \boldsymbol{u}_0, \quad \boldsymbol{u}'(0) = \boldsymbol{v}_0, \quad \theta(0) = \theta_0 \quad \text{in } \Omega. \tag{11.9}$$

Here, (11.1) represents the thermo-visco-elastic constitutive law of the material in which $\mathcal{A}$ is the viscosity tensor, $\mathcal{B}$ is the elasticity operator and C_e denotes the thermal expansion tensor. Equation (11.2) represents the equation of motion in which we assume the mass density $\varrho \equiv 1$. Condition (11.3) represents the traction boundary condition. Next, relation (11.4) represents the normal compliance contact condition in which σ_ν denotes the normal stress, c_ν is a positive constant related to the hardness of the foundation, u_ν represents the normal displacement and g is the initial gap between the foundation and the body. Here, the term $u_\nu(t) - g$ represents, when it is positive, the penetration of the surface asperities in those of the foundation. Conditions (11.5) represent a version of Coulomb's dry friction law, where $\boldsymbol{\sigma}_\tau$ is the tangential stress, μ_τ represents the coefficient of friction and, finally, $\boldsymbol{u}'_\tau$ denotes tangential velocity. The differential equation (11.6) describes the evolution of the temperature field, where K_c represents the thermal conductivity tensor and q is the density of volume heat sources. The associated temperature boundary condition is given by (11.7) and (11.8), where θ_R is the temperature of the foundation, and k_e is the heat exchange coefficient between the body and the obstacle. Finally, the data $\boldsymbol{u}_0$, $\boldsymbol{v}_0$, θ_0 in (11.9) represent the initial displacement, the initial velocity, and the initial temperature, respectively.

In order to derive the variational formulation of the mechanical problem (11.1)–(11.9), we need additional notation. Let $V = H_1$ be the space of admissible displacement fields, endowed with the inner product given by

$$(\boldsymbol{u}, \boldsymbol{v})_V = (\boldsymbol{\varepsilon}(\boldsymbol{u}), \boldsymbol{\varepsilon}(\boldsymbol{v}))_{\mathcal{H}} + (\boldsymbol{u}, \boldsymbol{v})_H \quad \forall \, \boldsymbol{u}, \, \boldsymbol{v} \in V,$$

and let $\|\cdot\|_V$ be the associated norm, i.e.

$$\|\boldsymbol{v}\|_V^2 = \|\boldsymbol{\varepsilon}(\boldsymbol{v})\|_{\mathcal{H}}^2 + \|\boldsymbol{v}\|_H^2 \quad \forall\, \boldsymbol{v} \in V.$$

It follows that $\|\cdot\|_{H_1}$ and $\|\cdot\|_V$ are equivalent norms on V and, therefore, $(V, \|\cdot\|_V)$ is a real Hilbert space. Moreover, by the Sobolev's trace theorem, there exists a constant $c_0 > 0$ depending on Ω such that

$$\|\boldsymbol{v}\|_{L^2(\Gamma_C)} \leq c_0\, \|\boldsymbol{v}\|_V \quad \forall\, \boldsymbol{v} \in V. \tag{11.10}$$

Next, let

$$E = \{\eta \in H^1(\Omega) \mid \eta = 0 \text{ on } \Gamma_F\}$$

be the space of admissible temperature fields, endowed with the canonical inner product of $H^1(\Omega)$. We also need two Gelfand evolution triples (see e.g. [20] II/A, p. 416), given by

$$V \subset H \equiv H' \subset V', \quad E \subset L^2(\Omega) \equiv (L^2(\Omega))' \subset E',$$

where the inclusions are dense and continuous, and we denote by $\langle\cdot,\cdot\rangle_{V'\times V}$, $\langle\cdot,\cdot\rangle_{E'\times E}$ the corresponding duality pairing mappings.

In the study of the mechanical problem (11.1)–(11.9), we assume that the tensor $\mathcal{A} = (a_{ijkh}) : \Omega \times \mathbb{S}^d \to \mathbb{S}^d$ satisfies the usual properties of symmetry and ellipticity, i.e.

$$\left.\begin{array}{l} \text{(i) } a_{ijkh} = a_{khij} = a_{ijhk} \in W^{1,\infty}(\Omega); \\ \text{(ii) there exists } m_{\mathcal{A}} > 0 \text{ such that} \\ \qquad \mathcal{A}\boldsymbol{\tau}\cdot\boldsymbol{\tau} \geq m_{\mathcal{A}}\, \|\boldsymbol{\tau}\|^2 \quad \forall\, \boldsymbol{\tau} \in \mathbb{S}^d, \text{ a.e. in } \Omega. \end{array}\right\} \tag{11.11}$$

We also suppose that the elasticity operator $\mathcal{B} : \Omega \times \mathbb{S}^d \to \mathbb{S}^d$ and the thermal tensor $C_e = (c_{ij}) : \Omega \times \mathbb{S}^d \to \mathbb{S}^d$ satisfy the following conditions.

$$\left.\begin{array}{ll} \text{(i)} & \text{there exists } L_{\mathcal{B}} > 0 \text{ such that} \\ & \|\mathcal{B}(\boldsymbol{x},\boldsymbol{\varepsilon}_1) - \mathcal{B}(\boldsymbol{x},\boldsymbol{\varepsilon}_2)\| \leq L_{\mathcal{B}}\|\boldsymbol{\varepsilon}_1 - \boldsymbol{\varepsilon}_2\| \\ & \forall\, \boldsymbol{\varepsilon}_1,\, \boldsymbol{\varepsilon}_2 \in \mathbb{S}^d,\ a.e.\ \boldsymbol{x} \in \Omega\,; \\ \text{(ii)} & \boldsymbol{x} \mapsto \mathcal{B}(\boldsymbol{x},\boldsymbol{\varepsilon}) \text{ is Lebesgue measurable on } \Omega,\ \forall\, \boldsymbol{\varepsilon} \in \mathbb{S}^d; \\ \text{(iii)} & \text{the mapping } \boldsymbol{x} \mapsto \mathcal{B}(\boldsymbol{x},\boldsymbol{0}) \in \mathcal{H}. \end{array}\right\} \tag{11.12}$$

$$c_{ij} = c_{ji} \in L^\infty(\Omega). \tag{11.13}$$

In addition, the body forces, surface tractions and the heat sources density have the regularity

$$f_0 \in L^2(0,T;H), \qquad f_F \in L^2(0,T;L^2(\Gamma_F)^d), \tag{11.14}$$

$$q \in L^2(0,T;L^2(\Omega)) \tag{11.15}$$

where, here and below, we use the standard notation for functions defined on $[0,T]$ with values in a Hilbert space.

The coefficients c_ν and μ_τ verify

$$c_\nu \in L^\infty(\Gamma_C;\mathbb{R}^+), \quad \mu_\tau \in L^\infty(\Gamma_C;\mathbb{R}^+) \tag{11.16}$$

and, moreover, the boundary thermal data satisfy the regularity

$$k_e \in L^\infty(\Omega;\mathbb{R}^+), \quad \theta_R \in W^{1,2}(0,T;L^2(\Gamma_C)). \tag{11.17}$$

We also suppose that the thermal conductivity tensor $K_c = (k_{ij}) : \Omega \times \mathbb{R}^d \to \mathbb{S}^d$ verifies the usual properties of symmetry and ellipticity, i.e.

$$\left.\begin{array}{l} \text{(i) } k_{ij} = k_{ji} \in L^\infty(\Omega); \\ \text{(ii) there exists } c_k > 0 \text{ such that} \\ \quad k_{ij}\xi_i\xi_j \ge c_k\xi_i\xi_i \quad \forall\, \xi = (\xi_{ij}) \in \mathbb{R}^d, \text{ a.e. in } \Omega. \end{array}\right\} \tag{11.18}$$

Finally, we assume that the initial data satisfy the conditions

$$\boldsymbol{u}_0 \in V, \quad \boldsymbol{v}_0 \in V, \quad \theta_0 \in E. \tag{11.19}$$

Next, using Green's formula, we obtain the following weak formulation of the mechanical Problem $\mathcal{P}$.

Problem $\mathcal{P}^V$. *Find a displacement field $\boldsymbol{u} : [0,T] \to V$ and a temperature field $\theta : [0,T] \to E$ such that*

$$\begin{aligned} &\langle \boldsymbol{u}''(t) + A\boldsymbol{u}'(t) + \boldsymbol{B}\boldsymbol{u}(t) + C\theta(t), \boldsymbol{w} - \boldsymbol{u}'(t)\rangle_{V'\times V} \\ &\qquad + j_\nu(\boldsymbol{u}(t), \boldsymbol{w} - \boldsymbol{u}'(t)) + j_\tau(\boldsymbol{u}(t), \boldsymbol{w}) - j_\tau(\boldsymbol{u}(t), \boldsymbol{u}'(t)) \\ &\qquad\qquad \ge \langle \boldsymbol{f}(t), \boldsymbol{w} - \boldsymbol{u}'(t)\rangle_{V'\times V} \quad \forall\, \boldsymbol{w} \in V, \\ &\theta'(t) + K\theta(t) = R\boldsymbol{u}'(t) + Q(t), \end{aligned}$$

a.e. $t \in (0,T)$ and, moreover,

$$\boldsymbol{u}(0) = \boldsymbol{u}_0, \quad \boldsymbol{u}'(0) = \boldsymbol{v}_0, \quad \theta(0) = \theta_0. \tag{11.20}$$

Note that in the statement Problem $\mathcal{P}^V$ we use various operators and functions, which are defined as follows: $A, B : V \to V'$, $C : E \to V'$, $j_\nu, j_\tau : V \times V \to \mathbb{R}$, $K : E \to E'$, $R : V \to E'$, $\boldsymbol{f} : [0,T] \to V'$ and $Q : [0,T] \to E'$,

$$\langle Av, w\rangle_{V'\times V} = (\mathcal{A}\varepsilon(v), \varepsilon(w))_{\mathcal{H}},$$

$$\langle Bv, w\rangle_{V'\times V} = (\mathcal{B}\varepsilon(v), \varepsilon(w))_{\mathcal{H}},$$

$$\langle C\zeta, w\rangle_{V'\times V} = -(\zeta\, C_e,\, \varepsilon(w))_{\mathcal{H}},$$

$$j_\nu(v, w) = \int_{\Gamma_C} c_\nu(v_\nu - g)_+ w_\nu\, da,$$

$$j_\tau(v, w) = \int_{\Gamma_C} \mu_\tau c_\nu(v_\nu - g)_+ \|w_\tau\|\, da,$$

$$\langle f(t), w\rangle_{V'\times V} = (f_0(t), w)_H + (f_F(t), w)_{(L^2(\Gamma_F))^d},$$

$$\langle K\zeta, \eta\rangle_{E'\times E} = \sum_{i,j=1}^{d} \int_\Omega k_{ij} \frac{\partial \zeta}{\partial x_j}\frac{\partial \eta}{\partial x_i}\, dx + \int_{\Gamma_C} k_e \zeta\cdot\eta\, da,$$

$$\langle Rv, \eta\rangle_{E'\times E} = -\int_\Omega c_{ij}\frac{\partial v_i}{\partial x_j}\,\eta\, dx,$$

$$\langle Q(t), \eta\rangle_{E'\times E} = \int_{\Gamma_C} k_e\theta_R(t)\,\eta\, dx + \int_\Omega q(t)\,\eta\, dx,$$

$\forall\, v \in V, \forall\, w \in V, \forall\, \zeta \in E, \forall\, \eta \in E$, a.e. $t \in (0, T)$.

Our main existence and uniqueness result that we state here and prove in the next section is the following.

Theorem 11.1. *Assume that (11.11)–(11.19) hold. Then there exists a positive constant c_Ω depending on Ω such that there exists a unique solution $\{u, \theta\}$ to Problem $\mathcal{P}^V$, if $\|\mu_\tau\, c_\nu\|_{L^\infty(\Gamma_C)} < c_\Omega$. Moreover, the solution has the regularity*

$$\left.\begin{aligned} &u \in C^1(0, T; H) \cap W^{1,2}(0, T; V) \cap W^{2,2}(0, T; V');\\ &\theta \in C(0, T; L^2(\Omega)) \cap L^2(0, T; E) \cap W^{1,2}(0, T; E'). \end{aligned}\right\} \tag{11.21}$$

Note that Theorem 11.1 states the unique weak solvability of the thermo-mechanical Problem $\mathcal{P}$, under a smallness assumption on the coefficient of friction.

11.3 Proof of Theorem 11.1

The proof is based on monotonicity, convexity and fixed point arguments and will be carried out in several steps. Everywhere in this section we denote by $c > 0$ a generic constant which value may change from line to line. We start by introducing the velocity variable $v = u'$. Then, Problem $\mathcal{P}^V$ leads to the following problem.

Problem $\mathcal{Q}^V$**.** *Find a velocity field* $\boldsymbol{v} : [0,T] \to V$ *and a temperature field* $\theta : [0,T] \to E$ *such that*

$$\begin{aligned}
&\langle \boldsymbol{v}'(t) + A\boldsymbol{v}(t) + B\boldsymbol{u}(t) + C\theta(t), \boldsymbol{w} - \boldsymbol{v}(t)\rangle_{V'\times V} \\
&\qquad + j_\nu(\boldsymbol{u}, \boldsymbol{w} - \boldsymbol{v}(t)) + j_\tau(\boldsymbol{u}, \boldsymbol{w}) - j_\tau(\boldsymbol{u}, \boldsymbol{v}(t)) \\
&\qquad\qquad \geq \langle \boldsymbol{f}(t), \boldsymbol{w} - \boldsymbol{v}(t)\rangle_{V'\times V} \quad \forall\, \boldsymbol{w} \in V, \\
&\theta'(t) + K\,\theta(t) = R\,\boldsymbol{u}'(t) + Q(t),
\end{aligned}$$

a.e. $t \in (0,T)$ *and, moreover,*

$$\boldsymbol{v}(0) = \boldsymbol{v}_0, \quad \theta(0) = \theta_0. \tag{11.22}$$

Here, $\boldsymbol{u} : [0,T] \to V$ is the function defined by

$$\boldsymbol{u}(t) = \boldsymbol{u}_0 + \int_0^t \boldsymbol{v}(s)\,ds \quad \forall\, t \in [0,T].$$

We start with the following result.

Lemma 11.2. *For all* $\boldsymbol{\eta} \in L^2(0,T;V')$*, there exists a unique function*

$$\boldsymbol{v}_\eta \in C(0,T;H) \cap L^2(0,T;V) \cap W^{1,2}(0,T;V') \tag{11.23}$$

which satisfies

$$\left.\begin{aligned}
&\langle \boldsymbol{v}_\eta'(t) + A\boldsymbol{v}_\eta(t), \boldsymbol{w} - \boldsymbol{v}_\eta(t)\rangle_{V'\times V} + \langle \boldsymbol{\eta}(t), \boldsymbol{w} - \boldsymbol{v}_\eta(t)\rangle_{V'\times V} \\
&\quad + j_\tau(\boldsymbol{u}_\eta(t), \boldsymbol{w}) - j_\tau(\boldsymbol{u}_\eta(t), \boldsymbol{v}_\eta(t)) \geq \langle \boldsymbol{f}(t), \boldsymbol{w} - \boldsymbol{v}_\eta(t)\rangle_{V'\times V}, \\
&\qquad\qquad \forall\, \boldsymbol{w} \in V, \quad \text{a.e. } t \in (0,T); \\
&\boldsymbol{v}_\eta(0) = \boldsymbol{v}_0,
\end{aligned}\right\} \tag{11.24}$$

where

$$\boldsymbol{u}_\eta(t) = \boldsymbol{u}_0 + \int_0^t \boldsymbol{v}_\eta(s)\,ds.$$

Moreover, there exists a positive constant c_Ω*, which depends on* Ω*, with the following property: if* $\|\mu_\tau\, c_\nu\|_{L^\infty(\Gamma_C)} < c_\Omega$*, then there exists* $c > 0$ *such that*

$$\|\boldsymbol{v}_{\eta_2}(t) - \boldsymbol{v}_{\eta_1}(t)\|_H^2 + \int_0^t \|\boldsymbol{v}_{\eta_2}(s) - \boldsymbol{v}_{\eta_1}(s)\|_V^2 \leq c \int_0^t \|\eta_1 - \eta_2\|_{V'}^2.$$

$$\forall\, \boldsymbol{\eta}_1, \boldsymbol{\eta}_2 \in L^2(0,T;V'),\ \forall\, t \in [0,T]. \tag{11.25}$$

Proof. Given $\boldsymbol{\eta} \in L^2(0,T;V')$ and $\boldsymbol{\xi} \in C(0,T;V)$, by using a general result on parabolic variational inequalities (see e.g. [7, Chap. 3]), we obtain the existence of a unique function $\boldsymbol{v}_{\eta\xi} \in C(0,T;H) \cap L^2(0,T;V) \cap W^{1,2}(0,T;V')$ which satisfies

$$\left.\begin{aligned} &\langle \boldsymbol{v}'_{\eta\xi}(t) + A\boldsymbol{v}_{\eta\xi}(t), \boldsymbol{w} - \boldsymbol{v}_{\eta\xi}(t)\rangle_{V'\times V} + \langle \boldsymbol{\eta}(t), \boldsymbol{w} - \boldsymbol{v}_{\eta\xi}(t)\rangle_{V'\times V} \\ &\quad + j_\tau(\boldsymbol{\xi}(t), \boldsymbol{w}) - j_\tau(\boldsymbol{\xi}(t), \boldsymbol{v}_{\eta\xi}(t)) \geq \langle \boldsymbol{f}(t), \boldsymbol{w} - \boldsymbol{v}_{\eta\xi}(t)\rangle_{V'\times V}, \\ &\qquad\qquad \forall\, \boldsymbol{w} \in V, \quad \text{a.e. } t \in (0,T), \\ &\boldsymbol{v}_{\eta\xi}(0) = \boldsymbol{v}_0, \end{aligned}\right\} \tag{11.26}$$

Now let us fix $\boldsymbol{\eta} \in L^2(0,T;V')$ and consider the operator $\Lambda_\eta : C(0,T;V) \to C(0,T;V)$ defined by

$$\forall\, \boldsymbol{\xi} \in C(0,T;V), \quad \Lambda_\eta \xi(t) = \boldsymbol{u}_0 + \int_0^t \boldsymbol{v}_{\eta\xi}(s)\, ds.$$

We use some algebraic manipulation to see that

$$j_\tau(\boldsymbol{u}_1, \boldsymbol{w}_2) - j_\tau(\boldsymbol{u}_1, \boldsymbol{w}_1) + j_\tau(\boldsymbol{u}_2, \boldsymbol{w}_1) - j_\tau(\boldsymbol{u}_2, \boldsymbol{w}_2) \leq c\, \|\boldsymbol{u}_2 - \boldsymbol{u}_1\|_V \, \|\boldsymbol{w}_2 - \boldsymbol{w}_1\|_V,$$

for all $\boldsymbol{u}_1, \boldsymbol{u}_2, \boldsymbol{w}_1, \boldsymbol{w}_2 \in V$. Here $c > 0$ is a positive constant proportional to $c_0 \|\mu_\tau\, c_\nu\|_{L^\infty(\Gamma_C)}$ where c_0 is defined in (11.10).

Let $\boldsymbol{\xi}_1, \boldsymbol{\xi}_2 \in C(0,T;V)$ be given. We use inequality (11.26) with $\boldsymbol{\xi} = \boldsymbol{\xi}_1$ and $\boldsymbol{w} = \boldsymbol{v}_{\eta\xi_2}$, then with $\boldsymbol{\xi} = \boldsymbol{\xi}_2$ and $\boldsymbol{w} = \boldsymbol{v}_{\eta\xi_1}$, add the resulting inequalities and integrate the result over $[0,t]$, for all $t \in [0,T]$. In this way we obtain

$$\begin{aligned} &\|\boldsymbol{v}_{\eta\xi_2}(t) - \boldsymbol{v}_{\eta\xi_1}(t)\|_H^2 + \int_0^t \|\boldsymbol{v}_{\eta\xi_2}(s) - \boldsymbol{v}_{\eta\xi_1}(s)\|_V^2 ds \\ &\quad \leq c \int_0^t \|\boldsymbol{\xi}_2(s) - \boldsymbol{\xi}_1(s)\|_V^2 ds + c \int_0^t \|\boldsymbol{v}_{\eta\xi_2}(s) - \boldsymbol{v}_{\eta\xi_1}(s)\|_H^2 ds \end{aligned}$$

for all $t \in [0,T]$. Next, using Gronwall's inequality, we deduce that

$$\|\Lambda_\eta(\xi_2)(t) - \Lambda_\eta(\boldsymbol{\xi}_1)(t)\|_V^2 \leq c \int_0^t \|\boldsymbol{\xi}_2(s) - \boldsymbol{\xi}_1(s)\|_V^2 ds$$

for all $t \in [0,T]$. Thus, by Banach's fixed point principle we know that the operator Λ_η has a unique fixed point, denoted $\boldsymbol{\xi}_\eta$. We then verify that

$$\boldsymbol{v}_\eta = \boldsymbol{v}_{\eta\xi_\eta}$$

is the unique solution of (11.24) with regularity (11.23).

Now let $\boldsymbol{\eta}_1, \boldsymbol{\eta}_2 \in L^2(0,T;V')$. We use (11.24) with $\boldsymbol{\eta} = \boldsymbol{\eta}_1$ and $\boldsymbol{w} = \boldsymbol{v}_{\eta_2}$, then with $\boldsymbol{\eta} = \boldsymbol{\eta}_2$ and $\boldsymbol{w} = \boldsymbol{v}_{\eta_1}$. We add the resulting inequalities and integrate their sum to obtain

$$\|\boldsymbol{v}_{\eta_2}(t) - \boldsymbol{v}_{\eta_1}(t)\|_H^2 + \int_0^t \|\boldsymbol{v}_{\eta_2}(s) - \boldsymbol{v}_{\eta_1}(s)\|_V^2 ds$$

$$\leq c \int_0^t \|\boldsymbol{\eta}_2(s) - \boldsymbol{\eta}_1(s)\|_{V'}^2 ds + c \int_0^t \|\boldsymbol{u}_{\eta_2}(s) - \boldsymbol{u}_{\eta_1}(s)\|_V^2 ds$$

$$+ c \int_0^t \|\boldsymbol{v}_{\eta_2}(s) - \boldsymbol{v}_{\eta_1}(s)\|_H^2 ds.$$

for all $t \in [0,T]$. Here, again, $c > 0$ is a positive constant which is proportional to $c_0\|\mu_\tau c_\nu\|_{L^\infty(\Gamma_C)}$. Let $\delta > 0$ be a given constant and let $c_\Omega = \frac{\delta}{c_0}$. It is clear that c_Ω depends on Ω and, moreover, if $\|\mu_\tau c_\nu\|_{L^\infty(\Gamma_C)} \leq c_\Omega$, then $c_0\|\mu_\tau c_\nu\|_{L^\infty(\Gamma_C)} \leq \delta$. Therefore, choosing δ small enough we can assume that $2c < 1$. Then, using Gronwall's inequality we deduce (11.25), which concludes the proof. □

We proceed with the following result.

Lemma 11.3. *For all $\boldsymbol{\eta} \in L^2(0,T;V')$, there exists a function*

$$\theta_\eta \in C(0,T;L^2(\Omega)) \cap L^2(0,T;E) \cap W^{1,2}(0,T;E') \tag{11.27}$$

which satisfies

$$\left.\begin{aligned} &\theta_\eta'(t) + K\,\theta_\eta(t) = R\,\boldsymbol{v}_\eta(t) + Q(t) \quad \text{in } E',\ \text{a.e. } t \in (0,T),\\ &\theta_\eta(0) = \theta_0. \end{aligned}\right\} \tag{11.28}$$

Moreover, if $\|\mu_\tau c_\nu\|_{L^\infty(\Gamma_C)} < c_\Omega$, then there exists $c > 0$ such that for all $\boldsymbol{\eta}_1, \boldsymbol{\eta}_2 \in L^2(0,T;V')$ the following inequality holds:

$$\|\theta_{\eta_1}(t) - \theta_{\eta_2}(t)\|_{L^2(\Omega)}^2 \leq c \int_0^t \|\boldsymbol{\eta}_1 - \boldsymbol{\eta}_2\|_{V'}^2 \quad \forall\, t \in [0,T]. \tag{11.29}$$

Proof. We verify that the operator $K : E \to E'$ is linear continuous and strongly monotone, and from the expression of the operator R, we have

$$\boldsymbol{v}_\eta \in L^2(0,T;V) \implies R\,\boldsymbol{v}_\eta \in L^2(0,T;E').$$

Now, since $Q \in L^2(0,T;E')$ it follows that $R\,\boldsymbol{v}_\eta + Q \in L^2(0,T;E')$. Therefore, the existence part of the lemma follows from a classical result on first order evolution equation.

Now, to provide the estimate (11.29), consider $\boldsymbol{\eta}_1, \boldsymbol{\eta}_2 \in L^2(0,T;V')$. We have

$$\langle \theta'_{\eta_1}(t) - \theta'_{\eta_2}(t), \theta_{\eta_1}(t) - \theta_{\eta_2}(t)\rangle_{E'\times E}$$
$$+ \langle K\theta_{\eta_1}(t) - K\theta_{\eta_2}(t), \theta_{\eta_1}(t) - \theta_{\eta_2}(t)\rangle_{E'\times E}$$
$$= \langle R\boldsymbol{v}_{\eta_1}(t) - R\boldsymbol{v}_{\eta_2}(t), \theta_{\eta_1}(t) - \theta_{\eta_2}(t)\rangle_{E'\times E} \quad \text{a.e. } t \in (0;T).$$

We then integrate this inequality over $[0,t]$ and use the strong monotonicity of K and the Lipschitz continuity of $R : V \to E'$ to deduce that

$$\|\theta_{\eta_1}(t) - \theta_{\eta_2}(t)\|^2_{L^2(\Omega)} \le c\int_0^t \|\boldsymbol{v}_{\eta_1} - \boldsymbol{v}_{\eta_2}\|^2_V \quad \forall\, t \in [0,T].$$

Inequality (11.29) follows then from Lemma 11.2. □

Lemmas 11.2 and 11.3 allow to consider the operator $\Lambda : L^2(0,T;V') \to L^2(0,T;V')$ defined, for all $\boldsymbol{\eta} \in L^2(0,T;V')$, by the equality

$$\langle \Lambda\boldsymbol{\eta}(t), \boldsymbol{w}\rangle_{V'\times V} = \langle B\boldsymbol{u}_\eta(t) + C\theta_\eta(t), \boldsymbol{w}\rangle_{V'\times V} + j_\nu(\boldsymbol{u}_\eta(t), \boldsymbol{w}),$$
$$\forall\, \boldsymbol{w} \in V, \text{ a.e. } t \in (0,T).$$

Here

$$\boldsymbol{u}_\eta(t) = \boldsymbol{u}_0 + \int_0^t \boldsymbol{v}_\eta(s)\,ds \quad \forall\, t \in [0,T]$$

where $\boldsymbol{v}_\eta$ and θ_η are the functions defined in Lemmas 11.2 and 11.3.

We have the following result.

Lemma 11.4. *Assume that $\|\mu_\tau c_\nu\|_{L^\infty(\Gamma_C)} < c_\Omega$. Then Λ has a unique fixed point $\boldsymbol{\eta}^* \in L^2(0,T;V')$.*

Proof. Let $\boldsymbol{\eta}_1, \boldsymbol{\eta}_2 \in L^2(0,T;V')$ be given. Then, it is easy to check that

$$\|\Lambda\boldsymbol{\eta}_2(t) - \Lambda\boldsymbol{\eta}_1(t)\|_{V'} \le c\,\|B\boldsymbol{u}_{\eta_2}(t) - B\boldsymbol{u}_{\eta_1}(t)\|_{V'} + c\,\|\theta_{\eta_2}(t) - \theta_{\eta_1}(t)\|_{L^2(\Omega)}$$
$$+ c\,\|\boldsymbol{u}_{\eta_2}(t) - \boldsymbol{u}_{\eta_1}(t)\|_V$$

a.e. $t \in (0,T)$. We combine (11.12), (11.25) and (11.29) to deduce that there exists $c > 0$ such that

$$\|\Lambda\boldsymbol{\eta}_2(t) - \Lambda\boldsymbol{\eta}_1(t)\|^2_{V'} \le c\int_0^t \|\boldsymbol{\eta}_2(s) - \boldsymbol{\eta}_1(s)\|^2_{V'}ds \quad \forall\, t \in [0,T].$$

Lemma 11.4 is a consequence of the previous inequality combined with the Banach fixed point principle. □

We now have all the ingredients to prove Theorem 11.1.

Proof of Theorem 11.1. Let $\boldsymbol{u}$, $\boldsymbol{v}$ and θ the functions defined by

$$\boldsymbol{u}(t) := \boldsymbol{u}_0 + \int_0^t \boldsymbol{v}_{\eta^*}(s)\,ds, \quad \boldsymbol{v}(t) := \boldsymbol{v}_{\eta^*}(t), \quad \theta(t) := \theta_{\eta^*}(t) \quad \forall\, t \in [0, T].$$

Then, using (11.24) and (11.28), it is easy to see that the couple $(\boldsymbol{v}, \theta)$ is a solution to Problem $\mathcal{Q}^V$ and, therefore $(\boldsymbol{u}, \theta)$ is a solution to Problem $\mathcal{P}^V$. Moreover, the regularity (11.21) follows from the regularity of the functions $\boldsymbol{v}_\eta$ and θ_η in Lemmas 11.2 and 11.3, see (11.23) and (11.27), respectively. This proves the existence part of the theorem. The the uniqueness part follows from the uniqueness of the solution in Lemmas 11.2 and 11.3. □

11.4 Numerical Computations

In this section, we provide a fully-discrete numerical approximation scheme for the variational Problem $\mathcal{P}^V$, and the associated numerical simulations in the study of two dimensional tests by using MATLAB computation codes. To this end we denote by $\{\boldsymbol{u}, \theta\}$ the unique solution of the Problem $\mathcal{P}^V$ and consider the velocity variable defined by

$$\boldsymbol{v}(t) = \boldsymbol{u}'(t) \quad \forall\, t \in [0, T].$$

We make the following additional assumptions on the solution and data:

$$\boldsymbol{f} \in C([0,T]; V'); \quad Q \in C([0,T]; E');$$

$$\boldsymbol{v} \in C(0,T; V); \quad \boldsymbol{v}' \in C(0,T; H);$$

$$\theta \in C(0,T; E); \quad \theta' \in C(0,T; L^2(\Omega)).$$

Now let $V^h \subset V$ and $E^h \subset E$ be a family of finite dimensional subspaces, defined by finite elements spaces of piecewise linear functions, where $h > 0$ is a discretization parameter which may be the maximal diameter of the elements. We divide the time interval $[0, T]$ into N equal parts: $t_n = n\,k$, $n = 0, 1, \ldots, N$, with the uniform time step $k = T/N$. For a continuous function $\boldsymbol{w} \in C([0, T]; X)$ with values in a space X, we use the notation $\boldsymbol{w}_n = \boldsymbol{w}(t_n) \in X$.

Then, Problem $\mathcal{Q}^V$ implies

$$\left.\begin{array}{l} \langle \boldsymbol{v}'(t) + A\boldsymbol{v}(t) + B\boldsymbol{u}(t) + C\theta(t) + D\boldsymbol{u}(t), \boldsymbol{w} - \boldsymbol{v}(t)\rangle_{V'\times V} \\ \quad + j_\tau(\boldsymbol{u}, \boldsymbol{w}) - j_\tau(\boldsymbol{u}, \boldsymbol{v}(t)) \geq \langle \boldsymbol{f}(t), \boldsymbol{w} - \boldsymbol{v}(t)\rangle_{V'\times V} \quad \forall\, \boldsymbol{w} \in V, \\ \langle \theta'(t) + K\theta(t) - R\boldsymbol{v}(t) - Q(t), \eta\rangle_{E'\times E} = 0 \quad \forall\, \eta \in E, \\ \boldsymbol{v}(0) = \boldsymbol{v}_0, \quad \theta(0) = \theta_0, \end{array}\right\}$$

for all $t \in [0, T]$, where

$$\langle D\boldsymbol{u}(t), \boldsymbol{w}\rangle_{V'\times V} = j_\nu(\boldsymbol{u}(t), \boldsymbol{w}),$$

$$\langle \boldsymbol{v}'(t), \boldsymbol{w}\rangle_{V'\times V} = (\boldsymbol{v}'(t), \boldsymbol{w})_H,$$

$$\langle \theta'(t), \eta\rangle_{E'\times E} = (\theta'(t), \eta)_{L^2(\Omega)}.$$

This suggests to introduce the following fully-discrete scheme.

Problem $\mathcal{P}^{hk}$. *Find a discrete velocity field* $\boldsymbol{v}^{hk} = \{\boldsymbol{v}_n^{hk}\}_{n=0}^N \subset V^h$ *and a discrete temperature field* $\theta^{hk} = \{\theta_n^{hk}\}_{n=0}^N \subset E^h$ *such that*

$$\boldsymbol{v}_0^{hk} = \boldsymbol{v}_0^h, \quad \theta_0^{hk} = \theta_0^h, \tag{11.30}$$

and for $n = 1, \cdots, N$,

$$\begin{aligned}
&\langle \frac{\boldsymbol{v}_n^{hk} - \boldsymbol{v}_{n-1}^{hk}}{k} + A\boldsymbol{v}_n^{hk}, \boldsymbol{w}^h - \boldsymbol{v}_n^{hk}\rangle_{V'\times V} \\
&\quad + \langle B\boldsymbol{u}_{n-1}^{hk} + C\theta_{n-1}^{hk} + D\boldsymbol{u}_{n-1}^{hk}, \boldsymbol{w}^h - \boldsymbol{v}_n^{hk}\rangle_{V'\times V} \\
&\quad + j_\tau(\boldsymbol{u}_{n-1}^{hk}, \boldsymbol{w}^h) - j_\tau(\boldsymbol{u}_{n-1}^{hk}, \boldsymbol{v}_n^{hk}) \\
&\qquad \geq \langle \boldsymbol{f}_n, \boldsymbol{w}^h - \boldsymbol{v}_n^{hk}\rangle_{V'\times V} \quad \forall\, \boldsymbol{w}^h \in V^h,
\end{aligned} \tag{11.31}$$

$$\begin{aligned}
&\Big(\frac{\theta_n^{hk} - \theta_{n-1}^{hk}}{k}, \eta^h\Big)_{L^2(\Omega)} + \langle K\theta_n^{hk}, \eta^h\rangle_{E'\times E} \\
&\qquad = \langle R\boldsymbol{v}_n^{hk}, \eta^h\rangle_{E'\times E} + \langle Q_n, \eta^h\rangle_{E'\times E}, \quad \forall\, \eta^h \in E^h.
\end{aligned} \tag{11.32}$$

Here

$$\boldsymbol{u}_n^{hk} = \boldsymbol{u}_{n-1}^{hk} + k\, \boldsymbol{v}_n^{hk}, \quad \boldsymbol{u}_0^{hk} = \boldsymbol{u}_0^h. \tag{11.33}$$

Moreover, $\boldsymbol{u}_0^h \in V^h$, $\boldsymbol{v}_0^h \in V^h$ and $\theta_0^h \in E^h$ represent suitable approximations of the initial values $\boldsymbol{u}_0$, $\boldsymbol{v}_0$, θ_0, respectively.

For $n = 1, \ldots, N$, once $\boldsymbol{u}_{n-1}^{hk}$, $\boldsymbol{v}_{n-1}^{hk}$ and θ_{n-1}^{hk} are known, we compute $\boldsymbol{v}_n^{hk}$, θ_n^{hk} and $\boldsymbol{u}_n^{hk}$ by using (11.31)–(11.33) and classical result on variational inequality (see e.g. [10]). Therefore, the discrete scheme has a unique solution by starting with initial values on displacement, velocity and temperature fields. Moreover, under additional regularity of solution and using arguments similar as those used in [19], we can prove that the errors estimate order is proportional to the discretization parameters h and k.

In view of the numerical simulations, we consider the domain Ω, the partition of its boundary, the elasticity tensor and the viscosity operator as follows:

$$\Omega = (0, L_1) \times (0, L_2);$$

$$\Gamma_F = (\{0\} \times [0, L_2]) \cup ([0, L_1] \times \{L_2\}) \cup (\{L_1\} \times [0, L_2]);\ \Gamma_C = [0, L_1] \times \{0\};$$

$$(\mathcal{B}\boldsymbol{\tau})_{ij} = \frac{E\,\kappa}{1-\kappa^2}(\tau_{11} + \tau_{22})\,\delta_{ij} + \frac{E}{1+\kappa}\tau_{ij}, \quad 1 \le i,\ j \le 2,\ \boldsymbol{\tau} \in \mathbb{S}^2;$$

$$(\mathcal{A}\boldsymbol{\tau})_{ij} = \mu\,(\tau_{11} + \tau_{22})\,\delta_{ij} + \eta\,\tau_{ij}, \quad 1 \le i,\ j \le 2,\ \boldsymbol{\tau} \in \mathbb{S}^2.$$

Here E is the Young's modulus, κ is the Poisson's ratio of the material, δ_{ij} denotes the Kronecker symbol and μ and η are viscosity constants.

We refer to the previous numerical scheme, and use spaces of continuous piecewise affine functions $V^h \subset V$ and $E^h \subset E$ as families of approximating subspaces. For our computations, we consider also the following data (IS unity):

$$L_1 = L_2 = 1, \quad T = 1,$$

$$\mu = 10, \quad \eta = 10, \quad E = 2, \quad \kappa = 0.1,$$

$$c_{ij} = k_{ij} = k_e = 1,\ 1 \le i,\ j \le 2,$$

$$\boldsymbol{f}_0(\boldsymbol{x}, t) = (0, -1.5), \quad q(\boldsymbol{x}, t) = 1 \quad \forall\, \boldsymbol{x} \in \Omega,\ t \in [0, T],$$

$$\boldsymbol{f}_F(\boldsymbol{x}, t) = (0,\ 0), \quad \forall\, \boldsymbol{x} \in \{0\} \times [0, L_2],\ t \in [0, T],$$

$$\boldsymbol{f}_F(\boldsymbol{x}, t) = (0.5,\ 0.4), \quad \forall\, \boldsymbol{x} \in ([0, L_1] \times \{L_2\}) \cup (\{L_1\} \times [0, L_2]),\ t \in [0, T],$$

$$\boldsymbol{u}_0(\boldsymbol{x}) = (0, 0), \quad \boldsymbol{v}_0(\boldsymbol{x}) = (0, 0), \quad \theta_0(\boldsymbol{x}) = 0 \quad \forall\, \boldsymbol{x} \in \Omega.$$

In Figs. 11.1 and 11.2 we show the deformed configurations at final time, for two different values of the normal compliance coefficient. We see that for a larger coefficient, penetration is less important. In Figs. 11.3 and 11.4, we show the deformed configurations at final time, for two different values of coefficients of friction. We note that for a smaller coefficient the slip phenomenon appears on the contact surface. In Figs. 11.5 and 11.6, we plot the deformed configurations at final time, for two values of the gap. In Figs. 11.7, 11.8, 11.9, 11.10, 11.11 and 11.12 we represent the Von Mises norm of the stress, corresponding to the numerical values in Figs. 11.1, 11.2, 11.3, 11.4, 11.5 and 11.6, respectively. These figures show that when penetration is more important then the norm of the stress on the contact surface is larger. In particular, the norm of the stress is minimal in the case where there is loss of contact with the foundation. Finally, in Figs. 11.13 and 11.14, we show the influence of the different temperatures of the foundation on the temperature field of the body. We observe that a high temperature of the foundation leads to a high temperature in the neighborhood of the contact surface.

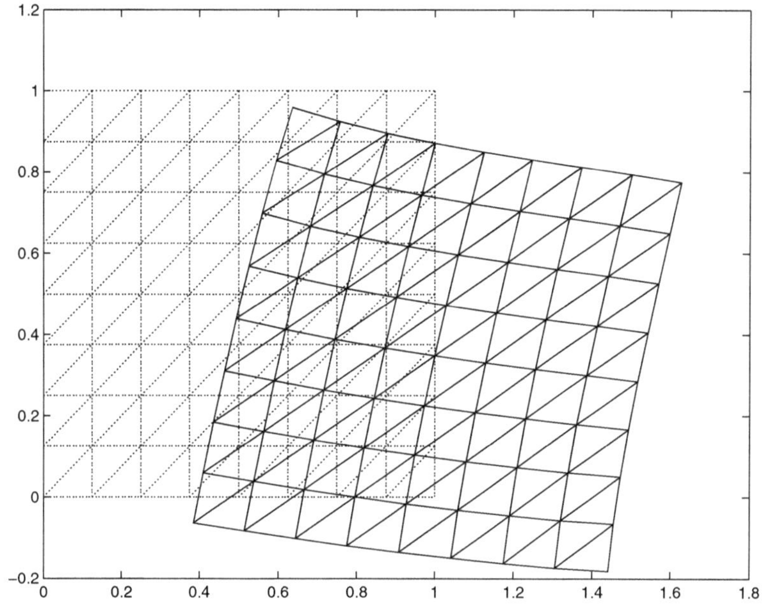

Fig. 11.1 Deformed configuration at final time, $\theta_R = 0$, $g = 0$, $\mu_\tau = 0.1$, $c_\nu = 10$

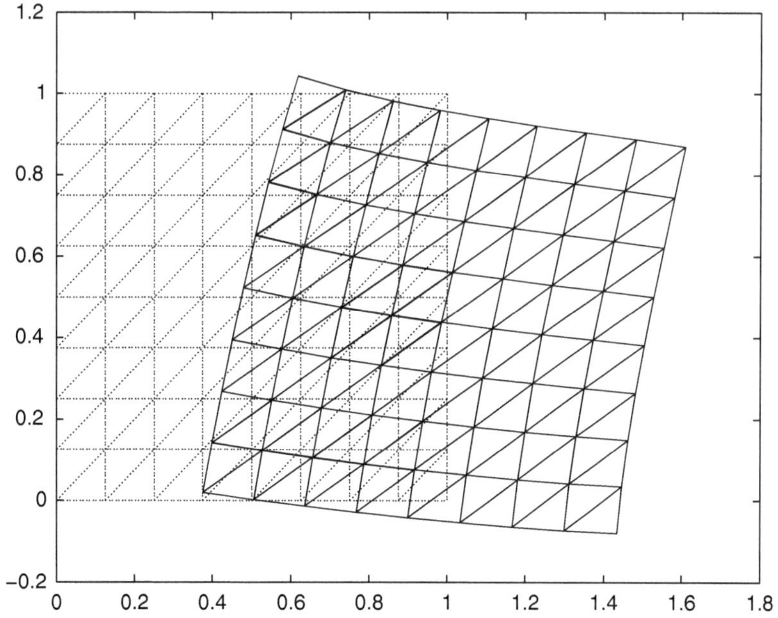

Fig. 11.2 Deformed configuration at final time, $\theta_R = 0$, $g = 0$, $\mu_\tau = 0.1$, $c_\nu = 20$

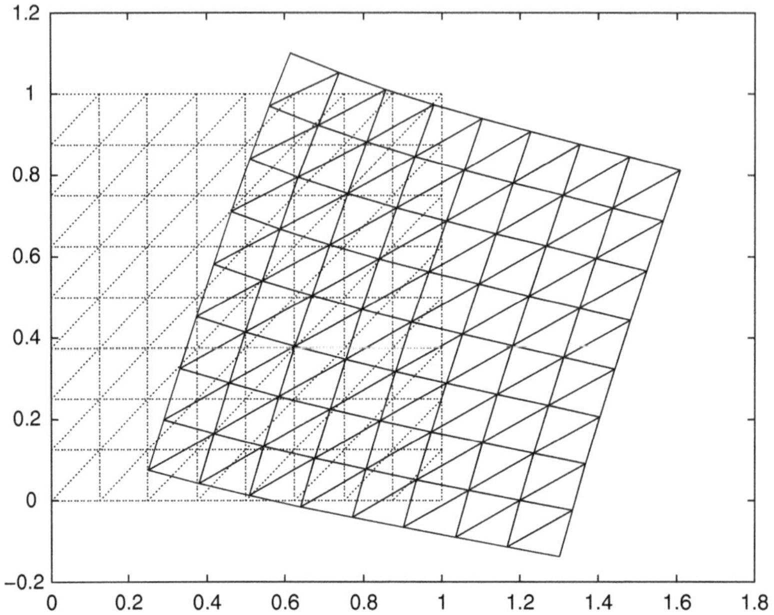

Fig. 11.3 Deformed configuration at final time, $\theta_R = 0$, $g = 0$, $c_\nu = 20$, $\mu_\tau = 0.30$

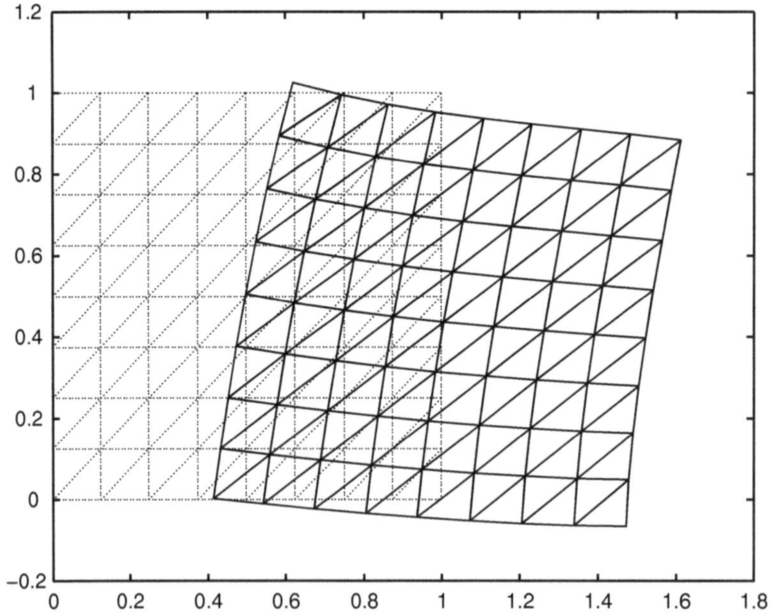

Fig. 11.4 Deformed configuration at final time, $\theta_R = 0$, $g = 0$, $c_\nu = 20$, $\mu_\tau = 0.05$

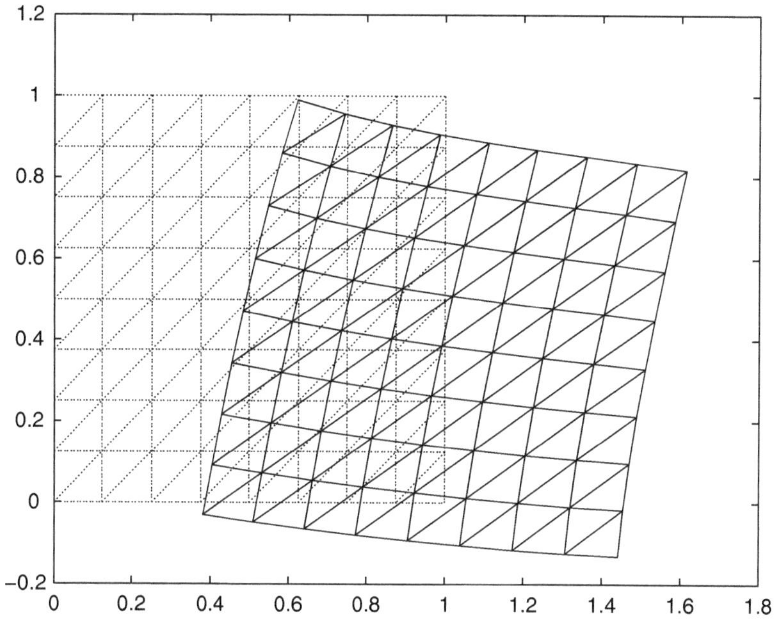

Fig. 11.5 Deformed configuration at final time, $\theta_R = 0$, $c_\nu = 20$, $\mu_\tau = 0.1$, $g = 0.03$

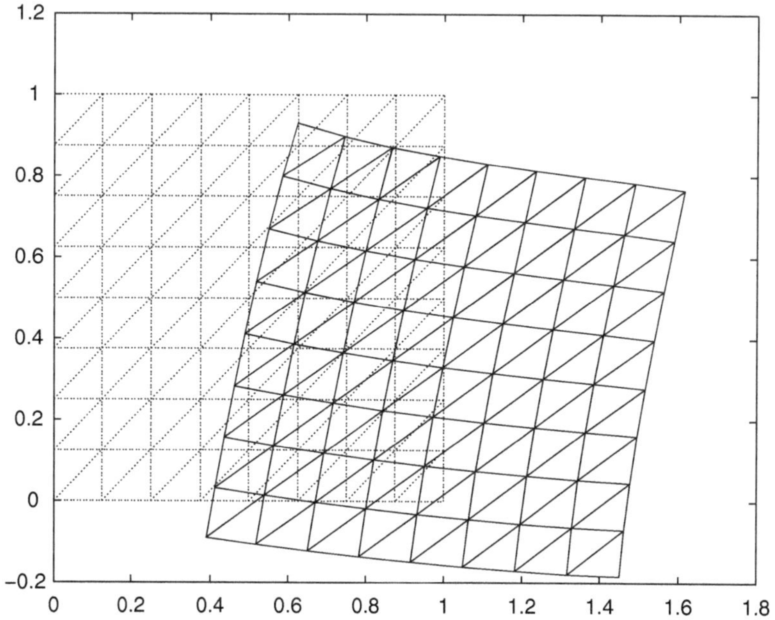

Fig. 11.6 Deformed configuration at final time, $\theta_R = 0$, $c_\nu = 20$, $\mu_\tau = 0.1$, $g = 0.06$

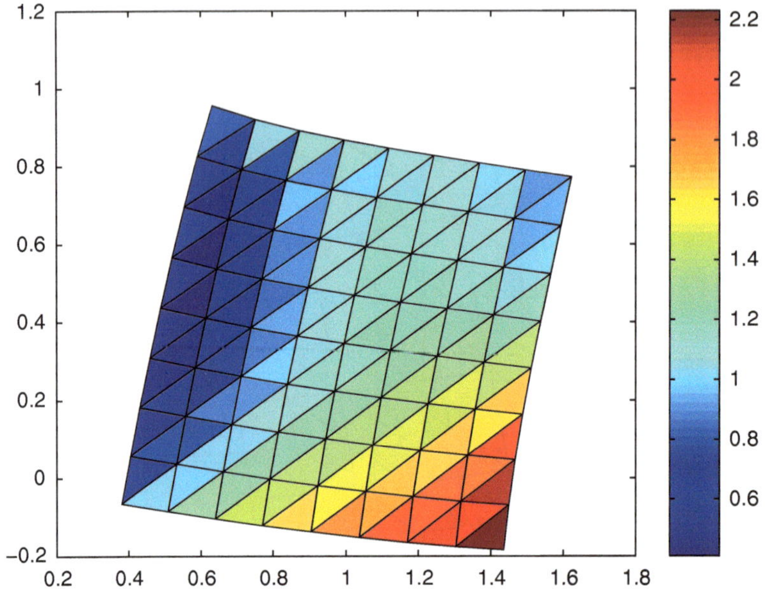

Fig. 11.7 Von Mises norm of the stress in deformed configurations, $\theta_R = 0$, $g = 0$, $\mu_\tau = 0.1$, $c_\nu = 10$

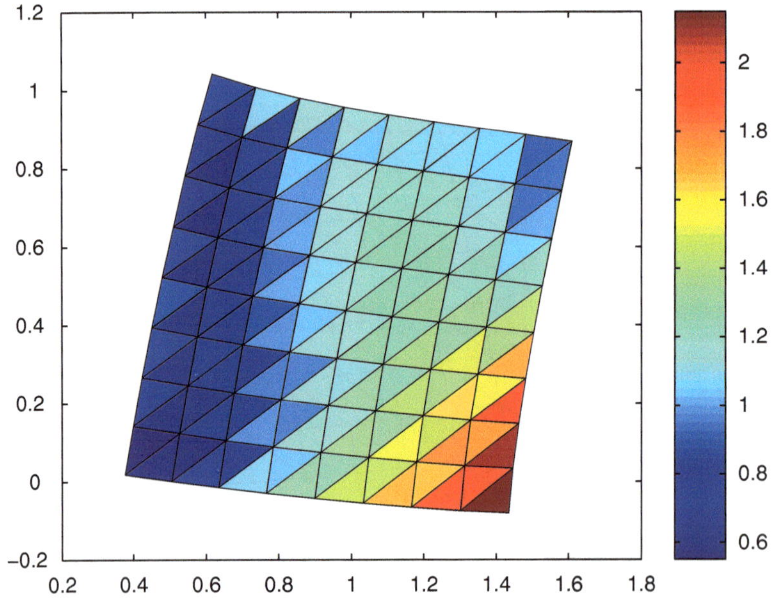

Fig. 11.8 Von Mises norm of the stress in deformed configuration, $\theta_R = 0$, $g = 0$, $\mu_\tau = 0.1$, $c_\nu = 20$

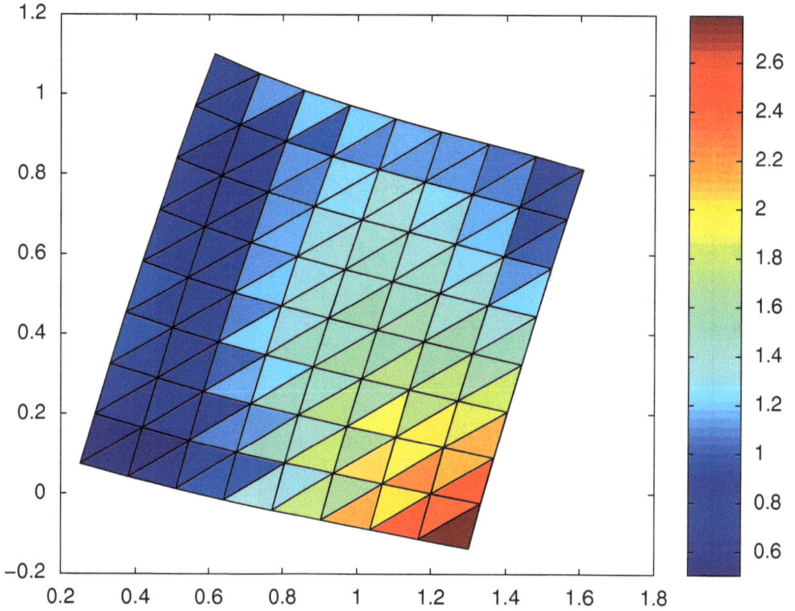

Fig. 11.9 Von Mises norm of the stress in deformed configuration, $\theta_R = 0$, $g = 0$, $c_\nu = 20$, $\mu_\tau = 0.3$

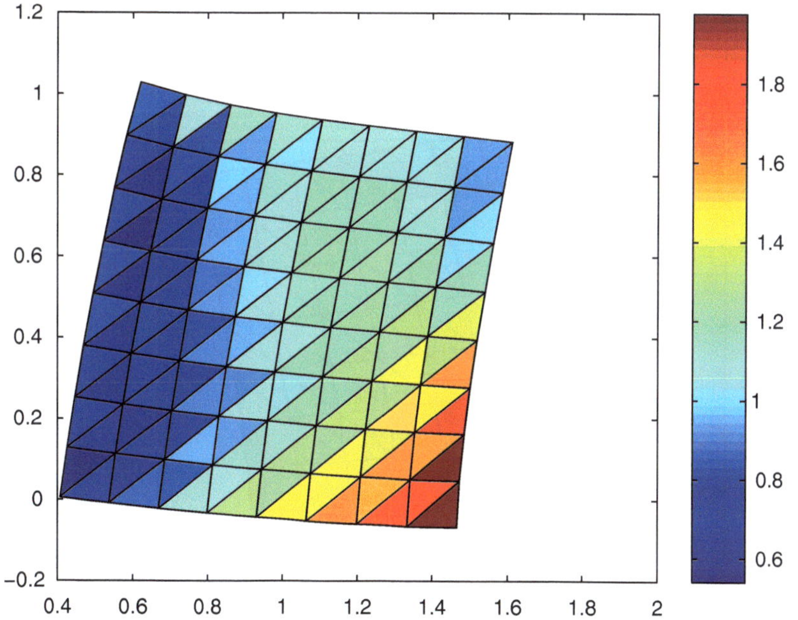

Fig. 11.10 Von Mises norm of the stress in deformed configuration, $\theta_R = 0$, $g = 0$, $c_\nu = 20$, $\mu_\tau = 0.05$

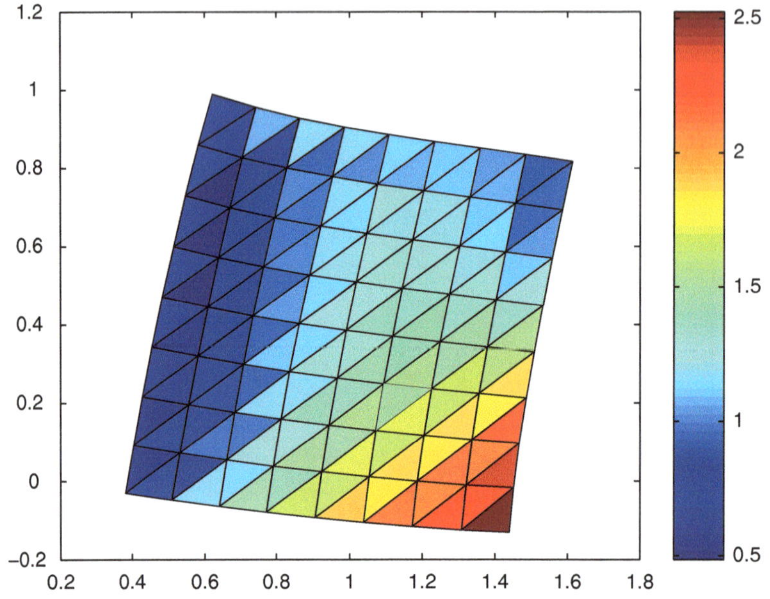

Fig. 11.11 Von Mises norm of the stress in deformed configuration, $\theta_R = 0$, $c_\nu = 20$, $\mu_\tau = 0.3$, $g = 0.03$

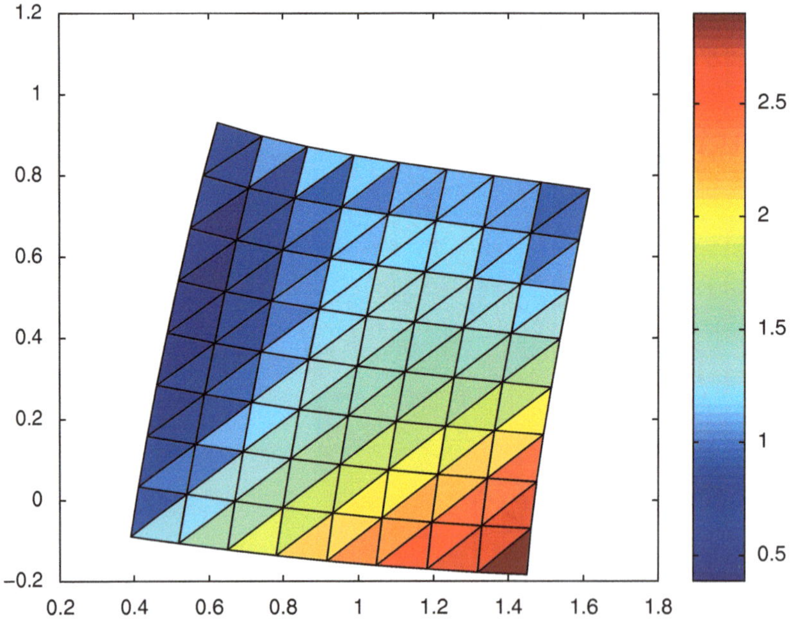

Fig. 11.12 Von Mises norm of the stress in deformed configuration, $\theta_R = 0$, $c_\nu = 20$, $\mu_\tau = 0.05$, $g = 0.06$

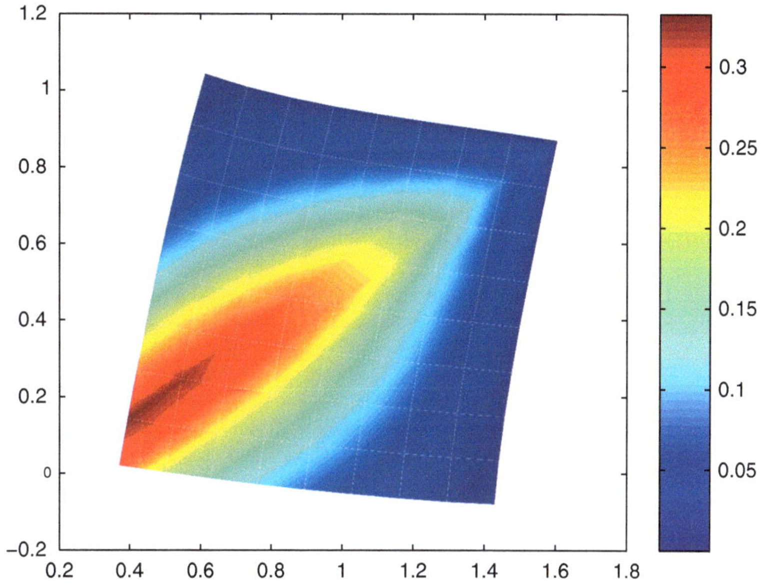

Fig. 11.13 Temperature field at final time, $c_v = 20$, $g = 0$, $\mu_\tau = 0.1$, $\theta_R = 0$

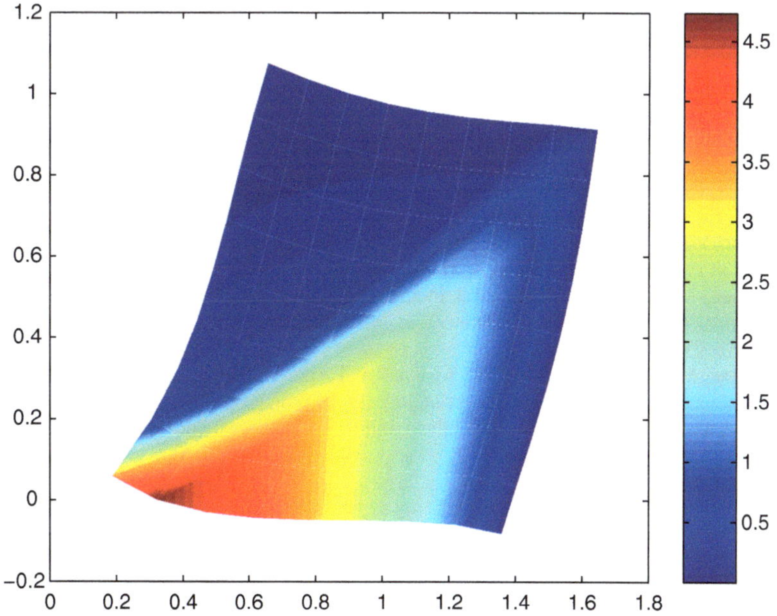

Fig. 11.14 Temperature field at final time, $c_v = 20$, $g = 0$, $\mu_\tau = 0.1$, $\theta_R = 10$

References

1. Adly, S., Chau, O.: On some dynamical thermal non clamped contact problems. Math. Program. Ser. B **139**, 5–26 (2013)
2. Brézis, H.: Problèmes unilatéraux. J. Math. Pures Appl. **51**, 1–168 (1972)
3. Brezis, H.: Operateurs Maximaux Monotones et Semigroups de Contractions dans les Espaces de Hilbert. North-Holland, Amsterdam (1973)
4. Brézis, H.: Analyse Fonctionnelle, Théorie et Application. Masson, Paris (1987)
5. Ciarlet, P.G.: The Finite Element Method for Elliptic Problems. North Holland, Amsterdam (1978)
6. Duvaut, G., Lions, J.L.: Les Inéquations en Mécanique et en Physique. Dunod, Paris (1972)
7. Glowinski, R.: Numerical Methods for Nonlinear Variational Problems. Springer, New York (2008)
8. Goeleven, D., Motreanu, D., Dumont, Y., Rochdi, M.: Variational and Hemivariational Inequalities, Theory, Methods and Applications. Volume I: Unilateral Analysis and Unilateral Mechanics. Kluwer Academic Publishers, Boston/New York (2003)
9. Han, W., Reddy, B.D.: Plasticity: Mathematical Theory and Numerical Analysis, 2nd edn. Springer, New York (2013)
10. Han, W., Sofonea, M.: Quasistatic Contact Problems in Viscoelasticity and Viscoplasticity. Studies in Advanced Mathematics, vol. 30. American Mathematical Society/International Press, Providence, RI/Somerville, MA (2002)
11. Kikuchi, N., Oden, J.T.: Contact Problems in Elasticity. SIAM, Philadelphia (1988)
12. Lions, J.L.: Quelques Méthodes de Résolution des Problèmes aux Limites Non linéaires. Dunod et Gauthier-Villars, Paris (1969)
13. Migórski, S., Ochal, A., Sofonea, M.: Nonlinear Inclusions and Hemivariational Inequalities. Models and Analysis of Contact Problems. Advances in Mechanics and Mathematics, vol. 26. Springer, New York (2013)
14. Nečas, J., Hlaváček, I.: Mathematical Theory of Elastic and Elastoplastic Bodies: An Introduction. Elsevier, Amsterdam (1981)
15. Panagiotopoulos, P.D.: Inequality Problems in Mechanics and Applications. Birkhäuser, Basel (1985)
16. Panagiotopoulos, P.D.: Hemivariational Inequalities, Applications in Mechanics and Engineering. Springer, New York (1993)
17. Shillor, M., Sofonea, M., Telega, J.J.: Models and Analysis of Quasistatic Contact. Lecture Notes in Physics, vol. 655. Springer, Berlin (2004)
18. Sofonea, M., Matei, A.: Variational Inequalities with Applications: A Study of Antiplane Frictional Contact Problems. Advances in Mechanics and Mathematics, vol. 18. Springer, New York (2009)
19. Sofonea, M., Han, W., Shillor, M.: Analysis and Approximation of Contact Problems with Adhesion or Damage. Chapman & Hall/CRC, Boca Raton (2006)
20. Zeidler, E.: Nonlinear Functional Analysis and Its Applications. Springer, New York (1997)

Chapter 12
On Large Time Asymptotics for Two Classes of Contact Problems

Piotr Kalita and Grzegorz Łukaszewicz

Abstract We consider two classes of evolution contact problems on two dimensional domains governed by first and second order evolution equations, respectively. The contact is represented by multivalued and nonmonotone boundary conditions that are expressed by means of Clarke subdifferentials of certain locally Lipschitz and semiconvex potentials. For both problems we study the existence and uniqueness of solutions as well as their asymptotic behavior in time. For the first order problem, that is governed by the Navier–Stokes equations with generalized Tresca law, we show the existence of global attractor of finite fractal dimension and existence of exponential attractor. For the second order problem, representing the frictional contact in antiplane viscoelasticity, we show that the global attractor exists, but both the global attractor and the set of stationary states are shown to have infinite fractal dimension.

Keywords Contact problem • Navier–Stokes equation • Antiplane viscoelasticity • Existence • Evolution inclusions • Global attractor • Fractal dimension • Exponential attractor

AMS Classification. 34G25, 35B41, 47J20, 76D05

P. Kalita
Faculty of Mathematics and Computer Science, Jagiellonian University, ul. Łojasiewicza 6, 30348 Krakow, Poland
e-mail: piotr.kalita@ii.uj.edu.pl

G. Łukaszewicz (✉)
Mathematics Department, University of Warsaw, ul. Banacha 2, 02957 Warsaw, Poland
e-mail: glukasz@mimuw.edu.pl

W. Han et al. (eds.), *Advances in Variational and Hemivariational Inequalities*, Advances in Mechanics and Mathematics 33, DOI 10.1007/978-3-319-14490-0_12

12.1 Introduction

Large time behaviour of solutions of problems in contact mechanics is an important though somewhat neglected part of the theory. In fact, remarking on future directions of research in the field of contact mechanics, in their book [18], the authors wrote: "The infinite-dimensional dynamical systems approach to contact problems is virtually nonexistent. (...) This topic certainly deserves further consideration".

From the mathematical point of view, a considerable difficulty in analysis of the contact problems, and dynamical ones in particular, comes from the presence of involved boundary constraints which are often modeled by multivalued boundary conditions of a subdifferential type and lead to a formulation of the considered problem in terms of a variational or hemivariational inequality with, frequently, nondifferentiable boundary functionals.

Our aim in this paper is to contribute to the topic of large time dynamics in problems of contact mechanics.

We consider examples belonging to two classes of evolutionary contact problems governed by equations of the form

$$u' + Au + Bu = f \tag{12.1}$$

or

$$u'' + Au' + Bu = f, \tag{12.2}$$

in a two-dimensional bounded domain Ω, with $\partial\Omega = \overline{\Gamma}_D \cup \overline{\Gamma}_C \cup \overline{\Gamma}_L$, where Γ_D, Γ_C, and Γ_L are the top, bottom and lateral parts of the boundary, respectively.

We supplement the equations with appropriate initial and boundary conditions. In particular, the boundary conditions on the bottom part of the boundary Γ_C are of subdifferential type, with semiconvex and locally Lipschitz superpotentials.

The problems can be written in the form of hemivariational inequalities or evolutionary differential inclusions. Since the superpotentials are semiconvex, the solutions (the existence of which we shall prove) are unique. Moreover, they exist on the whole time semiaxis $\mathbb{R}^+ = [0, +\infty)$.

Our aim is to study the large time asymptotics of solutions, that is the behavior of solutions after a large time of evolution, of chosen problems from the two above classes.

More precisely, we consider two specific examples of contact problems as representatives of the classes of problems governed by the first order and the second order equations, respectively. First, in Sect. 12.3, we consider the Navier–Stokes system and a corresponding problem that comes from the theory of lubrication, with a generalized Tresca law, where the friction bound depends on the tangential velocity. Then, in Sect. 12.4, we consider a second order model problem from the theory of antiplane viscoelasticity.

In both cases there exists a subset of the phase space of solutions (same as that of initial conditions) called *global attractor*. It is a compact, invariant set which attracts all solution trajectories as time goes to infinity. Existence of the global attractors follows from the dissipativity of the considered problems.

For the first order problem the attractor has an additional important property, namely, its fractal dimension is finite. Moreover, there exists an object, called *exponential attractor*. It contains the global attractor, has finite fractal dimension and attracts the trajectories exponentially fast in time. This property allows, among other things, to locate the exponential and thus the global attractor in the phase space by using numerical analysis.

Finally, in Sect. 12.4.4 we present a special case of our second order problem where there are no exponential attractors. For this case the global attractor has infinite fractal dimension and contains an infinite dimensional set of stationary states. This situation might have the following physical interpretation, namely, each stationary state is such that the body does not move and friction force equilibrates the inner stresses originating from the displacement u. The set of displacements corresponding to these stationary states must be relatively compact (since the attractor is compact). Intuitively with the presence of friction the body can stop in a large number of configurations—here we show that the set of equilibrium configurations can have infinite dimension.

The spatial domain Ω in the problems we consider is two-dimensional not only for the sake of simplicity. For the first order problem governed by the Navier–Stokes equation, where we need the uniqueness of solution, it is a crucial assumption.

Of course, one can extend our results to many other problems of contact mechanics governed by Eq. (12.1) or (12.2) under similar assumptions, and several extensions are also possible to problems without uniqueness of solutions, cf. e.g., [8, 20].

12.2 Preliminaries from the Theory of Dynamical Systems

Let us consider an abstract autonomous evolutionary problem

$$\begin{aligned} \frac{dv(t)}{dt} &= F(v(t)) \quad \text{in } Z \text{ for a.e. } t \in \mathbb{R}^+, \\ v(0) &= v_0. \end{aligned} \tag{12.3}$$

where Z is a Banach space, F is a nonlinear operator, and v_0 is in a Banach space X that embeds in Z. We assume that the above problem has a global in time unique solution $\mathbb{R}^+ \ni t \mapsto v(t) \in X$ for every $v_0 \in X$. In this case one can associate with the problem a semigroup $\{S(t)\}_{t\geq 0}$ of (nonlinear) operators $S(t) : X \to X$ setting $S(t)v_0 = v(t)$, where $v(t)$, $t > 0$, is the unique solution of (12.3).

From properties of the semigroup of operators $\{S(t)\}_{t\geq 0}$ we may then conclude the basic features of the behavior of solutions of problem (12.3), in particular, their time asymptotics. One of the objects existence of which characterize the asymptotic behavior of solutions is the global attractor. It is a compact and invariant with respect to operators $S(t)$ subset of the phase space X (in general, a metric space) that uniformly attracts all bounded subsets of X.

Definition 12.1. A global attractor for a semigroup $\{S(t)\}_{t\geq 0}$ in a Banach space X is a subset $\mathcal{A}$ of X such that

- $\mathcal{A}$ is compact in X.
- $\mathcal{A}$ is invariant, i.e., $S(t)\mathcal{A} = \mathcal{A}$ for every $t \geq 0$.
- For every $\varepsilon > 0$ and every bounded set B in X there exists $t_0 = t_0(B, \varepsilon)$ such that $\bigcup_{t\geq t_0} S(t)B$ is a subset of the ε-neighborhood of the attractor $\mathcal{A}$ (uniform attraction property).

The global attractor defined above is uniquely determined by the semigroup $\{S(t)\}_{t\geq 0}$. Moreover, it is connected and also has the following properties: it is the maximal compact invariant set and the minimal set that attracts all bounded sets. The global attractor may have a very complex structure. However, as a compact set (in an infinite dimensional Banach space) its interior is empty.

We provide a theorem that guarantees the global attractor existence for a semigroup $\{S(t)\}_{t\geq 0}$ (see for example Theorem 18 in [3] where more general, multivalued, case is considered).

Theorem 12.2. *Let $\{S(t)\}_{t\geq 0}$ be a semigroup of operators in a Banach space X such that*

- *For all $t \geq 0$ the operator $S(t) : X \to X$ is continuous.*
- *$\{S(t)\}_{t\geq 0}$ is dissipative, i.e. there exists a bounded set $B_0 \subset X$ such that for every bounded set $B \subset X$ there exists $t_0 = t_0(B)$ such that $\bigcup_{t\geq t_0} S(t)B \subset B_0$.*
- *$\{S(t)\}_{t\geq 0}$ is asymptotically compact, i.e. for every bounded set $B \subset X$ and every sequences $t_n \to \infty$ and $y_n \in S(t_n)B$, the sequence $\{y_n\}$ is relatively compact in X.*

Then $\{S(t)\}_{t\geq 0}$ has a global attractor $\mathcal{A}$.

For many dynamical systems the global attractor has a finite fractal dimension (defined below) which has a number of important consequences for the behaviour of the flow generated by the semigroup [15, 16].

Definition 12.3. The fractal dimension of a compact set K in a Banach space X is defined as

$$d_f^X(K) = \limsup_{\varepsilon\to 0} \frac{\log N_\varepsilon^X(K)}{\log(\frac{1}{\varepsilon})}$$

where $N_\varepsilon^X(K)$ is the minimal number of balls of radius ε in X needed to cover K.

Another important property that holds for many dynamical systems is the existence of an exponential attractor.

Definition 12.4. An exponential attractor for a semigroup $\{S(t)\}_{t\geq 0}$ in a Banach space X is a subset $\mathcal{M}$ of X such that

- $\mathcal{M}$ is compact in X.
- $\mathcal{M}$ is positively invariant, i.e., $S(t)\mathcal{M} \subset \mathcal{M}$ for every $t \geq 0$.
- Fractal dimension of $\mathcal{M}$ is finite, i.e., $d_f^X(\mathcal{M}) < \infty$.
- $\mathcal{M}$ attracts exponentially all bounded subsets of X, i.e., there exist a universal constant c_1 and a monotone function Φ such that for every bounded set B in X, its image $S(t)B$ is a subset of the $\varepsilon(t)$-neighborhood of $\mathcal{M}$ for all $t \geq t_0$, where $\varepsilon(t) = \Phi(||B||_X)e^{-c_1 t}$ (exponential attraction property).

Since the global attractor $\mathcal{A}$ is the minimal compact attracting set it follows that if both global and exponential attractors exist, then $\mathcal{A} \subset \mathcal{M}$ and the fractal dimension of $\mathcal{A}$ must be finite. Moreover, in contrast to the global attractor, an exponential attractor does not have to be unique.

In the following we will remind the abstract framework of [11] (see also [10]) that uses the so called method of l-trajectories (or short trajectories) to prove the existence of global and exponential attractors.

Let X, Y, and Z be three Banach spaces such that

$$Y \subset X \quad \text{with compact imbedding} \quad \text{and} \quad X \subset Z.$$

We assume, moreover, that X is reflexive and separable.

For $\tau > 0$, let

$$X_\tau = L^2(0, \tau; X),$$

and

$$Y_\tau = \left\{ u \in L^2(0, \tau; Y),\ \frac{du}{dt} \in L^2(0, \tau; Z) \right\}.$$

By $\mathcal{C}([0, \tau]; X_w)$ we denote the space of weakly continuous functions from the interval $[0, \tau]$ to the Banach space X, and we assume that the solutions of (12.3) are at least in $\mathcal{C}([0, T]; X_w)$ for all $T > 0$. Then by an l-trajectory we mean the restriction of any solution to the time interval $[0, l]$. If $v = v(t), t \geq 0$, is the solution of (12.3) then both $\chi = v|_{[0,l]}$ as well as all shifts $L_t(v)|_{[0,l]} = v|_{[t,l+t]}$ for $t > 0$ are l-trajectories. Note that the mapping L_t is defined as $L_t(v)(\tau) = v(\tau + t)$ for $t > 0$ and $\tau \in [0, l]$.

We can now formulate a theorem which gives criteria for the existence of a global attractor $\mathcal{A}$ for the semigroup $\{S(t)\}_{t\geq 0}$ in X and its finite dimensionality. These criteria are stated as assumptions $(A1)$–$(A8)$ in [11].

$(A1)$ For any $v_0 \in X$ and arbitrary $T > 0$ there exists (not necessarily unique) $v \in \mathcal{C}([0,T]; X_w) \cap Y_T$, a solution of the evolutionary problem on $[0,T]$ with $v(0) = v_0$. Moreover, for any solution v, the estimates of $||v||_{Y_T}$ are uniform with respect to $||v_0||_X$.

$(A2)$ There exists a bounded set $B^0 \subset X$ with the following properties: if v is an arbitrary solution with initial condition $v_0 \in X$ then (i) there exists $t_0 = t_0(||v_0||_X)$ such that $v(t) \in B^0$ for all $t \geq t_0$ and (ii) if $v_0 \in B^0$ then $v(t) \in B^0$ for all $t \geq 0$.

$(A3)$ Each l-trajectory has among all solutions a unique continuation which means that from an end point of an l-trajectory there starts at most one solution.

$(A4)$ For all $t > 0$, $L_t : X_l \to X_l$ is continuous on $\mathcal{B}_0^l$—the set of all l-trajectories starting at any point of B^0 from $(A2)$.

$(A5)$ For some $\tau > 0$, the closure in X_l of the set $L_\tau(\mathcal{B}_0^l)$ is included in $\mathcal{B}_0^l$.

$(A6)$ There exists a space W_l such that $W_l \subset X_l$ with compact embedding, and $\tau > 0$ such that $L_\tau : X_l \to W_l$ is Lipschitz continuous on $\mathcal{B}_l^1$—the closure of $L_\tau(\mathcal{B}_0^l)$ in X_l.

$(A7)$ The map $e : X_l \to X$, $e(\chi) = \chi(l)$ is continuous on $\mathcal{B}_l^1$.

$(A8)$ The map $e : X_l \to X$ is Hölder-continuous on $\mathcal{B}_l^1$.

Theorem 12.5. *Let the assumptions* $(A1)$–$(A5)$ *and* $(A7)$ *hold. Then there exists a global attractor* $\mathcal{A}$ *for the semigroup* $\{S(t)\}_{t\geq 0}$ *in* X. *Moreover, if the assumptions* $(A6)$, $(A8)$ *are satisfied then the fractal dimension of the attractor is finite.*

For the existence of an exponential attractor we need two additional properties to hold, where now X is a Hilbert space (cf. [11]).

$(A9)$ For all $\tau > 0$ the operators $L_t : X_l \to X_l$ are (uniformly with respect to $t \in [0,\tau]$) Lipschitz continuous on $\mathcal{B}_l^1$.

$(A10)$ For all $\tau > 0$ there exists $c > 0$ and $\beta \in (0,1]$ such that for all $\chi \in \mathcal{B}_l^1$ and $t_1, t_2 \in [0,\tau]$ it holds that

$$||L_{t_1}\chi - L_{t_2}\chi||_{X_l} \leq c|t_1 - t_2|^\beta. \tag{12.4}$$

Theorem 12.6. *Let* X *be a separable Hilbert space and let the assumptions* $(A1)$–$(A6)$ *and* $(A8)$–$(A10)$ *hold. Then there exists an exponential attractor* $\mathcal{M}$ *for the semigroup* $\{S(t)\}_{t\geq 0}$ *in* X.

For the proofs of Theorems 12.5 and 12.6 we refer the readers to corresponding theorems in [11].

12.3 Exponential Attractor for Planar Shear Flow with Generalized Tresca Type Friction Law

The problem considered in this section is based on the results of [10]. As we show, arguments of [10] can be generalized to a class of problems with the friction coefficient dependent on the slip rate.

12.3.1 *Classical Formulation of the Problem*

We will consider the planar flow of an incompressible viscous fluid governed by the equation of momentum

$$v_t - \eta \Delta v + (v \cdot \nabla)v + \nabla p = f \quad \text{in } \Omega \times \mathbb{R}^+ \tag{12.5}$$

and the incompressibility condition

$$\nabla \cdot v = 0 \quad \text{in } \Omega \times \mathbb{R}^+, \tag{12.6}$$

in domain Ω given by

$$\Omega = \{x = (x_1, x_2) \mid 0 < x_1 < L,\ 0 < x_2 < h(x_1)\},$$

with boundary $\partial\Omega = \overline{\Gamma}_D \cup \overline{\Gamma}_C \cup \overline{\Gamma}_L$, where $\Gamma_D = \{(x_1, h(x_1)) \mid x_1 \in (0, L)\}$, $\Gamma_C = (0, L) \times \{0\}$, and $\Gamma_L = \{0, L\} \times (0, h(0))$ and are the top, bottom and lateral parts of $\partial\Omega$, respectively. The function h is smooth and L-periodic such that $h(x_1) \geq \varepsilon > 0$ for all $x_1 \in \mathbb{R}$ with a constant $\varepsilon > 0$. We will use the notation $e_1 = (1, 0)$ and $e_2 = (0, 1)$ for the canonical basis of $\mathbb{R}^2$. Note, that on Γ_C the outer normal unit vector is given by $\nu = -e_2$.

The unknowns are the velocity field $v : \Omega \times \mathbb{R}^+ \to \mathbb{R}^2$ and the pressure $p : \Omega \times \mathbb{R}^+ \to \mathbb{R}$, $\eta > 0$ is the viscosity coefficient and $f : \Omega \to \mathbb{R}^2$ is the density of volume forces.

We are interested in the solutions of (12.5)–(12.6) such that $v(0, x_2, t) = v(L, x_2, t)$, $\frac{\partial v_2(0,x_2,t)}{\partial x_1} = \frac{\partial v_2(L,x_2,t)}{\partial x_1}$, and $p(0, x_2, t) = p(L, x_2, t)$ for $x_2 \in [0, h(0)]$ and $t \in \mathbb{R}^+$. The first condition represents the L-periodicity of velocities, while the latter two ones, the L-periodicity of normal stresses in the space of divergence free functions. Moreover we assume that

$$v = 0 \quad \text{on } \Gamma_D \times \mathbb{R}^+. \tag{12.7}$$

On the contact boundary Γ_C we decompose the velocity into the normal component $v_\nu = v \cdot \nu$, where ν is the unit outward normal vector and the tangential one $v_\tau = v \cdot e_1$. Note, that since the domain Ω is two dimensional it is possible to consider the tangential components as scalars, for the sake of the ease of notation. Likewise, we decompose the stress on the boundary Γ_C into its normal component $\sigma_\nu = \sigma\nu \cdot \nu$ and the tangential one $\sigma_\tau = \sigma\nu \cdot e_1$. The stress tensor is related to the velocity and pressure through the linear constitutive law $\sigma_{ij} = -p\delta_{ij} + \eta(v_{i,j} + v_{j,i})$.

We assume that there is no flux across Γ_C and hence we have

$$v_\nu = 0 \ \text{ at } \Gamma_C \times \mathbb{R}^+. \tag{12.8}$$

The boundary Γ_C is assumed to be moving with the constant velocity $U_0 e_1 = (U_0, 0)$ which, together with the mass force, produces the driving force of the flow. The friction bound k is assumed to be related to the slip rate through the relation $k = k(|v_\tau - U_0|)$, where $k : \mathbb{R}^+ \to \mathbb{R}^+$. If there is no slip between the fluid and the boundary then the friction force magnitude is bounded by friction threshold $k(0)$

$$v_\tau = U_0 \Rightarrow |\sigma_\tau| \le k(0) \ \text{ at } \Gamma_C \times \mathbb{R}^+, \tag{12.9}$$

while if there is a slip, then the friction force is given by the expression

$$v_\tau \ne U_0 \Rightarrow -\sigma_\tau = k(|v_\tau - U_0|)\frac{v_\tau - U_0}{|v_\tau - U_0|} \ \text{ at } \Gamma_C \times \mathbb{R}^+. \tag{12.10}$$

Note that (12.9)–(12.10) generalize the Tresca law considered in [10], where k was assumed to be a constant. Here k depends of the slip rate, this dependence represents the fact that the kinetic friction is less then the static one, which holds if k is a decreasing function. Similar friction law is used for example in the study of the motion of tectonic plates, see [6, 14, 17] and the references therein. We make the following assumptions on the friction coefficient k,

$H(k)(i) : \ k \in C([0,\infty); [0,\infty))$,
$H(k)(ii) : \ k(s) \le \alpha(1+s)$ for all $s \in \mathbb{R}^+$ with $\alpha > 0$,
$H(k)(iii) : \ k(s) - k(r) \ge -\mu(s-r)$ for all $s > r \ge 0$ with $\mu > 0$.

Note that the assumption that k has values in $\mathbb{R}^+$ has a clear physical interpretation, namely it means that the friction force is dissipative.

Finally, the initial condition for the velocity field is

$$v(x,0) = v_0(x) \ \text{ for } x \in \Omega. \tag{12.11}$$

In the next section we present a weak formulation of the problem in the form of an evolutionary differential inclusion with a suitable superpotential corresponding to the generalized Tresca condition.

12.3.2 Weak Formulation of the Problem

To be able to define a weak solution of problem (12.5)–(12.11) we have to introduce some notation.

Let

$$\tilde{V} = \{v \in \mathcal{C}^\infty(\Omega)^2 \mid \operatorname{div} v = 0 \text{ in } \Omega,\ v \text{ is L-periodic in } x_1, \\ v = 0 \text{ at } \Gamma_D,\ v_\nu = 0 \text{ at } \Gamma_C\}$$

and

$$V = \text{closure of } \tilde{V} \text{ in } H^1(\Omega)^2, \qquad H = \text{closure of } \tilde{V} \text{ in } L^2(\Omega)^2.$$

We define scalar products in H and V, respectively, by

$$(u, v) = \int_\Omega u(x)\, v(x)\, dx \quad \text{and} \quad (\nabla u, \nabla v) = \int_\Omega \nabla u(x) \cdot \nabla v(x)\, dx,$$

and their associated norms by

$$\|v\|_H = (v, v)^{\frac{1}{2}} \quad \text{and} \quad \|v\| = (\nabla v, \nabla v)^{\frac{1}{2}}.$$

We denote by V^* the dual space to V and the duality pairing between V^* and V will be denoted by $\langle \cdot, \cdot \rangle$. We denote the trace $\gamma : V \to L^2(\Gamma_C; \mathbb{R}^2)$ and $\|\gamma\| = \|\gamma\|_{\mathcal{L}(V;L^2(\Gamma_C;\mathbb{R}^2))}$. Moreover, let, for u, v and w in V,

$$a(u, v) = (\nabla u, \nabla v) \quad \text{and} \quad b(u, v, w) = ((u \cdot \nabla)v, w).$$

Finally, we define the functional $j : \mathbb{R} \to \mathbb{R}$ corresponding to the generalized Tresca condition (12.9)–(12.10) by

$$j(v) = \int_0^{|v - U_0|} k(s)\, ds. \tag{12.12}$$

The variational formulation of problem (12.5)–(12.11) can be derived by the calculation analogous to the proof of Proposition 2.1 in [10].

Problem 12.7. Given $v_0 \in H$ and $f \in H$, find $v : \mathbb{R}^+ \to H$ such that:

(i) $v \in \mathcal{C}(\mathbb{R}^+; H) \cap L^2_{\text{loc}}(\mathbb{R}^+; V)$, with $v_t \in L^2_{\text{loc}}(\mathbb{R}^+; V^*)$.
(ii) for all Θ in V and for almost all $t \in \mathbb{R}^+$, the following equality holds

$$\langle v_t(t), \Theta \rangle + \eta a(v(t), \Theta) + b(v(t), v(t), \Theta) + (\xi(t), \Theta_\tau)_{L^2(\Gamma_C)} = (f, \Theta), \tag{12.13}$$

with $\xi(t) \in S^2_{\partial j(v_\tau(t))}$ for a.e. $t \in \mathbb{R}^+$, where for $w \in L^2(\Gamma_C)$ we denote by $S^2_{\partial j(w)}$ the set of all functions $\eta \in L^2(\Gamma_C)$ such that $\eta(x) \in \partial j(w(x))$ for a.e. $x \in \Gamma_C$.

(iii) the following initial condition holds

$$v(0) = v_0. \tag{12.14}$$

The above definition is justified by part (i) of the following lemma.

Lemma 12.8. *Under assumptions* $H(k)$ *the functional* j *defined by* (12.12) *satisfies the following properties:*

(i) j *is locally Lipschitz and conditions* (12.9)–(12.10) *are equivalent to*

$$-\sigma_\tau \in \partial j(v_\tau), \tag{12.15}$$

where ∂j *denotes the Clarke subdifferential of the functional* j*.*

(ii) $|\xi| \leq \tilde{\alpha}(1 + |w|)$ *for all* $w \in \mathbb{R}$ *and* $\xi \in \partial j(w)$, $\tilde{\alpha} > 0$.

(iii) $m(\partial j) \geq -\mu$, *where* $m(\partial j)$ *is the one sided Lipschitz constant defined by*

$$m(\partial j) = \inf_{\substack{u,v \in \mathbb{R}, u \neq v \\ \xi \in \partial j(u), \eta \in \partial j(v)}} \frac{(\xi - \eta) \cdot (u - v)}{|u - v|^2}.$$

(iv) for all $w \in \mathbb{R}$ *and* $\xi \in \partial j(w)$ *we have* $\xi \cdot w \geq -\beta(1 + |w|)$ *with a constant* $\beta > 0$.

Proof. (i) The fact that j is locally Lipschitz follows from the continuity of k and the direct calculation. To see that (12.9)–(12.10) is equivalent to (12.15) observe that for $v \neq U_0$ the functional j is strictly differentiable at v and its derivative is given as $j'(v) = k(|v - U_0|)\frac{v - U_0}{|v - U_0|}$, whence (12.10) is equivalent to (12.15) by the Proposition 5.6.15 (b) from [4]. Now, if $v = U_0$, (12.9) is equivalent to (12.15) by the characterization of the Clarke subgradient (see [4, Proposition 5.6.17])

$$\partial j(v) = conv\{\lim_{n \to \infty} j'(v_n) \mid v_n \to v, v_n \notin S \cup N_j\},$$

where S is any Lebesgue-null set and $N_j = \{U_0\}$ is the set of points where j is not differentiable.

Assertion (ii) follows in a straightforward way from $H(k)(ii)$.

Assertion (iii) can be obtained by a following computation. Let $\xi \in \partial j(u)$, $\eta \in \partial j(v)$, where $u \neq U_0$ and $v \neq U_0$. Then

$$\begin{aligned}(\xi - \eta) \cdot (u - v) &= \left(k(|u - U_0|)\frac{u - U_0}{|u - U_0|} - k(|v - U_0|)\frac{v - U_0}{|v - U_0|}\right) \cdot (u - v) \\ &= k(|u - U_0|)|u - U_0| - k(|u - U_0|)\frac{(u - U_0) \cdot (v - U_0)}{|u - U_0|}\end{aligned}$$

$$
\begin{aligned}
&- k(|v-U_0|)\frac{(u-U_0)\cdot(v-U_0)}{|v-U_0|} + k(|v-U_0|)|v-U_0| \\
&\geq (k(|u-U_0|) - k(|v-U_0|))\,(|u-U_0| - |v-U_0|) \\
&\geq -\mu(|u-U_0| - |v-U_0|)^2 \geq -\mu|u-v|^2.
\end{aligned}
$$

The calculation in the case when either $u = U_0$ or $v = U_0$ is analogous.

Proof of assertion (iv) is also straightforward. Indeed, for $w = 0$ the assertion obviously holds. Let $w \neq 0$ and $\xi \in \partial j(w)$. We have, by $H(k)(i)$–(ii)

$$
\begin{aligned}
\xi \cdot w &= k(|w-U_0|)\frac{w-U_0}{|w-U_0|}(w-U_0+U_0) \\
&\geq k(|w-U_0|)|w-U_0| - k(|w-U_0|)|U_0| \geq -\alpha(1+|w|+|U_0|)|U_0|,
\end{aligned}
$$

and (iv) follows, which completes the proof. □

Remark 12.9. Observe that the assertion (iii) of Lemma 12.8 is equivalent to the claim that the functional j is semiconvex, i.e. the functional $s \to j(s) + \frac{\mu s^2}{2}$ is in fact convex (see [2, Definition 10]).

12.3.3 Existence and Properties of Solutions

We begin with some estimates that are satisfied by the solutions of Problem 12.7. We define the auxiliary operators $A : V \to V^*$ and $B : V \to V^*$ by $\langle Au, v\rangle = a(u, v)$ and $\langle Bu, v\rangle = b(u, u, v)$.

Lemma 12.10. *Let v be a solution of Problem 12.7. Then for all $t \geq 0$ we have*

$$
\max_{s\in[0,t]} \|v(s)\|_H^2 + \int_0^t \|v(s)\|^2 ds \leq C(t, \|v_0\|_H), \tag{12.16}
$$

$$
\int_0^t \|v'(s)\|_{V^*}^2 ds \leq C(t, \|v_0\|_H), \tag{12.17}
$$

where $C(t, \|v_0\|_H)$ is a nondecreasing function of t and $\|v_0\|_H$.

Proof. Take $\Theta = v(t)$ in (12.13). Since $b(v, v, v) = 0$ for $v \in V$, we get, with an arbitrary $\varepsilon > 0$,

$$
\begin{aligned}
\frac{1}{2}\frac{d}{dt}\|v(t)\|_H^2 + \eta\|v(t)\|^2 &\leq \|f\|_H\|v(t)\|_H + \|\xi(t)\|_{L^2(\Gamma_C)}\|v_\tau(t)\|_{L^2(\Gamma_C)} \\
&\leq \varepsilon\|v(t)\|_H^2 + \frac{\|f\|_H^2}{4\varepsilon} + \frac{3}{2}\tilde{\alpha}\|v(t)\|_{L^2(\Gamma_C)}^2 + \tilde{\alpha}\mu(\Gamma_C),
\end{aligned}
$$

where we used (ii) of Lemma 12.8 and $\mu(\Gamma_C) = L$ is the one dimensional boundary measure of Γ_C.

Denoting $Z = V \cap H^{1-\delta}(\Omega;\mathbb{R}^2)$ with some small $\delta \in \left(0, \frac{1}{2}\right)$ equipped with the norm $H^{1-\delta}(\Omega;\mathbb{R}^2)$ observe that the embedding $V \subset Z$ is compact and the trace map $\gamma_C : Z \to L^2(\Gamma_C;\mathbb{R}^2)$ is continuous. We denote $\|\gamma_C\| = \|\gamma_C\|_{\mathcal{L}(Z;L^2(\Gamma_C;\mathbb{R}^2))}$. From the Ehrling lemma we get, for an arbitrary $\varepsilon > 0$ independent on $v \in V$ and $C(\varepsilon) > 0$ that

$$\|v\|_Z^2 \leq \varepsilon\|v\|^2 + C(\varepsilon)\|v\|_H^2.$$

Hence, noting that the Poincaré inequality

$$\lambda_1\|v\|_H^2 \leq \|v\|^2$$

is valid for $v \in V$, we get

$$\frac{d}{dt}\|v(t)\|_H^2 + 2\eta\|v(t)\|^2 \leq \frac{2\varepsilon}{\lambda_1}\|v(t)\|^2 + \frac{\|f\|_H^2}{2\varepsilon} + 3\tilde{\alpha}\|\gamma_C\|^2\varepsilon\|v(t)\|^2$$
$$+ 3\tilde{\alpha}\|\gamma_C\|^2 C(\varepsilon)\|v(t)\|_H^2 + 2\tilde{\alpha}L.$$

It is clear that we can choose ε small enough such that

$$\frac{d}{dt}\|v(t)\|_H^2 + \eta\|v(t)\|^2 \leq C_1 + C_2\|v(t)\|_H^2,$$

with $C_1, C_2 > 0$. Using the Gronwall lemma we get (12.16).

For the proof of (12.17) observe, that proceeding like in Proposition 9.2 in [15] we get $\|Bu\|_{V^*} \leq C\|u\|_H\|u\|$. Hence (12.17) follows by (12.16) and the straightforward computation that uses the assertion (ii) of Lemma 12.8 to estimate the multivalued term. □

We formulate and prove the theorem on existence of solutions to Problem 12.7.

Theorem 12.11. *For any $u_0 \in H$, Problem 12.7 has at least one solution.*

Proof. Since the proof of solution existence, based on the Galerkin method, is standard and quite long, we provide only its main steps.

Let $\varrho \in C_0^\infty(-1,1)$ be a mollifier such that $\int_{-1}^{1}\varrho(s)\,ds = 1$ and $\varrho(s) \geq 0$. We define $\varrho_n : \mathbb{R} \to \mathbb{R}$ by $\varrho_n(x) = n\varrho(nx)$ for $n \in \mathbb{N}$ and $x \in \mathbb{R}$. Then $\operatorname{supp}\varrho_n \subset \left(-\frac{1}{n}, \frac{1}{n}\right)$. We consider $j_n : \mathbb{R} \to \mathbb{R}$ defined by the convolution

$$j_n(r) = \int_{\mathbb{R}} \varrho_n(s)j(r-s)\,ds \quad \text{for } r \in \mathbb{R}. \tag{12.18}$$

Note that $j_n \in C^\infty(\mathbb{R})$. From the proof of $H(j)(iii)$ in Lemma 5 in [7] it follows that for all $n \in \mathbb{N}$ and $r \in \mathbb{R}$ we have $|j_n'(r)| \le \overline{\alpha}(1+|r|)$, where $\overline{\alpha}$ is different from $\tilde{\alpha}$ in (ii) of Lemma 12.8 above but independent of n.

Let us furthermore take the sequence V_n of finite dimensional spaces such that $V_n = \text{span}\{z_1, \dots, z_n\}$ where $\{z_i\}$ are orthonormal in H eigenfunctions of the Stokes operator with the Dirichlet and periodic boundary conditions given in the definition of the space V. Then $\{V_n\}_{n=1}^\infty$ approximate V from inside, i.e. $\overline{\bigcup_{n=1}^\infty V_n} = V$. We denote by $P_n : V^* \to V_n$ the orthogonal projection on V_n defined as $P_n u = \sum_{i=1}^n \langle u, z_i \rangle z_i$.

We formulate the regularized Galerkin problems for $n \in \mathbb{N}$.

Find $v_n \in \mathcal{C}(\mathbb{R}^+; V_n)$ such that v_n is differentiable for a.e. $t \in \mathbb{R}^+$ and

$$\langle v_n'(t), \Theta \rangle + \eta a(v_n(t), \Theta) + b(v_n(t), v_n(t), \Theta) + (j_n'(v_{n\tau}(t)(\cdot)), \Theta_\tau)_{L^2(\Gamma_C)} = (f, \Theta), \tag{12.19}$$

$$v_n(0) = P_n v_0 \tag{12.20}$$

for a.e. $t \in \mathbb{R}^+$ and for all $\Theta \in V_n$. Note, that due to the fact that j_n', a, b are smooth, the solution of (12.19)–(12.20), if it exists, is a continuously differentiable function of time variable, with values in V_n.

We first show that if v_n solves (12.19) then estimates analogous to the ones in Lemma 12.10 hold.

Taking $v_n(t)$ in place of Θ in (12.19) and using the argument similar to that in the proof of (12.16) in Lemma 12.10 we obtain that, for a given $T > 0$,

$$\|v_n\|_{L^\infty(0,T;H)} \le \text{const}, \tag{12.21}$$

$$\|v_n\|_{L^2(0,T;V)} \le \text{const}. \tag{12.22}$$

Now denoting $\iota : V \to L^2(\Gamma_C)$ as $\iota v = (\gamma v)_\tau$ and ι^* adjoint to ι we observe that (12.19) is equivalent to the following equation in V^*

$$v_n'(t) + \eta A v_n(t) + P_n B(v_n(t)) + P_n \iota^* j_n'(v_{n\tau}(t)(\cdot)) = P_n f. \tag{12.23}$$

Since (see Lemma 7.5 in [15]) for $w \in V^*$ we have $\|P_n w\|_{V^*} \le \|w\|_{V^*}$, then from Eq. (12.23) by using the estimate $\|Bw\|_{V^*} \le C\|w\|_H \|w\|$ valid for $w \in V$, the growth condition on j_n' as well as (12.21) and (12.22), we get

$$\|v_n'\|_{L^2(0,T;V^*)} \le \text{const}. \tag{12.24}$$

The existence of the solution to the Galerkin problem (12.19) follows by the Caratheodory theorem and estimates (12.21)–(12.22).

Since the bounds (12.21), (12.22), and (12.24) are independent of n then, for a subsequence, we get

$$v_n \to v \ \text{weakly} - * \ \text{in } L^\infty(0,T;H), \tag{12.25}$$

$$v_n \to v \ \text{weakly in } L^2(0,T;V), \tag{12.26}$$

$$v_n' \to v' \ \text{weakly in } L^2(0,T;V^*), \tag{12.27}$$

$$v_{n\tau} \to v_\tau \ \text{strongly in } L^2(0,T;L^2(\Gamma_C)), \tag{12.28}$$

$$v_n \to v \ \text{strongly in } L^2(0,T;H), \tag{12.29}$$

where (12.28) is a consequence of the Lions–Aubin Lemma used for the spaces $V \subset Z \subset V^*$ and (12.29) is a consequence of the Lions–Aubin Lemma used for the spaces $V \subset H \subset V^*$.

From (12.22) and the growth condition on j_n' it follows that

$$\|j_n'(v_{n\tau})\|_{L^2(0,T;L^2(\Gamma_C))} \leq \text{const.}$$

Hence, perhaps for another subsequence, we have

$$j_n'(v_{n\tau}) \to \xi \ \text{weakly in } L^2(0,T;L^2(\Gamma_C)).$$

In a standard way (see for example Sections 7.4.3 and 9.4 in [15]) we can pass to the limit in (12.23) and obtain that v and ξ satisfy (12.13) for a.e. $t \in (0,T)$. The proof that v satisfies the initial condition $v(0) = v_0$ is analogous to the one in Theorem 3.1 in [8].

The proof that $\xi(t) \in S^2_{\partial j(v_\tau(t))}$ is also analogous to the proof of Theorem 3.1 in [8].

Finally, the solution can be extended from $[0,T]$ to the whole $\mathbb{R}^+$ by concatenating the solutions on intervals $[0,T]$ taking the value at the endpoint of the previous interval as the initial condition for the following one (see for example [5, Theorem 2 in Section 9.2.1]). □

Remark 12.12. The proof of Theorem 12.11 uses only assertions (i) and (ii) of Lemma 12.8 and, therefore, the theorem holds for a class of potentials j that satisfy these two conditions and are not necessarily given by (12.12). This is a new result of independent interest since it weakens the conditions required for the existence of solutions provided, for example, in [12]. Indeed, the sign condition (see H(j)(iv) p. 583 in [12]) is not needed here for the existence proof.

Next theorem shows that assertion (iii) of Lemma 12.8 gives the Lipschitz continuity of the solution map on bounded sets as well as the solution uniqueness.

Lemma 12.13. *The solution of Problem 12.7 is unique and if v, w are two solutions of Problem 12.7 with the initial conditions v_0, w_0, respectively, then for any $t > s \geq 0$ we have*

$$\|v(t) - w(t)\|_H \leq D(t, \|w_0\|_H)\|v(s) - w(s)\|_H, \tag{12.30}$$

where $D(t, \|w_0\|_H) > 0$ is a nondecreasing function of t, $\|w_0\|_H$.

Proof. Let v and w be two solutions of Problem 12.7 with the initial conditions v_0 and w_0, respectively. We denote $u(t) = v(t) - w(t)$. Subtracting (12.13) for v and w and taking $u(t)$ as a test function we get, for a.e. $t \in \mathbb{R}^+$,

$$\frac{1}{2}\frac{d}{dt}\|u(t)\|_H^2 + \eta\|u(t)\|^2 + b(u(t), w(t), u(t)) + (\xi(t) - \eta(t), v_\tau(t) - w_\tau(t))_{L^2(\Gamma_C)} = 0,$$

where $\xi(t) \in S^2_{\partial j(v_\tau(t))}$ and $\eta(t) \in S^2_{\partial j(w_\tau(t))}$.

Using the Ladyzhenskaya inequality [10],

$$||v||_{L^4(\Omega)} \leq ||v||_H^{1/2}||v||^{1/2} \quad \text{for} \quad v \in V, \tag{12.31}$$

(cf. also [15, Proposition 9.2]) to estimate the convective term and assertion (iii) in Lemma 12.8 to estimate the multivalued one we get, for a.e. $t \in \mathbb{R}^+$,

$$\frac{1}{2}\frac{d}{dt}\|u(t)\|_H^2 + \eta\|u(t)\|^2 - \mu\|u_\tau(t)\|^2_{L^2(\Gamma_C)} \leq C\,\|u(t)\|_H\,\|u(t)\|\,\|w(t)\|. \tag{12.32}$$

As in the proof of Lemma 12.10 we use the fact that the embedding $V \subset Z$ is compact and the trace $\gamma_C : Z \to L^2(\Gamma_C; \mathbb{R}^2)$ is continuous. We get, for a.e. $t \in \mathbb{R}^+$,

$$\frac{1}{2}\frac{d}{dt}\|u(t)\|_H^2 + \eta\|u(t)\|^2 \leq \varepsilon\|u(t)\|^2 + C(\varepsilon)\|u(t)\|_H^2\|w(t)\|^2$$
$$+ \mu\|\gamma_C\|^2(\varepsilon\|u(t)\|^2 + C(\varepsilon)\|u(t)\|_H^2), \tag{12.33}$$

with an arbitrary $\varepsilon > 0$ and the constant $C(\varepsilon) > 0$ independent on u, w and t. By choosing $\varepsilon > 0$ small enough we get, for a.e. $t \in \mathbb{R}^+$,

$$\frac{d}{dt}\|u(t)\|_H^2 \leq \|u(t)\|_H^2(C_1\|w(t)\|^2 + C_2)$$

with the constants $C_1, C_2 > 0$. The Gronwall lemma gives

$$\|u(t)\|_H^2 \leq \|u(s)\|_H^2 e^{C_2(t-s)+C_1\int_s^t \|w(\tau)\|^2 d\tau} \leq \|u(s)\|_H^2 e^{C_2 t+C_1\int_0^t \|w(\tau)\|^2 d\tau}.$$

Using (12.16) we get

$$\|v(t) - w(t)\|_H \leq \|v(s) - w(s)\|_H e^{\frac{1}{2}C_2 t+\frac{1}{2}C_1 C(t,R)}, \tag{12.34}$$

and the proof of (12.30) is complete. Taking $s = 0$ and $v(0) = w(0)$ in (12.30) we obtain the uniqueness. □

Now we will prove that assertion (iv) of Lemma 12.8 gives additional, dissipative, a priori estimates.

Lemma 12.14. *If v is a solution of Problem 12.7 with the initial condition v_0, then for all $t \geq 0$ we have*

$$\|v(t)\|_H \leq \|v_0\|_H e^{-C_1 t} + C_2,$$

with positive constants C_1, C_2 independent of t, v_0.

Proof. We proceed as in the proof of (12.16) in Lemma 12.8. Taking the test function $\Theta = v(t)$ in (12.13) we get

$$\frac{1}{2}\frac{d}{dt}\|v(t)\|_H^2 + \eta\|v(t)\|^2 + (\xi(t), v_\tau(t))_{L^2(\Gamma_C)} \leq \|f\|_H \|v(t)\|_H,$$

for a.e. $t \in \mathbb{R}^+$. We use assertion (iv) of Lemma 12.8 and the Poincaré inequality $\lambda_1\|v\|_H^2 \leq \|v\|^2$ to get

$$\frac{1}{2}\frac{d}{dt}\|v(t)\|_H^2 + \eta\|v(t)\|^2 - \beta L\left(1 + \frac{1}{4\varepsilon_1}\right) - \beta\varepsilon_1\|v_\tau(t)\|_{L^2(\Gamma_C)}^2$$
$$\leq \frac{\|f\|_H^2}{4\varepsilon_2} + \frac{\varepsilon_2}{\lambda_1}\|v(t)\|^2,$$

for a.e. $t \in \mathbb{R}^+$, with an arbitrary $\varepsilon_1 > 0$ and $\varepsilon_2 > 0$. Using the trace inequality and the Poincaré inequality, again, we can choose ε_1 and ε_2 small enough to get

$$\frac{d}{dt}\|v(t)\|_H^2 + \eta\lambda_1\|v(t)\|_H^2 \leq C,$$

for a.e. $t \in \mathbb{R}^+$, where $C > 0$ is a constant. Finally, lemma follows from the Gronwall inequality. □

12.3.4 Existence of Finite Dimensional Global Attractor

In this section we prove the following theorem.

Theorem 12.15. *The semigroup $\{S(t)\}_{t\geq 0}$ associated with Problem 12.7 has a global attractor $\mathcal{A} \subset H$ of finite fractal dimension in the space H.*

Proof. In view of Theorem 12.5 we need to verify assumptions $(A1)$–$(A8)$. We set $Y = V$, $X = H$, $Z = V^*$, $l > 0$ $X_l = L^2(0, l; H)$, and for $T > 0$, $Y_T = \{u \in L^2(0, T; V) \mid u' \in L^2(0, T; V^*)\}$.

Assumption $(A1)$ From Theorem 12.11 it follows that for any $u_0 \in H$ there exists at least one solution of Problem 12.7. This solution, restricted to the interval $(0, T)$, belongs to Y_T and $\mathcal{C}([0, T]; H)$. Uniform bounds in Y_T follow directly from Lemma 12.10.

Assumption $(A2)$ In view of Lemma 12.14 the closed ball $B_H(0, C_2+1)$ absorbs in finite time all bounded sets in H. Let $B^0 = \overline{\bigcup_{t\geq 0} S(t)B_H(0, C_2+1)}^H$. Obviously this set is closed in H (we will need this fact to verify the assumption $(A5)$ below). From Lemma 12.14 it follows that B^0 is bounded. Let us assume that $u \in B^0$ and $v = S(t)u$. Then, for a sequence $u_n \in S(t_n)B_H(0, C_2+1)$ with certain t_n we have $u_n \to u$ in H. From Lemma 12.13 it follows that $S(t)u_n \to S(t)u = v$ in H. But $S(t)u_n \in S(t+t_n)B_H(0, C_2+1) \subset B^0$. Hence, from the closedness of B^0 it follows that $v \in B^0$ and the assertion is proved.

In the rest of the proof the generic constant dependent on B^0, l and the problem data will be denoted by C_{l,B^0}.

Assumption $(A3)$ Recall that the l-trajectory is a restriction of any solution of Problem 12.7 to the time interval $[0, l]$. The fact that assumption $(A3)$ holds follows immediately from Lemma 12.13.

Assumption $(A4)$ Define the shift operator $L_t : X_l \to X_l$ by $(L_t\chi)(s) = u(s+t)$ for $s \in [0, l]$, where u is the unique solution on $[0, l+t]$ such that $u|_{[0,l]} = \chi$. We will prove that this operator is continuous on $\mathcal{B}_0^l$ for all $t \geq 0$ and $l > 0$, where $\mathcal{B}_0^l$ is the set of all l-trajectories with the initial condition in the set B^0 from assumption $(A2)$. Let u, v be two solutions with the initial conditions $u_0, v_0 \in B^0$. By Lemma 12.13 we have

$$\int_0^l \|v(s+t) - u(s+t)\|_H^2 ds \leq D^2(t+l, \|B^0\|_H) \int_0^l \|v(s) - u(s)\|_H^2 ds,$$

where $\|B^0\|_H = \sup_{v\in B_0} \|v\|_H$. Hence, denoting $\chi_1 = u|_{[0,l]}$ and $\chi_2 = v|_{[0,l]}$, we get

$$\|L_t\chi_1 - L_t\chi_2\|_{X_l}^2 \leq D^2(t+l, \|B^0\|_H)\|\chi_1 - \chi_2\|_{X_l}^2, \tag{12.35}$$

and the proof of $(A4)$ is complete.

Assumption $(A5)$ We need to prove that, for some $\tau > 0$, $\overline{L_\tau(\mathcal{B}_0^l)}^{X_l} \subset \mathcal{B}_0^l$. Let $u_0 \in B^0$. From the fact that for all $\tau > 0$ we have $S(\tau)u_0 \in B^0$ it follows that $L_\tau(\mathcal{B}_0^l) \subset \mathcal{B}_0^l$. To obtain the assertion we must show that $\mathcal{B}_0^l$ is closed in X_l. To this end assume that u_n is a sequence of solutions to Problem 12.7 such that $u_n(0) \in B^0$ and

$$\chi_n \to \chi \text{ strongly in } L^2(0, l; H), \tag{12.36}$$

where $\chi_n = u_n|_{[0,l]}$. From $(A1)$ it follows that

$$\chi_n \to \chi \text{ weakly in } L^2(0, l; V), \tag{12.37}$$

$$\chi_n' \to \chi' \ \text{ weakly in } L^2(0,l;V^*). \tag{12.38}$$

We must show that $\chi \in \mathcal{B}_0^l$. Let us first prove that $\chi(0) \in B^0$. Indeed from (12.36) it follows that, for a subsequence, $\chi_n(t) \to \chi(t)$ strongly in H for a.e. $t \in (0,l)$. But, since $\chi_n(t) \in B^0$ for all $n \in \mathbb{N}$ and all $t \in [0,l]$, then $\chi(t)$ must belong to the closed set B^0 for a.e. $t \in (0,l)$. The closedness of B^0 and the fact that $\chi \in \mathcal{C}([0,l];H)$ imply that $\chi(t) \in B^0$ for all $t \in [0,l]$ and, in particular $\chi(0) \in B^0$. We need to prove that χ satisfies (12.13) for a.e. $t \in (0,l)$. In view of Lemma 7.4 in [15] it is enough to show that for all $w \in L^2(0,l;V^*)$ we have

$$\int_0^l \langle \chi'(t), v(t)\rangle\, dt + \eta \int_0^l \langle A\chi(t), w(t)\rangle\, dt$$
$$+ \int_0^l \langle B\chi(t), w(t)\rangle\, dt + \int_0^l (\xi(t), w_\tau(t))_{L^2(\Gamma_C)}\, dt = \int_0^l (f, w(t))\, dt, \tag{12.39}$$

with $\xi(t) \in S^2_{\partial j(\chi_\tau(t))}$ for a.e. $t \in (0,l)$. From the fact that χ_n satisfies (12.13) for a.e. $t \in (0,l)$ we have

$$\int_0^l \langle \chi_n'(t), v(t)\rangle\, dt + \eta \int_0^l \langle A\chi_n(t), w(t)\rangle\, dt$$
$$+ \int_0^l \langle B\chi_n(t), w(t)\rangle\, dt + \int_0^l (\xi_n(t), w_\tau(t))_{L^2(\Gamma_C)}\, dt = \int_0^l (f, w(t))\, dt, \tag{12.40}$$

with $\xi_n(t) \in S^2_{\partial j(\chi_{n\tau}(t))}$ for a.e. $t \in (0,l)$. From the growth condition (ii) in Lemma 12.8 it follows that

$$\|\xi_n\|_{L^2(0,l;L^2(\Gamma_C))} \le \sqrt{2lL}\tilde{\alpha} + \sqrt{2}\tilde{\alpha}\|\chi_{n\tau}\|_{L^2(0,l;L^2(\Gamma_C))}.$$

But, since χ_n is bounded in $L^2(0,l;V)$, then $\chi_{n\tau}$ must be bounded in $L^2(0,l;L^2(\Gamma_C))$ and hence ξ_n is bounded in the same space. For a subsequence we have

$$\xi_n \to \xi \ \text{ weakly in } L^2(0,l;L^2(\Gamma_C)). \tag{12.41}$$

Now (12.36), (12.37), and (12.41) are sufficient to pass to the limit in all the terms in (12.40) and thus to prove (12.39). It remains to show that $\xi(t) \in S^2_{\partial j(\chi_{n\tau}(t))}$ for a.e. $t \in (0,l)$. Let $Z = V \cap H^{1-\delta}(\Omega;\mathbb{R}^2)$ be equipped with $H^{1-\delta}$ topology, where $\delta \in \left(0,\frac{1}{2}\right)$. Then for the triple $V \subset Z \subset V^*$ we can use the Lions–Aubin Compactness lemma and conclude from (12.36) and (12.37) that $\chi_n \to \chi$ strongly in $L^2(0,l;Z)$.

Now, since the trace operator $\gamma_C : Z \to L^2(\Gamma_C;\mathbb{R}^2)$ is linear and continuous, and so is the corresponding Nemytskii operator, we have

$$\chi_{n\tau} \to \chi_\tau \text{ strongly in } L^2(0,l;L^2(\Gamma_C)). \tag{12.42}$$

From (12.41), (12.42) and the fact that for a.e. $(x,t) \in \Gamma_C \times (0,l)$ we have $\xi_n(x,t) \in \partial j(\chi_{n\tau}(x,t))$ we can use the Convergence Theorem of Aubin and Cellina (see [1, Theorem 7.2.2]) to deduce that for a.e. $(x,t) \in \Gamma_C \times (0,l)$ we have $\xi(x,t) \in \partial j(\chi_\tau(x,t))$. This completes the proof of $(A5)$.

Assumption $(A6)$ We define $W = H_0^1(\Omega;\mathbb{R}^2) \cap V$, where we equip W with the norm of V. Then $V^* \subset W^*$. By the Lions–Aubin Compactness lemma the space

$$W_l = \{\chi \in L^2(0,l;V) \mid \chi' \in L^1(0,T;W^*)\}$$

is embedded compactly in X_l. We must show that the shift operator $L_\tau : X_l \to W_l$ is Lipschitz continuous on $\mathcal{B}_l^1 = \overline{L_\tau(\mathcal{B}_0^l)}^{X_l}$ for some $\tau > 0$. We will prove that L_τ is Lipschitz continuous on the larger set $\mathcal{B}_0^l$ for $\tau = l$. Indeed, take w, v to be two solutions of Problem 12.7 starting from B^0. Denote $u = v - w$. Then, by (12.33), for a.e. $t \in \mathbb{R}^+$,

$$\frac{d}{dt}\|u(t)\|_H^2 + \eta\|u(t)\|^2 \le C\|u(t)\|_H^2(1 + \|w(t)\|^2),$$

where the constant C depends only on the problem data. We fix $s \in (0,l)$ and integrate the last inequality over interval $(s,2l)$. We obtain

$$\|u(2l)\|_H^2 + \eta\int_s^{2l}\|u(t)\|^2 dt \le C\int_s^{2l}\|u(t)\|_H^2(1 + \|w(t)\|^2)\,dt + \|u(s)\|_H^2.$$

Using (12.34) we get

$$\eta\int_l^{2l}\|u(t)\|^2 \le \|u(s)\|_H^2\left(2C_{l,B^0}l + C_{l,B^0}\int_0^{2l}\|w(t)\|^2 dt + 1\right).$$

Since, by $(A1)$ we have

$$\int_0^{2l}\|w(t)\|^2 dt \le C_{l,B^0},$$

it follows that

$$\int_l^{2l}\|u(t)\|^2 dt \le C_{l,B^0}\|u(s)\|_H^2.$$

Integrating this inequality over the interval $(0, l)$ with respect to the variable s it follows that

$$l \int_l^{2l} \|u(t)\|^2 dt \leq C_{l,B^0} \int_0^l \|u(s)\|_H^2 ds$$

and, therefore,

$$\|L_l \chi_1 - L_l \chi_2\|_{L^2(0,l;V)} \leq \sqrt{\frac{C_{l,B^0}}{l}} \|\chi_1 - \chi_2\|_{X_l}. \tag{12.43}$$

It remains to prove that

$$\|(L_l \chi_1 - L_l \chi_2)'\|_{L^1(0,l;W^*)} \leq C_{l,B^0} \|\chi_1 - \chi_2\|_{X_l}.$$

In view of (12.43) it is sufficient to prove that

$$\|(\chi_1 - \chi_2)'\|_{L^1(0,l;W^*)} \leq C_{l,B^0} \|\chi_1 - \chi_2\|_{L^2(0,l;V)}.$$

To this end, we subtract (12.13) for w and v, and take as a test function $\Theta \in W$. We get

$$\langle u'(t), \Theta \rangle + \eta \langle Au(t), \Theta \rangle + \langle Bv(t) - Bw(t), \Theta \rangle = 0.$$

We have, for a constant $k > 0$ dependent only on the problem data,

$$\begin{aligned} \langle Bv(t) - Bw(t), \Theta \rangle &= b(u(t), w(t), \Theta) + b(w(t), u(t), \Theta) + b(u(t), u(t), \Theta) \\ &\leq k(\|u(t)\| \|w(t)\| + \|u(t)\|^2) \|\Theta\| \\ &\leq k(2\|w(t)\| + \|v(t)\|) \|u(t)\| \|\Theta\|, \end{aligned}$$

whence, for a constant $C > 0$ dependent only on the problem data,

$$\begin{aligned} \|u'(t)\|_{W^*} &= \sup_{\Theta \in W, \|\Theta\|=1} |\eta \langle Au(t), \Theta \rangle + \langle Bv(t) - Bw(t), \Theta \rangle| \\ &\leq \eta \|Au(t)\|_{V^*} + \|Bv(t) - Bw(t)\|_{V^*} \\ &\leq C(1 + \|w(t)\| + \|v(t)\|) \|u(t)\|. \end{aligned}$$

It follows that

$$\|u'\|_{L^1(0,l;W^*)} \leq C \|u\|_{L^2(0,l;V)} \sqrt{\int_0^l (1 + \|w(t)\| + \|v(t)\|)^2 \, dt}.$$

The assertion follows from $(A1)$.

Assumption $(A7)$ *and* $(A8)$ Since $(A8)$ implies $(A7)$, we prove only $(A8)$. We will show that the mapping $e : X_l \to X$ defined as $e(\chi) = \chi(l)$ is Lipschitz on $\mathcal{B}_l^0$. Indeed, by (12.34), for any two solutions v, w of Problem 12.7 such that their initial conditions belong to B^0, we get

$$\|w(l) - v(l)\|_H \le C_{l,B^0} \|w(s) - v(s)\|_H$$

for all $s \in (0, l)$. Integrating the square of this inequality over $s \in (0, l)$ we obtain

$$l \|w(l) - v(l)\|_H^2 \le C_{l,B^0}^2 \|w - v\|_{X_l}^2,$$

whence

$$\|w(l) - v(l)\|_H \le \frac{C_{l,B^0}}{\sqrt{l}} \|w - v\|_{X_l},$$

and $(A8)$ holds. □

Theorem 12.15 is about the existence of the global attractor of a finite fractal dimension. In the following subsection we shall prove the existence of an exponential attractor.

12.3.5 Existence of an Exponential Attractor

Our goal is to prove the following result.

Theorem 12.16. *The semigroup* $\{S(t)\}_{t\ge 0}$ *associated with Problem 12.7 has an exponential attractor in the space* H.

Proof. Since we have shown, in the proof of Theorem 12.15, that assumptions $(A1)$–$(A8)$ hold, in view of Theorem 12.6 it is enough to prove $(A9)$ and $(A10)$.

Assumption $(A9)$ Lipschitz continuity of $L_t : X_l \to X_l$, uniform with respect to $t \in [0, \tau]$ for all $\tau > 0$, follows immediately from inequality (12.35).

Assumption $(A10)$ It remains to prove that the inequality

$$\|L_{t_1}\chi - L_{t_2}\chi\|_{X_l} \le c|t_1 - t_2|^\delta$$

holds for all $t_0 > 0$, $0 \le t_2 \le t_1 \le t_0$ and $\chi \in \mathcal{B}_l^1$ with $\delta \in (0, 1]$ and $c > 0$. Since $\mathcal{B}_l^1 = \overline{L_\tau(\mathcal{B}_l^0)}^{X_l}$ for certain $\tau > 0$, it is enough to prove the assertion for $\chi \in L_\tau(\mathcal{B}_l^0)$. Let v be the unique solution of Problem 12.7 on $[0, \tau + l + t_0]$ with $v(0) \in B^0$ such that $v|_{[\tau,\tau+l]} = \chi$. It is enough to obtain the uniform bound on v' in the space $L^\infty(\tau, \tau + l + t_0; H)$. Indeed,

$$\begin{aligned}\|L_{t_1}\chi - L_{t_2}\chi\|_{X_l} &= \sqrt{\int_0^l \|v(\tau+s+t_2) - v(\tau+s+t_1)\|_H^2 ds} \\ &\le \sqrt{\int_0^l \left(\int_{\tau+s+t_1}^{\tau+s+t_2} \|v'(r)\|_H \, dr\right)^2 ds} \\ &\le |t_1 - t_2|\sqrt{l}\|v'\|_{L^\infty(\tau,\tau+l+t_0;H)}\end{aligned}$$

The a priori estimate will be computed for the smooth solutions of the regularized Galerkin problem (12.19)–(12.20) formulated in the proof of Theorem 12.11. We will show that, for the Galerkin approximations v_n, for any $T > \tau$, the functions v_n' are bounded in $L^\infty(\tau, T; H)$ independently of n and, in consequence, the desired estimate will be preserved by their limit v, a solution of Problem 12.7.

The argument follows the lines of the proof of Theorem 6.1 in [10] (compare [9]). First we take $\Theta = v_n'(t)$ in (12.19) which gives us

$$\begin{aligned}\|v_n'(t)\|_H^2 + \eta\frac{1}{2}\frac{d}{dt}\|v_n(t)\|^2 + b(v_n(t), v_n(t), v_n'(t)) \\ + \frac{d}{dt}\int_{\Gamma_C} j_n(v_{n\tau}(x,t))\, d\Gamma = (f, v_n'(t)).\end{aligned}$$

Estimating the convective term by using the Ladyzhenskaya inequality (12.31) we get

$$\begin{aligned}\|v_n'(t)\|_H^2 + \eta\frac{d}{dt}\|v_n(t)\|^2 + 2\frac{d}{dt}\int_{\Gamma_C} j_n(v_{n\tau}(x,t))\, d\Gamma \\ \le \|f\|_H^2 + C\,\|v_n(t)\|_H^{\frac{1}{2}}\|v_n(t)\|^{\frac{3}{2}}\|v_n'(t)\|_H^{\frac{1}{2}}\|v_n'(t)\|^{\frac{1}{2}},\end{aligned} \tag{12.44}$$

where $C > 0$ is a constant.

Now a straightforward and technical calculation that uses the Fatou lemma, the Lebourg mean value theorem and the conditions (ii) and (iii) in Lemma 12.8 shows that, for all $r_1, r_2 \in \mathbb{R}$, we have

$$(j_n'(r_2) - j_n'(r_1))(r_2 - r_1) \ge -\mu|r_2 - r_2|^2,$$

where the constant μ is the same as in (iii) of Lemma 12.8. Hence, since v_n is a continuously differentiable function of time variable, we have for all $(x,t) \in \Omega \times (0,T)$,

$$\left(\frac{d}{dt}j_n'(v_{n\tau}(x,t))\right)v_{n\tau}'(x,t) \ge -\mu|v_{n\tau}'(x,t)|^2. \tag{12.45}$$

We differentiate both sides of (12.19) with respect to time and take $\Theta = v_n'(t)$, which gives us

$$\frac{1}{2}\frac{d}{dt}\|v_n'(t)\|_H^2 + \eta\|v_n'(t)\|^2 + \int_{\Gamma_C}\left(\frac{d}{dt}j_n'(v_{n\tau}(x,t))\right)v_{n\tau}'(x,t)\,d\Gamma + b(v_n'(t),v_n(t),v_n'(t)) + b(v_n(t),v_n'(t),v_n'(t)) = 0.$$

Using (12.45) and the fact that $b(v_n(t),v_n'(t),v_n'(t)) = 0$ we get

$$\frac{1}{2}\frac{d}{dt}\|v_n'(t)\|_H^2 + \eta\|v_n'(t)\|^2 - \mu\|v_{n\tau}'(t)\|_{L^2(\Gamma_C)}^2 + b(v_n'(t),v_n(t),v_n'(t)) \le 0.$$

Proceeding exactly as in the proof of (12.32) we get, for an arbitrary $\varepsilon > 0$ and a constant $C(\varepsilon) > 0$,

$$\frac{1}{2}\frac{d}{dt}\|v_n'(t)\|_H^2 + \eta\|v_n'(t)\|^2 \le C(\varepsilon)\|v_n(t)\|^2\|v_n'(t)\|_H^2 + \varepsilon\|v_n'(t)\|^2 + C(\varepsilon)\|v_n'(t)\|_H^2,$$

whence

$$\frac{d}{dt}\|v_n'(t)\|_H^2 + \eta\|v_n'(t)\|^2 \le C\|v_n'(t)\|_H^2(\|v_n(t)\|^2 + 1).$$

Multiplying this inequality by t^2 we get

$$\frac{d}{dt}(t^2\|v_n'(t)\|_H^2) + \eta t^2\|v_n'(t)\|^2 \le Ct^2\|v_n'(t)\|_H^2(\|v_n(t)\|^2 + 1) + 2t\|v_n'(t)\|_H^2.$$

We add this inequality to inequality (12.44) multiplied by $2t$ to get, after simple calculations that use the fact that $\int_{\Gamma_C} j_n(v_{n\tau}(x,t))\,d\Gamma \le C(1 + \|v_n(t)\|^2)$,

$$\frac{d}{dt}\left(t^2\|v_n'(t)\|_H^2 + 2t\|v_n'(t)\|^2 + 4t\int_{\Gamma_C} j_n(v_{n\tau}(x,t))\,d\Gamma\right) + \eta t^2\|v_n'(t)\|^2$$
$$\le C_1\|v_n(t)\|^2(t\|v_n'(t)\|_H)^2 + C_2(t\|v_n'(t)\|_H)^2 + C_3 + C_4\|v_n(t)\|^2$$
$$+ C_5\|v_n(t)\|_H^{\frac{1}{2}}\|v_n(t)\|^{\frac{3}{2}}(t\|v_n'(t)\|_H)^{\frac{1}{2}}(t\|v_n'(t)\|)^{\frac{1}{2}},$$

with $C_i > 0$ for $i = 1,\ldots,5$ independent on n and initial data. Denoting $y_n(t) = t^2\|v_n'(t)\|_H^2 + 2t\|v_n'(t)\|^2 + 4t\int_{\Gamma_C} j_n(v_{n\tau}(x,t))\,d\Gamma$, using the bound of Lemma 12.14 for $\|v_n(t)\|_H$, Young inequality, and the fact that j_n assumes nonnegative values, we get

$$\frac{d}{dt}y_n(t) + \frac{\eta}{2}t^2\|v_n'(t)\|^2 \leq \left(C_6\|v_n(t)\|^2 + C_2\right) y_n(t) + C_3 + C_7\|v_n(t)\|^2,$$

where $C_6, C_7 > 0$. Finally, an application of the Gronwall lemma as well as the bound in (12.16) of Lemma 12.10 for $\int_0^T \|v_n(s)\|^2\, ds$ yields the uniform boundedness of v_n' in $L^\infty(\eta, T; H) \cap L^2(\eta, T; V)$ for all intervals $[\eta, T], 0 < \eta < T$. This completes the proof of $(A10)$ and shows the existence of an exponential attractor for the semigroup $\{S(t)\}_{t\geq 0}$ associated with Problem 12.7. □

12.4 Global Attractor for Dynamic Antiplane Contact Problem

In this section we deal with the antiplane contact problem in dynamic Kelvin–Voigt viscoelasticity formulated in [13].

12.4.1 *Problem Formulation*

The classical formulation of the problem under consideration will be the following

$$\begin{aligned}
&u''(x,t) - \Delta u'(x,t) - \Delta u(x,t) = f_0(x) \text{ on } \Omega \times \mathbb{R}^+,\\
&u(x,0) = u_0(x) \text{ on } \Omega,\\
&u'(x,0) = u_1(x) \text{ on } \Omega,\\
&u(x,t) = 0 \text{ on } \Gamma_D \times \mathbb{R}^+,\\
&\frac{\partial u(x,t)}{\partial \nu} + \frac{\partial u'(x,t)}{\partial \nu} = f_2(x) \text{ on } \Gamma_I \times \mathbb{R}^+,\\
&-\frac{\partial u(x,t)}{\partial \nu} - \frac{\partial u'(x,t)}{\partial \nu} \in \partial j(u'(x,t)) \text{ on } \Gamma_C \times \mathbb{R}^+.
\end{aligned}$$

The mechanical interpretation of the equations and the boundary conditions present in the problem can be found in [13].

Define $V = \{v \in H^1(\Omega) \mid v = 0 \text{ on } \Gamma_D\}$, $H = L^2(\Omega)$, $X = V \times H$, $U = L^2(\Gamma_C)$ and $\gamma_C : V \to U$ as the trace operator. We will denote $\|\gamma_C\| := \|\gamma_C\|_{\mathcal{L}(V;U)}$. The norm in V will be denoted by $\|\cdot\|$, while all other norms will be denoted by corresponding subscripts. Note that, for $v \in V$, we have the Poincaré inequality $\lambda_1\|v\|_H^2 \leq \|v\|^2$, and the inequality

$$\|v\|_U^2 \leq \varepsilon\|v\|^2 + c(\varepsilon)\|v\|_H^2, \tag{12.46}$$

with an arbitrary $\varepsilon > 0$ and a constant $c(\varepsilon) > 0$. The last inequality is a consequence of the Ehrling lemma and the fact that if we define $Z = V \cap H^{1-\delta}(\Omega)$ for $\delta \in \left(0, \frac{1}{2}\right)$ endowed with $H^{1-\delta}$ topology, then the trace $\gamma_C^Z : Z \to U$ is continuous and $V \subset Z \subset H$ is an evolution triple with compact embedding $V \subset Z$. The duality pairing between V and V^* will be denoted by $\langle \cdot, \cdot \rangle$, while the scalar product in H by $(\cdot, \cdot)$. Moreover, we denote $\mathcal{V}(\mathbb{R}^+) = L^2_{loc}(\mathbb{R}^+; V)$, $\mathcal{V}^*(\mathbb{R}^+) = L^2_{loc}(\mathbb{R}^+; V^*)$, and

$$\mathcal{W}(\mathbb{R}^+) = \{u \in \mathcal{V}(\mathbb{R}^+) \mid u' \in \mathcal{V}^*(\mathbb{R}^+)\}.$$

We define the operator $A : V \to V^*$ as

$$\langle Av, w \rangle = \int_\Omega \nabla v \cdot \nabla w \, dx \quad \text{for } v, w \in V.$$

Finally, we assume that $f_0 \in L^2(\Omega)$ and $f_2 \in L^2(\Gamma_N)$, which makes possible to define the functional $f \in V^*$ as

$$\langle f, v \rangle = \int_\Omega f_0 v \, dx + \int_{\Gamma_L} f_2 v \, d\Gamma \quad \text{for } v \in V.$$

Weak form of the above problem will be the following

Problem 12.17. Find $u \in \mathcal{V}(\mathbb{R}^+)$ with $u' \in \mathcal{W}(\mathbb{R}^+)$ such that for all $v \in V$ and a.e. $t \in \mathbb{R}^+$,

$$\langle u''(t) + Au'(t) + Au(t), v \rangle + (\eta(t), \gamma_C v)_U = \langle f, v \rangle, \tag{12.47}$$

with $\eta(t) \in S^2_{\partial j(\gamma_C u'(t))}$ for a.e. $t \in \mathbb{R}^+$ and $u'(0) = u_1 \in H$, $u(0) = u_0 \in V$.

We will need the following assumptions on the potential $j : \mathbb{R} \to \mathbb{R}$

$H(j)$ (i) $j : \mathbb{R} \to \mathbb{R}$ is locally Lipschitz,
(ii) ∂j satisfies the growth condition $|\xi| \le c_1 + c_2|s|$ for all $\xi \in \partial j(s)$ for all $s \in \mathbb{R}$ with $c_1, c_2 > 0$,
(iii) ∂j satisfies the dissipativity condition $\inf_{\xi \in \partial j(s)} \xi s \ge d_1 - d_2|s|^2$ for all $s \in \mathbb{R}$ with $d_1 \in \mathbb{R}$ and $d_2 \ge 0$ is such that $d_2 \|\gamma_C\|^2 < 1$,
(iv) $m(\partial j) > -\infty$, where $m(\partial j)$ is the one sided Lipschitz constant defined by

$$m(\partial j) = \inf_{\substack{r,s \in \mathbb{R}, r \neq s \\ \xi \in \partial j(r), \eta \in \partial j(s)}} \frac{(\xi - \eta) \cdot (r - s)}{|r - s|^2}.$$

In the following, we will show that $H(j)(i)$–(ii) guarantee the solution existence, while, adding the assumption $H(j)(iv)$, we obtain uniqueness. In fact the existence of weak solution for Problem 12.17 has been proved in [13]. We sketch the proof here, however, since we use different technique than in [13] and, likewise, our assumptions on j are different. Moreover, we prove that all assumptions $H(j)(i)$–(iv) guarantee the existence of global attractor for associated semiflow. We remark that global attractor existence can be proved without the condition $H(j)(iv)$ and hence without solution uniqueness, using the method of *multivalued semiflows* (see [19,20]).

12.4.2 Existence and Uniqueness of Solutions

Lemma 12.18. *The operator $A : V \to V^*$ is bilinear, symmetric, continuous and coercive, with $\langle Av, w\rangle \le \|v\|\|w\|$ for $v, w \in V$ and $\langle Av, v\rangle = \|v\|^2$ for $v \in V$.*

For $(u, v) \in X$ we define the energy

$$E(u, v) = \frac{1}{2}\left(\|u\|^2 + \|v\|_H^2\right)$$

and the auxiliary energy

$$E_\varepsilon(u, v) = E(u, v) + \varepsilon(u, v)_H,$$

where $\varepsilon > 0$ is a given parameter. Note that, for $(u, v) \in X$,

$$|(u, v)_H| \le \frac{1}{\sqrt{\lambda_1}} E(u, v)$$

and hence we have, for all $(u, v) \in X$,

$$\left(1 - \frac{\varepsilon}{\sqrt{\lambda_1}}\right) E(u, v) \le E_\varepsilon(u, v) \le \left(1 + \frac{\varepsilon}{\sqrt{\lambda_1}}\right) E(u, v). \tag{12.48}$$

Theorem 12.19. *Under assumptions $H(j)(i)$–(ii), Problem 12.17 has at least one solution for any $u_0 \in V$ and $u_1 \in H$.*

Proof. The proof proceeds by the Galerkin method and mollification of the non-smooth term, similarly to that of Theorem 12.11. We outline only main points of the proof here. We pick a family of mollifier kernels $\{\varrho_n\}$ and define j_n by (12.18). We furthermore define $V_n = \text{span}\{z_1, \ldots, z_n\}$ where $\{z_i\}$ are orthonormal in H and orthogonal in V eigenfunctions of the operator A with the Dirichlet boundary conditions given in the definition of the space V. Then $\{V_n\}_{n=1}^\infty$ approximate V from inside, i.e. $\overline{\bigcup_{n=1}^\infty V_n} = V$. We define the orthogonal projection $P_n : V^* \to V_n$ by $P_n w = \sum_{i=1}^n \langle w, z_i\rangle z_i$ for $w \in V^*$.

We can now define the regularized Galerkin problems for $n \in \mathbb{N}$ as follows.

Find $u_n \in C^1(\mathbb{R}^+; V_n)$ such that u_n' is differentiable for a.e. $t \in \mathbb{R}^+$ and

$$\langle u_n''(t) + Au_n'(t) + Au_n(t) - f, \Theta\rangle + (j_n'(u_n'(\cdot,t)), \gamma_C\Theta)_U = 0, \quad (12.49)$$

$$u_n(0) = P_n u_0, \ u_n'(0) = P_n u_1 \quad (12.50)$$

for a.e. $t \in \mathbb{R}^+$ and for all $\Theta \in V_n$. Note that, due to the fact that j_n' are smooth, the solution of (12.49)–(12.50), if it exists, is a twice continuously differentiable function of time variable, with values in V_n. Taking $u_n'(t)$ in place of Θ in (12.49) we obtain

$$\frac{1}{2}\frac{d}{dt}\left(\|u_n'(t)\|_H^2 + \|u_n(t)\|^2\right) + \|u_n'(t)\|^2$$
$$\leq \varepsilon\|u_n'(t)\|^2 + \frac{1}{4\varepsilon}\|f\|_{V^*}^2 + \int_{\Gamma_C} |j_n'(u_n'(x,t))||u_n'(x,t)|\, d\Gamma. \quad (12.51)$$

Since, by $H(j)(ii)$, $|j_n'(s)| \leq \tilde{c}_1 + \tilde{c}_2|s|$ for all $s \in \mathbb{R}$ with $\tilde{c}_1, \tilde{c}_2 > 0$ independent of n, from (12.46) we get, for an arbitrary $\varepsilon > 0$ and some $C(\varepsilon), C_1, C_2 > 0$,

$$\int_{\Gamma_C} |j_n'(u_n'(x,t))||u_n'(x,t)|\, d\Gamma \leq C_1 + C_2\|\gamma_C u_n'(t)\|_U^2$$
$$\leq C_1 + \varepsilon\|u_n'(t)\|^2 + C(\varepsilon)\|u_n'(t)\|_H^2. \quad (12.52)$$

From (12.51) and (12.52) with $\varepsilon = \frac{1}{4}$ and from the Gronwall lemma we get, for a given $T > 0$,

$$\|u_n\|_{L^\infty(0,T;V)} \leq \text{const}, \quad (12.53)$$
$$\|u_n'\|_{L^2(0,T;V)} \leq \text{const}, \quad (12.54)$$
$$\|u_n'\|_{L^\infty(0,T;H)} \leq \text{const}, \quad (12.55)$$

where the constants depend on T and the initial conditions but are independent on n. Now, if $\gamma_C^* : U^* \to V^*$ is the mapping adjoint to γ_C, we have that (12.49) is equivalent to the following equation in V^*,

$$u_n''(t) + Au_n'(t) + Au_n(t) + P_n\gamma_C^* j_n'(\gamma_C u_n'(t)(\cdot)) = P_n f,$$

whence

$$\|u_n''\|_{L^2(0,T;V^*)} \leq \text{const}. \quad (12.56)$$

The existence of the solution to the Galerkin problem (12.49)–(12.50) follows from the Carathéodory theorem and estimates (12.53)–(12.56).

Since the bounds (12.53)–(12.56) are independent of n then, for a subsequence, we get

$$u_n \to u \;\text{ weakly} - * \;\text{ in } L^\infty(0,T;V) \text{ and weakly in } L^2(0,T;V), \tag{12.57}$$

$$u_n' \to u' \;\text{ weakly} - * \;\text{ in } L^\infty(0,T;H) \text{ and weakly in } L^2(0,T;V), \tag{12.58}$$

$$u_n'' \to u'' \;\text{ weakly }\; \text{in } L^2(0,T;V^*), \tag{12.59}$$

$$\gamma_C u_n' \to \gamma_C u' \;\text{ strongly }\; \text{in } L^2(0,T;U), \tag{12.60}$$

$$u_n' \to u' \;\text{ strongly }\; \text{in } L^2(0,T;H), \tag{12.61}$$

where (12.60) is a consequence of the Lions–Aubin lemma used for the spaces $V \subset Z \subset V^*$ and (12.61) is a consequence of the Lions–Aubin lemma used for the spaces $V \subset H \subset V^*$. Note that in (12.60) we use the symbol γ_C for the Nemytskii trace operator $\gamma_C : L^2(0,T;V) \to L^2(0,T;U)$.

From (12.54) and the growth condition on j_n' it follows that

$$\|j_n'(\gamma_C u_n')\|_{L^2(0,T;U))} \le \text{const.}$$

Hence, perhaps for another subsequence, we have

$$j_n'(\gamma_C u_n') \to \eta \;\text{ weakly in }\; L^2(0,T;U).$$

In a standard way we can pass to the limit in (12.49) and obtain that u and η satisfy (12.47) for a.e. $t \in (0,T)$. The proof that u satisfies the initial conditions $u(0) = u_0$ and $u'(0) = u_1$ is standard and it follows from the fact that from (12.57)–(12.59) we must have $u_n(0) \to u(0)$ weakly in V and $u_n'(0) \to u'(0)$ weakly in H. The proof that $\eta(t) \in S^2_{\partial j(\gamma_C u'(t))}$ is analogous to the argument in the proof of Theorem 3.1 in [8]. The solution can be extended from $[0,T]$ to the whole positive semiaxis by taking the values of u, u' at T as initial conditions and concatenating the solutions obtained in this way. □

Theorem 12.20. *Under assumptions $H(j)(i)$–(ii), (iv), Problem 12.17 has exactly one solution for any $(u_0,u_1) \in X$. Moreover the map $(u(t_1),u'(t_1)) \to (u(t_2),u'(t_2))$ is Lipschitz continuous in X for all $t_2 > t_1 \ge 0$.*

Proof. The existence part follows from Theorem 12.19. For the proof of uniqueness and Lipschitz continuity, assume that u_1, u_2 solve Problem 12.17. Denote $w(t) = u_1(t) - u_2(t)$. Let the functions η such that (12.47) for u_1 and u_2 holds, be denoted by η_1, η_2, respectively. Subtracting (12.47) written for u_1 and u_2 we obtain for $v \in V$ and a.e. $t \in \mathbb{R}^+$

$$\langle w''(t) + Aw'(t) + Aw(t), v\rangle + (\eta_1(t) - \eta_2(t), \gamma_C v)_U = 0.$$

Taking $v = w'(t)$ and using Lemma 12.18 as well as $H(j)(iv)$ we get, for a.e. $t \in \mathbb{R}^+$,

$$\frac{1}{2}\frac{d}{dt}\|w'(t)\|_H^2 + \|w'(t)\|^2 + \frac{1}{2}\frac{d}{dt}\|w(t)\|^2 + m(\partial j)\|\gamma_C w'(t)\|_U^2 \le 0.$$

Using (12.46) and denoting $M = |m(\partial j)|$ we obtain

$$\frac{1}{2}\frac{d}{dt}\left(\|w'(t)\|_H^2 + \|w(t)\|^2\right) + (1 - M\varepsilon)\|w'(t)\|^2 \le Mc(\varepsilon)\|w'(t)\|_H^2.$$

It suffices to take $\varepsilon = M^{-1}$ (if $M = |m(\partial j)| = 0$ then any choice of ε is fine) to get, for some $C_1 > 0$,

$$\frac{d}{dt}E(w(t), w'(t)) \le C_1 E(w(t), w'(t)).$$

As $E(w(t), w'(t))$ is a continuous function of time, we can apply the Gronwall lemma to obtain, for $t_2 > t_1 \ge 0$,

$$E(w(t_2), w'(t_2)) \le e^{C_1(t_2 - t_1)} E(w(t_1), w'(t_1)),$$

whence the Lipschitz continuity follows. In particular if $t_1 = 0$, $u_1(0) = u_2(0)$ and $u_1'(0) = u_2'(0)$ then for all $t > 0$ we have $u_1(t) = u_2(t)$ and $u_1'(t) = u_2'(t)$, and the proof is complete. □

Observe that from Theorem 12.20 it follows that we can define a family $\{S(t)\}_{t\ge 0}$ of Lipschitz continuous mappings $S(t) : X \to X$ by $S(t)(u_0, u_1) = (u(t), u'(t))$, where u solves Problem 12.17 with $u(0) = u_0$ and $u'(0) = u_1$. As $S(t_1 + t_2) = S(t_1)S(t_2)$ for all $t_1, t_1 \in \mathbb{R}^+$, $\{S(t)\}_{t\ge 0}$ is a semigroup.

12.4.3 Existence of Global Attractor

The existence of global attractor, being the compact set in X, that is invariant and attracts all bounded sets in X follows from the results of [19, 20] (see [19, Theorem 2] and [20, Theorem 2.9, p. 88]). We repeat here some steps of the proof for the completeness of exposition.

Lemma 12.21. *Assume $H(j)(i)$–(iv). There exist constants $D_1, D_2, D_3 > 0$ dependent only on the problem data such that if u solves Problem 12.17 then for all $t \in \mathbb{R}^+$ we have*

$$E(u(t), u'(t)) \le E(u(0), u'(0))D_1 e^{-D_2 t} + D_3$$

Proof. The proof follows the lines of the proof of Lemma 2.15 in [20]. Let u solve Problem 12.17. We have, for some $\eta(t) \in S^2_{\partial j(\gamma_C u'(t))}$ and for a.e. $t \in \mathbb{R}^+$,

$$
\begin{aligned}
\frac{dE_\varepsilon(u(t),u'(t))}{dt} &= \langle u''(t),u'(t)\rangle + \langle Au(t),u'(t)\rangle + \varepsilon\langle u''(t),u(t)\rangle + \varepsilon\|u'(t)\|_H^2 \\
&= \langle f,u'(t)\rangle - \langle Au'(t),u'(t)\rangle - (\eta(t),\gamma_C u'(t))_U \\
&\quad + \varepsilon\langle f,u(t)\rangle - \varepsilon\langle Au'(t),u(t)\rangle - \varepsilon(\eta(t),\gamma_C u(t))_U \\
&\quad - \varepsilon\langle Au(t),u(t)\rangle + \varepsilon\|u'(t)\|_H^2 \\
&\le \|f\|_{V^*}\|u'(t)\| - \|u'(t)\|^2 + d_2\|\gamma_C\|^2\|u'(t)\|^2 - d_1 \\
&\quad + \varepsilon\|f\|_{V^*}\|u(t)\| - \varepsilon\|u(t)\|^2 + \varepsilon\|u(t)\|\|u'(t)\| \\
&\quad + \varepsilon\|\gamma_C\|\|u(t)\|(c_1\sqrt{\mu(\Gamma_C)} + c_2\|\gamma_C\|\|u'(t)\|) + \varepsilon\|u'(t)\|_H^2.
\end{aligned}
$$

Using the relations $xy \le \frac{x^2}{2} + \frac{y^2}{2}$ and $xy \le \varepsilon_1 x^2 + \frac{y^2}{4\varepsilon_1}$ valid for nonnegative x, y with any constant $\varepsilon_1 > 0$ we obtain

$$
\begin{aligned}
\frac{dE_\varepsilon(u(t),u'(t))}{dt} \le &- \left(1 - d_2\|\gamma_C\|^2 - 3\varepsilon_1\right)\|u'(t)\|^2 + \varepsilon\|u'(t)\|_H^2 \\
&- \varepsilon\left(1 - \varepsilon - \frac{\varepsilon}{4\varepsilon_1}(1 + c_2^2\|\gamma_C\|^4)\right)\|u(t)\|^2 \\
&+ \frac{\|f\|_{V^*}^2}{4\varepsilon_1} + \frac{\|f\|_{V^*}^2}{2} - d_1 + \frac{\|\gamma_C\|^2 c_1^2\mu(\Gamma_C)}{2}
\end{aligned}
$$

valid for any $\varepsilon_1 > 0$. It is enough to take $\varepsilon_1 = \frac{1-d_2\|\gamma_C\|^2}{6}$ to obtain

$$
\frac{dE_\varepsilon(u(t),u'(t))}{dt} \le -\left(C_1 - \varepsilon\right)\|u'(t)\|_H^2 - \varepsilon\left(1 - \varepsilon C_2\right)\|u(t)\|^2 + C_3,
$$

where $C_1, C_2, C_3 > 0$ are constants dependent only on the problem data. We choose ε small enough to get

$$
\frac{dE_\varepsilon(u(t),u'(t))}{dt} \le -C_4 E(u(t),u'(t)) + C_3.
$$

with a constant $C_4 > 0$ dependent only on the problem data. We can use (12.48) to obtain

$$
\frac{dE_\varepsilon(u(t),u'(t))}{dt} \le -C_5 E_\varepsilon(u(t),u'(t)) + C_3,
$$

with a constant $C_5 > 0$, the latter inequality being valid for a.e. $t \in \mathbb{R}^+$. From the Gronwall lemma we get

$$E_\varepsilon(u(t), u'(t)) \le E_\varepsilon(u(0), u'(0))e^{-C_5 t} + \frac{C_3}{C_5}.$$

Hence, from (12.48) we obtain the assertion. □

Lemma 12.22. *Assume $H(j)(i)$–(iv). The semigroup $\{S(t)\}_{t\ge 0}$ is asymptotically compact, i.e. if $B \subset X$ is bounded and $t_n \to \infty$, then any sequence $\{(u_n, v_n)\}_n \subset X$ such that $(u_n, v_n) \in S(t_n)B$ is relatively compact in the strong topology of X.*

We do not provide the proof of the lemma, since, on one hand it is quite involved and, on the other hand, it exactly follows the lines of the proof of Theorem 2.9 in [20] (also compare the proof of Theorem 2 in [19]).

Theorem 12.2 together with Theorem 12.20, and Lemmata 12.21, 12.22 yield the following theorem.

Theorem 12.23. *Assume $H(j)(i)$–(iv). Then the semigroup $\{S(t)\}_{t\ge 0}$ associated with Problem 12.17 has a global attractor.*

12.4.4 Infinite Dimensionality of Stationary Set

Our aim in this subsection is to analyze the global attractor of a special case of the problem considered in the previous subsection and prove that its fractal dimension is infinite.

We present the analysis for a very simple case: $\Omega = (0,1)^2$, $\Gamma_D = (0,1)\times\{1\}$, $\Gamma_C = (0,1)\times\{0\}$ and $\Gamma_L = \{0,1\}\times(0,1)$. In addition, we set $j(s) = |s|$ for $s \in \mathbb{R}$. Then j satisfies $H(j)(i)$–(iv). Moreover j is convex and $\partial j(0) = [-1,1]$. This leads to the Tresca friction law (see [13, Example 5.1]).

One of the ways to prove that the global attractor has infinite fractal dimension is to show that the stationary trajectories of the system, which are all contained in the global attractor, constitute a set of infinite dimensionality.

Thus, let us look at stationary trajectories. They must satisfy the condition $u'(t) = 0_H$, $u(t) = u = const$. Define the set $K = \{\xi \in U \mid |\xi(x)| \le 1 \text{ a.e. } x \in \Gamma_C\}$. If there exists $\eta \in K$ such that for all $v \in V$ we have

$$(\nabla u, \nabla v)_{L^2(\Omega;\mathbb{R}^2)} + (\eta, v)_U = \langle f, v\rangle, \tag{12.62}$$

then the solution $u \in V$ defines a stationary trajectory. We denote the set of all such stationary points by $\mathcal{S} \subset V$.

Now define an $N-1$ dimensional simplex as

$$C^{N-1} = \left\{\xi \in \mathbb{R}^N \mid \xi_k \ge 0, \sum_{k=1}^{N} \xi_k = 1\right\}$$

and $C_{\mathbb{N}}^{N-1} = C^{N-1} \cap \{\xi \in \mathbb{R}^N \mid \forall k \in \{1,\dots,N\}\ \xi_k = \frac{j}{N} \text{ for some } j \in \{0,\dots,N\}\}$. If $\xi \in C^{N-1}$, then the function $\eta_\xi : \Gamma_C \to \mathbb{R}$ defined as $\eta_\xi(x_1,0) = \sum_{k=1}^N \xi_k \sin(k\pi x_1)$ belongs to K. Moreover for $\xi \in C^{N-1}$ we define $v_\xi : \Omega \to \mathbb{R}$ as $v_\xi(x_1,x_2) = (1-x_2)\eta_\xi(x_1,0)$. Note that $v_\xi \in V$ and $\gamma_C v_\xi = \eta_\xi$.

By the Lax–Milgram lemma, for each $\xi \in C^{N-1}$ we can find uniquely defined $u_\xi \in \mathcal{S}$, the solution of (12.62) with $\eta_\xi \in K$ in place of η. We subtract (12.62) for ξ^1 and ξ^2 and test it by $v_{\xi^1} - v_{\xi^2}$. We get

$$\begin{aligned}\|\eta_{\xi^1} - \eta_{\xi^2}\|_U^2 &= (\nabla(u_{\xi^2} - u_{\xi^1}), \nabla(v_{\xi^1} - v_{\xi^2}))_{L^2(\Omega;\mathbb{R}^2)} \\ &\le \|u_{\xi^1} - u_{\xi^2}\| \|\nabla(v_{\xi^1} - v_{\xi^2})\|_{L^2(\Omega;\mathbb{R}^2)}. \end{aligned} \tag{12.63}$$

As it is easy to compute we have $\|\eta_{\xi^1} - \eta_{\xi^2}\|_{L_2(\Gamma_C)}^2 = \frac{1}{2}|\xi^1 - \xi^2|^2$ (where $|\cdot|$ is Euclidean norm in $\mathbb{R}^N$), and

$$\nabla(v_{\xi^1} - v_{\xi^2}) = \begin{pmatrix} (1-x_2)\sum_{k=1}^N (\xi_k^1 - \xi_k^2)\pi k \cos(\pi k x_1) \\ \sum_{k=1}^N (\xi_k^2 - \xi_k^1)\sin(k\pi x_1) \end{pmatrix}.$$

Hence

$$\|\nabla(v_{\xi^1} - v_{\xi^2})\|_{L^2(\Omega;\mathbb{R}^2)}^2 = \frac{1}{2}\sum_{k=1}^N (\xi_k^1 - \xi_k^2)^2 \left(1 + \frac{k^2\pi^2}{3}\right) \le \frac{3 + N^2\pi^2}{6}|\xi^1 - \xi^2|^2.$$

For $N \ge 5$ we have $3 + N^2\pi^2 \le 10N^2$. In consequence, in this case, from (12.63) we obtain

$$\|u_{\xi^1} - u_{\xi^2}\|_V \ge \frac{1}{3N}|\xi^1 - \xi^2|.$$

Now assume that $\xi^1, \xi^2 \in C_{\mathbb{N}}^{N-1}$ and $\xi^1 \ne \xi^2$. Then ξ^1 and ξ^2 must differ on at least one coordinate by at least $\frac{1}{N}$. Hence $|\xi^1 - \xi^2| \ge \frac{1}{N}$. In consequence, if $\xi^1, \xi^2 \in C_{\mathbb{N}}^{N-1}$ with $\xi^1 \ne \xi^2$, then

$$\|u_{\xi^1} - u_{\xi^2}\|_V \ge \frac{1}{3N^2} > \frac{1}{4N^2}.$$

We observe that the cardinality of $C_{\mathbb{N}}^{N-1}$ is equal to the number of partitions of N into the sum of N components, being the natural numbers (with zero). This number of partitions is equal to $\binom{2N-1}{N}$. Note that $\binom{2N-1}{N} = \frac{(N+1)(N+2)\dots(2N-1)}{1\cdot 2\cdot\dots(N-1)} > 2^{N-1}$.

Hence we have at least 2^{N-1} distinct points $u_\xi \in \mathcal{S}$ such that the distance between each pair of them is greater then $\frac{1}{4N^2}$.

Hence, denoting by $N_\varepsilon^V(\mathcal{S})$ the minimal number of balls in V of radius ε that cover $\mathcal{S}$ we observe that $N^V_{1/(8N^2)}(\mathcal{S}) \geq 2^{N-1}$ and we have

$$d_f^V(\mathcal{S}) = \limsup_{\varepsilon\to 0} \frac{\ln N_\varepsilon^V(\mathcal{S})}{\ln\left(\frac{1}{\varepsilon}\right)} \geq \limsup_{N\to\infty} \frac{\ln N^V_{1/(8N^2)}(\mathcal{S})}{\ln(8N^2)}$$
$$\geq \limsup_{N\to\infty} \frac{\ln(2^{N-1})}{\ln(8N^2)} = \infty.$$

Denoting the global attractor for our problem by $\mathcal{A} \subset X$ being the compact set in the strong topology of $X = V \times H$ it must hold that $\{0_H\} \times \mathcal{S} \subset \mathcal{A}$. Hence, since $d_f^V(\mathcal{S}) = \infty$ it must be that $d_f^X(\mathcal{A}) = \infty$. The proof is complete. In particular the considered problem does not have exponential attractors.

To fully understand the dynamics of considered problem one needs to answer the following open problem. It is easy to verify that the problem formulated in this section has a Lyapunov function. Does the global attractor consist of the stationary points only or does it also contain complete trajectories that connect different levels of Lyapunov function?

Our analysis provided in this chapter shows that in a way similar contact problems governed by the equations of the first and second order with respect to the time variable, respectively, may exhibit very different dynamical characteristics.

Acknowledgements The research of first author was supported by the Marie Curie International Research Staff Exchange Scheme Fellowship within the 7th European Community Framework Programme under Grant Agreement No. 295118, the National Science Center of Poland under grant no. N N201 604640, the International Project co-financed by the Ministry of Science and Higher Education of Republic of Poland under grant no. W111/7.PR/2012, the National Science Center of Poland under Maestro Advanced Project no. DEC-2012/06/A/ST1/00262, and the project Polonium "Mathematical and Numerical Analysis for Contact Problems with Friction" 2014/15 between the Jagiellonian University in Krakow and Université de Perpignan Via Domitia.

References

1. Aubin, J.P., Frankowska, H.: Set-Valued Analysis. Birkhäuser, Boston (1990)
2. Bolte, J., Daniliidis, A., Ley, O., Mazet, L.: Characterizations of Łojasiewicz inequalities: subgradient flows, talweg, convexity. Trans. Am. Math. Soc. **362**, 3319–3363 (2010)
3. Caraballo, T., Marin-Rubio, P., Robinson, J.C.: A comparison between two theories for multi-valued semiflows and their asymptotic behaviour. Set-Valued Anal. **11**, 297–322 (2003)
4. Denkowski, Z., Migórski, S., Papageorgiou, N.S.: An Introduction to Nonlinear Analysis: Theory. Kluwer Academic/Plenum Publishers, Boston/Dordrecht/London/New York (2003)
5. Evans, L.C.: Partial Differential Equations, 2nd edn. American Mathematical Society, Providence (2010)
6. Ionescu, I.R., Nguyen, Q.L.: Dynamic contact problems with slip dependent friction in viscoelasticity. Int. J. Appl. Math. Comput. Sci. **12**, 71–80 (2002)
7. Kalita, P.: Decay of energy for second-order boundary hemivariational inequalities with coercive damping. Nonlinear Anal. Theory Methods Appl. **74**, 1164–1181 (2011)

8. Kalita, P., Łukaszewicz, G.: Attractors for Navier–Stokes flows with multivalued and non-monotone subdifferential boundary conditions. Nonlinear Anal. Real World Appl. **19**, 75–88 (2004)
9. Ladyzhenskaya, O., Seregin, G.: On semigroups generated by initial-boundary problems describing two dimensional visco-plastic flows. In: Nonlinear Evolution Equations. American Mathematical Society Translations: Series 2, vol. 164, pp. 99–123. American Mathematical Society, Providence (1995)
10. Łukaszewicz, G.: On the existence of an exponential attractor for a planar shear flow with the Tresca friction condition. Nonlinear Anal. Real World Appl. **14**, 1585–1600 (2013)
11. Málek, J., Pražák, D.: Large time behavior via the method of l-trajectories. J. Differ. Equ. **181**, 243–279 (2002)
12. Migórski, S., Ochal, A.: Boundary hemivariational inequality of parabolic type. Nonlinear Anal. Theory Methods Appl. **57**, 579–596 (2004)
13. Migórski, S., Ochal, A., Sofonea, M.: Solvability of dynamic antiplane frictional contact problems for viscoelastic cylinders. Nonlinear Anal. Theory Methods Appl. **70**, 3738–3748 (2009)
14. Perrin, G.J., Rice, J.R., Zheng, G.: Self-healing slip pulse on a frictional surface. J. Mech. Phys. Solids **43**, 1461–1495 (1995)
15. Robinson, J.C.: Infinite-dimensional Dynamical Systems. Cambridge University Press, Cambridge (2001)
16. Robinson, J.C.: Dimensions, Embeddings, and Attractors. Cambridge University Press, Cambridge (2011)
17. Scholz, C.H.: The Mechanics of Earthquakes and Faulting. Cambridge University Press, Cambridge (1990)
18. Shillor, M., Sofonea, M., Telega, J.J.: Models and Analysis of Quasistatic Contact: Variational Methods. Springer, Berlin/Heidelberg (2004)
19. Zadoianchuk, N.V., Kasyanov, P.O.: Dynamics of solutions of a class of second-order autonomous evolution inclusions. Cybern. Syst. Anal. **48**, 414–428 (2012)
20. Zgurovsky, M.Z., Kasyanov, P.O., Kapustyan, O.V., Valero, J., Zadoianchuk, N.V.: Evolution Inclusions and Variation Inequalities for Earth Data Processing III. Long-Time Behavior of Evolution Inclusions Solutions in Earth Data Analysis. Springer, Heidelberg/New York (2012)

Chapter 13
Hemivariational Inequalities for Dynamic Elastic-Viscoplastic Contact Problems

Anna Kulig

Abstract In this chapter we consider two mathematical models which describe the contact between a body and a foundation. The contact is frictional whereas the body is deformable and the process is dynamic. In both models the constitutive law is elastic-viscoplastic and the frictional contact is modeled with subdifferential boundary conditions. For the two problems we present their classical and variational formulations. The latter has the form of a system which couples an evolutionary hemivariational inequality with an integro-differential equation. Finally, we prove the existence of unique weak solutions to both models.

Keywords Evolutionary inclusion • Viscoplastic contact • Clarke subgradient • Multifunction • Hyperbolic • Contact problem • Hemivariational inequality

AMS Classification. 34G25, 35L90, 35R70, 45P05, 74C10, 47H04, 74H20, 74H25, 74M15

13.1 Introduction

In this chapter we consider two mechanical models which describe the frictional contact between a body and a foundation. In both models the process is dynamic and the body is deformable. We model the material behaviour with a nonlinear elastic-viscoplastic constitutive law of the from

$$\begin{aligned}\boldsymbol{\sigma}(t) &= \mathcal{A}(t, \boldsymbol{\varepsilon}(\boldsymbol{u}'(t))) + \mathcal{E}(t, \boldsymbol{\varepsilon}(\boldsymbol{u}(t))) \\ &\quad + \int_0^t \mathcal{G}(s, \boldsymbol{\sigma}(s) - \mathcal{A}(s, \boldsymbol{\varepsilon}(\boldsymbol{u}'(s))), \boldsymbol{\varepsilon}(\boldsymbol{u}(s)))\, ds, \qquad (13.1)\end{aligned}$$

A. Kulig (✉)
Faculty of Mathematics and Computer Science, Jagiellonian University, ul. Łojasiewicza 6, 30348 Krakow, Poland
e-mail: anna.kulig@ii.uj.edu.pl

W. Han et al. (eds.), *Advances in Variational and Hemivariational Inequalities*, Advances in Mechanics and Mathematics 33, DOI 10.1007/978-3-319-14490-0_13

where $\boldsymbol{u}$ denotes the displacement field and $\boldsymbol{u}'$ is the velocity field. The function $\boldsymbol{\sigma}$ denotes the stress tensor whereas $\boldsymbol{\varepsilon}$ denotes the linearized strain tensor. The operator $\mathcal{A}$ describes the purely viscous properties of the material, $\mathcal{E}$ is a nonlinear elasticity operator and $\mathcal{G}$ is a nonlinear constitutive function which describes the viscoplastic behavior of the material. Here and below, for simplicity, we skip the dependence of various functions on the spatial variable $\boldsymbol{x}$. Constitutive laws of the form (13.1) have been used in the literature in order to describe the mechanical properties of various materials such as rubber, metals, pastes and rocks.

One-dimensional constitutive laws of the form (13.1) can be constructed by using rheological arguments. Indeed, for example, consider a dashpot connected in parallel with a Maxwell model. In this case an additive formula holds

$$\sigma = \sigma^V + \sigma^R, \tag{13.2}$$

where σ, σ^V and σ^R denote the total stress, the stress in the dashpot and the stress in the Maxwell model, respectively. Moreover, we have

$$\sigma^V = A\varepsilon(u') \tag{13.3}$$

and

$$(\sigma^R)' = E\varepsilon(u') - \frac{E}{\eta}\sigma^R, \tag{13.4}$$

where $E > 0$ is the Young modulus of the Maxwell material, A and η are positive viscosity coefficients and ε denotes the strain. For $t \in [0, T]$, we integrate (13.4) on $[0, t]$. We use (13.2), (13.3) and the initial conditions $\sigma^R(0) = 0$, $\varepsilon(u(0)) = 0$, and we obtain

$$\sigma(t) = A\varepsilon(u'(t)) + E\varepsilon(u(t)) - \frac{E}{\eta}\int_0^t (\sigma(s) - A\boldsymbol{\varepsilon}(u'(s)))\, ds,$$

which represents a constitutive equation of the form (13.1). More details on the one-dimensional models and the construction of rheological models can be found in [5] and [6].

Note also that when $G \equiv 0$, the constitutive law (13.1) reduces to a nonlinear viscoelastic constitutive law with short memory

$$\boldsymbol{\sigma}(t) = \mathcal{A}(t, \boldsymbol{\varepsilon}(\boldsymbol{u}'(t))) + \mathcal{E}(t, \boldsymbol{\varepsilon}(\boldsymbol{u}(t))). \tag{13.5}$$

Quasistatic contact problems for materials of the form (13.5) with time independent elasticity and viscosity operators were investigated in a large number of papers, see [6, 19] for a survey. There, both the variational analysis and the numerical approach of the problems, including the study of semi-discrete and fully discrete schemes, were provided. Existence results for the dynamic problems with viscoelastic materials of the form (13.5) can be found in [7, 9–14]. The related existence and uniqueness results for hemivariational inequalities are provided in [12, 13, 16]. Quasistatic contact problems with elastic-viscoplastic materials were treated in [1,2,6,19], for instance.

In the present work we consider two frictional contact problems with subdifferential boundary conditions of the form

$$\begin{aligned} -\sigma_\nu(t) &\in \partial j_1(t, \boldsymbol{u}(t), \boldsymbol{u}'(t), u_\nu(t)), \\ -\boldsymbol{\sigma}_\tau(t) &\in \partial j_3(t, \boldsymbol{u}(t), \boldsymbol{u}'(t), \boldsymbol{u}_\tau(t)), \end{aligned} \tag{13.6}$$

or

$$\begin{aligned} -\sigma_\nu(t) &\in \partial j_2(t, \boldsymbol{u}(t), \boldsymbol{u}'(t), u'_\nu(t)), \\ -\boldsymbol{\sigma}_\tau(t) &\in \partial j_4(t, \boldsymbol{u}(t), \boldsymbol{u}'(t), \boldsymbol{u}'_\tau(t)) \end{aligned} \tag{13.7}$$

where σ_ν and $\boldsymbol{\sigma}_\tau$, u_ν and $\boldsymbol{u}_\tau$, u'_ν and $\boldsymbol{u}'_\tau$ denote the normal and the tangential components of the stress tensor, the displacement and the velocity, respectively, and ∂j_k, $k = 1, \ldots, 4$ stands for the Clarke subdifferential of functions j_k taken with respect to their last arguments. Similar boundary conditions were studied in [7, 8]. The boundary conditions (13.6) and (13.7) are more general than those treated in [15], due to the additional dependency of the functions j_k with respect to $\boldsymbol{u}$ and $\boldsymbol{u}'$. Examples of such boundary conditions will be presented in Sect. 13.6.

The rest of the chapter is structured as follows. In Sect. 13.2 we introduce some notation and recall an abstract existence result for evolutionary inclusions of second order. The mathematical models of frictional contact are presented in Sect. 13.3, together with the hypotheses on the data. Then, in Sect. 13.4 we present variational formulations of the models whereas in Sect. 13.5 we state and prove our main existence and uniqueness result, Theorem 13.4. Finally, in Sect. 13.6 we present examples of subdifferential boundary conditions for which our main result works.

13.2 Preliminaries

In this section we recall the basic notations and a result on existence of solutions for evolutionary inclusion of second order that will be needed in following sections.

Let $(X, \|\cdot\|_X)$ be a reflexive Banach space; the symbol w-X denotes the space X endowed with the weak topology; if $U \subset X$, then we write $\|U\|_X = \sup\{\, \|x\|_X \mid x \in U\,\}$. Let $T\colon X \to 2^{X^*}$ be a multivalued operator and let $\langle\cdot,\cdot\rangle$ be the duality pairing for (X^*, X). We say that operator T is *upper semicontinuous (usc)* if set $T^-(C) = \{x \in X \mid Tx \cap C \neq \emptyset\}$ is closed in X for any closed subset $C \subset X^*$. We say it is *coercive* if there exists a function $c\colon \mathbb{R}_+ \to \mathbb{R}$ with $\lim_{r\to+\infty} c(r) = +\infty$

such that for all $x \in X$ and $x^* \in Tx$, we have $\langle x^*, x\rangle \geq c\,(\|x\|_X)\|x\|_X$. A single valued operator $A\colon X \to X^*$ is called hemicontinuous if the real-valued function $\lambda \mapsto \langle A(u+\lambda v), w\rangle$ is continuous on $[0,1]$ for all u, v, $w \in X$.

The generalized directional derivative of Clarke of a locally Lipschitz function $h\colon X \to \mathbb{R}$ at $x \in X$ in the direction $v \in X$, denoted by $h^0(x;v)$, is defined as follows (cf. [3]):

$$h^0(x;v) = \limsup_{y\to x,\lambda\downarrow 0} \frac{h(y+\lambda v) - h(y)}{\lambda}.$$

The generalized gradient of the function $h\colon X \to \mathbb{R}$ at $x \in X$, denoted by $\partial h(x)$, is a subset of a dual space X^* given by $\partial h(x) = \{\zeta \in X^* \mid h^0(x;v) \geq \langle \zeta, v\rangle_{X^*\times X}$ for all $v \in X\}$.

Let V and Z be a separable and reflexive Banach spaces with the duals V^* and Z^*, respectively. Let H denote a separable Hilbert space and we identify H with its dual. We suppose that $V \subset H \subset V^*$ and $Z \subset H \subset Z^*$ are evolution triples of spaces where all embeddings are continuous, dense and compact (see e.g. Chap. 23.4 of [20], Chap. 3.4 of [4]). Let $\|\cdot\|$ and $|\cdot|$ denote the norms in V and H, respectively, and let $\langle\cdot,\cdot\rangle$ be the duality pairing between V^* and V. We also introduce the following spaces

$$\mathcal{V} = L^2(0,T;V), \quad \mathcal{Z} = L^2(0,T;Z), \quad \hat{\mathcal{H}} = L^2(0,T;H),$$
$$\mathcal{Z}^* = L^2(0,T;Z^*), \quad \mathcal{V}^* = L^2(0,T;V^*), \quad \mathcal{W} = \{v \in \mathcal{V} \mid v' \in \mathcal{V}^*\}.$$

The linear space of second order symmetric tensors on $\mathbb{R}^d$ will be denoted by $\mathbb{S}^d$. The inner product and the corresponding norm on $\mathbb{S}^d$ is defined similarly to the inner product on $\mathbb{R}^d$, i.e.

$$\boldsymbol{u}\cdot\boldsymbol{v} = u_i v_i, \qquad \|\boldsymbol{v}\|_{\mathbb{R}^d} = (\boldsymbol{v}\cdot\boldsymbol{v})^{1/2} \qquad \text{for all } \boldsymbol{u},\boldsymbol{v} \in \mathbb{R}^d,$$
$$\boldsymbol{\sigma}:\boldsymbol{\tau} = \sigma_{ij}\tau_{ij}, \qquad \|\boldsymbol{\tau}\|_{\mathbb{S}^d} = (\boldsymbol{\tau}:\boldsymbol{\tau})^{1/2} \qquad \text{for all } \boldsymbol{\sigma},\boldsymbol{\tau} \in \mathbb{S}^d.$$

We recall now a result on the following evolutionary inclusion.

Problem $\mathcal{P}$. *Find $u \in \mathcal{V}$ such that $u' \in \mathcal{W}$ and*

$$\begin{cases} u''(t) + A(t,u'(t)) + F(t,u(t),u'(t)) \ni f(t) \ \text{ a.e. } t \in (0,T), \\ u(0) = u_0, \ \ u'(0) = u_1. \end{cases}$$

A solution to Problem $\mathcal{P}$ is understood as follows.

Definition 13.1. A function $u \in \mathcal{V}$ is a solution of Problem $\mathcal{P}$ if and only if $u' \in \mathcal{W}$ and there exists $z \in \mathcal{Z}^*$ such that

$$\begin{cases} u''(t) + A(t, u'(t)) + z(t) = f(t) \text{ a.e. } t \in (0, T), \\ z(t) \in F(t, u(t), u'(t)) \quad \text{a.e. } t \in (0, T), \\ u(0) = u_0, \quad u'(0) = u_1. \end{cases}$$

We will need the following hypotheses.

$H(A)$: The operator $A: (0, T) \times V \to V^*$ is such that

(i) $A(\cdot, v)$ is measurable on $(0, T)$ for all $v \in V$.
(ii) $A(t, \cdot)$ is hemicontinuous for a.e. $t \in (0, T)$.
(iii) $\|A(t, v)\|_{V^*} \leq a_0(t) + a_1\|v\|$ for all $v \in V$, a.e. $t \in (0, T)$ with $a_0 \in L^2(0, T)$, $a_0 \geq 0$ and $a_1 > 0$.
(iv) $\langle A(t, v), v\rangle \geq \alpha\|v\|^2$ for all $v \in V$, a.e. $t \in (0, T)$ with $\alpha > 0$.
(v) $A(t, \cdot)$ is strongly monotone for a.e. $t \in (0, T)$, i.e. there exists $m_1 > 0$ such that $\langle A(t, v) - A(t, u), v - u\rangle \geq m_1\|v - u\|^2$ for all $u, v \in V$, a.e. $t \in (0, T)$.

$H(F)$: The multifunction $F: (0, T) \times V \times V \to 2^{Z^*}$ has nonempty closed and convex values and it is such that

(i) $F(\cdot, u, v)$ is measurable on $(0, T)$ for all $u, v \in V$.
(ii) $F(t, \cdot, \cdot)$ is upper semicontinuous from $V \times V$ into w-Z^* for a.e. $t \in (0, T)$, where $V \times V$ is endowed with $(Z \times Z)$-topology.
(iii) $\|F(t, u, v)\|_{Z^*} \leq d_0(t) + d_1\|u\| + d_2\|v\|$ for all $u, v \in V$, a.e. $t \in (0, T)$ with $d_0 \in L^2(0, T)$ and $d_0, d_1, d_2 \geq 0$.
(iv) $\langle F(t, u_1, v_1) - F(t, u_2, v_2), v_1 - v_2\rangle_{Z^* \times Z} \geq -m_2\|v_1 - v_2\|^2 - m_3\|v_1 - v_2\|\|u_1 - u_2\|$ for all $u_i, v_i \in V$, $i = 1, 2$, a.e. $t \in (0, T)$ with $m_2, m_3 \geq 0$.

(H_0) : $f \in \mathcal{V}^*$, $u_0 \in V$, $u_1 \in H$.
(H_1) : $\alpha > 2\sqrt{3}c_e(d_1 T + d_2)$, where $c_e > 0$ is the embedding constant of V into Z, i.e. $\|\cdot\|_Z \leq c_e\|\cdot\|$.
(H_2) : $m_1 > m_2 + \frac{1}{\sqrt{2}} m_3 T$.

Then, we can state the following existence and uniqueness result.

Theorem 13.2. *Assume that hypotheses $H(A)$, $H(F)$, (H_0), (H_1) and (H_2) hold. Then Problem $\mathcal{P}$ has a unique solution which satisfies*

$$\|u\|_{C(0,T;V)} + \|u'\|_{\mathcal{W}} \leq C\Big(1 + \|u_0\| + |u_1| + \|f\|_{\mathcal{V}^*}\Big).$$

with a constant $C > 0$.

The proof of Theorem 13.2 can be found in [9].

13.3 Problems Statement

In this section we introduce two mathematical models of contact and present the hypotheses on the data. Let Ω be a bounded domain in $\mathbb{R}^d$, $d = 2, 3$, which is occupied, in its reference configuration, by an elastic-viscoplastic body. The boundary Γ of Ω is assumed to be Lipschitz continuous and it is divided into three mutually disjoint measurable parts Γ_D, Γ_N and Γ_C in such a way that $meas(\Gamma_D) > 0$. Let $\boldsymbol{\nu}$ denote the unit outward normal vector to Γ. We put $Q = \Omega \times (0, T)$, where $0 < T < +\infty$. Volume forces of density $\boldsymbol{f}_0$ act in Ω and surface tractions of density $\boldsymbol{f}_N$ are applied on Γ_N. The body is clamped on Γ_D and, therefore, the displacement field vanishes there. The part Γ_C is where the body may come in contact with a foundation. By $\boldsymbol{u} = (u_1, \ldots, u_d)$, $\boldsymbol{\sigma} = (\sigma_{ij})$ and $\boldsymbol{\varepsilon}(\boldsymbol{u}) = (\boldsymbol{\varepsilon}_{ij}(\boldsymbol{u}))$, $i, j = 1, \ldots, d$, we denote the displacement vector, the stress tensor and linearized strain tensor, respectively. We assume that the mass density is equal to 1 and re recall that sometimes we skip the dependence of various function with respect to the spatial variable $\boldsymbol{x}$. Moreover, we use the elastic-viscoplastic constitutive law of the form (13.1).

In the models under consideration, the frictional contact is described by subdifferential boundary conditions of the form (13.6) or (13.7). We note that functions j_k depend on both $\boldsymbol{u}$ and $\boldsymbol{u}'$; moreover, in the first problem the subdifferential is considered with respect to the displacement and in the second problem is considered with respect to the velocity. All these ingredients represent traits of novelty of our frictional contact models. Finally, the initial displacement and the initial velocity are denoted by $\boldsymbol{u}_0$ and $\boldsymbol{u}_1$, respectively.

With this preliminaries, the classical formulation of the two contact problems can be stated as follows.

Problem C_1**.** *Find a displacement field* $\boldsymbol{u}\colon Q \to \mathbb{R}^d$ *and a stress field* $\boldsymbol{\sigma}\colon Q \to \mathbb{S}^d$ *such that*

$$\boldsymbol{u}''(t) - \mathrm{Div}\,\boldsymbol{\sigma}(t) = \boldsymbol{f}_0(t) \quad \text{in } \Omega, \tag{13.8}$$

$$\boldsymbol{\sigma}(t) = \mathcal{A}(t, \boldsymbol{\varepsilon}(\boldsymbol{u}'(t))) + \mathcal{E}(t, \boldsymbol{\varepsilon}(\boldsymbol{u}(t))) + \int_0^t \mathcal{G}(s, \boldsymbol{\sigma}(s) - \mathcal{A}(s, \boldsymbol{\varepsilon}(\boldsymbol{u}'(s))), \boldsymbol{\varepsilon}(\boldsymbol{u}(s)))\, ds \quad \text{in } \Omega, \tag{13.9}$$

$$\boldsymbol{u}(t) = \boldsymbol{0} \quad \text{on } \Gamma_D, \tag{13.10}$$

$$\boldsymbol{\sigma}(t)\,\boldsymbol{\nu} = \boldsymbol{f}_N \quad \text{on } \Gamma_N, \tag{13.11}$$

$$-\sigma_\nu(t) \in \partial j_1(t, \boldsymbol{u}(t), \boldsymbol{u}'(t), u_\nu(t)) \quad \text{on } \Gamma_C, \tag{13.12}$$

$$-\boldsymbol{\sigma}_\tau(t) \in \partial j_3(t, \boldsymbol{u}(t), \boldsymbol{u}'(t), \boldsymbol{u}_\tau(t)) \quad \text{on } \Gamma_C, \tag{13.13}$$

$$\boldsymbol{u}(0) = \boldsymbol{u}_0, \ \ \boldsymbol{u}'(0) = \boldsymbol{u}_1 \quad \text{in } \Omega, \tag{13.14}$$

for all $t \in (0, T)$.

Problem C_2. *Find a displacement field $\boldsymbol{u}\colon Q \to \mathbb{R}^d$ and a stress field $\boldsymbol{\sigma}\colon Q \to \mathbb{S}^d$ such that*

$$\boldsymbol{u}''(t) - \mathrm{Div}\,\boldsymbol{\sigma}(t) = \boldsymbol{f}_0(t) \qquad \text{in } \Omega, \tag{13.15}$$

$$\boldsymbol{\sigma}(t) = \mathcal{A}(t, \boldsymbol{\varepsilon}(\boldsymbol{u}'(t))) + \mathcal{E}(t, \boldsymbol{\varepsilon}(\boldsymbol{u}(t))) + \int_0^t \mathcal{G}(s, \boldsymbol{\sigma}(s) - \mathcal{A}(s, \boldsymbol{\varepsilon}(\boldsymbol{u}'(s))), \boldsymbol{\varepsilon}(\boldsymbol{u}(s)))\, ds \qquad \text{in } \Omega, \tag{13.16}$$

$$\boldsymbol{u}(t) = \boldsymbol{0} \qquad \text{on } \Gamma_D, \tag{13.17}$$

$$\boldsymbol{\sigma}(t)\,\boldsymbol{\nu} = \boldsymbol{f}_N \qquad \text{on } \Gamma_N, \tag{13.18}$$

$$-\,\sigma_\nu(t) \in \partial j_2(t, \boldsymbol{u}(t), \boldsymbol{u}'(t), u'_\nu(t)) \qquad \text{on } \Gamma_C, \tag{13.19}$$

$$-\,\boldsymbol{\sigma}_\tau(t) \in \partial j_4(t, \boldsymbol{u}(t), \boldsymbol{u}'(t), \boldsymbol{u}'_\tau(t)) \qquad \text{on } \Gamma_C, \tag{13.20}$$

$$\boldsymbol{u}(0) = \boldsymbol{u}_0, \;\; \boldsymbol{u}'(0) = \boldsymbol{u}_1 \qquad \text{in } \Omega, \tag{13.21}$$

for all $t \in (0, T)$.

In the study of Problems C_1 and C_2 we need the following hypotheses on the data.

$H(\mathcal{A})$: The viscosity operator $\mathcal{A}\colon Q \times \mathbb{S}^d \to \mathbb{S}^d$ is such that

(i) $\mathcal{A}(\cdot, \cdot, \boldsymbol{\varepsilon})$ is measurable on Q for all $\boldsymbol{\varepsilon} \in \mathbb{S}^d$.
(ii) $\mathcal{A}(\boldsymbol{x}, t, \cdot)$ is continuous on $\mathbb{S}^d$ for a.e. $(\boldsymbol{x}, t) \in Q$.
(iii) $\|\mathcal{A}(\boldsymbol{x}, t, \boldsymbol{\varepsilon})\|_{\mathbb{S}^d} \le \widetilde{a_1}(\boldsymbol{x}, t) + \widetilde{a_2}\, \|\boldsymbol{\varepsilon}\|_{\mathbb{S}^d}$ for all $\boldsymbol{\varepsilon} \in \mathbb{S}^d$, a.e. $(\boldsymbol{x}, t) \in Q$ with $\widetilde{a_1} \in L^2(Q)$, $\widetilde{a_1}, \widetilde{a_2} \ge 0$.
(iv) $(\mathcal{A}(\boldsymbol{x}, t, \boldsymbol{\varepsilon}_1) - \mathcal{A}(\boldsymbol{x}, t, \boldsymbol{\varepsilon}_2)) : (\boldsymbol{\varepsilon}_1 - \boldsymbol{\varepsilon}_2) \ge \widetilde{a_4}\, \|\boldsymbol{\varepsilon}_1 - \boldsymbol{\varepsilon}_2\|^2_{\mathbb{S}^d}$ for all $\boldsymbol{\varepsilon}_1$, $\boldsymbol{\varepsilon}_2 \in \mathbb{S}^d$, a.e. $(\boldsymbol{x}, t) \in Q$ with $\widetilde{a_4} > 0$.
(v) $\mathcal{A}(\boldsymbol{x}, t, \boldsymbol{\varepsilon}) : \boldsymbol{\varepsilon} \ge \widetilde{a_3}\, \|\boldsymbol{\varepsilon}\|^2_{\mathbb{S}^d}$ for all $\boldsymbol{\varepsilon} \in \mathbb{S}^d$, a.e. $(\boldsymbol{x}, t) \in Q$ with $\widetilde{a_3} > 0$.

$H(\mathcal{E})$: The elasticity operator $\mathcal{E}\colon Q \times \mathbb{S}^d \to \mathbb{S}^d$ is such that

(i) $\|\mathcal{E}(\boldsymbol{x}, t, \boldsymbol{\varepsilon}_1) - \mathcal{E}(\boldsymbol{x}, t, \boldsymbol{\varepsilon}_2)\|_{\mathbb{S}^d} \le L_{\mathcal{E}}\|\boldsymbol{\varepsilon}_1 - \boldsymbol{\varepsilon}_2\|_{\mathbb{S}^d}$ for all $\boldsymbol{\varepsilon}_1, \boldsymbol{\varepsilon}_2 \in \mathbb{S}^d$, a.e. $(\boldsymbol{x}, t) \in Q$.
(ii) $\mathcal{E}(\cdot, \cdot, \boldsymbol{\varepsilon})$ is measurable on Q for all $\boldsymbol{\varepsilon} \in \mathbb{S}^d$.
(iii) $\mathcal{E}(\cdot, \cdot, \boldsymbol{0}) \in L^2(Q, \mathbb{S}^d)$.

$H(\mathcal{G})$: The operator $\mathcal{G}\colon Q \times \mathbb{S}^d \times \mathbb{S}^d \to \mathbb{S}^d$ is such that

(i) $\|\mathcal{G}(\boldsymbol{x}, t, \boldsymbol{\sigma}_1, \boldsymbol{\varepsilon}_1) - \mathcal{G}(\boldsymbol{x}, t, \boldsymbol{\sigma}_2, \boldsymbol{\varepsilon}_2)\|_{\mathbb{S}^d} \le L_{\mathcal{G}}\big(\|\boldsymbol{\sigma}_1 - \boldsymbol{\sigma}_2\|_{\mathbb{S}^d} + \|\boldsymbol{\varepsilon}_1 - \boldsymbol{\varepsilon}_2\|_{\mathbb{S}^d}\big)$ for all $\boldsymbol{\sigma}_1, \boldsymbol{\sigma}_2, \boldsymbol{\varepsilon}_1, \boldsymbol{\varepsilon}_2 \in \mathbb{S}^d$, a.e. $(\boldsymbol{x}, t) \in Q$.
(ii) $\mathcal{G}(\cdot, \cdot, \boldsymbol{\sigma}, \boldsymbol{\varepsilon})$ is measurable on Q for all $\boldsymbol{\sigma}, \boldsymbol{\varepsilon} \in \mathbb{S}^d$.
(iii) $\mathcal{G}(\cdot, \cdot, \boldsymbol{0}, \boldsymbol{0}) \in L^2(Q, \mathbb{S}^d)$.

Remark 13.3. Under the conditions $H(\mathcal{E})$(i), (iii) and $H(\mathcal{G})$(i), (iii), we have

$$\|\mathcal{E}(\boldsymbol{x},t,\boldsymbol{\varepsilon})\|_{\mathbb{S}^d} \le L_{\mathcal{E}}\|\boldsymbol{\varepsilon}\|_{\mathbb{S}^d} + e(\boldsymbol{x},t) \text{ for all } \boldsymbol{\varepsilon} \in \mathbb{S}^d, \text{ a.e. } (\boldsymbol{x},t) \in Q,$$

$$\|\mathcal{G}(\boldsymbol{x},t,\boldsymbol{\sigma},\boldsymbol{\varepsilon})\|_{\mathbb{S}^d} \le L_{\mathcal{G}}(\|\boldsymbol{\sigma}\|_{\mathbb{S}^d} + \|\boldsymbol{\varepsilon}\|_{\mathbb{S}^d}) + g(\boldsymbol{x},t) \text{ for all } \boldsymbol{\varepsilon},\boldsymbol{\sigma} \in \mathbb{S}^d, \text{ a.e. } (\boldsymbol{x},t) \in Q,$$

where $e(\boldsymbol{x},t) = \|\mathcal{E}(\boldsymbol{x},t,\boldsymbol{0})\|_{\mathbb{S}^d}$, $e \in L^2(Q)$, $e \ge 0$ and $g(\boldsymbol{x},t) = \|\mathcal{G}(\boldsymbol{x},t,\boldsymbol{0},\boldsymbol{0})\|_{\mathbb{S}^d}$, $g \in L^2(Q)$, $g \ge 0$.

$H(f)$: $\boldsymbol{f}_0 \in L^2(0,T;H)$, $\boldsymbol{f}_N \in L^2(0,T;L^2(\Gamma_N;\mathbb{R}^d))$, $\boldsymbol{u}_0 \in V$, $\boldsymbol{u}_1 \in H$.

The functions j_k for $k = 1, 2$ satisfy the following conditions.

$H(j_k)$: The function $j_k : \Gamma_C \times (0,T) \times (\mathbb{R}^d)^2 \times \mathbb{R} \to \mathbb{R}$ is such that

(i) $j_k(\cdot,\cdot,\boldsymbol{\zeta},\boldsymbol{\rho},r)$ is measurable for all $\boldsymbol{\zeta}$, $\boldsymbol{\rho} \in \mathbb{R}^d$, $r \in \mathbb{R}$, $j_k(\cdot,\cdot,v(\cdot),w(\cdot),0) \in L^1(\Gamma_C \times (0,T))$ for all $v, w \in L^2(\Gamma_C;\mathbb{R}^d)$.

(ii) $j_k(\boldsymbol{x},t,\cdot,\cdot,r)$ is continuous for all $r \in \mathbb{R}$, a.e. $(\boldsymbol{x},t) \in \Gamma_C \times (0,T)$, $j_k(\boldsymbol{x},t,\boldsymbol{\zeta},\boldsymbol{\rho},\cdot)$ is locally Lipschitz for all $\boldsymbol{\zeta}$, $\boldsymbol{\rho} \in \mathbb{R}^d$, a.e. $(\boldsymbol{x},t) \in \Gamma_C \times (0,T)$.

(iii) $|\partial j_k(\boldsymbol{x},t,\boldsymbol{\zeta},\boldsymbol{\rho},r)| \le c_{k0} + c_{k1}\|\boldsymbol{\zeta}\|_{\mathbb{R}^d} + c_{k2}\|\boldsymbol{\rho}\|_{\mathbb{R}^d} + c_{k3}|r|$ for all $\boldsymbol{\zeta}$, $\boldsymbol{\rho} \in \mathbb{R}^d$, $r \in \mathbb{R}$, a.e. $(\boldsymbol{x},t) \in \Gamma_C \times (0,T)$ with $c_{ki} \ge 0$, $i = 0, 1, 2, 3$, where ∂j_k denotes the Clarke subdifferential of $j_k(\boldsymbol{x},t,\boldsymbol{\zeta},\boldsymbol{\rho},\cdot)$.

(iv) $j_k^0(\boldsymbol{x},t,\cdot,\cdot,\cdot;s)$ is upper semicontinuous on $(\mathbb{R}^d)^2 \times \mathbb{R}$ for all $s \in \mathbb{R}$, a.e. $(\boldsymbol{x},t) \in \Gamma_C \times (0,T)$, where j_k^0 denotes the generalized directional derivative of Clarke of $j_k(\boldsymbol{x},t,\boldsymbol{\zeta},\boldsymbol{\rho},\cdot)$ in the direction s.

(v) For $k = 1$ we have, $|\partial j_k(\boldsymbol{x},t,\boldsymbol{\zeta}_1,\boldsymbol{\rho}_1,r_1) - \partial j_k(\boldsymbol{x},t,\boldsymbol{\zeta}_2,\boldsymbol{\rho}_2,r_2)| \le L_k\left(\|\boldsymbol{\zeta}_1 - \boldsymbol{\zeta}_2\|_{\mathbb{R}^d} + \|\boldsymbol{\rho}_1 - \boldsymbol{\rho}_2\|_{\mathbb{R}^d} + |r_1 - r_2|\right)$ and for $k = 2$, we have $|\left(\partial j_k(\boldsymbol{x},t,\boldsymbol{\zeta}_1,\boldsymbol{\rho}_1,r_1) - \partial j_k(\boldsymbol{x},t,\boldsymbol{\zeta}_2,\boldsymbol{\rho}_2,r_2)\right)(r_1 - r_2) \ge -L_k\left(\|\boldsymbol{\zeta}_1 - \boldsymbol{\zeta}_2\|_{\mathbb{R}^d} + \|\boldsymbol{\rho}_1 - \boldsymbol{\rho}_2\|_{\mathbb{R}^d} + |r_1 - r_2|\right)|r_1 - r_2|$ for all $\boldsymbol{\zeta}_1$, $\boldsymbol{\zeta}_2$, $\boldsymbol{\rho}_1$, $\boldsymbol{\rho}_2 \in \mathbb{R}^d$, r_1, $r_2 \in \mathbb{R}$, a.e. $(\boldsymbol{x},t) \in \Gamma_C \times (0,T)$ with constants $L_k \ge 0$.

The functions j_k for $k = 3, 4$ satisfy conditions $H(j_k)$(i)–(v) with the last variable being in $\mathbb{R}^d$.

Moreover, we need the following hypothesis.

$H(j)_{reg}$: The functions j_k for $k = 1, \ldots, 4$ are such that for all $\boldsymbol{\zeta}$, $\boldsymbol{\rho} \in \mathbb{R}^d$, a.e. $(\boldsymbol{x},t) \in \Gamma_C \times (0,T)$, either all $j_k(\boldsymbol{x},t,\boldsymbol{\zeta},\boldsymbol{\rho},\cdot)$ are regular or all $-j_k(\boldsymbol{x},t,\boldsymbol{\zeta},\boldsymbol{\rho},\cdot)$ are regular for $k = 1, \ldots, 4$.

According to the comments and the mechanical interpretations presented in the books [6, 18, 19] it follows that the above hypotheses are realistic from the physical point of view.

13.4 Variational Formulation

We now turn to the variational formulations of Problems C_1 and C_2. Let V be the closed subspace of $H^1(\Omega;\mathbb{R}^d)$ given by

$$V = \{\, \boldsymbol{v} \in H_1 \mid \boldsymbol{v} = \boldsymbol{0} \text{ on } \Gamma_D \,\}.$$

The space V is a closed subspace of H_1 which means it is a Hilbert space with the inner product and the corresponding norm defined by $(\boldsymbol{u}, \boldsymbol{v})_V = (\boldsymbol{\varepsilon}(\boldsymbol{u}), \boldsymbol{\varepsilon}(\boldsymbol{v}))_{\mathcal{H}}$ and $\|\boldsymbol{v}\| = \|\boldsymbol{\varepsilon}(\boldsymbol{v})\|_{\mathcal{H}}$ for $\boldsymbol{u}, \boldsymbol{v} \in V$. Due to Korn's inequality

$$\|\boldsymbol{v}\|_{H_1} \le c\, \|\boldsymbol{\varepsilon}(\boldsymbol{v})\|_{\mathcal{H}} \qquad \text{for all } \boldsymbol{v} \in V \text{ with } c > 0$$

(cf. Sect. 6.3 of [17]), we know that $\|\cdot\|_{H_1}$ and $\|\cdot\|$ are the equivalent norms on V. We identify $H = L^2(\Omega;\mathbb{R}^d)$ with its dual space and we obtain an evolution triple of spaces (V, H, V^*) in which all embeddings are dense, continuous and compact. The duality pairing between V^* and V is denoted by $\langle \cdot, \cdot \rangle$. We define the spaces

$$\mathcal{V} = L^2(0,T;V), \quad \mathcal{H} = L^2(0,T;\mathbb{S}^d), \quad \hat{\mathcal{H}} = L^2(0,T;H),$$
$$\mathcal{V}^* = L^2(0,T;V^*), \quad \mathcal{W} = \{v \in \mathcal{V} \mid v' \in \mathcal{V}^*\}.$$

Next, let $\boldsymbol{v} \in V$. We define $\boldsymbol{f} \in \mathcal{V}^*$ by

$$\langle \boldsymbol{f}(t), \boldsymbol{v} \rangle = \langle \boldsymbol{f}_0(t), \boldsymbol{v} \rangle_H + \langle \boldsymbol{f}_N(t), \boldsymbol{v} \rangle_{L^2(\Gamma_N;\mathbb{R}^d)} \text{ for a.e. } t \in (0,T).$$

Assuming that the functions involved in Problem C_1 are sufficiently smooth, using the equation of motion (8.37) and the Green formula, we obtain

$$\langle \boldsymbol{u}''(t), \boldsymbol{v} \rangle + \langle \boldsymbol{\sigma}(t), \boldsymbol{\varepsilon}(\boldsymbol{v}) \rangle_{\mathcal{H}} - \int_\Gamma \boldsymbol{\sigma}(t)\boldsymbol{\nu} \cdot \boldsymbol{v}\, d\Gamma = \langle \boldsymbol{f}_0(t), \boldsymbol{v} \rangle_H$$

for a.e. $t \in (0,T)$. From the boundary conditions (8.39) and (8.40), we have

$$\int_\Gamma \boldsymbol{\sigma}(t)\boldsymbol{\nu} \cdot \boldsymbol{v}\, d\Gamma = \int_{\Gamma_N} \boldsymbol{f}_N(t) \cdot \boldsymbol{v}\, d\Gamma + \int_{\Gamma_C} (\boldsymbol{\sigma}_\tau(t) \cdot \boldsymbol{v}_\tau + \sigma_\nu(t) v_\nu)\, d\Gamma.$$

On the other hand, the subdifferential boundary conditions (8.41) and (13.13) imply

$$-\sigma_\nu(t)\, r \le j_1^0(t, \boldsymbol{u}(t), \boldsymbol{u}'(t), u_\nu(t); r) \ \text{ for all } r \in \mathbb{R},$$
$$-\boldsymbol{\sigma}_\tau(t) \cdot \boldsymbol{\xi} \le j_3^0(t, \boldsymbol{u}(t), \boldsymbol{u}'(t), \boldsymbol{u}_\tau(t); \boldsymbol{\xi}) \ \text{ for all } \boldsymbol{\xi} \in \mathbb{R}^d,$$

a.e. on Γ_3. Then, using the constitutive law (8.38) and the above relations, we obtain the following variational formulation of Problem C_1.

Problem (HVI_1). *Find* $\boldsymbol{u}\colon (0,T) \to V$ *such that* $\boldsymbol{u} \in \mathcal{V}$, $\boldsymbol{u}' \in \mathcal{W}$ *and*

$$\begin{aligned}\boldsymbol{\sigma}(t) = \mathcal{A}(t, \boldsymbol{\varepsilon}(\boldsymbol{u}'(t))) + \mathcal{E}(t, \boldsymbol{\varepsilon}(\boldsymbol{u}(t))) \\ + \int_0^t \mathcal{G}(s, \boldsymbol{\sigma}(s) - \mathcal{A}(s, \boldsymbol{\varepsilon}(\boldsymbol{u}'(s))), \boldsymbol{\varepsilon}(\boldsymbol{u}(s)))\, ds \text{ for a.e. } t \in (0,T),\end{aligned} \tag{13.22}$$

$$\begin{aligned}&\langle \boldsymbol{u}''(t), \boldsymbol{v}\rangle + \langle \boldsymbol{\sigma}(t), \boldsymbol{\varepsilon}(\boldsymbol{v})\rangle_{\mathcal{H}} \\ &\quad + \int_{\Gamma_C} \left(j_1^0(t, \boldsymbol{u}(t), \boldsymbol{u}'(t), u_\nu(t); v_\nu) + j_3^0(t, \boldsymbol{u}(t), \boldsymbol{u}'(t), \boldsymbol{u}_\tau(t); \boldsymbol{v}_\tau)\right) d\Gamma \\ &\quad \geq \langle \boldsymbol{f}(t), \boldsymbol{v}\rangle \text{ for all } \boldsymbol{v} \in V, \text{ a.e. } t \in (0,T),\end{aligned} \tag{13.23}$$

$$\boldsymbol{u}(0) = \boldsymbol{u}_0,\ \boldsymbol{u}'(0) = \boldsymbol{u}_1, \tag{13.24}$$

In a similar way we find the following variational formulation of the Problem C_2.

Problem (HVI_2). *Find* $\boldsymbol{u}\colon (0,T) \to V$ *such that* $\boldsymbol{u} \in \mathcal{V}$, $\boldsymbol{u}' \in \mathcal{W}$ *and*

$$\begin{aligned}\boldsymbol{\sigma}(t) = \mathcal{A}(t, \boldsymbol{\varepsilon}(\boldsymbol{u}'(t))) + \mathcal{E}(t, \boldsymbol{\varepsilon}(\boldsymbol{u}(t))) \\ + \int_0^t \mathcal{G}(s, \boldsymbol{\sigma}(s) - \mathcal{A}(s, \boldsymbol{\varepsilon}(\boldsymbol{u}'(s))), \boldsymbol{\varepsilon}(\boldsymbol{u}(s)))\, ds \text{ for a.e. } t \in (0,T),\end{aligned} \tag{13.25}$$

$$\begin{aligned}&\langle \boldsymbol{u}''(t), \boldsymbol{v}\rangle + \langle \boldsymbol{\sigma}(t), \boldsymbol{\varepsilon}(\boldsymbol{v})\rangle_{\mathcal{H}} \\ &\quad + \int_{\Gamma_C} \left(j_2^0(t, \boldsymbol{u}(t), \boldsymbol{u}'(t), u'_\nu(t); v_\nu) + j_4^0(t, \boldsymbol{u}(t), \boldsymbol{u}'(t), \boldsymbol{u}'_\tau(t); \boldsymbol{v}_\tau)\right) d\Gamma \\ &\quad \geq \langle \boldsymbol{f}(t), \boldsymbol{v}\rangle \text{ for all } \boldsymbol{v} \in V, \text{ a.e. } t \in (0,T),\end{aligned} \tag{13.26}$$

$$\boldsymbol{u}(0) = \boldsymbol{u}_0,\ \boldsymbol{u}'(0) = \boldsymbol{u}_1. \tag{13.27}$$

Note that variational formulations (HVI_1) and (HVI_2) consist of a system coupling the integro-differential equation (13.22) or (13.25) with second order hemivariational inequality (13.23) or (13.26), and with the initial conditions (13.24) or (13.27), respectively.

13.5 Existence and Uniqueness Results

In this section we prove the existence and uniqueness theorem for variational problems (HVI_1) and (HVI_2). Our main result is the following.

Theorem 13.4. *Assume that the hypotheses* $H(\mathcal{A})$, $H(\mathcal{G})$, $H(\mathcal{E})$, $H(f)$, $H(j_k)$ *for* $k = 1, \ldots, 4$, $H(j)_{reg}$ *hold. Moreover, assume that*

$$\tilde{a}_3 > 4\sqrt{15}\, c_e^2\, \|\gamma\|^2 \Big(T \max\{ \max_{1\le k\le 4} c_{k1}, c_{13}, c_{33}\} + \max\{ \max_{1\le k\le 4} c_{k2}, c_{23}, c_{43}\} \Big)$$

and

$$\tilde{a}_4 > c_e \|\gamma\|^2 \Big(\max\{L_1, 2L_2, L_3, 2L_4\} + \frac{T}{\sqrt{2}} \max\{2L_1, L_2, 2L_3, L_4\} \Big).$$

Then, Problems (HVI_1) *and* (HVI_2) *have unique solutions.*

The proof of Theorem 13.4 is carried out in two main steps and it is based on Theorem 13.2 and a fixed point argument. First we focus on Problem (HVI_1).

Step 1. Let $\boldsymbol{\eta} \in L^2(0, T; \mathcal{H})$ be fixed and consider the following auxiliary problem.

Problem $(HVI_{1\eta})$. *Find* $\boldsymbol{u}\colon (0, T) \to V$ *such that* $\boldsymbol{u} \in \mathcal{V}$, $\boldsymbol{u}' \in \mathcal{W}$ *and*

$$\boldsymbol{\sigma}(t) = \mathcal{A}(t, \boldsymbol{\varepsilon}(\boldsymbol{u}_\eta'(t))) + \langle \boldsymbol{\eta}(t), \boldsymbol{\varepsilon}(t)\rangle_{\mathcal{H}} \text{ for a.e. } t \in (0, T), \tag{13.28}$$

$$\begin{aligned} &\langle \boldsymbol{u}_\eta''(t), \boldsymbol{v}\rangle + \langle \boldsymbol{\sigma}(t), \boldsymbol{\varepsilon}(v)\rangle_{\mathcal{H}} + \int_{\Gamma_C} \Big(j_1^0(t, \boldsymbol{u}_\eta(t), \boldsymbol{u}_\eta'(t), u_{\eta\nu}(t); v_\nu) \\ &\quad + j_3^0(t, \boldsymbol{u}_\eta(t), \boldsymbol{u}_\eta'(t), \boldsymbol{u}_{\eta\tau}(t); \boldsymbol{v}_\tau) \Big)\, d\Gamma \\ &\quad \ge \langle \boldsymbol{f}(t), \boldsymbol{v}\rangle \text{ for all } \boldsymbol{v} \in V, \text{ a.e. } t \in (0, T), \end{aligned} \tag{13.29}$$

$$\boldsymbol{u}_\eta(0) = \boldsymbol{u}_0,\ \boldsymbol{u}_\eta'(0) = \boldsymbol{u}_1. \tag{13.30}$$

To solve Problem $(HVI_{1\eta})$ we substitute $\boldsymbol{\sigma}_\eta$ from (13.28) in (13.29) and use (13.30) to obtain the following hemivariational inequality: find $\boldsymbol{u}_\eta\colon (0, T) \to V$ such that

$$\begin{aligned} &\langle \boldsymbol{u}_\eta''(t), \boldsymbol{v}\rangle + \langle \mathcal{A}(t, \boldsymbol{\varepsilon}(\boldsymbol{u}_\eta'(t))), \boldsymbol{\varepsilon}(\boldsymbol{v})\rangle_{\mathcal{H}} + \langle \eta(t), \boldsymbol{\varepsilon}(v)\rangle_{\mathcal{H}} \\ &+ \int_{\Gamma_C} \Big(j_1^0(t, \boldsymbol{u}_\eta(t), \boldsymbol{u}_\eta'(t), u_{\eta\nu}(t); v_\nu) + j_3^0(t, \boldsymbol{u}_\eta(t), \boldsymbol{u}_\eta'(t), \boldsymbol{u}_{\eta\tau}(t); \boldsymbol{v}_\tau) \Big)\, d\Gamma \\ &\ge \langle \boldsymbol{f}(t), \boldsymbol{v}\rangle \text{ for all } \boldsymbol{v} \in V \text{ a.e. } t \in (0, T), \end{aligned} \tag{13.31}$$

$$\boldsymbol{u}_\eta(0) = \boldsymbol{u}_0,\ \ \boldsymbol{u}_\eta'(0) = \boldsymbol{u}_1. \tag{13.32}$$

In what follows we need additional notation. Let the function $g\colon \Gamma_C \times (0,T) \times (\mathbb{R}^d)^4 \to \mathbb{R}$ be defined by

$$g(\boldsymbol{x},t,\boldsymbol{\zeta},\boldsymbol{\rho},\boldsymbol{\xi},\boldsymbol{\eta}) = j_1(\boldsymbol{x},t,\boldsymbol{\zeta},\boldsymbol{\rho},\xi_\nu) + j_3(\boldsymbol{x},t,\boldsymbol{\zeta},\boldsymbol{\rho},\boldsymbol{\xi}_\tau)$$

for $\boldsymbol{\zeta}$, $\boldsymbol{\rho}$, $\boldsymbol{\xi}$, $\boldsymbol{\eta} \in \mathbb{R}^d$ and a.e. $(\boldsymbol{x},t) \in \Gamma_C \times (0,T)$ and let the functional $G\colon (0,T) \times L^2(\Gamma_C;\mathbb{R}^d)^4 \to \mathbb{R}$ be given by

$$G(t,\boldsymbol{w},\boldsymbol{z},\boldsymbol{u},\boldsymbol{v}) = \int_{\Gamma_C} g(\boldsymbol{x},t,\boldsymbol{w}(\boldsymbol{x}),\boldsymbol{z}(\boldsymbol{x}),\boldsymbol{u}(\boldsymbol{x}),\boldsymbol{v}(\boldsymbol{x}))\, d\Gamma$$

for $\boldsymbol{w}, \boldsymbol{z}, \boldsymbol{u}, \boldsymbol{v} \in L^2(\Gamma_C;\mathbb{R}^d)$ and $t \in (0,T)$. Note, that in this setup G is independent of the variable $\boldsymbol{v}$. Let $Z = H^{1/2}(\Omega;\mathbb{R}^d)$. We introduce the operators $R\colon Z \times Z \to L^2(\Gamma_C;\mathbb{R}^d)^2$ and $S\colon Z^* \times Z^* \to Z^*$ defined by

$$R(z_1,z_2) = (\gamma z_1, \gamma z_2), \quad S(z_1^*,z_2^*) = z_1^* + z_2^*$$

for all $z_1, z_2 \in Z$ and $z_1^*, z_2^* \in Z^*$. Then, the adjoint operator of R is the operator $R^*\colon L^2(\Gamma_C;\mathbb{R}^d)^2 \to Z^* \times Z^*$ given by

$$R^*(\boldsymbol{u},\boldsymbol{v}) = (\gamma^*\boldsymbol{u}, \gamma^*\boldsymbol{v}) \quad \text{for all } \boldsymbol{u}, \boldsymbol{v} \in L^2(\Gamma_C;\mathbb{R}^d).$$

Next, we define the multivalued mapping $F\colon (0,T) \times V \times V \to 2^{Z^*}$ by

$$F(t,\boldsymbol{u},\boldsymbol{v}) = S\, R^*\, \partial G(t, R(\boldsymbol{u},\boldsymbol{v}), R(\boldsymbol{u},\boldsymbol{v})) \tag{13.33}$$

for $\boldsymbol{u}$, $\boldsymbol{v} \in V$ and a.e. $t \in (0,T)$, where ∂G denotes the Clarke subdifferential of the functional $G = G(t,\boldsymbol{w},\boldsymbol{z},\boldsymbol{u},\boldsymbol{v})$ with respect to the pair $(\boldsymbol{u},\boldsymbol{v})$.

From [8, Lemma 17] we have the following result.

Lemma 13.5. *Assume that the hypotheses $H(j_k)$ hold for $k = 1,\ldots,4$ and moreover,*

$$\text{either } j_k(\boldsymbol{x},t,\boldsymbol{\zeta},\boldsymbol{\rho},\cdot) \text{ are regular and } j_k \text{ satisfy } H(j_k)_1 \tag{13.34}$$

$$\text{or } -j_k(\boldsymbol{x},t,\boldsymbol{\zeta},\boldsymbol{\rho},\cdot) \text{ are regular and } -j_k \text{ satisfy } H(j_k)_1 \tag{13.35}$$

for $k = 1,\ldots,4$. Then the multifunction $F\colon (0,T) \times V \times V \to 2^{Z^*}$ defined by (13.33) *satisfies the condition $H(F)$ with $m_2 = c_e k_1 \|\gamma\|^2$ and $m_3 = c_e k_2 \|\gamma\|^2$.*

Next, we derive the following result.

Lemma 13.6. *Assume that the hypotheses $H(\mathcal{A})$, $H(f)$, $H(j_k)$ for $k = 1,\ldots,4$, $H(j)_{reg}$ hold and, in addition, assume that either* (13.34) *or* (13.35) *hold. Moreover, assume the following conditions*

$$\tilde{a}_3 > 4\sqrt{15}\, c_e^2 \|\gamma\|^2 \Big(T \max\{ \max_{1\le k\le 4} c_{k1}, c_{13}, c_{33}\} + \max\{ \max_{1\le k\le 4} c_{k2}, c_{23}, c_{43}\} \Big) \tag{13.36}$$

and

$$\tilde{a}_4 > c_e \|\gamma\|^2 \Big(\max\{L_1, 2L_2, L_3, 2L_4\} + \frac{T}{\sqrt{2}} \max\{2L_1, L_2, 2L_3, L_4\} \Big). \tag{13.37}$$

Then, Problem $(HVI_{1\eta})$ *has a unique solution which satisfies*

$$\|\boldsymbol{u}\|_{C(0,T;V)} + \|\boldsymbol{u}'\|_{\mathcal{W}} \le C\Big(1 + \|\boldsymbol{u}_0\| + |\boldsymbol{u}_1| + \|\boldsymbol{f}\|_{\mathcal{V}^*}\Big) \text{ with } C > 0.$$

Proof. We define the operator $A: (0,T) \times V \to V^*$ and the function $\boldsymbol{f}_\eta: (0,T) \to V^*$ by

$$\langle A(t, \boldsymbol{u}), \boldsymbol{v} \rangle = (\mathcal{A}(\boldsymbol{x}, t, \boldsymbol{\varepsilon}(\boldsymbol{u})), \boldsymbol{\varepsilon}(\boldsymbol{v}))_{\mathcal{H}}, \tag{13.38}$$

$$\langle \boldsymbol{f}_\eta(t), \boldsymbol{v} \rangle = \langle f(t), \boldsymbol{v} \rangle - \langle \boldsymbol{\eta}(t), \boldsymbol{v} \rangle \tag{13.39}$$

for $\boldsymbol{u}, \boldsymbol{v} \in V$, a.e. $t \in (0,T)$. We associate to (13.31) and (13.32) the following evolution inclusion: find $\boldsymbol{u}_\eta \in \mathcal{V}$ such that $\boldsymbol{u}'_\eta \in \mathcal{W}$ and

$$\begin{cases} \boldsymbol{u}''_\eta(t) + A(t, \boldsymbol{u}'_\eta(t)) + F(t, \boldsymbol{u}_\eta(t), \boldsymbol{u}'_\eta(t)) \ni \boldsymbol{f}_\eta(t) \ \text{ a.e. } t \in (0,T), \\ \boldsymbol{u}_\eta(0) = \boldsymbol{u}_0, \ \ \boldsymbol{u}'_\eta(0) = \boldsymbol{u}_1, \end{cases} \tag{13.40}$$

where A is given by (13.38), and F and $\boldsymbol{f}_\eta$ are defined by (13.33) and (13.39), respectively. It is easy to check that under the assumption $H(\mathcal{A})$ the operator A satisfies $H(A)$ with $a_0(t) = \sqrt{2}\, \|\widetilde{a_1}(t)\|_{L^2(\Omega)}$, $a_1 = \sqrt{2}\, \widetilde{a_2}$, $\alpha = \widetilde{a_3}$, and $m_1 = \widetilde{a_4}$ (cf. Lemma 8 in [8]). Due to $H(f)$, it is clear that $\boldsymbol{f}_\eta \in \mathcal{V}^*$ and

$$\|\boldsymbol{f}_\eta\|_{\mathcal{V}^*} \le \sqrt{2}(\|\boldsymbol{f}\|_{\mathcal{V}^*} + \|\boldsymbol{\eta}\|_{L^2(0,T;\mathcal{H})}).$$

Next, $H(0)$ follows from (H_0) whereas (H_1) and (H_2) are consequences of (13.36) and (13.37), respectively. The assumption $H(F)$ is satisfied due to Lemma 13.5. Applying Theorem 13.2, we deduce that the problem (13.40) has a unique solution $\boldsymbol{u}_\eta \in \mathcal{V}$ such that $\boldsymbol{u}'_\eta \in \mathcal{W}$ which satisfies

$$\begin{aligned} \|\boldsymbol{u}\|_{C(0,T;V)} + \|\boldsymbol{u}'\|_{\mathcal{W}} &\le C(1 + \|\boldsymbol{u}_0\| + |\boldsymbol{u}_1| + \|\boldsymbol{f}_\eta\|_{\mathcal{V}^*}) \\ &\le C(1 + \|\boldsymbol{u}_0\| + |\boldsymbol{u}_1| + \|\boldsymbol{f}\|_{\mathcal{V}^*} + \|\boldsymbol{\eta}\|_{L^2(0,T;\mathcal{H})}). \end{aligned} \tag{13.41}$$

Applying Theorems 19 and 20 of [8] we deduce that $\boldsymbol{u}_\eta \in \mathcal{V}$, $\boldsymbol{u}'_\eta \in \mathcal{W}$ is the unique solution to $(HVI_{1\eta})$, which completes the proof of the lemma. □

Step 2. From Step 1 we know that for every $\boldsymbol{\eta} \in L^2(0,T;\mathcal{H})$, Problem $(HVI_{1\eta})$ has a unique solution $\boldsymbol{u}_\eta \in \mathcal{V}$ such that $\boldsymbol{u}'_\eta \in \mathcal{W}$. We consider the operator $\Lambda\colon L^2(0,T;\mathcal{H}) \to L^2(0,T;\mathcal{H})$ defined by

$$(\Lambda\boldsymbol{\eta})(t) = \mathcal{E}(t, \boldsymbol{\varepsilon}(\boldsymbol{u}_\eta(t))) + \int_0^t \mathcal{G}(s, \eta(s), \boldsymbol{\varepsilon}(\boldsymbol{u}_\eta(s)))\, ds \tag{13.42}$$

for $\boldsymbol{\eta} \in L^2(0,T;\mathcal{H})$, a.e. $t \in (0,T)$, where $\boldsymbol{u}_\eta \in \mathcal{V}$ is the unique solution to $(HVI_{1\eta})$.

Lemma 13.7. *Under the hypotheses of Theorem 13.4, the operator Λ has a unique fixed point $\boldsymbol{\eta}^* \in L^2(0,T;\mathcal{H})$.*

Proof. First, we observe that the operator Λ is well defined. Indeed, from Remark 13.3, we have

$$\|\mathcal{E}(t, \boldsymbol{\varepsilon}(\boldsymbol{u}_\eta(t)))\|_{\mathcal{H}} \le \sqrt{2}\,(L_{\boldsymbol{\varepsilon}}\|\boldsymbol{u}_\eta(t)\| + |e(t)|),$$

$$\|\mathcal{G}(t, \boldsymbol{\eta}(t), \boldsymbol{\varepsilon}(\boldsymbol{u}_\eta(t)))\|_{\mathcal{H}} \le 2L_{\mathcal{G}}(\|\boldsymbol{u}_\eta(t)\| + \|\boldsymbol{\eta}(t)\|_{\mathcal{H}} + \sqrt{2}|g(t)|)$$

for a.e. $t \in (0,T)$. Hence

$$\begin{aligned}\|(\Lambda\boldsymbol{\eta})(t)\|_{\mathcal{H}} &\le \|\mathcal{E}(t, \boldsymbol{\varepsilon}(\boldsymbol{u}_\eta(t)))\|_{\mathcal{H}} + \int_0^t \|\mathcal{G}(s, \boldsymbol{\eta}(s), \boldsymbol{\varepsilon}(\boldsymbol{u}_\eta(s)))\|_{\mathcal{H}} ds\\ &\le \sqrt{2}\,(L_{\mathcal{E}}\|\boldsymbol{u}_\eta(t)\| + |e(t)|)\\ &\quad + L_G \int_0^t (\|\boldsymbol{u}_\eta(s)\| + \|\boldsymbol{\eta}(s)\|_{\mathcal{H}})\, ds + \sqrt{2}\int_0^t |g(s)|\, ds\end{aligned}$$

for a.e. $t \in (0,T)$ and, subsequently,

$$\|\Lambda\boldsymbol{\eta}\|^2_{L^2(0,T;\mathcal{H})} = \int_0^T \|(\Lambda\boldsymbol{\eta})(t)\|^2_{\mathcal{H}}\, dt \le \overline{c}\left(\|\boldsymbol{u}_\eta\|^2_{\mathcal{V}} + \|\boldsymbol{\eta}\|^2_{L^2(0,T;\mathcal{H})} + 1\right),$$

where $\overline{c} > 0$. Keeping in mind the estimate (13.41), we deduce that the integral (13.42) is well defined and Λ takes values in $L^2(0,T;\mathcal{H})$. Next, we will show that the operator Λ has a unique fixed point. Let $\boldsymbol{\eta}_1, \boldsymbol{\eta}_2 \in L^2(0,T;\mathcal{H})$ and let $\boldsymbol{u}_1 = \boldsymbol{u}_{\eta_1}$ and $\boldsymbol{u}_2 = \boldsymbol{u}_{\eta_2}$ be the corresponding solutions to $(HVI_{1\eta})$ such that $\boldsymbol{u}_i \in \mathcal{V}$ and $\boldsymbol{u}'_i \in \mathcal{W}$ for $i = 1, 2$. We have

$$\boldsymbol{u}''_1(t) + A(t, \boldsymbol{u}'_1(t)) + z_1(t) = \boldsymbol{f}(t) - \boldsymbol{\eta}_1(t) \text{ a.e. } t \in (0,T), \tag{13.43}$$

$$\boldsymbol{u}''_2(t) + A(t, \boldsymbol{u}'_2(t)) + z_2(t) = \boldsymbol{f}(t) - \boldsymbol{\eta}_2(t) \text{ a.e. } t \in (0,T), \tag{13.44}$$

$$z_1(t) \in F(t, \boldsymbol{u}_1(t), \boldsymbol{u}_1'(t)), \quad z_2(t) \in F(t, \boldsymbol{u}_2(t), \boldsymbol{u}_2'(t)) \text{ a.e. } t \in (0, T),$$

$$\boldsymbol{u}_1(0) = \boldsymbol{u}_2(0) = \boldsymbol{u}_0, \quad \boldsymbol{u}_1'(0) = \boldsymbol{u}_2'(0) = \boldsymbol{u}_1.$$

We subtract (13.44) from (13.43) and multiply the result by $\boldsymbol{u}_1'(t) - \boldsymbol{u}_2'(t)$. Using the integration by parts formula, we obtain

$$\frac{1}{2} |\boldsymbol{u}_1'(t) - \boldsymbol{u}_2'(t)|^2 + \int_0^t \langle A(s, \boldsymbol{u}_1'(s)) - A(s, \boldsymbol{u}_2'(s)), \boldsymbol{u}_1'(s) - \boldsymbol{u}_2'(s)\rangle \, ds$$
$$+ \int_0^t \langle z_1(s) - z_2(s), \boldsymbol{u}_1'(s) - \boldsymbol{u}_2'(s)\rangle \, ds$$
$$= \int_0^t \langle \eta_1(s) - \eta_2(s), \boldsymbol{u}_1'(s) - \boldsymbol{u}_2'(s)\rangle \, ds \tag{13.45}$$

for all $t \in [0, T]$. Since $\boldsymbol{u}_1, \boldsymbol{u}_2 \in H^1(0, T; V)$ and V is reflexive, by Theorem 8.4.11 of [4], we know that $\boldsymbol{u}_1$ and $\boldsymbol{u}_2$ may be identified with absolutely continuous functions with values in V and

$$\boldsymbol{u}_1(t) = \boldsymbol{u}_1(0) + \int_0^t \boldsymbol{u}_1'(s) \, ds, \quad \boldsymbol{u}_2(t) = \boldsymbol{u}_2(0) + \int_0^t \boldsymbol{u}_2'(s) \, ds \tag{13.46}$$

for all $t \in [0, T]$. This implies

$$\|\boldsymbol{u}_1(t) - \boldsymbol{u}_2(t)\| \le \int_0^t \|\boldsymbol{u}_1'(s) - \boldsymbol{u}_2'(s)\| \, ds$$

for all $t \in [0, T]$. Hence, by the Jensen inequality, we obtain

$$\int_0^t \|\boldsymbol{u}_1(s) - \boldsymbol{u}_2(s)\|^2 \, ds \le \int_0^t \left(\int_0^s \|\boldsymbol{u}_1'(\tau) - \boldsymbol{u}_2'(\tau)\| \, d\tau \right)^2 ds$$
$$\le \int_0^t s \, \|\boldsymbol{u}_1' - \boldsymbol{u}_2'\|^2_{L^2(0,T;V)} \, ds$$
$$\le \frac{T^2}{2} \|\boldsymbol{u}_1' - \boldsymbol{u}_2'\|^2_{L^2(0,T;V)}$$

for all $t \in [0, T]$. Therefore, exploiting $H(F)$(iv) and the Hölder inequality, we have

$$\int_0^t \langle z_1(s) - z_2(s), \boldsymbol{u}_1'(s) - \boldsymbol{u}_2'(s)\rangle_{Z^* \times Z} \, ds$$
$$\ge - m_2 \|\boldsymbol{u}_1' - \boldsymbol{u}_2'\|^2_{L^2(0,t;V)} - m_3 \|\boldsymbol{u}_1' - \boldsymbol{u}_2'\|_{L^2(0,t;V)} \left(\int_0^t \|\boldsymbol{u}_1(s) - \boldsymbol{u}_2(s)\|^2 \, ds \right)^2$$
$$= - \left(m_2 + \frac{m_3 T}{\sqrt{2}} \right) \|\boldsymbol{u}_1' - \boldsymbol{u}_2'\|^2_{L^2(0,t;V)}. \tag{13.47}$$

Hence, using (13.45), (13.47) and $H(A)$(vi), we obtain

$$\frac{1}{2}\,|\boldsymbol{u}_1'(t)-\boldsymbol{u}_2'(t)|^2+\Big(m_1-m_2-\frac{m_3\,T}{\sqrt{2}}\Big)\|\boldsymbol{u}_1'-\boldsymbol{u}_2'\|^2_{L^2(0,t;V)}\le 0$$

for all $t\in[0,T]$ which, together with (H_2), implies

$$\|\boldsymbol{u}_1'-\boldsymbol{u}_2'\|_{L^2(0,t;V)}\le\frac{1}{c}\,\|\boldsymbol{\eta}_1-\boldsymbol{\eta}_2\|_{L^2(0,t;V^*)},\tag{13.48}$$

where $c=m_1-m_2-\frac{m_3T}{\sqrt{2}}>0$. We use (13.46) again and we obtain

$$\|\boldsymbol{u}_1(t)-\boldsymbol{u}_2(t)\|\le\int_0^t\|\boldsymbol{u}_1'(s)-\boldsymbol{u}_2'(s)\|ds\le\sqrt{t}\|\boldsymbol{u}_1'-\boldsymbol{u}_2'\|_{L^2(0,t;V)}\tag{13.49}$$

for all $t\in[0,T]$. Combining (13.48) and (13.49), we deduce

$$\|\boldsymbol{u}_1(t)-\boldsymbol{u}_2(t)\|\le\frac{\sqrt{t}}{\hat{c}}\,\|\boldsymbol{\eta}_1-\boldsymbol{\eta}_2\|_{L^2(0,t;\mathcal{H})}\ \text{ for all } t\in[0,T]\tag{13.50}$$

with $\hat{c}>0$. On the other hand, from $H(\mathcal{E})(i)$ and $H(\mathcal{G})(i)$ it follows that

$$\|\mathcal{E}(t,\boldsymbol{\varepsilon}_1)-\mathcal{E}(t,\boldsymbol{\varepsilon}_2)\|_{\mathcal{H}}\le L_{\mathcal{E}}\|\boldsymbol{\varepsilon}_1-\boldsymbol{\varepsilon}_2\|_{\mathcal{H}}\ \text{for all } \boldsymbol{\varepsilon}_1,\boldsymbol{\varepsilon}_2\in\mathcal{H} \text{ and } t\in[0,T],$$

$$\|\mathcal{G}(t,\boldsymbol{\sigma}_1,\boldsymbol{\varepsilon}_1)-\mathcal{G}(t,\boldsymbol{\sigma}_2,\boldsymbol{\varepsilon}_2)\|_{\mathcal{H}}\le\sqrt{2}\,L_{\mathcal{G}}(\|\boldsymbol{\sigma}_1-\boldsymbol{\sigma}_2\|_{\mathcal{H}}+\|\boldsymbol{\varepsilon}_1-\boldsymbol{\varepsilon}_2\|_{\mathcal{H}})$$

for all $\boldsymbol{\sigma}_1,\boldsymbol{\sigma}_2,\boldsymbol{\varepsilon}_1,\boldsymbol{\varepsilon}_2\in\mathcal{H}$, $t\in[0,T]$. Then, from (13.50), we have

$$\begin{aligned}
&\|(\Lambda\boldsymbol{\eta}_1)(t)-(\Lambda\boldsymbol{\eta}_2)(t)\|_{\mathcal{H}}\le\|\mathcal{E}(t,\boldsymbol{\varepsilon}(\boldsymbol{u}_1(t)))-\mathcal{E}(t,\boldsymbol{\varepsilon}(\boldsymbol{u}_2(t)))\|_{\mathcal{H}}\\
&\quad+\int_0^t\|\mathcal{G}(s,\boldsymbol{\eta}_1(s),\boldsymbol{\varepsilon}(\boldsymbol{u}_1(t)))-\mathcal{G}(s,\boldsymbol{\eta}_2(s),\boldsymbol{\varepsilon}(\boldsymbol{u}_2(t)))\|_{\mathcal{H}}\\
&\quad\le L_{\mathcal{E}}\|\boldsymbol{\varepsilon}(\boldsymbol{u}_1(t)))-\boldsymbol{\varepsilon}(\boldsymbol{u}_2(t)))\|_{\mathcal{H}}\\
&\qquad+\sqrt{2}\,L_{\mathcal{G}}\int_0^t\|\boldsymbol{\eta}_1(s)-\boldsymbol{\eta}_2(s)\|_{\mathcal{H}}\,ds+\sqrt{2}\,L_{\mathcal{G}}\int_0^t\|\boldsymbol{\varepsilon}(\boldsymbol{u}_1(s))-\boldsymbol{\varepsilon}(\boldsymbol{u}_2(s))\|_{\mathcal{H}}\,ds\\
&\quad=L_{\mathcal{E}}\|\boldsymbol{u}_1(t)-\boldsymbol{u}_2(t)\|+\sqrt{2}L_{\mathcal{G}}\int_0^t(\|\boldsymbol{\eta}_1(t)-\boldsymbol{\eta}_2(t)\|_{\mathcal{H}}+\|\boldsymbol{u}_1(s)-\boldsymbol{u}_2(s)\|)\,ds\\
&\quad\le\overline{c}\,\|\boldsymbol{\eta}_1-\boldsymbol{\eta}_2\|_{L^2(0,t;\mathcal{H})}
\end{aligned}$$

for all $t \in [0, T]$, where $\overline{c} = (\sqrt{2}L_{\mathcal{G}} + \frac{L_{\mathcal{E}}+2L_{\mathcal{G}}T}{\hat{c}})\sqrt{T}$. This implies

$$\|(\Lambda\boldsymbol{\eta}_1)(t) - (\Lambda\boldsymbol{\eta}_2)(t)\|^2_{\mathcal{H}} \leq \overline{c} \int_0^t \|\boldsymbol{\eta}_1(s) - \boldsymbol{\eta}_2(s)\|_{\mathcal{H}}\, ds$$

for all $t \in [0, T]$. From the above and Lemma 7 of [9], we deduce that Λ has a unique fixed point, which completes the proof of Lemma 13.7. □

Now we are in a position to complete the proof of Theorem 13.4. Let $\boldsymbol{\eta}^* \in L^2(0, T; \mathcal{H})$ be the fixed point of the operator Λ and let $(\boldsymbol{u}_{\eta^*}, \boldsymbol{\sigma}_{\eta^*})$ be a solution to Problem $(HVI_{1\eta^*})$. Since

$$\boldsymbol{\sigma}_{\eta^*}(t) = \mathcal{A}(t, \boldsymbol{\varepsilon}(\boldsymbol{u}'_{\eta^*}(t))) + \boldsymbol{\eta}^*(t) \text{ for } t \in [0, T]$$

and

$$\boldsymbol{\eta}^*(t) = \mathcal{E}(t, \boldsymbol{\varepsilon}(\boldsymbol{u}_{\eta^*}(t))) + \int_0^t \mathcal{G}(s, \boldsymbol{\eta}^*(s), \boldsymbol{\varepsilon}(\boldsymbol{u}_{\eta^*}(s)))\, ds,$$

we have

$$\boldsymbol{\sigma}_{\eta^*}(t) = \mathcal{A}(t, \boldsymbol{\varepsilon}(\boldsymbol{u}'_{\eta^*}(t))) + \mathcal{E}(t, \boldsymbol{\varepsilon}(\boldsymbol{u}_{\eta^*}(t))) + \int_0^t \mathcal{G}(s, \boldsymbol{\eta}^*(s), \boldsymbol{\varepsilon}(\boldsymbol{u}_{\eta^*}(s)))\, ds,$$

for a.e. $t \in (0, T)$. This concludes the proof of Theorem 13.4 in the case of Problem (HVI_1).

To prove Theorem 13.4 for Problem (HVI_2) we proceed in the similar way. The difference arises in Step 1 of the proof, where the function g is now defined by

$$g(\boldsymbol{x}, t, \boldsymbol{\zeta}, \boldsymbol{\rho}, \boldsymbol{\xi}, \boldsymbol{\eta}) = j_2(\boldsymbol{x}, t, \boldsymbol{\zeta}, \boldsymbol{\rho}, \eta_\nu) + j_4(\boldsymbol{x}, t, \boldsymbol{\zeta}, \boldsymbol{\rho}, \boldsymbol{\eta}_\tau)$$

for $\boldsymbol{\zeta}, \boldsymbol{\rho}, \boldsymbol{\xi}, \boldsymbol{\eta} \in \mathbb{R}^d$ and a.e. $(\boldsymbol{x}, t) \in \Gamma_C \times (0, T)$. Then, it follows that the function G is independent of the variable $\boldsymbol{u}$. The rest of the proof is the same and we conclude from here the proof of Theorem 13.4 in the case of Problem (HVI_2).

13.6 Examples of Boundary Conditions

In this section we present three examples of boundary conditions of types (13.6) and (13.7) for which our main existence and uniqueness result works.

1. **Contact with nonmonotone normal compliance.** This multivalued contact condition describes reactive foundation assigning a reactive normal traction or pressure that depends on the interpenetration of the asperities on the body surface and those on the foundation. We comment on it in a simple case when

$$-\sigma_\nu(t) \in \partial j_1(u_\nu(t)) \text{ on } \Gamma_C \times (0,T) \tag{13.51}$$

with $j_1 : \mathbb{R} \to \mathbb{R}$ defined by

$$j_1(r) = \int_0^r p(s)\, ds \text{ for } r \in \mathbb{R}.$$

We assume the following hypothesis on the integrand of j_1.

$H(p)$: $p : \mathbb{R} \to \mathbb{R}$ is a function such that

$$p \in L^\infty_{loc}(\mathbb{R}),\ |p(s)| \le p_1(1+|s|) \text{ for } s \in \mathbb{R} \text{ with } p_1 > 0.$$

Then it is well known that $\partial j_1(s) = \hat{p}(s)$ for $s \in \mathbb{R}$, where the multivalued function $\hat{p} : \mathbb{R} \to 2^{\mathbb{R}}$ is given by $\hat{p}(s) = [p^{(1)}(s), p^{(2)}(s)]$ ($[\cdot,\cdot]$ denotes an interval in $\mathbb{R}$) and

$$p^{(1)}(r) = \lim_{\varepsilon \to 0+} \operatorname*{ess\,inf}_{|\tau - r| \le \varepsilon} p(\tau), \qquad p^{(2)}(r) = \lim_{\varepsilon \to 0+} \operatorname*{ess\,sup}_{|\tau - r| \le \varepsilon} p(\tau).$$

Under the hypothesis $H(p)$, the function j_1 is a locally Lipschitz and $|\partial j_1(r)| \le p_1(1+|r|)$ for $r \in \mathbb{R}$. Therefore, it is easy to see that the function j_1 satisfies assumption $H(j_1)$.

2. **Friction contact between reinforcement and concrete.** We consider boundary conditions of the form

$$-\boldsymbol{\sigma}_\tau(t) \in \partial j_3(t, \boldsymbol{u}(t), \boldsymbol{u}'(t), \boldsymbol{u}_\tau(t)) \text{ on } \Gamma_C \times (0,T). \tag{13.52}$$

This relation describes the tangential contact law between reinforcement and concrete in a concrete structure. In literature, cf. Chap. 2.4 in Panagiotopoulos [18] (Fig. 2.4.1), Chap. 1.4 in Naniewicz and Panagiotopoulos [16] (Fig. 1.4.3), one can find a couple of examples of the superpotential j_3 which describes such type of contact. We give an example of nonconvex function which appears in this type of contact condition. The superpotential $j_3 : \mathbb{R} \to \mathbb{R}$ is of the following form.

$$j_3(r) = \begin{cases} 0 & \text{if } r < 0, \\ 2r^2 & \text{if } 0 \le r < 1, \\ -\frac{1}{3}r^3 + r^2 + 3r - \frac{5}{3} & \text{if } 1 \le r < 3, \\ \frac{22}{3} & \text{if } r \ge 3, \end{cases}$$

It is easy to check that the function j_3 satisfies $H(j_3)$ with $c_{30} = 4$ and $c_{31} = c_{32} = c_{33} = 0$.

3. **Nonmonotone friction depending on slip and slip rate.** We consider the nonmonotone friction conditions which depend on both the slip and the slip rate. This is the case when the superpotential $j_4 = j_4(\boldsymbol{x}, t, \boldsymbol{\zeta}, \boldsymbol{\rho}, \boldsymbol{\theta})$ depends on $\boldsymbol{\zeta}$ and $\boldsymbol{\rho}$, and it is nonconvex in $\boldsymbol{\theta}$. As an example of this function we choose

$$j_4(\boldsymbol{x}, t, \boldsymbol{\zeta}, \boldsymbol{\rho}, \boldsymbol{\theta}) = \psi(\boldsymbol{x}, t, \boldsymbol{\zeta}, \boldsymbol{\rho})\, h(\boldsymbol{\theta}) \tag{13.53}$$

for $\boldsymbol{\zeta}$, $\boldsymbol{\rho}$, $\boldsymbol{\theta} \in \mathbb{R}^d$, a.e. $(\boldsymbol{x}, t) \in \Gamma_C \times (0, T)$, where $\psi \colon \Gamma_C \times (0, T) \times (\mathbb{R}^d)^2 \to \mathbb{R}$ satisfies

$$\begin{cases} \psi(\cdot, \cdot, \boldsymbol{\zeta}, \boldsymbol{\rho}) \text{ is measurable for all } \boldsymbol{\zeta}, \boldsymbol{\rho} \in \mathbb{R}^d, \\ \psi(\boldsymbol{x}, t, \cdot, \cdot) \text{ is continuous for a.e. } (\boldsymbol{x}, t) \in \Gamma_C \times (0, T), \\ 0 \le \psi(\boldsymbol{x}, t, \boldsymbol{\zeta}, \boldsymbol{\rho}) \le \psi_0(1 + \|\boldsymbol{\zeta}\| + \|\boldsymbol{\rho}\|) \text{ for all } \boldsymbol{\zeta}, \boldsymbol{\rho} \in \mathbb{R}^d, \\ \qquad \qquad \text{a.e. } (\boldsymbol{x}, t) \in \Gamma_C \times (0, T) \text{ with } \psi_0 > 0 \end{cases}$$

and $h \colon \mathbb{R}^d \to \mathbb{R}$ is a locally Lipschitz function such that $h(0) = 0$ and

$$\|\partial h(\boldsymbol{\theta})\| \le h_0 \quad \text{for } \boldsymbol{\theta} \in \mathbb{R}^d \text{ with } h_0 > 0.$$

Under these hypotheses on ψ and h, the function j_4 given by (13.53) satisfies $H(j_4)$(i)–(iii) with $c_{40} = c_{41} = c_{42} = h_0\, \psi_0$ and $c_{43} = 0$. The friction law (13.7) takes now the form

$$-\boldsymbol{\sigma}_\tau(t) \in \psi(t, \boldsymbol{u}(t), \boldsymbol{u}'(t))\, \partial h(\boldsymbol{u}'_\tau(t)) \quad \text{on } \Gamma_C \times (0, T). \tag{13.54}$$

It is clear that $j_4(\boldsymbol{x}, t, \boldsymbol{\zeta}, \boldsymbol{\rho}, \cdot)$ is regular if and only if h is regular. Next, let $(\boldsymbol{\zeta}_n, \boldsymbol{\rho}_n, \boldsymbol{\theta}_n) \in (\mathbb{R}^d)^3$, $(\boldsymbol{\zeta}_n, \boldsymbol{\rho}_n, \boldsymbol{\theta}_n) \to (\boldsymbol{\zeta}, \boldsymbol{\rho}, \boldsymbol{\theta})$ and $\boldsymbol{\sigma} \in \mathbb{R}^d$. We have

$$\begin{aligned} \limsup j_4^0(\boldsymbol{x}, t, \boldsymbol{\zeta}_n, \boldsymbol{\rho}_n, \boldsymbol{\theta}_n; \boldsymbol{\sigma}) &= \limsup \psi(\boldsymbol{x}, t, \boldsymbol{\zeta}_n, \boldsymbol{\rho}_n)\, h^0(\boldsymbol{\theta}_n; \boldsymbol{\sigma}) \\ &= \limsup\, [(\psi(\boldsymbol{x}, t, \boldsymbol{\zeta}_n, \boldsymbol{\rho}_n) - \psi(\boldsymbol{x}, t, \boldsymbol{\zeta}, \boldsymbol{\rho}))\, h^0(\boldsymbol{\theta}_n; \boldsymbol{\sigma}) + \psi(\boldsymbol{x}, t, \boldsymbol{\zeta}, \boldsymbol{\rho})\, h^0(\boldsymbol{\theta}_n; \boldsymbol{\sigma})] \\ &\le h_0 \|\boldsymbol{\sigma}\|_{\mathbb{R}^d} \lim\, (\psi(\boldsymbol{x}, t, \boldsymbol{\zeta}_n, \boldsymbol{\rho}_n) - \psi(\boldsymbol{x}, t, \boldsymbol{\zeta}, \boldsymbol{\rho})) + \psi(\boldsymbol{x}, t, \boldsymbol{\zeta}, \boldsymbol{\rho}) \limsup h^0(\boldsymbol{\theta}_n; \boldsymbol{\sigma}) \\ &\le \psi(\boldsymbol{x}, t, \boldsymbol{\zeta}, \boldsymbol{\rho})\, h^0(\boldsymbol{\theta}; \boldsymbol{\sigma}) = j_4^0(\boldsymbol{x}, t, \boldsymbol{\zeta}, \boldsymbol{\rho}, \boldsymbol{\theta}; \boldsymbol{\sigma}) \end{aligned}$$

for a.e. $(\boldsymbol{x},t) \in \Gamma_C \times (0,T)$. Hence $H(j_4)$(iv) holds. Moreover, for instance, if $\psi(\boldsymbol{x},t,\cdot,\cdot)$ is Lipschitz continuous for a.e. $(\boldsymbol{x},t) \in \Gamma_C \times (0,T)$ (i.e. $|\psi(\boldsymbol{x},t,\boldsymbol{\zeta}_1,\boldsymbol{\rho}_1) - \psi(\boldsymbol{x},t,\boldsymbol{\zeta}_2,\boldsymbol{\rho}_2)| \le L_\psi\,(\|\boldsymbol{\zeta}_1 - \boldsymbol{\zeta}_2\|_{\mathbb{R}^d} + \|\boldsymbol{\rho}_1 - \boldsymbol{\rho}_2\|_{\mathbb{R}^d})$ for all $\boldsymbol{\zeta}_1$, $\boldsymbol{\zeta}_2$, $\boldsymbol{\rho}_1$, $\boldsymbol{\rho}_2 \in \mathbb{R}^d$, a.e. $(\boldsymbol{x},t) \in \Gamma_C \times (0,T)$) and h is convex, then

$$
\begin{aligned}
&(\partial j_4(\boldsymbol{x},t,\boldsymbol{\zeta}_1,\boldsymbol{\rho}_1,\boldsymbol{\theta}_1) - \partial j_4(\boldsymbol{x},t,\boldsymbol{\zeta}_2,\boldsymbol{\rho}_2,\boldsymbol{\theta}_2), \boldsymbol{\theta}_1 - \boldsymbol{\theta}_2) \\
&\quad = ((\psi(\boldsymbol{x},t,\boldsymbol{\zeta}_1,\boldsymbol{\rho}_1) - \psi(\boldsymbol{x},t,\boldsymbol{\zeta}_2,\boldsymbol{\rho}_2))\,\partial h(\boldsymbol{\theta}_1), \boldsymbol{\theta}_1 - \boldsymbol{\theta}_2) \\
&\qquad + \psi(\boldsymbol{x},t,\boldsymbol{\zeta}_2,\boldsymbol{\rho}_2)\,(\partial h(\boldsymbol{\theta}_1) - \partial h(\boldsymbol{\theta}_2), \boldsymbol{\theta}_1 - \boldsymbol{\theta}_2) \\
&\quad \ge -L_\psi\, h_0\,(\|\boldsymbol{\zeta}_1 - \boldsymbol{\zeta}_2\|_{\mathbb{R}^d} + \|\boldsymbol{\rho}_1 - \boldsymbol{\rho}_2\|_{\mathbb{R}^d})\,\|\boldsymbol{\theta}_1 - \boldsymbol{\theta}_2\|_{\mathbb{R}^d}
\end{aligned}
$$

for all $\boldsymbol{\zeta}_1, \boldsymbol{\zeta}_2, \boldsymbol{\rho}_1, \boldsymbol{\rho}_2 \in \mathbb{R}^d$, a.e. $(\boldsymbol{x},t) \in \Gamma_C \times (0,T)$, which implies that the condition $H(j_4)(v)$ is satisfied with $L_4 = L_\psi h_0$.

References

1. Amassad, A., Fabre, C., Sofonea, M.: A quasistatic viscoplastic contact problem with normal compliance and friction. IMA J. Appl. Math. **69**, 463–482 (2004)
2. Campo, M., Fernandez, J.R., Kuttler, K.L.: An elastic-viscoplastic quasistatic contact problem with damage. Comput. Methods Appl. Mech. Eng. **196**, 3219–3229 (2007)
3. Clarke, F.H.: Optimization and Nonsmooth Analysis. Wiley Interscience, New York (1983)
4. Denkowski, Z., Migórski, S., Papageorgiou, N.S.: An Introduction to Non-linear Analysis: Applications. Kluwer Academic/Plenum, Boston/Dordrecht/London/New York (2003)
5. Drozdov, A.D.: Finite Elasticity and Viscoelasticity: A Course in the Nonlinear Mechanics of Solids. World Scientific, Singapore (1996)
6. Han, W., Sofonea, M.: Quasistatic Contact Problems in Viscoelasticity and Viscoplasticity. American Mathematical Society/International Press, Providence (2002)
7. Kulig, A.: Hemivariational inequality approach to the dynamic viscoelastic contact problem with nonmonotone normal compliance and slip-dependent friction. Nonlinear Anal. Real World Appl. **9**, 1741–1755 (2008)
8. Kulig, A.: Hyperbolic hemivariational inequalities for dynamic viscoelastic contact problems. J. Elasticity **110**, 1–31 (2013)
9. Kulig, A., Migórski, S.: Nonlinear evolution inclusions and hemivariational inequalities for nonsmooth problems in contact mechanics. Nonlinear Anal. Theory Methods Appl. **75**, 4729–4746 (2012)
10. Kuttler, K.L.: Dynamic friction contact problem with general normal and friction laws. Nonlinear Anal. Theory Methods Appl. **28**, 559–575 (1997)
11. Kuttler, K.L., Shillor, M.: Set-valued pseudomonotone maps and degenerate evolution inclusions. Commun. Contemp. Math. **1**, 87–123 (1999)
12. Migórski, S.: Dynamic hemivariational inequality modeling viscoelastic contact problem with normal damped response and friction. Appl. Anal. **84**, 669–699 (2005)
13. Migórski, S., Ochal, A.: Hemivariational inequality for viscoelastic contact problem with slip dependent friction. Nonlinear Anal. Theory Methods Appl. **61**, 135–161 (2005)
14. Migórski, S., Ochal, A.: A unified approach to dynamic contact problems in viscoelasticity. J. Elasticity **83**, 247–275 (2006)
15. Migórski, S., Ochal, A., Sofonea, M.: Analysis of a dynamic elastic-viscoplastic contact problem with friction. Discrete Continuous Dyn. Syst. Ser. B **10**, 887–902 (2008)

16. Naniewicz, Z., Panagiotopoulos, P.D.: Mathematical Theory of Hemivariational Inequalities and Applications. Marcel Dekker, New York/Basel/Hong Kong (1995)
17. Nečas, J., Hlaváček, I.: Mathematical Theory of Elastic and Elasto-Plastic Bodies: An Introduction. Elsevier, Amsterdam/Oxford (1981)
18. Panagiotopoulos, P.D.: Hemivariational Inequalities, Applications in Mechanics and Engineering. Springer, Berlin (1993)
19. Shillor, M., Sofonea, M., Telega, J.J.: Models and Analysis of Quasistatic Contact. Springer, Berlin (2004)
20. Zeidler, E.: Nonlinear Functional Analysis and Applications, II A/B. Springer, New York (1990)

Chapter 14
Two History-Dependent Contact Problems

Mircea Sofonea, Stanisław Migórski, and Anna Ochal

Abstract We consider two initial boundary value problems which describe the evolution of a viscoelastic and viscoplastic body, respectively, in contact with a piston or a device. In both problems the contact process is assumed to be dynamic and the friction is described with a subdifferential boundary condition. Both the constitutive laws and the contact conditions we use involve memory terms. For each problem we derive a variational formulation which is in the form of a system coupling a nonlinear integral equation with a history-dependent hemivariational inequality. Then, we prove the existence of a weak solution and, under additional assumptions, its uniqueness. The proofs are based on results for history-dependent hemivariational inequalities presented in Chap. 2.

Keywords Dynamic frictional contact • Viscoelastic material • Viscoplastic material • History-dependent operator • Hemivariational inequality

AMS Classification. 35L86, 47J20, 47J22, 74M15, 74H20, 74H25

14.1 Preliminaries

In this section we recall the notation and a result on history-dependent hemivariational inequalities of second order presented in Chap. 2.

Given a normed space $(E, \|\cdot\|_E)$ we denote by E^* its dual space and $\langle\cdot,\cdot\rangle_{E^*\times E}$ will represent the duality pairing of E and E^*. The space of all linear and continuous operators from a normed space E to a normed space F is denoted by $\mathcal{L}(E, F)$. Let

M. Sofonea (✉)
Laboratoire de Mathématiques et Physique (LAMPS), Université de Perpignan Via Domitia, 52 Avenue Paul Alduy, 66860 Perpignan, France
e-mail: sofonea@univ-perp.fr

S. Migórski • A. Ochal
Faculty of Mathematics and Computer Science, Jagiellonian University, ul. Łojasiewicza 6, 30348 Krakow, Poland
e-mail: stanislaw.migorski@uj.edu.pl; ochal@ii.uj.edu.pl

W. Han et al. (eds.), *Advances in Variational and Hemivariational Inequalities*, Advances in Mechanics and Mathematics 33, DOI 10.1007/978-3-319-14490-0_14

$h: E \to \mathbb{R}$ be a locally Lipschitz function. The generalized directional derivative of h at $x \in E$ in the direction $v \in E$ will be denoted by $h^0(x; v)$ and the generalized gradient of h at $x \in E$ will be denoted by $\partial h(x)$. Recall also that a locally Lipschitz function h is called regular (in the sense of Clarke) at $x \in E$ if for all $v \in E$ the one-sided directional derivative $h'(x; v)$ exists and satisfies $h^0(x; v) = h'(x; v)$ for all $v \in E$. Also, everywhere below we use bold face letters for vectors and tensors, as it is usual in Contact Mechanics.

Let $\Omega \subset \mathbb{R}^d$ be a bounded domain with a Lipschitz boundary Γ and let Γ_C be a measurable part of Γ, $\Gamma_C \subseteq \Gamma$. Let V be a closed subspace of $H^1(\Omega; \mathbb{R}^d)$ and $H = L^2(\Omega; \mathbb{R}^d)$. It is well known that $V \subset H \subset V^*$ form an evolution triple of spaces, cf. e.g., Section 3.4 of [5]. We introduce the trace operator $\gamma: V \to L^2(\Gamma; \mathbb{R}^d)$ and its adjoint $\gamma^*: L^2(\Gamma; \mathbb{R}^d) \to V^*$. We set $\mathcal{V} = L^2(0, T; V)$ and introduce the space $\mathcal{W} = \{w \in \mathcal{V} \mid w' \in \mathcal{V}^*\}$, where $\mathcal{V}^* = L^2(0, T; V^*)$ is the dual space to $\mathcal{V}$ and the time derivative $w' = \partial w / \partial t$ is understood in the sense of vector-valued distributions. The space $\mathcal{W}$ endowed with the graph norm $\|w\|_{\mathcal{W}} = \|w\|_{\mathcal{V}} + \|w'\|_{\mathcal{V}^*}$ is a Banach space which is separable and reflexive due to the separability and reflexivity of $\mathcal{V}$ and $\mathcal{V}^*$.

We consider the following hemivariational inequality of second order.

$$
\left.\begin{aligned}
&\text{Find } u \in \mathcal{V} \text{ such that } u' \in \mathcal{W} \text{ and} \\
&\langle u''(t) + A(t, u'(t)) + (\mathcal{S}u')(t), v \rangle_{V^* \times V} \\
&\qquad + \int_{\Gamma_C} j^0(x, t, \gamma u'(t); \gamma v)\, d\Gamma \geq \langle f(t), v \rangle_{V^* \times V} \\
&\qquad\qquad\qquad\qquad \text{for all } v \in V \text{ and a.e. } t \in (0, T), \\
&u(0) = u_0, \;\; u'(0) = v_0.
\end{aligned}\right\} \qquad (14.1)
$$

In the study of the inequality (14.1) we need the following hypotheses on the data.

$$
\left.\begin{aligned}
&A: (0, T) \times V \to V^* \text{ is such that} \\
&\text{(a) } A(\cdot, v) \text{ is measurable on } (0, T) \text{ for all } v \in V. \\
&\text{(b) } A(t, \cdot) \text{ is pseudomonotone on } V \text{ for a.e. } t \in (0, T). \\
&\text{(c) } \|A(t, v)\|_{V^*} \leq a_0(t) + a_1 \|v\|_V \text{ for all } v \in V, \text{ a.e. } t \in (0, T) \\
&\qquad \text{with } a_0 \in L^2(0, T), a_0 \geq 0 \text{ and } a_1 > 0. \\
&\text{(d) } \langle A(t, v), v \rangle_{V^* \times V} \geq \alpha \|v\|_V^2 \text{ for all } v \in V, \text{ a.e. } t \in (0, T) \\
&\qquad \text{with } \alpha > 0. \\
&\text{(e) } A(t, \cdot) \text{ is strongly monotone for a.e. } t \in (0, T), \text{ i.e., there} \\
&\qquad \text{is } m_1 > 0 \text{ such that for all } v_1, v_2 \in V, \text{ a.e. } t \in (0, T) \\
&\qquad \langle A(t, v_1) - A(t, v_2), v_1 - v_2 \rangle_{V^* \times V} \geq m_1 \|v_1 - v_2\|_V^2.
\end{aligned}\right\} \qquad (14.2)
$$

$$\left.\begin{array}{l}\mathcal{S}\colon \mathcal{V} \to \mathcal{V}^* \text{ is such that} \\ \qquad \|(\mathcal{S}v_1)(t) - (\mathcal{S}v_2)(t)\|_{V^*} \le L_{\mathcal{S}} \displaystyle\int_0^t \|v_1(s) - v_2(s)\|_V \, ds \\ \text{for all } v_1,\ v_2 \in \mathcal{V},\ \text{a.e. } t \in (0,T) \text{ with } L_{\mathcal{S}} > 0.\end{array}\right\} \tag{14.3}$$

$$\left.\begin{array}{l} j\colon \Gamma_C \times (0,T) \times \mathbb{R}^d \to \mathbb{R} \text{ is such that} \\ \text{(a) } j(\cdot,\cdot,\xi) \text{ is measurable for all } \xi \in \mathbb{R}^d \text{ and there exists} \\ \qquad e \in L^2(\Gamma_C;\mathbb{R}^d) \text{ such that } j(\cdot,\cdot,e(\cdot)) \in L^1(\Gamma_C \times (0,T)). \\ \text{(b) } j(x,t,\cdot) \text{ is locally Lipschitz for a.e. } (x,t) \in \Gamma_C \times (0,T). \\ \text{(c) } \|\partial j(x,t,\xi)\|_{\mathbb{R}^d} \le b_0(x,t) + b_1 \|\xi\|_{\mathbb{R}^d} \text{ for all } \xi \in \mathbb{R}^d, \\ \qquad \text{a.e. } (x,t) \in \Gamma_C \times (0,T) \text{ with } b_0 \in L^2(\Gamma_C \times (0,T)), \\ \qquad b_0, b_1 \ge 0. \\ \text{(d) } (\zeta_1 - \zeta_2, \xi_1 - \xi_2)_{\mathbb{R}^d} \ge -\overline{m}_2 \|\xi_1 - \xi_2\|^2_{\mathbb{R}^d} \text{ for all } \xi_i \in \mathbb{R}^d, \\ \qquad \zeta_i \in \partial j(x,t,\xi_i), i = 1,2 \text{ with } \overline{m}_2 \ge 0. \end{array}\right\} \tag{14.4}$$

$$f \in \mathcal{V}^*,\ u_0 \in V,\ v_0 \in V. \tag{14.5}$$

$$m_1 \ge \overline{m}_2 \|\gamma\|^2. \tag{14.6}$$

$$\left.\begin{array}{l}\text{One of the following conditions is satisfied} \\ \text{(a) } \alpha > 2\sqrt{2}\, b_1 \|\gamma\|^2, \text{ where } \|\gamma\| = \|\gamma\|_{\mathcal{L}(V,L^2(\Gamma;\mathbb{R}^d))}. \\ \text{(b) } j^0(x,t,\xi;-\xi) \le \overline{d}_0 (1 + \|\xi\|_{\mathbb{R}^d}) \text{ for all } \xi \in \mathbb{R}^d, \text{ a.e.} \\ \qquad (x,t) \in \Gamma_C \times (0,T) \text{ with } \overline{d}_0 \ge 0.\end{array}\right\} \tag{14.7}$$

$$\left.\begin{array}{l} j\colon \Gamma_C \times (0,T) \times \mathbb{R}^d \to \mathbb{R} \text{ is such that} \\ \qquad \text{either } j(x,t,\cdot) \text{ or } -j(x,t,\cdot) \text{ is regular on } \mathbb{R}^d \text{ for a.e.} \\ \qquad (x,t) \in \Gamma_C \times (0,T). \end{array}\right\} \tag{14.8}$$

The following result on the existence and uniqueness of solution to the history-dependent hemivariational inequality (14.1) was proved in Corollary 2.20, see page 61.

Theorem 14.1. *Assume that* (14.2)–(14.7) *hold. Then the hemivariational inequality* (14.1) *has at least one solution. If, in addition* (14.8) *holds, then the solution to* (14.1) *is unique.*

Using the terminology introduced in Chap. 2 on page 52 we refer to the operators which satisfy (14.3) as *history-dependent operators*. For this reason, we say that a hemivariational inequality of the form (14.1) represents a *history-dependent hemivariational inequality*.

14.2 Modeling of Contact Problems

The physical setting is as follows. A deformable body occupies a domain Ω of $\mathbb{R}^d$, $d = 2, 3$ in applications. The body is acted upon by volume forces and surface tractions and, as a result, its state is evolving. The boundary Γ of Ω is supposed to be Lipschitz continuous and, therefore, the unit outward normal vector exists a.e. on Γ. It is supposed that Γ is divided into three mutually disjoint measurable parts Γ_D, Γ_N and Γ_C such that the measure of Γ_D is positive. We assume that the body is clamped on Γ_D, so the displacement field vanishes there. Volume forces of density $\boldsymbol{f}_0$ act in Ω and surface tractions of density $\boldsymbol{f}_N$ are applied on Γ_N. On Γ_C the body is or could arrive in contact with an obstacle, the so-called foundation.

We are interested in mathematical models which describe the evolution of the mechanical state of the body, in the physical setting above. To this end we use the notation $\boldsymbol{x} = (x_i)$ for a point in $\Omega \cup \Gamma$ and we denote by $\boldsymbol{\nu} = (\nu_i)$ the outward unit normal at Γ. Here and below, the indices i, j, k, l run between 1 and d and, unless stated otherwise, the summation convention over repeated indices is used. Also, the index that follows a comma indicates a partial derivative with the corresponding component of the spatial variable $\boldsymbol{x}$. We denote by $\boldsymbol{u} = (u_i)$, $\boldsymbol{\sigma} = (\sigma_{ij})$, and $(\boldsymbol{u}) = (\varepsilon_{ij}(\boldsymbol{u}))$ the displacement vector, the stress tensor, and linearized strain tensor, respectively. These are functions which depend on the spatial variable $\boldsymbol{x}$ and on the time variable t. Nevertheless, in what follows we do not indicate explicitly the dependence of these quantities on $\boldsymbol{x}$ and t and for instance, we write $\boldsymbol{\sigma}(t)$ instead of $\boldsymbol{\sigma}(\boldsymbol{x})$ or $\boldsymbol{\sigma}(\boldsymbol{x}, t)$. We recall that the components of the linearized strain tensor $\boldsymbol{\varepsilon}(\boldsymbol{u})$ are given by

$$\varepsilon_{ij}(\boldsymbol{u}) = \frac{1}{2}(u_{i,j} + u_{j,i}).$$

Everywhere below T is a positive constant, $[0, T]$ denotes the time interval of interest and primes will represent the derivatives with respect to time, i.e., $\boldsymbol{u}' = \partial \boldsymbol{u}/\partial t$ and $\boldsymbol{u}'' = \partial^2 \boldsymbol{u}/\partial t^2$.

We also use $\mathbb{R}^d$ for the d-dimensional real linear space and the symbol $\mathbb{S}^d$ stands for the space of second order symmetric tensors on $\mathbb{R}^d$ or, equivalently, the space of symmetric matrices of order d. The canonical inner products and the corresponding norms on $\mathbb{R}^d$ and $\mathbb{S}^d$ are given by

$$\boldsymbol{u} \cdot \boldsymbol{v} = u_i v_i, \quad \|\boldsymbol{v}\|_{\mathbb{R}^d} = (\boldsymbol{v} \cdot \boldsymbol{v})^{1/2} \quad \text{for all } \boldsymbol{u} = (u_i),\ \boldsymbol{v} = (v_i) \in \mathbb{R}^d,$$

$$\boldsymbol{\sigma} \cdot \boldsymbol{\tau} = \sigma_{ij} \tau_{ij}, \quad \|\boldsymbol{\tau}\|_{\mathbb{S}^d} = (\boldsymbol{\tau} \cdot \boldsymbol{\tau})^{1/2} \quad \text{for all } \boldsymbol{\sigma} = (\sigma_{ij}),\ \boldsymbol{\tau} = (\tau_{ij}) \in \mathbb{S}^d,$$

respectively. For a vector field, we use the notation v_ν and $\boldsymbol{v}_\tau$ for the normal and tangential components of $\boldsymbol{v}$ on Γ given by $v_\nu = \boldsymbol{v} \cdot \boldsymbol{\nu}$ and $\boldsymbol{v}_\tau = \boldsymbol{v} - v_\nu \boldsymbol{\nu}$. Finally, we recall that the normal and tangential components of the stress field $\boldsymbol{\sigma}$ on the boundary are defined by $\sigma_\nu = (\boldsymbol{\sigma}\boldsymbol{\nu}) \cdot \boldsymbol{\nu}$ and $\boldsymbol{\sigma}_\tau = \boldsymbol{\sigma}\boldsymbol{\nu} - \sigma_\nu \boldsymbol{\nu}$, respectively.

To present a mathematical model for a specific contact process there is a need to precise the constitutive law, the balance equation, the boundary conditions, the contact conditions and, eventually, the initial conditions. A constitutive law represents a relationship between the stress $\boldsymbol{\sigma}$, the strain $\boldsymbol{\varepsilon}$ and their derivatives, eventually, which characterizes a specific material. It describes the deformations of the body resulting from the action of forces and tractions. Though the constitutive laws must satisfy some basic axioms and invariance principles, they originate mostly from experiments. We refer the reader to [4, 9, 10] for a general description of several diagnostic experiments which provide information needed in constructing constitutive laws for specific materials.

One of the most popular constitutive law for viscoelastic materials is the Kelvin-Voigt constitutive law, i.e.,

$$\boldsymbol{\sigma}(t) = \mathcal{A}\boldsymbol{\varepsilon}(\boldsymbol{u}'(t)) + \mathcal{B}\boldsymbol{\varepsilon}(\boldsymbol{u}(t)). \tag{14.9}$$

Here $\mathcal{A}$ and $\mathcal{B}$ represent the viscosity and the elasticity operators, respectively. Quasistatic contact problems for viscoelastic materials of the form (14.9) have been considered in [9, 16, 19, 20] and the references therein. There, the contact was modeled with normal compliance with or without unilateral constraint and with subdifferential boundary conditions, as well. Friction was described by versions of Coulomb's law of dry friction and its regularization. The unique weak solvability of the corresponding problems was proved by using arguments of variational and hemivariational inequalities with monotone operators and fixed point. The numerical analysis of part of these models can be found in [9]. There, semi-discrete and fully discrete scheme were considered, error estimates and convergence results were proved and numerical simulation in the study of two-dimensional test problems were presented. The analysis of various dynamic problems for materials of the form (14.9), including existence and uniqueness results, has been carried out in [7] and the references therein.

A large number of viscoplastic constitutive laws for viscoplastic materials used in the literature can be cast on the general form

$$\boldsymbol{\sigma}'(t) = \mathcal{B}\boldsymbol{\varepsilon}(\boldsymbol{u}'(t)) + \mathcal{G}(\boldsymbol{\sigma}(t), \boldsymbol{\varepsilon}(\boldsymbol{u}(t))). \tag{14.10}$$

Here $\mathcal{B}$ and $\mathcal{G}$ represent the elasticity operator and the viscoplastic potential, respectively. Existence and uniqueness results for quasistatic contact problems for viscoelastic materials of the form (14.10) have been considered in [9, 19, 20] and the references therein. The numerical analysis of part of these models can be found in [9].

The boundary conditions on the contact surface are divided naturally into conditions in the normal direction (called also contact conditions) and those in the tangential directions (called also friction laws). The so-called *normal compliance* contact condition describes a deformable foundation. It assigns a reactive normal pressure that depends on the interpenetration of the asperities on the body surface and those of the foundation. A general expression for the normal reactive pressure is

$$-\sigma_\nu = p_\nu(u_\nu), \tag{14.11}$$

where p_ν is a prescribed nonnegative function which vanishes for negative argument. Indeed, when $u_\nu < 0$ there is no contact and the normal pressure vanishes. When there is contact then u_ν is positive and it represents a measure of the interpenetration of the asperities. Then, condition (14.11) shows that the foundation exerts a pressure on the body which depends on the penetration. The normal compliance contact condition was first introduced in [17] and since then used in many publications, see e.g. [7,9,11–14] and the references therein. The term *normal compliance* was first used in [12, 13].

In the case when the friction force $\boldsymbol{\sigma}_\tau$ does not vanish on the contact surface, the contact is *frictional*. Frictional contact is usually modeled with the *Coulomb law of dry friction* or its variants. It states that the magnitude of the friction force is bounded by a function, the so-called friction bound, which is the maximal frictional resistance that the surface can generate; also, once slip starts, the friction force opposes the direction of the motion and its magnitude reaches the friction bound. Thus,

$$\|\boldsymbol{\sigma}_\tau\|_{\mathbb{S}^d} \le F_b, \quad \boldsymbol{\sigma}_\tau = -\,F_b\,\frac{\boldsymbol{u}'_\tau}{\|\boldsymbol{u}'_\tau\|_{\mathbb{R}^d}} \quad \text{if } \boldsymbol{u}'_\tau \neq \boldsymbol{0} \text{ on } \Gamma_C, \tag{14.12}$$

where $\boldsymbol{u}'_\tau$ represents the tangential velocity or slip rate and F_b is the friction bound. On a nonhomogeneous surface F_b depends explicitly on the position $\boldsymbol{x}$ on the surface. It could also depend on the process variables. Nevertheless, when F_b depends only on the spatial variable $\boldsymbol{x}$, we refer to (14.12) as the Tresca friction law.

Note that the Coulomb law (14.12) is characterized by the existence of stick-slip zones on the contact boundary. Indeed, it follows from (14.12) that, if in a point $\boldsymbol{x} \in \Gamma_C$ the inequality $\|\boldsymbol{\sigma}_\tau(\boldsymbol{x})\|_{\mathbb{S}^d} < F_b(\boldsymbol{x})$ holds, then $\boldsymbol{u}'_\tau(\boldsymbol{x}) = \boldsymbol{0}$ and the material point $\boldsymbol{x}$ is in the so-called *stick zone*; if $\|\boldsymbol{\sigma}_\tau(\boldsymbol{x})\|_{\mathbb{S}^d} = F_b(\boldsymbol{x})$ then the point $\boldsymbol{x}$ is in the so-called *slip zone*. We conclude that the Coulomb friction law (14.12) models the phenomenon that slip may occur only when the friction force reaches a critical value, the friction bound F_b.

In variational formulation, frictional contact problems with Coulomb's law lead to variational inequalities involving nondifferentiable functionals and, for this reason, their numerical analysis could present some difficulties. To avoid these difficulties, several regularizations of Coulomb's law (14.12) are used in the literature. A simple example is given by

$$\boldsymbol{\sigma}_\tau = -F_b\,\frac{\boldsymbol{u}'_\tau}{\sqrt{\|\boldsymbol{u}'_\tau\|^2_{\mathbb{R}^d} + \rho^2}} \quad \text{on } \Gamma_C, \tag{14.13}$$

where $\rho > 0$ is a regularization parameter. Note that the friction law (14.13) describes situation when slip appears even for small tangential shears which is the case when the surfaces are lubricated by a thin layer of non-Newtonian fluid. We remark that the Coulomb law (14.12) is obtained, formally, from the friction law (14.13) in the limit as $\rho \to 0$.

Everywhere in what follows we use the standard notation for Lebesgue and Sobolev spaces associated to Ω and Γ. We recall that if $\boldsymbol{\sigma}$ denotes a smooth stress tensor then the following Green formula holds

$$\int_{\Omega} \boldsymbol{\sigma} \cdot \boldsymbol{\varepsilon}(\boldsymbol{v})\, dx + \int_{\Omega} \mathrm{Div}\, \boldsymbol{\sigma} \cdot \boldsymbol{v}\, dx = \int_{\Gamma} \boldsymbol{\sigma}\boldsymbol{\nu} \cdot \boldsymbol{v}\, d\Gamma \tag{14.14}$$

for all $\boldsymbol{v} \in H^1(\Omega;\mathbb{R}^d)$, where Div denotes the divergence operator given by $\mathrm{Div}\, \boldsymbol{\sigma} = (\sigma_{ij,j})$.

Next, we introduce the spaces V and $\mathcal{H}$, defined by

$$V = \{\, \boldsymbol{v} = (v_i) \in H^1(\Omega;\mathbb{R}^d) \mid \boldsymbol{v} = \mathbf{0} \text{ a.e. on } \Gamma_D \,\},$$

$$\mathcal{H} = \{\, \boldsymbol{\tau} = (\tau_{ij}) \mid \tau_{ij} = \tau_{ji} \in L^2(\Omega)\} = L^2(\Omega;\mathbb{S}^d).$$

The space $\mathcal{H}$ is a Hilbert space with the canonical inner product given by

$$(\boldsymbol{\sigma}, \boldsymbol{\tau})_{\mathcal{H}} = \int_{\Omega} \sigma_{ij}(\boldsymbol{x})\, \tau_{ij}(\boldsymbol{x})\, dx, \quad \boldsymbol{\sigma}, \boldsymbol{\tau} \in \mathcal{H}$$

and the associated norm $\|\cdot\|_{\mathcal{H}}$. Also, since meas $(\Gamma_D) > 0$, it is well known that V is a Hilbert space with the inner product

$$(\boldsymbol{u}, \boldsymbol{v})_V = (\boldsymbol{\varepsilon}(\boldsymbol{u}), \boldsymbol{\varepsilon}(\boldsymbol{v}))_{\mathcal{H}} = \int_{\Omega} \varepsilon_{ij}(\boldsymbol{u})\varepsilon_{ij}(\boldsymbol{v})\, dx, \quad \boldsymbol{u}, \boldsymbol{v} \in V$$

and the associated norm $\|\cdot\|_V$. Finally, denote

$$Q = \Omega \times (0,T), \qquad \Sigma_D = \Gamma_D \times (0,T),$$

$$\Sigma_N = \Gamma_N \times (0,T), \qquad \Sigma_C = \Gamma_C \times (0,T).$$

We shall use the notations above in the next sections of this chapter.

14.3 A Viscoelastic Contact Problem

In the first problem of contact we consider a viscoelastic body which is attached to a piston or a device over the surface Γ_C. Following the terminology used in Contact Mechanics we refer to Γ_C as the contact surface and, below, the device will be refereed as the obstacle or the foundation. Note that, according to the physical setting, no separation between the body and the obstacle is allowed, which represents one of the novelties of the model we introduce in this section. Then, the classical formulation of the problem is as follows.

Problem $\mathcal{P}$. *Find a displacement field $\boldsymbol{u}\colon Q \to \mathbb{R}^d$ and a stress field $\boldsymbol{\sigma}\colon Q \to \mathbb{S}^d$ such that*

$$\boldsymbol{\sigma}(t) = \mathcal{A}(t, \boldsymbol{\varepsilon}(\boldsymbol{u}'(t))) + \mathcal{B}(t, \boldsymbol{\varepsilon}(\boldsymbol{u}(t))) + \int_0^t \mathcal{C}(t-s)\boldsymbol{\varepsilon}(\boldsymbol{u}'(s))\,ds \quad \text{in } Q, \tag{14.15}$$

$$\boldsymbol{u}''(t) = \mathrm{Div}\,\boldsymbol{\sigma}(t) + \boldsymbol{f}_0(t) \quad \text{in } Q, \tag{14.16}$$

$$\boldsymbol{u}(t) = \boldsymbol{0} \quad \text{on } \Sigma_D, \tag{14.17}$$

$$\boldsymbol{\sigma}(t)\boldsymbol{\nu} = \boldsymbol{f}_N(t) \quad \text{on } \Sigma_N, \tag{14.18}$$

$$-\sigma_\nu(t) = p(t, u_\nu(t)) + \int_0^t b(t-s)\,u_\nu(s)\,ds \quad \text{on } \Sigma_C, \tag{14.19}$$

$$-\boldsymbol{\sigma}_\tau(t) \in \partial j_\tau(t, \boldsymbol{u}'_\tau(t)) \quad \text{on } \Sigma_C, \tag{14.20}$$

$$\boldsymbol{u}(0) = \boldsymbol{u}_0, \quad \boldsymbol{u}'(0) = \boldsymbol{v}_0 \quad \text{in } \Omega. \tag{14.21}$$

We proceed with some explanations and comments on equations and boundary conditions in (14.15)–(14.21).

First, Eq. (14.15) represents the viscoelastic constitutive law in which $\mathcal{A}$ is a nonlinear operator which describes the viscous properties of the material, $\mathcal{B}$ is a nonlinear operator which describes its elastic behavior, and $\mathcal{C}$ represents the relaxation tensor. Various results, examples and mechanical interpretations in the study of such kind of constitutive laws, can be found in [2] and the references therein. Such kind of laws were used in the literature in order to model the behavior of real materials like rubbers, rocks, metals, pastes and polymers. In particular, Eq. (14.15) was employed in [1, 3] in order to model the hysteresis damping in elastomers. Note that in the case when the relaxation tensor vanishes and the viscosity and elasticity operators do not depend on time, Eq. (14.15) is reduced to the Kelvin-Voigt constitutive Eq. (14.9).

Equation (14.16) represents the equation of motion in which, for simplicity, we supposed that the mass density is equal to one. Conditions (14.17) and (14.18) are the displacement and the traction boundary conditions, respectively. They model the situation when the body is fixed on the part Γ_D of its boundary and the Cauchy stress vector is prescribed on Γ_N, respectively.

Equation (14.19) is the contact condition in which σ_ν denotes the normal stress, u_ν is the normal displacement, and p and b are given functions which describe the instantaneous and the memory reaction of the obstacle, respectively. It follows from (14.19) that at each moment t, the reaction of the obstacle depends both on the current value of the normal displacement (represented by the term $p(t, u_\nu(t))$) as well as on the history of the normal displacement (represented by the integral term in (14.19)). This reaction could be towards the body (when the obstacle is in compression) or towards the obstacle (when the latter is in extension). Condition (14.19) models situations when the memory effects of the obstacle are taken into account. A similar contact condition was considered in [8,21] in the study of frictionless contact problems for rate-type viscoplastic materials and viscoelastic materials with long memory, respectively. Moreover, note that when the memory function b vanishes, condition (14.19) reduces to the normal compliance contact condition (14.11).

Condition (14.20) represents the friction law, where j_τ is a given function and symbol ∂j_τ denotes the Clarke subdifferential of j_τ with respect to its last variable. Concrete examples of frictional conditions which lead to subdifferential boundary conditions of the form (14.20) with the function j_τ satisfying assumptions (14.27) below can be found in [9, 15, 16]. Here, we only remark that these examples include the nonmonotone friction law, the power-law friction, the Tresca friction law (14.12) as well as its regularization (14.13).

Finally, (14.21) represents the initial conditions in which $\boldsymbol{u}_0$ and $\boldsymbol{v}_0$ denote the initial displacement and the initial velocity, respectively.

In the study of problem (14.15)–(14.21) we consider the following assumptions on the viscosity operator $\mathcal{A}$, the elasticity operator $\mathcal{B}$ and the relaxation operator $\mathcal{C}$.

$$\left.\begin{array}{l}
\mathcal{A}\colon Q \times \mathbb{S}^d \to \mathbb{S}^d \text{ is such that} \\
\quad \text{(a) } \mathcal{A}(\cdot,\cdot,\boldsymbol{\varepsilon}) \text{ is measurable on } Q \text{ for all } \boldsymbol{\varepsilon} \in \mathbb{S}^d. \\
\quad \text{(b) } \mathcal{A}(\boldsymbol{x},t,\cdot) \text{ is continuous on } \mathbb{S}^d \text{ for a.e. } (\boldsymbol{x},t) \in Q. \\
\quad \text{(c) } (\mathcal{A}(\boldsymbol{x},t,\boldsymbol{\varepsilon}_1) - \mathcal{A}(\boldsymbol{x},t,\boldsymbol{\varepsilon}_2)) \cdot (\boldsymbol{\varepsilon}_1 - \boldsymbol{\varepsilon}_2) \geq m_{\mathcal{A}} \|\boldsymbol{\varepsilon}_1 - \boldsymbol{\varepsilon}_2\|^2_{\mathbb{S}^d} \\
\qquad\qquad \text{for all } \boldsymbol{\varepsilon}_1, \boldsymbol{\varepsilon}_2 \in \mathbb{S}^d, \text{ a.e. } (\boldsymbol{x},t) \in Q \text{ with } m_{\mathcal{A}} > 0. \\
\quad \text{(d) } \|\mathcal{A}(\boldsymbol{x},t,\boldsymbol{\varepsilon})\|_{\mathbb{S}^d} \leq \overline{a}_0(\boldsymbol{x},t) + \overline{a}_1 \|\boldsymbol{\varepsilon}\|_{\mathbb{S}^d} \text{ for all } \boldsymbol{\varepsilon} \in \mathbb{S}^d, \\
\qquad\qquad \text{a.e. } (\boldsymbol{x},t) \in Q \text{ with } \overline{a}_0 \in L^2(Q), \overline{a}_0 \geq 0 \text{ and } \overline{a}_1 > 0. \\
\quad \text{(e) } \mathcal{A}(\boldsymbol{x},t,\boldsymbol{0}) = \boldsymbol{0} \text{ for a.e. } (\boldsymbol{x},t) \in Q.
\end{array}\right\} \qquad (14.22)$$

$$\left.\begin{array}{l}
\mathcal{B}\colon Q \times \mathbb{S}^d \to \mathbb{S}^d \text{ is such that} \\
\quad \text{(a) } \mathcal{B}(\cdot,\cdot,\boldsymbol{\varepsilon}) \text{ is measurable on } Q \text{ for all } \boldsymbol{\varepsilon} \in \mathbb{S}^d. \\
\quad \text{(b) } \|\mathcal{B}(\boldsymbol{x},t,\boldsymbol{\varepsilon}_1) - \mathcal{B}(\boldsymbol{x},t,\boldsymbol{\varepsilon}_2)\|_{\mathbb{S}^d} \leq L_{\mathcal{B}} \|\boldsymbol{\varepsilon}_1 - \boldsymbol{\varepsilon}_2\|_{\mathbb{S}^d} \\
\qquad\qquad \text{for all } \boldsymbol{\varepsilon}_1, \boldsymbol{\varepsilon}_2 \in \mathbb{S}^d, \text{ a.e. } (\boldsymbol{x},t) \in Q \text{ with } L_{\mathcal{B}} > 0. \\
\quad \text{(c) } \mathcal{B}(\cdot,\cdot,\boldsymbol{0}) \in L^2(Q;\mathbb{S}^d).
\end{array}\right\} \qquad (14.23)$$

$$\left.\begin{array}{l}\mathcal{C}: Q \times \mathbb{S}^d \to \mathbb{S}^d \text{ is such that} \\ \quad \text{(a) } \mathcal{C}(\boldsymbol{x}, t, \boldsymbol{\varepsilon}) = (c_{ijkl}(\boldsymbol{x}, t)\varepsilon_{kl}) \\ \qquad\qquad \text{for all } \boldsymbol{\varepsilon} = (\varepsilon_{ij}) \in \mathbb{S}^d, \text{ a.e. } (\boldsymbol{x}, t) \in Q. \\ \quad \text{(b) } c_{ijkl}(\cdot, t) = c_{jikl}(\cdot, t) = c_{lkij}(\cdot, t) \in L^\infty(\Omega) \text{ a.e. } t \in (0, T). \\ \quad \text{(c) } t \mapsto c_{ijkl}(\cdot, t) \in L^\infty(0, T; L^\infty(\Omega)).\end{array}\right\} \tag{14.24}$$

The contact function p, the memory function b and the friction potential j_τ satisfy the following hypotheses.

$$\left.\begin{array}{l}p: \Sigma_C \times \mathbb{R} \to \mathbb{R} \text{ is such that} \\ \text{(a) } p(\cdot, \cdot, r) \text{ is measurable on } \Sigma_C \text{ for all } r \in \mathbb{R}. \\ \text{(b) } |p(\boldsymbol{x}, t, r_1) - p(\boldsymbol{x}, t, r_2)| \le L_p |r_1 - r_2| \\ \qquad \text{for all } r_1, r_2 \in \mathbb{R}, \text{ a.e. } (\boldsymbol{x}, t) \in \Sigma_C \text{ with } L_p > 0. \\ \text{(c) } p(\cdot, \cdot, 0) \in L^2(\Sigma_C).\end{array}\right\} \tag{14.25}$$

$$\left.\begin{array}{l}b: \Sigma_C \to \mathbb{R} \text{ is such that} \\ \quad \text{(a) } b \in L^1(0, T; L^\infty(\Gamma_C)). \\ \quad \text{(b) } b(\boldsymbol{x}, t) \ge 0 \text{ for a.e. } (\boldsymbol{x}, t) \in \Sigma_C.\end{array}\right\} \tag{14.26}$$

$$\left.\begin{array}{l}j_\tau: \Sigma_C \times \mathbb{R}^d \to \mathbb{R} \text{ is such that} \\ \quad \text{(a) } j_\tau(\cdot, \cdot, \boldsymbol{\xi}) \text{ is measurable on } \Sigma_C \text{ for all } \boldsymbol{\xi} \in \mathbb{R}^d \text{ and there} \\ \qquad \text{exists } \mathbf{e} \in L^2(\Gamma_C; \mathbb{R}^d) \text{ such that } j_\tau(\cdot, \cdot, \mathbf{e}(\cdot)) \in L^1(\Sigma_C). \\ \quad \text{(b) } j_\tau(\boldsymbol{x}, t, \cdot) \text{ is locally Lipschitz on } \mathbb{R}^d \text{ for a.e. } (\boldsymbol{x}, t) \in \Sigma_C. \\ \quad \text{(c) } \|\partial j_\tau(\boldsymbol{x}, t, \boldsymbol{\xi})\|_{\mathbb{R}^d} \le c_0 + c_1 \|\boldsymbol{\xi}\|_{\mathbb{R}^d} \text{ for all } \boldsymbol{\xi} \in \mathbb{R}^d, \\ \qquad \text{a.e. } (\boldsymbol{x}, t) \in \Sigma_C \text{ with } c_0, c_1 \ge 0. \\ \quad \text{(d) } (\boldsymbol{\zeta}_1 - \boldsymbol{\zeta}_2) \cdot (\boldsymbol{\xi}_1 - \boldsymbol{\xi}_2) \ge -m_\tau \|\boldsymbol{\xi}_1 - \boldsymbol{\xi}_2\|^2_{\mathbb{R}^d} \text{ for all} \\ \qquad \boldsymbol{\zeta}_i \in \partial j_\tau(\boldsymbol{x}, t, \boldsymbol{\xi}_i), \boldsymbol{\xi}_i \in \mathbb{R}^d, i = 1, 2, \text{ a.e. } (\boldsymbol{x}, t) \in \Sigma_C \\ \qquad \text{with } m_\tau \ge 0. \\ \quad \text{(e) } j^0_\tau(\boldsymbol{x}, t, \boldsymbol{\xi}; -\boldsymbol{\xi}) \le d_0 \, (1 + \|\boldsymbol{\xi}\|_{\mathbb{R}^d}) \text{ for all } \boldsymbol{\xi} \in \mathbb{R}^d, \\ \qquad \text{a.e. } (\boldsymbol{x}, t) \in \Sigma_C \text{ with } d_0 \ge 0.\end{array}\right\} \tag{14.27}$$

We also assume that the densities of the body forces and tractions have the regularity

$$\boldsymbol{f}_0 \in L^2(0, T; L^2(\Omega; \mathbb{R}^d)), \qquad \boldsymbol{f}_N \in L^2(0, T; L^2(\Gamma_N; \mathbb{R}^d)), \tag{14.28}$$

and, finally, the initial data are such that

$$\boldsymbol{u}_0 \in V, \qquad \boldsymbol{v}_0 \in V. \tag{14.29}$$

We now turn to the variational formulation of Problem $\mathcal{P}$ and, to this end, we assume in what follows that $(\boldsymbol{u}, \boldsymbol{\sigma})$ is a couple of sufficiently smooth functions which solve (14.16)–(14.20). Let $\boldsymbol{v} \in V$ and $t \in [0, T]$. Then, using (14.14) and (14.16) we deduce that

$$\int_\Omega \boldsymbol{u}''(t) \cdot \boldsymbol{v}\, dx + \int_\Omega \boldsymbol{\sigma}(t) \cdot \boldsymbol{\varepsilon}(\boldsymbol{v})\, dx = \int_\Omega \boldsymbol{f}_0(t) \cdot \boldsymbol{v}\, dx + \int_\Gamma \boldsymbol{\sigma}(t)\boldsymbol{\nu} \cdot \boldsymbol{v}\, d\Gamma.$$

We now split the surface integral to three integrals on Γ_D, Γ_N and Γ_C, then we use the boundary conditions (14.17) and (14.18) to obtain

$$\int_\Omega \boldsymbol{u}''(t) \cdot \boldsymbol{v}\, dx + \int_\Omega \boldsymbol{\sigma}(t) \cdot \boldsymbol{\varepsilon}(\boldsymbol{v})\, dx = \int_\Omega \boldsymbol{f}_0(t) \cdot \boldsymbol{v}\, dx + \int_{\Gamma_N} \boldsymbol{f}_N(t) \cdot \boldsymbol{v}\, d\Gamma$$
$$+ \int_{\Gamma_C} \boldsymbol{\sigma}(t)\boldsymbol{\nu} \cdot \boldsymbol{v}\, d\Gamma. \qquad (14.30)$$

Next, we use the identity $\boldsymbol{\sigma}(t)\boldsymbol{\nu} \cdot \boldsymbol{v} = \sigma_\nu(t) v_\nu + \boldsymbol{\sigma}_\tau(t) \cdot \boldsymbol{v}_\tau$ a.e. on Γ, the frictional contact conditions (14.19) and (14.20), and the definition of the subdifferential to see that

$$\boldsymbol{\sigma}(t)\boldsymbol{\nu} \cdot \boldsymbol{v} \geq -p(t, u_\nu(t)) v_\nu - \Big(\int_0^t b(t-s)\, u_\nu(s)\, ds \Big) v_\nu - j_\tau^0(t, \boldsymbol{u}_\tau'(t); \boldsymbol{v}_\tau)$$

a.e. on Γ_C, which implies that

$$\int_{\Gamma_C} \boldsymbol{\sigma}(t)\boldsymbol{\nu} \cdot \boldsymbol{v}\, d\Gamma + \int_{\Gamma_C} p(t, u_\nu(t)) v_\nu d\Gamma + \int_{\Gamma_C} \Big(\int_0^t b(t-s) u_\nu(s)\, ds \Big) v_\nu d\Gamma$$
$$+ \int_{\Gamma_C} j_\tau^0(t, \boldsymbol{u}_\tau'(t); \boldsymbol{v}_\tau)\, d\Gamma \geq 0. \qquad (14.31)$$

We consider also the function $\boldsymbol{f} \colon (0, T) \to V^*$ given by

$$\langle \boldsymbol{f}(t), \boldsymbol{v} \rangle_{V^* \times V} = (\boldsymbol{f}_0(t), \boldsymbol{v})_{L^2(\Omega;\mathbb{R}^d)} + (\boldsymbol{f}_N(t), \boldsymbol{v})_{L^2(\Gamma_N;\mathbb{R}^d)} \qquad (14.32)$$

for all $\boldsymbol{v} \in V$ and a.e. $t \in (0, T)$. Then, exploiting (14.30)–(14.32), we infer that

$$\langle \boldsymbol{u}''(t), \boldsymbol{v} \rangle_{V^* \times V} + (\boldsymbol{\sigma}(t), \boldsymbol{\varepsilon}(\boldsymbol{v}))_{\mathcal{H}}$$
$$+ \int_{\Gamma_C} p(t, u_\nu(t)) v_\nu\, d\Gamma + \int_{\Gamma_C} \Big(\int_0^t b(t-s) u_\nu(s)\, ds \Big) v_\nu\, d\Gamma$$
$$+ \int_{\Gamma_C} j_\tau^0(t, \boldsymbol{u}_\tau'(t); \boldsymbol{v}_\tau)\, d\Gamma \geq \langle \boldsymbol{f}(t), \boldsymbol{v} \rangle_{V^* \times V}. \qquad (14.33)$$

We now combine inequality (14.33) with the constitutive law (14.15) and the initial conditions (14.21) to obtain the following variational formulation of Problem $\mathcal{P}$.

Problem $\mathcal{P}^V$**.** *Find a displacement field* $\boldsymbol{u}\colon Q \to \mathbb{R}^d$ *and a stress field* $\boldsymbol{\sigma}\colon Q \to \mathbb{S}^d$ *such that*

$$\boldsymbol{\sigma}(t) = \mathcal{A}(t, \boldsymbol{\varepsilon}(\boldsymbol{u}'(t))) + \mathcal{B}(t, \boldsymbol{\varepsilon}(\boldsymbol{u}(t))) + \int_0^t \mathcal{C}(t-s)\boldsymbol{\varepsilon}(\boldsymbol{u}'(s))\, ds \quad \text{for a.e. } t \in (0,T), \tag{14.34}$$

$$\begin{aligned} &\langle \boldsymbol{u}''(t), \boldsymbol{v}\rangle_{V^*\times V} + (\boldsymbol{\sigma}(t), \boldsymbol{\varepsilon}(\boldsymbol{v}))_{\mathcal{H}} + \int_{\Gamma_C} p(t, u_\nu(t)) v_\nu \, d\Gamma \\ &\quad + \int_{\Gamma_C} \Big(\int_0^t b(t-s) u_\nu(s)\, ds \Big) v_\nu \, d\Gamma + \int_{\Gamma_C} j_\tau^0(t, \boldsymbol{u}_\tau'(t); \boldsymbol{v}_\tau)\, d\Gamma \\ &\qquad \geq \langle \boldsymbol{f}(t), \boldsymbol{v}\rangle_{V^*\times V} \quad \text{for all } \boldsymbol{v} \in V, \text{ a.e. } t \in (0,T), \end{aligned} \tag{14.35}$$

$$\boldsymbol{u}(0) = \boldsymbol{u}_0, \qquad \boldsymbol{u}'(0) = \boldsymbol{v}_0. \tag{14.36}$$

Our main result in the study of Problem $\mathcal{P}^V$ is the following.

Theorem 14.2. *Assume that* (14.22)–(14.26), (14.28) *and* (14.29) *hold. If one of the following hypotheses*

i) (14.27)(a)–(d) *and* $m_{\mathcal{A}} > \max\{2\sqrt{2}c_1, m_\tau\}\, \|\gamma\|^2$

ii) (14.27) *and* $m_{\mathcal{A}} > m_\tau \|\gamma\|^2$

is satisfied, then Problem $\mathcal{P}^V$ *has at least one solution which satisfies*

$$\boldsymbol{u} \in W^{1,2}(0,T;V), \quad \boldsymbol{\sigma} \in L^2(0,T;\mathcal{H}), \quad \mathrm{Div}\,\boldsymbol{\sigma} \in L^2(0,T;V^*). \tag{14.37}$$

If, in addition,

$$\text{either } j_\tau(\boldsymbol{x}, t, \cdot) \text{ or } -j_\tau(\boldsymbol{x}, t, \cdot) \text{ is regular on } \mathbb{R}^d \text{ for a.e. } (\boldsymbol{x}, t) \in \Sigma_C, \tag{14.38}$$

then the solution of Problem $\mathcal{P}^V$ *is unique.*

Proof. First, we introduce the operator $\boldsymbol{\zeta}\colon \mathcal{V} \to \mathcal{V}$ defined by

$$\boldsymbol{\zeta}\boldsymbol{w}(t) = \int_0^t \boldsymbol{w}(s)\, ds + \boldsymbol{u}_0 \tag{14.39}$$

for all $w \in \mathcal{V}$ and $t \in (0,T)$. Recall that $\mathcal{V} = L^2(0,T;V)$ and $\mathcal{V}^*$ represents its dual. We denote by $\zeta_\nu w(t)$ and $\boldsymbol{\zeta}_\tau w(t)$ the normal and tangential components of the element $\boldsymbol{\zeta} w(t)$, i.e.,

$$\zeta_\nu \boldsymbol{w}(t) = (\boldsymbol{\zeta}\boldsymbol{w}(t))_\nu = \int_0^t w_\nu(s)\,ds + u_{0\nu}, \tag{14.40}$$

$$\boldsymbol{\zeta}_\tau \boldsymbol{w}(t) = (\boldsymbol{\zeta}\boldsymbol{w}(t))_\tau = \int_0^t \boldsymbol{w}_\tau(s)\,ds + \boldsymbol{u}_{0\tau},$$

where $u_{0\nu}$ and $\boldsymbol{u}_{0\tau}$ are the normal and the tangential components of the initial displacement $\boldsymbol{u}_0 \in V$.

In what follows we apply Theorem 14.1. To this end, we insert (14.34) into (14.35) to obtain the following problem.

$$\left.\begin{aligned}
&\text{Find } \boldsymbol{u} \in \mathcal{V} \text{ such that } \boldsymbol{u}' \in \mathcal{W} \text{ and}\\
&\langle \boldsymbol{u}''(t) + A(t,\boldsymbol{u}'(t)) + (\mathcal{S}\boldsymbol{u}')(t), \boldsymbol{v}\rangle_{V^*\times V}\\
&\qquad + \int_{\Gamma_C} j^0(x,t,\gamma\boldsymbol{u}'(t);\gamma\boldsymbol{v})\,d\Gamma \ge \langle \boldsymbol{f}(t),\boldsymbol{v}\rangle_{V^*\times V}\\
&\qquad\qquad\qquad\qquad \text{for all } \boldsymbol{v} \in V \text{ and a.e. } t \in (0,T),\\
&\boldsymbol{u}(0) = \boldsymbol{u}_0, \;\; \boldsymbol{u}'(0) = \boldsymbol{v}_0,
\end{aligned}\right\} \tag{14.41}$$

where $A\colon (0,T)\times V \to V^*$, $\mathcal{S}\colon \mathcal{V} \to \mathcal{V}^*$ and $j\colon \Sigma_C \times \mathbb{R}^d \to \mathbb{R}$ are defined by

$$\langle A(t,\boldsymbol{u}),\boldsymbol{v}\rangle_{V^*\times V} = (\mathcal{A}(t,\boldsymbol{\varepsilon}(\boldsymbol{u})),\boldsymbol{\varepsilon}(\boldsymbol{v}))_{\mathcal{H}} \tag{14.42}$$

for all $\boldsymbol{u}, \boldsymbol{v} \in V$, a.e. $t \in (0,T)$,

$$\langle (\mathcal{S}\boldsymbol{w})(t),\boldsymbol{v}\rangle_{V^*\times V} = \sum_{i=1}^{4} \langle (\mathcal{S}_i\boldsymbol{w})(t),\boldsymbol{v}\rangle_{V^*\times V} \tag{14.43}$$

for all $\boldsymbol{w} \in \mathcal{V}$, $\boldsymbol{v} \in V$, a.e. $t \in (0,T)$, with

$$\langle (\mathcal{S}_1\boldsymbol{w})(t),\boldsymbol{v}\rangle_{V^*\times V} = \Big(\mathcal{B}\big(t,\boldsymbol{\varepsilon}(\boldsymbol{\zeta}\boldsymbol{w}(t))\big),\boldsymbol{\varepsilon}(\boldsymbol{v})\Big)_{\mathcal{H}}, \tag{14.44}$$

$$\langle (\mathcal{S}_2\boldsymbol{w})(t),\boldsymbol{v}\rangle_{V^*\times V} = \Big(\int_0^t \mathcal{C}(t-s)\boldsymbol{\varepsilon}(\boldsymbol{w}(s))\,ds,\boldsymbol{\varepsilon}(\boldsymbol{v})\Big)_{\mathcal{H}}, \tag{14.45}$$

$$\langle (\mathcal{S}_3\boldsymbol{w})(t),\boldsymbol{v}\rangle_{V^*\times V} = \int_{\Gamma_C} p(t,\zeta_\nu\boldsymbol{w}(t))\,v_\nu\,d\Gamma, \tag{14.46}$$

$$\langle (\mathcal{S}_4\boldsymbol{w})(t),\boldsymbol{v}\rangle_{V^*\times V} = \int_{\Gamma_C} \Big(\int_0^t b(t-s)\zeta_\nu\boldsymbol{w}(s)\,ds\Big)\,v_\nu\,d\Gamma \tag{14.47}$$

for all $\boldsymbol{w} \in \mathcal{V}$, $\boldsymbol{v} \in V$, a.e. $t \in (0, T)$, and

$$j(\boldsymbol{x}, t, \boldsymbol{\xi}) = j_\tau(\boldsymbol{x}, t, \boldsymbol{\xi}_\tau) \ \text{ for all } \boldsymbol{\xi} \in \mathbb{R}^d, \ \text{a.e. } (\boldsymbol{x}, t) \in \Sigma_C. \tag{14.48}$$

We will check that the operators $A: (0, T) \times V \to V^*$, $\mathcal{S}: \mathcal{V} \to \mathcal{V}^*$ and the function $j: \Sigma_C \times \mathbb{R}^d \to \mathbb{R}$ satisfy the hypotheses of Theorem 14.1.

First, note that under assumption (14.22), the operator A satisfies (14.2) with $m_1 = m_{\mathcal{A}} > 0$, $a_0(t) = \sqrt{2}\|\overline{a}_0(t)\|_{L^2(\Omega)} \geq 0$ and $a_1 = \overline{a}_1\sqrt{2} > 0$. Indeed, it is clear that $A(\cdot, \boldsymbol{u})$ is measurable on $(0, T)$ for all $\boldsymbol{u} \in V$. By (14.22)(d) and the Hölder inequality, we obtain

$$\begin{aligned} |\langle A(t, \boldsymbol{u}), \boldsymbol{v}\rangle_{V^*\times V}| &\leq \int_\Omega \|\mathcal{A}(t, \boldsymbol{\varepsilon}(\boldsymbol{u}))\|_{\mathbb{S}^d} \|\boldsymbol{\varepsilon}(\boldsymbol{v})\|_{\mathbb{S}^d}\, dx \\ &\leq \left(\int_\Omega (\overline{a}_0(\boldsymbol{x}, t) + \overline{a}_1\|\boldsymbol{\varepsilon}(\boldsymbol{u})\|_{\mathbb{S}^d})^2\, dx\right)^{1/2} \|\boldsymbol{v}\|_V \end{aligned}$$

for all $\boldsymbol{u}$, $\boldsymbol{v} \in V$ and a.e. $t \in (0, T)$. Hence $\|A(t, \boldsymbol{u})\|_{V^*} \leq a_0(t) + a_1\|\boldsymbol{u}\|_V$ for all $\boldsymbol{u} \in V$, a.e. $t \in (0, T)$ with $a_0 \in L^2(0, T)$, $a_0 \geq 0$, $a_1 > 0$ which means that A satisfies (14.2)(c). The condition (14.2)(d) is an easy consequence of (14.22)(c) and (d). From (14.22)(c), we get

$$\begin{aligned} \langle A(t, \boldsymbol{u}_1) - A(t, \boldsymbol{u}_2), \boldsymbol{u}_1 - \boldsymbol{u}_2\rangle_{V^*\times V} &\geq m_{\mathcal{A}} \int_\Omega \|\boldsymbol{\varepsilon}(\boldsymbol{u}_1) - \boldsymbol{\varepsilon}(\boldsymbol{u}_2)\|^2_{\mathbb{S}^d} dx \\ &= m_{\mathcal{A}}\|\boldsymbol{u}_1 - \boldsymbol{u}_2\|^2_V \end{aligned}$$

for all $\boldsymbol{u}_1$, $\boldsymbol{u}_2 \in V$ which implies that condition (14.2) of the strong monotonicity of $A(t, \cdot)$ for a.e. $t \in (0, T)$ holds. Next, we show that $A(t, \cdot)$ is continuous for a.e. $t \in (0, T)$. Fix $t \in (0, T)$ and suppose that $\boldsymbol{u}_n \to \boldsymbol{u}$ in V, i.e., $\boldsymbol{\varepsilon}(\boldsymbol{u}_n) \to \boldsymbol{\varepsilon}(\boldsymbol{u})$ in $L^2(\Omega; \mathbb{S}^d)$. Using Proposition 2.2.41 of [5], we know that there exist a subsequence $\{\boldsymbol{u}_{n_k}\}$ and a function $z \in L^2(\Omega)$ such that $\boldsymbol{\varepsilon}(\boldsymbol{u}_{n_k})(\boldsymbol{x}) \to \boldsymbol{\varepsilon}(\boldsymbol{u})(\boldsymbol{x})$ in $\mathbb{S}^d$ and $\|\boldsymbol{\varepsilon}(\boldsymbol{u}_{n_k})(\boldsymbol{x})\|_{\mathbb{S}^d} \leq z(\boldsymbol{x})$ for a.e. $\boldsymbol{x} \in \Omega$. By hypothesis (14.22)(b), we have

$$\mathcal{A}(\boldsymbol{x}, t, \boldsymbol{\varepsilon}(\boldsymbol{u}_{n_k})(\boldsymbol{x})) \to \mathcal{A}(\boldsymbol{x}, t, \boldsymbol{\varepsilon}(\boldsymbol{u})(\boldsymbol{x})) \ \text{ in } \mathbb{S}^d \ \text{ for a.e. } \boldsymbol{x} \in \Omega.$$

Exploiting (14.22)(d), we get

$$\begin{aligned} &\|\mathcal{A}(\boldsymbol{x}, t, \boldsymbol{\varepsilon}(\boldsymbol{u}_{n_k})(\boldsymbol{x})) - \mathcal{A}(\boldsymbol{x}, t, \boldsymbol{\varepsilon}(\boldsymbol{u})(\boldsymbol{x}))\|^2_{\mathbb{S}^d} \\ &\leq 2\left(\overline{a}_0(\boldsymbol{x}, t) + \overline{a}_1\|\boldsymbol{\varepsilon}(\boldsymbol{u}_{n_k})(\boldsymbol{x})\|_{\mathbb{S}^d}\right)^2 + 2\left(\overline{a}_0(\boldsymbol{x}, t) + \overline{a}_1\|\boldsymbol{\varepsilon}(\boldsymbol{u})(\boldsymbol{x})\|_{\mathbb{S}^d}\right)^2 \\ &\leq 8\,\overline{a}_0^2(\boldsymbol{x}, t) + 4\,\overline{a}_1^2\left(z^2(\boldsymbol{x}) + \|\boldsymbol{\varepsilon}(\boldsymbol{u})(\boldsymbol{x})\|^2_{\mathbb{S}^d}\right). \end{aligned}$$

Hence, by the Lebesgue theorem, we get

$$\|\mathcal{A}(t,\boldsymbol{\varepsilon}(\boldsymbol{u}_{n_k}))-\mathcal{A}(t,\boldsymbol{\varepsilon}(\boldsymbol{u}))\|_{\mathcal{H}}^2=\int_\Omega\|\mathcal{A}(t,\boldsymbol{\varepsilon}(\boldsymbol{u}_{n_k}))-\mathcal{A}(t,\boldsymbol{\varepsilon}(\boldsymbol{u}))\|_{\mathbb{S}^d}^2\,dx\to 0$$

as $n_k\to\infty$. Hence, applying again the Hölder inequality, we have

$$\begin{aligned}\langle A(t,\boldsymbol{u}_{n_k})-A(t,\boldsymbol{u}),\boldsymbol{v}\rangle_{V^*\times V}&=\int_\Omega\big(\mathcal{A}(t,\boldsymbol{\varepsilon}(\boldsymbol{u}_{n_k}))-\mathcal{A}(t,\boldsymbol{\varepsilon}(\boldsymbol{u}))\big)\cdot\boldsymbol{\varepsilon}(\boldsymbol{v})\,dx\\&\le\|\mathcal{A}(t,\boldsymbol{\varepsilon}(\boldsymbol{u}_{n_k}))-\mathcal{A}(t,\boldsymbol{\varepsilon}(\boldsymbol{u}))\|_{\mathcal{H}}\|\boldsymbol{\varepsilon}(\boldsymbol{v})\|_{\mathcal{H}}\end{aligned}$$

for all $\boldsymbol{v}\in V$. Thus $A(t,\boldsymbol{u}_{n_k})\to A(t,\boldsymbol{u})$ in V^* and subsequently we deduce that the whole sequence $A(t,\boldsymbol{u}_n)$ converges to $A(t,\boldsymbol{u})$ in V^*, which shows that $A(t,\cdot)$ is continuous for a.e. $t\in(0,T)$. Since the operator $A(t,\cdot)$ for a.e. $t\in(0,T)$ is bounded [by (14.2)(c)], monotone (being strongly monotone) and hemicontinuous (being continuous), from Proposition 27.6 of [23], it follows that it is pseudomonotone, i.e., (14.2)(b) holds. We deduce that the operator $A\colon(0,T)\times V\to V^*$ given by (14.42) satisfies (14.2).

Next, we prove that under the hypotheses (14.23)–(14.26), the operators $\mathcal{S}_i$, $i=1,\ldots,4$ defined by (14.44)–(14.47), respectively, satisfy

$$\|(\mathcal{S}_i\boldsymbol{w}_1)(t)-(\mathcal{S}_i\boldsymbol{w}_2)(t)\|_{V^*}\le L_i\int_0^t\|\boldsymbol{w}_1(s)-\boldsymbol{w}_2(s)\|_V\,ds \tag{14.49}$$

for all $\boldsymbol{w}_1,\boldsymbol{w}_2\in\mathcal{V}$, a.e. $t\in(0,T)$ with $L_i>0$, which entails that the operator $\mathcal{S}$ satisfies (14.3) with $L_{\mathcal{S}}=\sum_{i=1}^4L_i$.

Let $\boldsymbol{w}_1,\boldsymbol{w}_2\in\mathcal{V}$, $\boldsymbol{v}\in V$ and $t\in(0,T)$. Using (14.23)(b) and the Hölder inequality, we have

$$\begin{aligned}&\langle(\mathcal{S}_1\boldsymbol{w}_1)(t)-(\mathcal{S}_1\boldsymbol{w}_2)(t),\boldsymbol{v}\rangle_{V^*\times V}\\&\quad=\Big(\mathcal{B}\big(t,\boldsymbol{\varepsilon}(\zeta\boldsymbol{w}_1(t))\big)-\mathcal{B}\big(t,\boldsymbol{\varepsilon}(\zeta\boldsymbol{w}_2(t))\big),\boldsymbol{\varepsilon}(\boldsymbol{v})\Big)_{\mathcal{H}}\\&\quad\le\|\mathcal{B}\big(t,\boldsymbol{\varepsilon}(\zeta\boldsymbol{w}_1(t))\big)-\mathcal{B}\big(t,\boldsymbol{\varepsilon}(\zeta\boldsymbol{w}_2(t))\big)\|_{\mathcal{H}}\,\|\boldsymbol{v}\|_V\\&\quad\le L_{\mathcal{B}}\|\boldsymbol{\varepsilon}(\zeta\boldsymbol{w}_1(t)-\zeta\boldsymbol{w}_2(t))\|_{\mathcal{H}}\,\|\boldsymbol{v}\|_V\\&\quad\le L_{\mathcal{B}}\Big(\int_0^t\|\boldsymbol{w}_1(s)-\boldsymbol{w}_2(s)\|_V\,ds\Big)\|\boldsymbol{v}\|_V.\end{aligned}$$

Hence, the operator $\mathcal{S}_1$ satisfies (14.49) with $L_1=L_{\mathcal{B}}$. Next, from (14.24) and by the Hölder inequality again, we obtain

$$\begin{aligned}&\langle(\mathcal{S}_2\boldsymbol{w}_1)(t)-(\mathcal{S}_2\boldsymbol{w}_2)(t),\boldsymbol{v}\rangle_{V^*\times V}\\&\quad=\Big(\int_0^t\mathcal{C}(t-s)\,\boldsymbol{\varepsilon}\big(\boldsymbol{w}_1(s)-\boldsymbol{w}_2(s)\big)\,ds,\boldsymbol{\varepsilon}(\boldsymbol{v})\Big)_{\mathcal{H}}\end{aligned}$$

$$\le \| \int_0^t \mathcal{C}(t-s)\boldsymbol{\varepsilon}\big(\boldsymbol{w}_1(s)-\boldsymbol{w}_2(s)\big)\,ds\|_{\mathcal{H}}\,\|\boldsymbol{v}\|_V$$

$$\le \Big(\int_0^t \|\mathcal{C}(t-s)\|_{L^\infty(\Omega;\mathbb{S}^d)}\|\boldsymbol{w}_1(s)-\boldsymbol{w}_2(s)\|_V\,ds\Big)\|\boldsymbol{v}\|_V,$$

which implies

$$\langle(\mathcal{S}_2\boldsymbol{w}_1)(t)-(\mathcal{S}_2\boldsymbol{w}_2)(t),\boldsymbol{v}\rangle_{V^*\times V}$$
$$\le \|\mathcal{C}\|_{L^\infty(Q;\mathbb{S}^d)}\Big(\int_0^t \|\boldsymbol{w}_1(s)-\boldsymbol{w}_2(s)\|_V\,ds\Big)\|\boldsymbol{v}\|_V.$$

So, the operator $\mathcal{S}_2$ satisfies (14.49) with $L_2=\|\mathcal{C}\|_{L^\infty(Q;\mathbb{S}^d)}$.

Using the continuity of the trace operator and the inequality $|\xi_\nu|\le\|\boldsymbol{\xi}\|_{\mathbb{R}^d}$ for $\boldsymbol{\xi}\in\mathbb{R}^d$, by the definition of the operator $\boldsymbol{\zeta}$, we have

$$\begin{aligned}\|\zeta_\nu\boldsymbol{w}_1(t)-\zeta_\nu\boldsymbol{w}_2(t)\|_{L^2(\Gamma_C)} &\le \|\boldsymbol{\zeta}\boldsymbol{w}_1(t)-\boldsymbol{\zeta}\boldsymbol{w}_2(t)\|_{L^2(\Gamma_C;\mathbb{R}^d)}\\ &\le \|\gamma\|\,\|\boldsymbol{\zeta}\boldsymbol{w}_1(t)-\boldsymbol{\zeta}\boldsymbol{w}_2(t)\|_V\\ &\le \|\gamma\|\int_0^t \|\boldsymbol{w}_1(s)-\boldsymbol{w}_2(s)\|_V ds. \end{aligned}\tag{14.50}$$

Now, making use of (14.25)(b) and (14.50), we have

$$\langle(\mathcal{S}_3\boldsymbol{w}_1)(t)-(\mathcal{S}_3\boldsymbol{w}_2)(t),\boldsymbol{v}\rangle_{V^*\times V}$$
$$=\int_{\Gamma_C}\big(p(t,\zeta_\nu\boldsymbol{w}_1(t))-p(t,\zeta_\nu\boldsymbol{w}_2(t))\big)\,v_\nu\,d\Gamma$$
$$\le \|p(t,\zeta_\nu\boldsymbol{w}_1(t))-p(t,\zeta_\nu\boldsymbol{w}_2(t))\|_{L^2(\Gamma_C)}\,\|\boldsymbol{v}\|_{L^2(\Gamma_C;\mathbb{R}^d)}$$
$$\le L_p\,\|\zeta_\nu\boldsymbol{w}_1(t)-\zeta_\nu\boldsymbol{w}_2(t)\|_{L^2(\Gamma_C)}\|\gamma\|\,\|\boldsymbol{v}\|_V$$
$$\le L_p\,\|\gamma\|^2\Big(\int_0^t \|\boldsymbol{w}_1(s)-\boldsymbol{w}_2(s)\|_V\,ds\Big)\|\boldsymbol{v}\|_V.$$

Thus, the operator $\mathcal{S}_3$ satisfies (14.49) with $L_3=L_p\|\gamma\|^2$. From (14.26), the Hölder inequality and (14.50), we get

$$\langle(\mathcal{S}_4\boldsymbol{w}_1)(t)-(\mathcal{S}_4\boldsymbol{w}_2)(t),\boldsymbol{v}\rangle_{V^*\times V}$$
$$=\int_{\Gamma_c}\Big(\int_0^t b(t-s)\big(\zeta_\nu\boldsymbol{w}_1(s)-\zeta_\nu\boldsymbol{w}_2(s)\big)\,ds\Big)\,v_\nu\,d\Gamma$$
$$\le \|\int_0^t b(t-s)\big(\zeta_\nu\boldsymbol{w}_1(s)-\zeta_\nu\boldsymbol{w}_2(s)\big)\,ds\|_{L^2(\Gamma_C)}\,\|\boldsymbol{v}\|_{L^2(\Gamma_C;\mathbb{R}^d)}$$

$$\leq \left(\int_0^t \|b(t-s)\|_{L^\infty(\Gamma_C)}\|\zeta_\nu \boldsymbol{w}_1(s) - \zeta_\nu \boldsymbol{w}_2(s)\|_{L^2(\Gamma_C)}\,ds\right)\|\gamma\|\,\|\boldsymbol{v}\|_V$$

$$\leq \|\gamma\|^2\left(\int_0^t \|b(t-s)\|_{L^\infty(\Gamma_C)}\left(\int_0^t \|\boldsymbol{w}_1(\tau) - \boldsymbol{w}_2(\tau)\|_V\,d\tau\right)ds\right)\|\boldsymbol{v}\|_V$$

$$\leq \|\gamma\|^2\|b\|_{L^1(0,T;L^\infty(\Gamma_C))}\left(\int_0^t \|\boldsymbol{w}_1(s) - \boldsymbol{w}_2(s)\|_V\,ds\right)\|\boldsymbol{v}\|_V.$$

This means that the operator $\mathcal{S}_4$ satisfies condition (14.49) with $L_4 = \|\gamma\|^2\ \|b\|_{L^1(0,T;L^\infty(\Gamma_C))}$.

Moreover, it follows from (14.23) and (14.25) that

$$\|\mathcal{B}(\boldsymbol{x},t,\boldsymbol{\varepsilon})\|_{\mathbb{S}^d} \leq \tilde{b}(\boldsymbol{x},t) + L_{\mathcal{B}}\|\boldsymbol{\varepsilon}\|_{\mathbb{S}^d} \text{ for all } \boldsymbol{\varepsilon} \in \mathbb{S}^d, \text{ a.e. } (\boldsymbol{x},t) \in Q, \tag{14.51}$$

$$|p(\boldsymbol{x},t,r)| \leq \tilde{p}(\boldsymbol{x},t) + L_p|r| \text{ for all } r \in \mathbb{R}, \text{ a.e. } (\boldsymbol{x},t) \in \Sigma_C, \tag{14.52}$$

where $\tilde{b}(\boldsymbol{x},t) = \|\mathcal{B}(\boldsymbol{x},t,\boldsymbol{0})\|_{\mathbb{S}^d}$, $\tilde{b} \in L^2(Q)$ and $\tilde{p}(\boldsymbol{x},t) = |p(\boldsymbol{x},t,0)|$, $\tilde{p} \in L^2(\Sigma_C)$. Therefore, exploiting (14.43)–(14.47), (14.51), (14.52) and condition (14.49), we have

$$\begin{aligned}\|(\mathcal{S}\boldsymbol{w})(t)\|_{V^*} &\leq L_{\mathcal{S}}\|\boldsymbol{w}\|_{L^1(0,t;V)} + \|(\mathcal{S}\,\boldsymbol{0})(t)\|_{V^*}\\ &\leq L_{\mathcal{S}}\|\boldsymbol{w}\|_{L^1(0,t;V)} + \tilde{b}(t) + L_{\mathcal{B}}\|\boldsymbol{u}_0\|_V + \tilde{p}(t)\\ &\quad + L_p\|\boldsymbol{u}_0\|_V + \|b\|_{L^1(0,T;L^\infty(\Gamma_c))}\|\boldsymbol{u}_0\|_V\end{aligned}$$

for all $\boldsymbol{w} \in \mathcal{V}$, a.e. $t \in (0,T)$. Hence, we deduce

$$\|\mathcal{S}\boldsymbol{w}\|_{\mathcal{V}^*} \leq c\left(\|\boldsymbol{w}\|_{\mathcal{V}} + \|\boldsymbol{u}_0\|_V + \|\tilde{b}\|_{L^2(Q)} + \|\tilde{p}\|_{L^2(\Sigma_C)}\right)$$

with $c > 0$. Therefore the operator $\mathcal{S}$ is well defined, takes values in $\mathcal{V}^*$ and satisfies the hypothesis (14.3).

Next, we observe that condition (14.5) follows easily from hypotheses (14.28) and (14.29). It is clear that (14.27)(a)–(d) entails hypothesis (14.4). Also, the conditions i) and ii) imply hypothesis (14.6) with $m_1 = m_{\mathcal{A}}$ and $\overline{m}_2 = m_\tau$. Conditions (a) and (b) of (14.7) are consequences of the hypotheses i) and ii), respectively.

Since conditions (14.2)–(14.7) hold, then applying Theorem 14.1, we infer that problem (14.41) has at least one solution $\boldsymbol{u} \in \mathcal{V}$ such that $\boldsymbol{u}' \in \mathcal{W}$. Furthermore, the regularity condition (14.38) implies (14.8). Therefore, we infer from the uniqueness part of Theorem 14.1 that under the additional hypothesis (14.38), the solution to problem (14.41) is unique.

Finally, using the regularity of $\boldsymbol{u}$ and hypotheses (14.22)–(14.24), from condition (14.34), we deduce that the pair $(\boldsymbol{u},\boldsymbol{\sigma})$ is a solution to Problem $\mathcal{P}^V$ with the regularity $\boldsymbol{u} \in W^{1,2}(0,T;V)$, $\boldsymbol{\sigma} \in L^2(0,T;\mathcal{H})$ and $\mathrm{Div}\,\boldsymbol{\sigma} \in L^2(0,T;V^*)$. This concludes the proof of the theorem. □

A couple of functions $(\boldsymbol{u},\boldsymbol{\sigma})$ which satisfies (14.34)–(14.36) is called a *weak solution* of the frictional contact problem (14.15)–(14.21). We conclude that, under the hypotheses of Theorem 14.2, the frictional contact problem (14.15)–(14.21) has at least one weak solution. If, in addition, the regularity condition (14.38) holds, then the weak solution of Problem $\mathcal{P}$ is unique.

14.4 A Viscoplastic Contact Problem

The second problem of contact we consider is viscoplastic. The physical setting is similar to that described in Sect. 14.3 but here we could have separation between the body and the foundation. The classical formulation of the problem is as follows.

Problem $\mathcal{Q}$. *Find a displacement field $\boldsymbol{u}\colon Q \to \mathbb{R}^d$ and a stress field $\boldsymbol{\sigma}\colon Q \to \mathbb{S}^d$ such that*

$$\boldsymbol{\sigma}(t) = \mathcal{A}(t, \boldsymbol{\varepsilon}(\boldsymbol{u}'(t))) + \mathcal{B}(t, \boldsymbol{\varepsilon}(\boldsymbol{u}(t)))$$

$$+ \int_0^t \mathcal{G}(s, \boldsymbol{\sigma}(s) - \mathcal{A}(s, \boldsymbol{\varepsilon}(\boldsymbol{u}'(s))), \boldsymbol{\varepsilon}(\boldsymbol{u}(s)))\, ds \quad \text{in } Q, \qquad (14.53)$$

$$\boldsymbol{u}''(t) = \mathrm{Div}\, \boldsymbol{\sigma}(t) + \boldsymbol{f}_0(t) \quad \text{in } Q, \qquad (14.54)$$

$$\boldsymbol{u}(t) = \boldsymbol{0} \quad \text{on } \Sigma_D, \qquad (14.55)$$

$$\boldsymbol{\sigma}(t)\boldsymbol{\nu} = \boldsymbol{f}_N(t) \quad \text{on } \Sigma_N, \qquad (14.56)$$

$$-\sigma_\nu(t) = k\Big(\int_0^t u_\nu(s)\, ds\Big)\, p(t, u_\nu(t)) \quad \text{on } \Sigma_C, \qquad (14.57)$$

$$-\boldsymbol{\sigma}_\tau(t) \in \partial j_\tau(t, \boldsymbol{u}'_\tau(t)) \quad \text{on } \Sigma_C, \qquad (14.58)$$

$$\boldsymbol{u}(0) = \boldsymbol{u}_0, \ \ \boldsymbol{u}'(0) = \boldsymbol{v}_0 \quad \text{in } \Omega. \qquad (14.59)$$

The equations and boundary conditions in Problem $\mathcal{Q}$ have a similar meaning as those in Problem $\mathcal{P}$. Nevertheless, there are two major differences between these two mathematical models. The first one arises in the fact that in Problem $\mathcal{Q}$ we use the viscoplastic constitutive law (14.53), instead of the viscoelastic constitutive law (14.15), used in Problem $\mathcal{P}$. The second one consists in the fact that in Problem $\mathcal{Q}$ we use a version of the normal compliance condition, (14.57), instead of the contact condition (14.19). A short description of these new ingredients used in Problem $\mathcal{Q}$ follows.

First, we note that concrete examples of constitutive laws of the form (14.53) can be constructed by using rheological arguments presented in [6] and [9], for instance. Here, we restrict ourselves to note that the stress field in (14.53) has an additive decomposition of the form

$$\boldsymbol{\sigma}(t) = \boldsymbol{\sigma}^V(t) + \boldsymbol{\sigma}^{VP}(t) \tag{14.60}$$

where

$$\boldsymbol{\sigma}^V(t) = \mathcal{A}(t, \boldsymbol{\varepsilon}(\boldsymbol{u}'(t))), \tag{14.61}$$

$$\boldsymbol{\sigma}^{VP}(t) = \mathcal{B}(t, \boldsymbol{\varepsilon}(\boldsymbol{u}(t))) + \int_0^t \mathcal{G}(s, \boldsymbol{\sigma}^{VP}(s), \boldsymbol{\varepsilon}(\boldsymbol{u}(s)))\, ds. \tag{14.62}$$

Equation (14.61) represents the constitutive law for a nonlinear time-dependent purely viscous material. Equation (14.62) represents a version of the constitutive law for a nonlinear rate-type viscoplastic materials, (14.10), written under the assumptions that the elasticity operator and the viscoplastic function depend explicitly on the time. We conclude from (14.60) that models of the form (14.53) are obtained by connecting in parallel a purely viscous time-dependent damper with a rate-type viscoplastic constitutive model. In addition, we note that when the operators $\mathcal{A}$ and $\mathcal{B}$ do not depend on time and the function $\mathcal{G}$ vanishes, then (14.53) reduces, again, to the Kelvin-Voigt constitutive law (14.9).

Next, we turn on the contact condition (14.57) which represents an extension of the normal compliance contact condition (14.11). Here, k is a given positive function, the stiffness coefficient of the obstacle and, again, p is a given function. If we allow separation between the body and the foundation, we have to assume that the function p vanishes when the second argument is negative. Indeed, when there is separation between the body and the foundation, then the normal stress vanishes. As the cycles of penetration proceed, the stiffness of the obstacle may be increasing or decreasing, making it a function of the history of the contact process. In this way we take into account the hardening or the softening of the foundation. Practical examples of surface hardening or softening abound, see, e.g., [18]. A contact condition similar to condition (14.57) was considered in [22] in the study of a frictionless contact problems for rate-type viscoplastic. Finally, note that (14.57) reduces to the normal compliance contact condition (14.11), when the stiffness coefficient k is a constant and p does not depend explicitly on time.

In the study of problem (14.53)–(14.59) we assume that the viscosity operator $\mathcal{A}$, the elasticity operator $\mathcal{B}$ and the friction potential j_τ satisfy conditions (14.22), (14.23) and (14.27), respectively. We also assume that the densities of the body forces and tractions have the regularity (14.28), and the initial displacement is such that (14.29) holds. Moreover, we assume that the constitutive function $\mathcal{G}$, the stiffness coefficient k and the contact function p satisfy the following conditions.

$$\left.\begin{array}{l} \mathcal{G}\colon Q\times\mathbb{S}^d\times\mathbb{S}^d\to\mathbb{S}^d \text{ is such that}\\ \quad\text{(a) } \mathcal{G}(\cdot,\cdot,\boldsymbol{\sigma},\boldsymbol{\varepsilon}) \text{ is measurable on } Q \text{ for all } \boldsymbol{\sigma},\boldsymbol{\varepsilon}\in\mathbb{S}^d.\\ \quad\text{(b) } \|\mathcal{G}(\boldsymbol{x},t,\boldsymbol{\sigma}_1,\boldsymbol{\varepsilon}_1)-\mathcal{G}(\boldsymbol{x},t,\boldsymbol{\sigma}_2,\boldsymbol{\varepsilon}_2)\|_{\mathbb{S}^d}\le\\ \qquad\qquad \le L_{\mathcal{G}}\,(\|\boldsymbol{\sigma}_1-\boldsymbol{\sigma}_2\|_{\mathbb{S}^d}+\|\boldsymbol{\varepsilon}_1-\boldsymbol{\varepsilon}_2\|_{\mathbb{S}^d})\\ \qquad\quad \text{for all } \boldsymbol{\sigma}_1,\boldsymbol{\sigma}_2,\boldsymbol{\varepsilon}_1,\boldsymbol{\varepsilon}_2\in\mathbb{S}^d,\ \text{a.e. } (\boldsymbol{x},t)\in Q \text{ with } L_{\mathcal{G}}>0.\\ \quad\text{(c) } \mathcal{G}(\cdot,\cdot,\mathbf{0},\mathbf{0})\in L^2(Q;\mathbb{S}^d). \end{array}\right\} \tag{14.63}$$

$$\left.\begin{array}{l} k\colon \Gamma_C\times\mathbb{R}\to\mathbb{R}_+ \text{ is such that}\\ \quad\text{(a) } k(\cdot,r) \text{ is measurable on } \Gamma_C \text{ for all } r\in\mathbb{R}.\\ \quad\text{(b) } |k(\boldsymbol{x},r_1)-k(\boldsymbol{x},r_2)|\le L_k\,|r_1-r_2|\\ \qquad\quad \text{for all } r_1,r_2\in\mathbb{R},\ \text{a.e. } \boldsymbol{x}\in\Gamma_C \text{ with } L_k>0.\\ \quad\text{(c) } |k(\boldsymbol{x},r)|\le\overline{k} \ \text{ for all } r\in\mathbb{R},\ \text{a.e. } \boldsymbol{x}\in\Gamma_C \text{ with } \overline{k}>0. \end{array}\right\} \tag{14.64}$$

$$\left.\begin{array}{l} p\colon \Sigma_C\times\mathbb{R}\to\mathbb{R} \text{ is such that}\\ \quad\text{(a) } p(\cdot,\cdot,r) \text{ is measurable on } \Sigma_C \text{ for all } r\in\mathbb{R}.\\ \quad\text{(b) } |p(\boldsymbol{x},t,r_1)-p(\boldsymbol{x},t,r_2)|\le L_p\,|r_1-r_2|\\ \qquad\quad \text{for all } r_1,r_2\in\mathbb{R},\ \text{a.e. } (\boldsymbol{x},t)\in\Sigma_C \text{ with } L_p>0.\\ \quad\text{(c) } |p(\boldsymbol{x},t,r)|\le\overline{p} \ \text{ for all } r\in\mathbb{R},\ \text{a.e. } (\boldsymbol{x},t)\in\Sigma_C \text{ with } \overline{p}>0. \end{array}\right\} \tag{14.65}$$

We use the function $\boldsymbol{f}\colon(0,T)\to V^*$ given by (14.32) and the Green formula (14.14) to obtain the following variational formulation of Problem $\mathcal{Q}$.

Problem $\mathcal{Q}^V$. *Find a displacement field $\boldsymbol{u}\colon Q\to\mathbb{R}^d$ and a stress field $\boldsymbol{\sigma}\colon Q\to\mathbb{S}^d$ such that*

$$\begin{aligned}\boldsymbol{\sigma}(t)&=\mathcal{A}(t,\boldsymbol{\varepsilon}(\boldsymbol{u}'(t)))+\mathcal{B}(t,\boldsymbol{\varepsilon}(\boldsymbol{u}(t)))\\ &\quad+\int_0^t\mathcal{G}(s,\boldsymbol{\sigma}(s)-\mathcal{A}(s,\boldsymbol{\varepsilon}(\boldsymbol{u}'(s))),\boldsymbol{\varepsilon}(\boldsymbol{u}(s)))\,ds \ \ \textit{a.e. } t\in(0,T),\end{aligned} \tag{14.66}$$

$$\begin{aligned}&\langle\boldsymbol{u}''(t),\boldsymbol{v}\rangle_{V^*\times V}+(\boldsymbol{\sigma}(t),\boldsymbol{\varepsilon}(\boldsymbol{v}))_{\mathcal{H}}+\int_{\Gamma_C}k\Big(\int_0^t u_\nu(s)\,ds\Big)\,p(t,u_\nu(t))v_\nu\,d\Gamma\\ &\quad+\int_{\Gamma_C}j_\tau^0(t,\boldsymbol{u}'_\tau(t);\boldsymbol{v}_\tau)\,d\Gamma\ge\langle\boldsymbol{f}(t),\boldsymbol{v}\rangle_{V^*\times V}\\ &\qquad\qquad\text{for all } \boldsymbol{v}\in V,\ \text{a.e. } t\in(0,T),\end{aligned} \tag{14.67}$$

$$\boldsymbol{u}(0)=\boldsymbol{u}_0,\qquad \boldsymbol{u}'(0)=\boldsymbol{v}_0. \tag{14.68}$$

Our main result in the study of Problem $\mathcal{Q}^V$ is the following.

Theorem 14.3. *Assume that* (14.22), (14.23), (14.28), (14.29) *and* (14.63)–(14.65) *hold. If one of the following hypotheses*

i) (14.27)(a)–(d) *and* $m_{\mathcal{A}} > \max\{2\sqrt{2}c_1, m_\tau\}\, \|\gamma\|^2$

ii) (14.27) *and* $m_{\mathcal{A}} > m_\tau\, \|\gamma\|^2$

is satisfied, then Problem $\mathcal{Q}^V$ has at least one solution which satisfies (14.37). *If, in addition,* (14.38) *holds, then the solution of Problem $\mathcal{Q}^V$ is unique.*

Proof. The proof will be made in several steps. To present it, we use operator $\zeta: \mathcal{V} \to \mathcal{V}$ defined by (14.39). Moreover, we need the following auxiliary result.

Lemma 14.4. *Assume that* (14.23) *and* (14.63) *hold. Then, for all* $\boldsymbol{u} \in L^2(0,T;V)$, *there exists a unique function* $\boldsymbol{\sigma}^I(\boldsymbol{u}) \in L^2(0,T;\mathcal{H})$ *such that*

$$\boldsymbol{\sigma}^I(\boldsymbol{u})(t) = \int_0^t \mathcal{G}(s, \boldsymbol{\sigma}^I(\boldsymbol{u})(s) + \mathcal{B}(s, \boldsymbol{\varepsilon}(\boldsymbol{u}(s))), \boldsymbol{\varepsilon}(\boldsymbol{u}(s)))\, ds \tag{14.69}$$

for a.e. $t \in (0,T)$. *Moreover, if* $\boldsymbol{u}_1, \boldsymbol{u}_2 \in L^2(0,T;V)$, *then*

$$\|\boldsymbol{\sigma}^I(\boldsymbol{u}_1)(t) - \boldsymbol{\sigma}^I(\boldsymbol{u}_2)(t)\|_{\mathcal{H}} \le c \int_0^t \|\boldsymbol{u}_1(s) - \boldsymbol{u}_2(s)\|_V\, ds \tag{14.70}$$

for a.e. $t \in (0,T)$ *with* $c > 0$.

Proof. Let $\boldsymbol{u} \in \mathcal{V}$ be given. We introduce the operator $\Lambda: L^2(0,T;\mathcal{H}) \to L^2(0,T;\mathcal{H})$ defined by

$$\Lambda\boldsymbol{\eta}(t) = \int_0^t \mathcal{G}(s, \boldsymbol{\eta}(s) + \mathcal{B}(s, \boldsymbol{\varepsilon}(\boldsymbol{u}(s))), \boldsymbol{\varepsilon}(\boldsymbol{u}(s)))\, ds$$

for $\boldsymbol{\eta} \in L^2(0,T;\mathcal{H})$, a.e. $t \in (0,T)$. The operator Λ depends on $\boldsymbol{u}$ and, for simplicity, we do not indicate explicitly this dependence. Let $\boldsymbol{\eta}_1$, $\boldsymbol{\eta}_2 \in L^2(0,T;\mathcal{H})$ and $t \in (0,T)$. Then

$$\|\Lambda\boldsymbol{\eta}_1(t) - \Lambda\boldsymbol{\eta}_2(t)\|_{\mathcal{H}} \le L_{\mathcal{G}} \int_0^t \|\boldsymbol{\eta}_1(s) - \boldsymbol{\eta}_2(s)\|_{\mathcal{H}}\, ds.$$

It is clear from the hypotheses on $\mathcal{G}$ and $\mathcal{B}$ that the operator Λ is well defined and takes values in $L^2(0,T;\mathcal{H})$. From Lemma 2.3 on page 42, we deduce that the operator Λ has a unique fixed point, denoted by $\boldsymbol{\sigma}^I(\boldsymbol{u}) \in L^2(0,T;\mathcal{H})$.

The proof of inequality (14.70) is a consequence of (14.69) combined with a Gronwall type argument. Indeed, let $\boldsymbol{u}_1, \boldsymbol{u}_2 \in \mathcal{V}$. We have

$$
\begin{aligned}
&\|\boldsymbol{\sigma}^I(\boldsymbol{u}_1)(t) - \boldsymbol{\sigma}^I(\boldsymbol{u}_2)(t)\|_{\mathcal{H}} \\
&\quad \le \sqrt{2}\, L_{\mathcal{G}} \int_0^t \Big(\|\boldsymbol{\sigma}^I(\boldsymbol{u}_1)(s) - \boldsymbol{\sigma}^I(\boldsymbol{u}_2)(s)\|_{\mathcal{H}} \\
&\qquad + \|\mathcal{B}(s, \boldsymbol{\varepsilon}(\boldsymbol{u}_1(s))) - \mathcal{B}(s, \boldsymbol{\varepsilon}(\boldsymbol{u}_2(s)))\|_{\mathcal{H}} + \|\boldsymbol{\varepsilon}(\boldsymbol{u}_1(s) - \boldsymbol{u}_2(s))\|_{\mathcal{H}} \Big)\, ds \\
&\quad \le \sqrt{2}\, L_{\mathcal{G}} \int_0^t \|\boldsymbol{\sigma}^I(\boldsymbol{u}_1)(s) - \boldsymbol{\sigma}^I(\boldsymbol{u}_2)(s)\|_{\mathcal{H}}\, ds \\
&\qquad + \sqrt{2}\, L_{\mathcal{G}}\, (1 + L_{\mathcal{B}}) \int_0^t \|\boldsymbol{u}_1(s) - \boldsymbol{u}_2(s)\|_V\, ds
\end{aligned}
$$

for a.e. $t \in (0, T)$. From the Gronwall inequality (cf. [16], Lemma 2.7), we infer that

$$
\|\boldsymbol{\sigma}^I(\boldsymbol{u}_1)(t) - \boldsymbol{\sigma}^I(\boldsymbol{u}_2)(t)\|_{\mathcal{H}} \le e^{\sqrt{2} L_{\mathcal{G}} T} \sqrt{2} L_{\mathcal{G}} (1 + L_{\mathcal{B}}) \int_0^t \|\boldsymbol{u}_1(s) - \boldsymbol{u}_2(s)\|_V\, ds
$$

for a.e. $t \in (0, T)$. This completes the proof of the lemma. □

We continue the proof of Theorem 14.3. In order to formulate an equivalent form of Problem $\mathcal{Q}^V$, we use Lemma 14.4 and the notation (14.39) and (14.40). We consider the following intermediate problem.

Problem $\tilde{\mathcal{Q}}^V$. *Find a displacement field $\boldsymbol{u}\colon Q \to \mathbb{R}^d$ and a stress field $\boldsymbol{\sigma}\colon Q \to \mathbb{S}^d$ such that*

$$
\boldsymbol{\sigma}(t) = \mathcal{A}(t, \boldsymbol{\varepsilon}(\boldsymbol{u}'(t))) + \mathcal{B}(t, \boldsymbol{\varepsilon}(\boldsymbol{u}(t))) + \boldsymbol{\sigma}^I(\boldsymbol{u})(t) \ \text{ a.e. } t \in (0, T), \tag{14.71}
$$

$$
\begin{aligned}
&\langle \boldsymbol{u}''(t), \boldsymbol{v} \rangle_{V^* \times V} + (\boldsymbol{\sigma}(t), \boldsymbol{\varepsilon}(\boldsymbol{v}))_{\mathcal{H}} + \int_{\Gamma_C} k\Big(\int_0^t u_\nu(s)\, ds \Big)\, p(t, u_\nu(t)) v_\nu\, d\Gamma \\
&\quad + \int_{\Gamma_C} j_\tau^0(t, \boldsymbol{u}_\tau'(t); \boldsymbol{v}_\tau)\, d\Gamma \ge \langle \boldsymbol{f}(t), \boldsymbol{v} \rangle_{V^* \times V} \\
&\qquad\qquad \text{for all } \boldsymbol{v} \in V, \ \text{a.e. } t \in (0, T),
\end{aligned} \tag{14.72}
$$

$$
\boldsymbol{u}(0) = \boldsymbol{u}_0, \qquad \boldsymbol{u}'(0) = \boldsymbol{v}_0,
$$

where $\boldsymbol{\sigma}^I(\boldsymbol{u}) \in L^2(0, T; \mathcal{H})$ is the unique function defined in Lemma 14.4.

Inserting (14.71) into (14.72), it is clear that Problem $\tilde{\mathcal{Q}}^V$ is of the form (14.41), where the operator $A\colon (0, T) \times V \to V^*$, the functions $j\colon \Sigma_C \times \mathbb{R}^d \to \mathbb{R}$ and $\boldsymbol{f}\colon (0, T) \to V^*$ are defined as before (see (14.42), (14.48) and (14.32), respectively), and the operator $\mathcal{S}\colon \mathcal{V} \to \mathcal{V}^*$ is now a sum of three operators $\mathcal{S}_1$, $\mathcal{S}_5$, $\mathcal{S}_6\colon \mathcal{V} \to \mathcal{V}^*$ given, respectively, by (14.44),

$$
\langle (\mathcal{S}_5 \boldsymbol{w})(t), \boldsymbol{v} \rangle_{V^* \times V} = \big(\boldsymbol{\sigma}^I(\zeta \boldsymbol{w})(t), \boldsymbol{\varepsilon}(\boldsymbol{v}) \big)_{\mathcal{H}},
$$

where $\boldsymbol{\sigma}^I(\boldsymbol{\zeta}\boldsymbol{w}) \in L^2(0,T;\mathcal{H})$ is the unique function defined in Lemma 14.4, and

$$\langle(\mathcal{S}_6\boldsymbol{w})(t),\boldsymbol{v}\rangle_{V^*\times V} = \int_{\Gamma_C} k\Big(\int_0^t \zeta_\nu \boldsymbol{w}(s)\,ds\Big)\, p(t,\zeta_\nu\boldsymbol{w}(t))\, v_\nu\, d\Gamma$$

for all $\boldsymbol{w} \in \mathcal{V}$, $\boldsymbol{v} \in V$, a.e. $t \in (0,T)$. It is enough to check that $\mathcal{S}_5$ and $\mathcal{S}_6$ satisfy inequality (14.49) with some positive constants L_5 and L_6, respectively.

Let $\boldsymbol{w}_1, \boldsymbol{w}_2 \in \mathcal{V}$, $\boldsymbol{v} \in V$ and $t \in (0,T)$. First, using (14.70) and the Hölder inequality, we obtain

$$\begin{aligned}\langle(\mathcal{S}_5\boldsymbol{w}_1)(t)-(\mathcal{S}_5\boldsymbol{w}_2)(t),\boldsymbol{v}\rangle_{V^*\times V} &= \big(\boldsymbol{\sigma}^I(\boldsymbol{\zeta}\boldsymbol{w}_1)(t)-\boldsymbol{\sigma}^I(\boldsymbol{\zeta}\boldsymbol{w}_2)(t),\boldsymbol{\varepsilon}(\boldsymbol{v})\big)_{\mathcal{H}}\\ &\le \|\boldsymbol{\sigma}^I(\boldsymbol{\zeta}\boldsymbol{w}_1)(t)-\boldsymbol{\sigma}^I(\boldsymbol{\zeta}\boldsymbol{w}_2)(t)\|_{\mathcal{H}}\|\boldsymbol{v}\|_V\\ &\le c\Big(\int_0^t \|\boldsymbol{\zeta}\boldsymbol{w}_1(s)-\boldsymbol{\zeta}\boldsymbol{w}_2(s)\|_V ds\Big)\|\boldsymbol{v}\|_V\\ &\le c\,T\Big(\int_0^t \|\boldsymbol{w}_1(s)-\boldsymbol{w}_2(s)\|_V ds\Big)\|\boldsymbol{v}\|_V.\end{aligned}$$

Hence, the operator $\mathcal{S}_5$ satisfies (14.49) with $L_5 = c\,T > 0$. Next, since the functions p and k are Lipschitz continuous and bounded by constants, by the Hölder inequality and inequality (14.50), we get

$$\begin{aligned}&\langle(\mathcal{S}_6\boldsymbol{w}_1)(t)-(\mathcal{S}_6\boldsymbol{w}_2)(t),\boldsymbol{v}\rangle_{V^*\times V}\\ &= \int_{\Gamma_c}\Big(k\Big(\int_0^t \zeta_\nu\boldsymbol{w}_1(s)\,ds\Big)\,p(t,\zeta_\nu\boldsymbol{w}_1(t)) - k\Big(\int_0^t \zeta_\nu\boldsymbol{w}_2(s)\,ds\Big)\,p(t,\zeta_\nu\boldsymbol{w}_2(t))\Big)\,v_\nu\,d\Gamma\\ &\le \|k\Big(\int_0^t \zeta_\nu\boldsymbol{w}_1(s)\,ds\Big)\,p(t,\zeta_\nu\boldsymbol{w}_1(t))\\ &\quad -k\Big(\int_0^t \zeta_\nu\boldsymbol{w}_2(s)\,ds\Big)\,p(t,\zeta_\nu\boldsymbol{w}_2(t))\|_{L^2(\Gamma_C)}\,\|\boldsymbol{v}\|_{L^2(\Gamma_C;\mathbb{R}^d)}\\ &\le \Big(\|k\Big(\int_0^t \zeta_\nu\boldsymbol{w}_1(s)\,ds\Big)\,\big(p(t,\zeta_\nu\boldsymbol{w}_1(t))-p(t,\zeta_\nu\boldsymbol{w}_2(t))\big)\|_{L^2(\Gamma_C)}\\ &\quad +\|\Big(k\Big(\int_0^t \zeta_\nu\boldsymbol{w}_1(s)\,ds\Big)-k\Big(\int_0^t \zeta_\nu\boldsymbol{w}_2(s)\,ds\Big)\Big)\,p(t,\zeta_\nu\boldsymbol{w}_2(t))\|_{L^2(\Gamma_C)}\Big)\,\|\gamma\|\,\|\boldsymbol{v}\|_V\\ &\le \Big(\overline{k}\,L_p\|\zeta_\nu\boldsymbol{w}_1(t)-\zeta_\nu\boldsymbol{w}_2(t))\|_{L^2(\Gamma_C)}\\ &\quad +\overline{p}\,L_k\,\|\int_0^t\big(\zeta_\nu\boldsymbol{w}_1(s)-\zeta_\nu\boldsymbol{w}_2(s)\big)\,ds\|_{L^2(\Gamma_C)}\Big)\,\|\gamma\|\,\|\boldsymbol{v}\|_V\end{aligned}$$

$$\leq \Big(\overline{k}\, L_p\, \|\gamma\| \int_0^t \|\boldsymbol{w}_1(s) - \boldsymbol{w}_2(s)\|_V\, ds +$$

$$+ \overline{p}\, L_k \int_0^t \|\gamma\| \int_0^s \|\boldsymbol{w}_1(\tau) - \boldsymbol{w}_2(\tau)\|_V\, d\tau\, ds\Big)\, \|\gamma\|\, \|\boldsymbol{v}\|_V$$

$$\leq \|\gamma\|^2 (\overline{k} L_p + \overline{p} L_k T) \Big(\int_0^t \|\boldsymbol{w}_1(s) - \boldsymbol{w}_2(s)\|_V\, ds \Big) \|\boldsymbol{v}\|_V.$$

This implies that the operator $\mathcal{S}_6$ satisfies (14.49) with constant $L_6 = \|\gamma\|^2(\overline{k} L_p + \overline{p} L_k T) > 0$. Hence, the operator $\mathcal{S} = \mathcal{S}_1 + \mathcal{S}_5 + \mathcal{S}_6$ satisfies (14.49) with $L_{\mathcal{S}} = L_1 + L_5 + L_6$.

Using similar arguments as in the proof of Theorem 14.2, we infer that

$$\|\mathcal{S}\boldsymbol{w}\|_{\mathcal{V}^*} \leq \overline{c} \left(\|\boldsymbol{w}\|_{\mathcal{V}} + \|\boldsymbol{u}_0\|_V + \|\tilde{b}\|_{L^2(Q)} + \|\boldsymbol{\sigma}^I(\boldsymbol{u}_0)\|_{L^2(0,T;\mathcal{H})} + \overline{k}\, \overline{p} \right)$$

for $\boldsymbol{w} \in \mathcal{V}$ with $\overline{c} > 0$, which means that the operator $\mathcal{S}$ is well defined and takes values in $\mathcal{V}^*$.

Since conditions (14.2)–(14.7) are satisfied, we are now in a position to apply Theorem 14.1 and we infer that problem (14.41) has at least one solution $\boldsymbol{u} \in \mathcal{V}$ such that $\boldsymbol{u}' \in \mathcal{W}$. Furthermore, the regularity condition (14.38) implies (14.8). Therefore, we infer from the uniqueness part of Theorem 14.1 that under the additional hypothesis (14.38), the solution to problem (14.41) is unique.

Finally, using the regularity of $\boldsymbol{u}$ and hypotheses (14.22)–(14.24), from condition (14.34), we deduce that the pair $(\boldsymbol{u}, \boldsymbol{\sigma})$ is a solution to Problem $\mathcal{Q}^V$ with the regularity $\boldsymbol{u} \in W^{1,2}(0, T; V)$, $\boldsymbol{\sigma} \in L^2(0, T; \mathcal{H})$ and $\mathrm{Div}\, \boldsymbol{\sigma} \in L^2(0, T; V^*)$. This concludes the proof of the theorem. □

A couple of functions $(\boldsymbol{u}, \boldsymbol{\sigma})$ which satisfies (14.66)–(14.68) is called a *weak solution* of the frictional contact problem (14.53)–(14.59). We conclude that, under the hypotheses of Theorem 14.3, the frictional contact problem (14.53)–(14.59) has at least one weak solution. If, in addition, the regularity condition (14.38) holds, then the weak solution of Problem $\mathcal{Q}$ is unique.

Acknowledgements This research was supported by the Marie Curie International Research Staff Exchange Scheme Fellowship within the 7th European Community Framework Programme under Grant Agreement No. 295118, the National Science Center of Poland under grant no. N N201 604640, the International Project co-financed by the Ministry of Science and Higher Education of Republic of Poland under grant no. W111/7.PR/2012, the National Science Center of Poland under Maestro Advanced Project no. DEC-2012/06/A/ST1/00262, and the project Polonium "Mathematical and Numerical Analysis for Contact Problems with Friction" 2014/15 between the Jagiellonian University in Krakow and Université de Perpignan Via Domitia.

References

1. Banks, H.T., Pinter, G.A., Potter, L.K., Munoz, B.C., Yanyo, L.C.: Estimation and control related issues in smart material structure and fluids. In: Caccetta, L., et al. (eds.) Optimization Techniques and Applications, pp. 19–34. Curtin University Press, Perth (1998)
2. Banks, H.T., Hu, S., Kenz, Z.R.: A brief review of elasticity and viscoelasticity for solids. Adv. Appl. Math. Mech. **3**, 1–51 (2011)
3. Banks, H.T., Pinter, G.A., Potter, L.K., Gaitens, J.M., Yanyo, L.C.: Modeling of quasistatic and dynamic load responses of filled viesoelastic materials, Chapter 11. In: Cumberbatch, E., Fitt, A. (eds.) Mathematical Modeling: Case Studies from Industry, pp. 229–252. Cambridge University Press, Cambridge (2011)
4. Cristescu, N., Suliciu, I.: Viscoplasticity. Martinus Nijhoff Publishers, Editura Tehnica, Bucharest (1982)
5. Denkowski, Z., Migórski, S., Papageorgiou, N.S.: An Introduction to Nonlinear Analysis: Applications. Kluwer Academic/Plenum, Boston (2003)
6. Drozdov, A.D.: Finite Elasticity and Viscoelasticity—A Course in the Nonlinear Mechanics of Solids. World Scientific, Singapore (1996)
7. Eck, C., Jarušek, J., Krbeč, M.: Unilateral Contact Problems: Variational Methods and Existence Theorems. Pure and Applied Mathematics, vol. 270. Chapman/CRC Press, New York (2005)
8. Farcaş, A., Pătrulescu, F., Sofonea, M.: A history-dependent contact problem with unilateral constraint. Math. Appl. **2**, 105–111 (2012)
9. Han, W., Sofonea, M.: Quasistatic Contact Problems in Viscoelasticity and Viscoplasticity. Studies in Advanced Mathematics, vol. 30. American Mathematical Society/International Press, Providence/Somerville (2002)
10. Ionescu, I.R., Sofonea, M.: Functional and Numerical Methods in Viscoplasticity. Oxford University Press, Oxford (1993)
11. Kikuchi, N., Oden, J.T.: Contact Problems in Elasticity: A Study of Variational Inequalities and Finite Element Methods. SIAM, Philadelphia (1988)
12. Klarbring, A., Mikelič, A., Shillor, M.: Frictional contact problems with normal compliance. Int. J. Eng. Sci. **26**, 811–832 (1988)
13. Klarbring, A., Mikelič, A., Shillor, M.: On friction problems with normal compliance. Nonlinear Anal. **13**, 935–955 (1989)
14. Martins, J.A.C., Oden, J.T.: Existence and uniqueness results for dynamic contact problems with nonlinear normal and friction interface laws. Nonlinear Anal. Theory Methods Appl. **11**, 407–428 (1987)
15. Migórski, S., Ochal, A.: A unified approach to dynamic contact problems in viscoelasticity. J. Elast. **83**, 247–275 (2006)
16. Migórski, S., Ochal, A., Sofonea, M.: Nonlinear Inclusions and Hemivariational Inequalities. Models and Analysis of Contact Problems. Advances in Mechanics and Mathematics, vol. 26. Springer, New York (2013)
17. Oden, J.T., Martins, J.A.C.: Models and computational methods for dynamic friction phenomena. Comput. Methods Appl. Mech. Eng. **52**, 527–634 (1985)
18. Piotrowski, J.: Smoothing dry friction damping by dither generated in rolling contact of wheel and rail and its influence on ride dynamics of freight wagons. Veh. Syst. Dyn. **48**, 675–703 (2010)
19. Shillor, M., Sofonea, M., Telega, J.J.: Models and Analysis of Quasistatic Contact. Springer, Berlin (2004)
20. Sofonea, M., Matei, A.: Mathematical Models in Contact Mechanics. London Mathematical Society Lecture Note Series, vol. 398. Cambridge University Press, Cambridge (2012)
21. Sofonea, M., Pătrulescu, F.: Analysis of a history-dependent frictionless contact problem. Math. Mech. Solids **18**, 409–430 (2013)

22. Sofonea, M., Shillor, M.: A viscoplastic contact problem with a normal compliance with limited penetration condition and history-dependent stiffness coefficient. Commun. Pure Appl. Anal. **13**, 371–387 (2014)
23. Zeidler, E.: Nonlinear Functional Analysis and its Applications, II/B: Nonlinear Monotone Operators. Springer, New York (1990)

Index

W. Han et al. (eds.), *Advances in Variational and Hemivariational Inequalities*,
Advances in Mechanics and Mathematics 33, DOI 10.1007/978-3-319-14490-0

MIX
Papier aus verantwortungsvollen Quellen
Paper from responsible sources
FSC® C105338

If you have any concerns about our products,
you can contact us on
ProductSafety@springernature.com

In case Publisher is established outside the EU,
the EU authorized representative is:
Springer Nature Customer Service Center GmbH
Europaplatz 3, 69115 Heidelberg, Germany

Printed by Libri Plureos GmbH
in Hamburg, Germany